GENETIC ALGORITHMS AND ENGINEERING OPTIMIZATION

WILEY SERIES IN ENGINEERING DESIGN AND AUTOMATION

Series Editor
HAMID R. PARSAEI

GENETIC ALGORITHMS AND ENGINEERING DESIGN
Mitsuo Gen and Runwei Cheng

ADVANCED TOLERANCING TECHNIQUES
Hong-Chao Zhang

INTEGRATED PRODUCT AND PROCESS DEVELOPMENT: METHODOLOGIES, TOOLS, AND TECHNOLOGIES
John M. Usher, Utpal Roy, and Hamid R. Parsaei

GENETIC ALGORITHMS AND ENGINEERING OPTIMIZATION
Mitsuo Gen and Runwei Cheng

GENETIC ALGORITHMS AND ENGINEERING OPTIMIZATION

MITSUO GEN
RUNWEI CHENG
Ashikaga Institute of Technology
Ashikaga, Japan

A Wiley-Interscience Publication
JOHN WILEY & SONS, INC.
New York • Chichester • Weinheim • Brisbane • Singapore • Toronto

This book is printed on acid-free paper. ♾

Published simultaneously in Canada.

Library of Congress Cataloging-in-Publication Data:
Gen, Mitsuo, 1944–
Genetic algorithms and engineering optimization / by Mitsuo Gen, Runwei Cheng.
p. cm.
"A Wiley-Interscience publication."
Includes bibliographical references and index.
ISBN 0-471-31531-1 (alk. paper)
1. Industrial engineering—Mathematical models. 2. Genetic algorithms. 3. Mathematical optimization. I. Cheng, Runwie. II. Title.
T56.24.G46 2000
670′.285′51—dc21 99-16023

Printed in the United States of America

10 9 8 7 6 5 4 3 2 1

To Our Fathers, Mothers, and Families

CONTENTS

Preface xiii

1 Foundations of Genetic Algorithms **1**

1.1 Introduction 1

1.1.1 Encoding Issue 2

1.1.2 Genetic Operators 7

1.1.3 Selection 9

1.1.4 Genetic Local Search 11

1.2 Adaptation of Genetic Algorithms 14

1.2.1 Structure Adaptation 15

1.2.2 Parameter Adaptation 16

1.2.3 Fuzzy Logic Controller 18

1.3 Genetic Optimizations 27

1.3.1 Global Optimizations 27

1.3.2 Constrained Optimizations 34

1.3.3 Combinatorial Optimizations 38

1.3.4 Multiobjective Optimizations 39

1.4 Recent Genetic Algorithm Dissertations 40

2 Combinatorial Optimization Problems **53**

2.1 Introduction 53

2.2 Set-Covering Problem 53

2.2.1 Airline Crew Scheduling Problems 56

2.2.2 Genetic Representation 56

2.2.3 Genetic Operators 58

2.2.4 Genetic Algorithm 60

2.2.5 Computational Experience 61

2.3 Bin-Packing Problem 61

2.3.1 Heuristic Algorithms 63

2.3.2 Genetic Representation 65
2.3.3 Genetic Operators 68
2.3.4 Fitness Function 69
2.3.5 Initial Population 69
2.3.6 Computational Experience 70
2.4 Knapsack Problem 71
2.4.1 Multiple-Choice Knapsack Problem 72
2.4.2 Multiconstraint Knapsack Problem 77
2.5 Minimum Spanning Tree Problem 81
2.5.1 Quadratic Minimum Spanning Tree Problem 82
2.5.2 Degree-Constrained Minimum Spanning Tree Problem 85
2.5.3 Bicriteria Minimum Spanning Tree Problem 90
3 Multiobjective Optimization Problems 97
3.1 Introduction 97
3.2 Basic Concepts of Multiobjective Optimizations 97
3.2.1 Nondominated Solutions 98
3.2.2 Preference Structures 101
3.2.3 Basic Solution Approaches 102
3.2.4 Structures and Properties of Problems 106
3.3 Genetic Mulitobjective Optimization 106
3.3.1 Features of Genetic Search 106
3.3.2 Fitness Assignment Mechanism 107
3.3.3 Fitness Sharing and Population Diversity 111
3.3.4 The Concept of Pareto Solution 113
3.4 Vector-Evaluated Genetic Algorithms 115
3.5 Pareto Ranking and Tournament Methods 118
3.5.1 Pareto Ranking Method 118
3.5.2 Pareto Tournament Method 122
3.6 Weighted-Sum Approach 124
3.6.1 Random-Weight Approach 125
3.6.2 Adaptive Weight Approach 127
3.7 Distance Method 131
3.7.1 General Idea of the Distance Method 131
3.7.2 Calculation of Distance Measure 133
3.7.3 Application of the Distance Method 136
3.8 Compromise Approach 136
3.9 Goal Programming Approach 138

4 Fuzzy Optimization Problems 142

4.1 Introduction 142

4.2 Fuzzy Linear Programming 143

4.2.1 Fuzzy Linear Programming Model 143

4.2.2 Genetic Algorithm Approach 149

4.2.3 Interactive Approach 152

4.2.4 Numerical Example 154

4.3 Fuzzy Nonlinear Programming 156

4.3.1 Nonlinear Programming Model 157

4.3.2 Inexact Approach to FO/RNP-1 161

4.3.3 Interactive Approach 163

4.3.4 Numerical Example 164

4.4 Fuzzy Nonlinear Mixed-Integer Goal Programming 165

4.4.1 Fuzzy Nonlinear Mixed-Integer Goal Programming Model 168

4.4.2 Genetic Algorithm Approach 170

4.4.3 Numerical Examples 173

4.5 Fuzzy Multiobjective Integer Programming 178

4.5.1 Problem Formulation 181

4.5.2 Augmented Minimax Problems 184

4.5.3 Genetic Algorithm Approach 185

4.5.4 Interactive Fuzzy Satisfaction Method 189

4.5.5 Numerical Example 190

5 Reliability Design Problems 194

5.1 Introduction 194

5.2 Network Reliability Design 195

5.2.1 Problem Formulation 196

5.2.2 Dengiz, Altiparmak, and Smith's Approach 198

5.2.3 Deeter and Smith's Approach 204

5.3 Tree-Based Network Reliability and LAN Design 211

5.3.1 Bicriteria Network Topology Design 212

5.3.2 Numerical Examples 219

5.4 Multiobjective Reliability Design 221

5.4.1 Bicriteria Reliability Design 221

5.4.2 Genetic Algorithm Approach 224

5.4.3 Hybrid Genetic Algorithm Approach 226

5.4.4 Reliability Design with Fuzzy Goals 230

6 Scheduling Problems **235**

6.1 Introduction 235

6.2 Job-Shop Scheduling 235

6.2.1 Basic Approaches 236

6.2.2 Encodings 236

6.2.3 Adapted Genetic Operators 238

6.2.4 Heuristic-Featured Genetic Operators 242

6.2.5 Hybrid Genetic Algorithms 244

6.2.6 Discussion 251

6.3 Grouped Job Scheduling Problem 253

6.3.1 Problem Description and Necessary Condition 253

6.3.2 Fundamental Runs 255

6.3.3 Representation 257

6.3.4 Evaluation 259

6.3.5 Genetic Operators 260

6.3.6 Overall Procedure 260

6.3.7 Numerical Example 261

6.4 Resource-Constrained Project Scheduling 263

6.4.1 Priority-Based Encoding 265

6.4.2 Genetic Operators 270

6.4.3 Evaluation and Selection 272

6.4.4 Experimental Results 274

6.5 Parallel Machine Scheduling 278

6.5.1 Dominance Condition 280

6.5.2 Memetic Algorithms 284

6.5.3 Experimental Results 287

6.6 Multiprocessor Scheduling 288

6.6.1 Problem Description and Assumptions 289

6.6.2 Genetic Algorithm for MSP 290

6.6.3 Numerical Example 295

7 Advanced Transportation Problems **297**

7.1 Introduction 297

7.1.1 Transportation Model 297

7.1.2 Formulation of Transportation Problems 299

7.2 Spanning Tree–Based Approach 304

7.2.1 Tree Representation 304

7.2.2 Initialization 306

7.2.3 Genetic Operators 308

7.2.4 Evaluation and Selection 308
7.2.5 Overall Procedure 311
7.3 Multiobjective Transporation Problem 311
7.3.1 Problem Description 312
7.3.2 Spanning Tree–Based Genetic Algorithm for Multiobjective Transportation Problem 313
7.3.3 Numerical Examples 316
7.4 Fixed-Charge Transportation Problem 320
7.4.1 Mathematical Model 321
7.4.2 Difficulty of fcTP Instances 321
7.4.3 Solution Method for the fcTP 322
7.4.4 Implementation of the Genetic Algorithm 322
7.4.5 Numerical Examples 323
7.5 Capacitated Plant Location Problem 325
7.5.1 Mathematical Model 326
7.5.2 Spanning Tree–Based Genetic Algorithm for Plant Location Problems 328
7.5.3 Numerical Examples 329
7.6 Bicriteria Transportation Problem with Fuzzy Coefficients 330
7.6.1 Problem Description 331
7.6.2 Ranking Fuzzy Numbers 332
7.6.3 Implementation of the Genetic Algorithm 333
7.6.4 Numerical Examples 336
8 Network Design and Routing 341
8.1 Introduction 341
8.2 Shortest Path Problem 341
8.2.1 Problem Formulation 342
8.2.2 Genetic Algorithm Approach 343
8.2.3 Numerical Examples 351
8.3 Adaptive Network Routing 352
8.3.1 Genetic-Based Adaptive Routing 353
8.3.2 Representation of Chromosomes 354
8.3.3 Evaluation of Chromosomes 355
8.3.4 Genetic Operators 355
8.3.5 Numerical Examples 360
8.4 Centralized Network Design 363
8.4.1 Problem Formulation 364
8.4.2 Genetic Algorithm Approach 365
8.4.3 Numerical Example 366

8.5 Computer Network Expansion 366
8.5.1 Problem Description 368
8.5.2 Kummar, Pathak, and Gupta's Approach 369
8.5.3 Numerical Example 373
8.6 Multistage Process Planning 373
8.6.1 Problem Description 374
8.6.2 Genetic Algorithm Approach 376
8.6.3 Numerical Examples 377
8.7 *M/G/s* Queuing Facility Location on a Network 378
8.7.1 Problem Description 379
8.7.2 Evolutionary Computation Approach 383
8.7.3 Numerical Examples 387
9 Manufacturing Cell Design 390
9.1 Introduction 390
9.2 Manufacturing Cell Design 391
9.3 Traditional MCD Approaches 393
9.3.1 Similarity Coefficient Methods 394
9.3.2 Array-Based Methods 395
9.3.3 Mathematical Programming Methods 395
9.3.4 Graph and Network Methods 396
9.4 Genetic Algorithm Approaches 396
9.4.1 Representative and Genetic Operators 396
9.4.2 Joines' Order-Based Approach 399
9.4.3 Moon and Kim's Approach 405
9.4.4 Joines' Integer Programming Formulation Approach 410
9.4.5 Other Approaches 419
9.5 Cell Design with Alternative Process Plans 419
9.5.1 Incorporation of Alternative Operations and Machine Redundancy 421
9.5.2 Incorporation of Alternative Routings 424
9.5.3 Moon, Gen, and Kim's Approach to Independent Cells 433
9.6 Designing Independent Cells 438
9.6.1 Family Formation for Minimizing Number of Machine Types 439
9.6.2 Determining Number of Families 443
9.6.3 Minimizing Number of Machines 448
9.6.4 Other Considerations 449

References 451

Index 491

PREFACE

In the past decade, the study of how to apply genetic algorithms to problems in the industrial engineering world has been a subject engaging the curiosity of many researchers and practitioners in the area of management science, operations research, and industrial and systems engineering. A major reason for this interest is that genetic algorithms are powerful and broadly applicable stochastic search and optimization techniques that really work for many problems that are very difficult to solve by conventional techniques. Most engineering problems are optimization problems subject to complex constraints. Simple genetic algorithms usually do not produce successful applications for these thorny engineering optimization problems. Therefore, a method to tailor genetic algorithms to meet the nature of these problems is one of the major focuses of research into industrial engineering-oriented genetic algorithms. This book is intended to cover the major research topics of the application of genetic algorithms to industrial engineering optimization problems.

Since the spring of 1993, we have devoted our efforts to the study of industrial engineering-oriented genetic algorithms. In our group, six students in the doctoral course have finished their studies on applying genetic algorithms to scheduling problems, spanning tree problems, transportation problems, reliability optimization problems, location and allocation problems, and production plan problems. One of the authors has been invited to give a tutorial on the topic "Genetic Algorithms and Applications" at the 21st *International Conference on Computers and Industrial Engineering,* San Juan, Puerto Rico, 1997, the topic "Genetic Algorithms for Engineering Design" at *the Symposium on Management Science and Industrial Engineering,* Beijing, China, 1997, and the topic "Hybrid Genetic Algorithms: Networks and Engineering Applications," at *the Artificial Neural Networks in Engineering Conference,* St. Louis, Missouri, 1998. Also one of the authors has served as the guest editor of several special issues of international journals, including the special issue on "Genetic Algorithms and Industrial Engineering," *International Journal of Computers and Industrial Engineering,* Vol. 30, No. 4, 1996, the special issue on "Intelligent Engineering Design," *International*

Journal of Engineering Design and Automation, Vol. 3, No. 2, 1997, the special issue on "Evolutionary Computation for Engineering Valuation," *International Journal of Engineering Valuation and Cost Analysis,* Vol. 2, No. 3, 1999, and the special issue on "Computational Intelligence for Industrial Engineering," *International Journal of Computers and Industrial Engineering,* Vol. 36, No. 2, 1999.

We summarized the results of related genetic algorithm studies from 1992 to early 1996 in our book *Genetic Algorithms and Engineering Design* (New York: John Wiley & Sons, 1997), including the following major topics: constrained optimization problems, combinatorial optimization problems, reliability optimization problems, flow-shop scheduling problems, job-shop scheduling problems, machine scheduling problems, transportation problems, facility layout design problems, and other topics in engineering design. The major motivations for writing this book are that due to size limitations, some interesting studies were not included in our first book, and also, some new and interesting results have been reported since then.

The book is suitable for a course in the genetic algorithms and applications at the upper-level undergraduate or beginning graduate level in industrial and system engineering, management science, operations researches, computer science, and related areas. The book is also useful as a comprehensive reference text for system analysts, operations researchers, management scientists, engineers, and other specialists who face challenging optimization problems inherent in industrial engineering/operations research.

The book is organized as follows: Chapters 1 through 4 provide fundamental knowledge for readers. The remaining chapters can be covered nearly independently, except that they all use the basic material represented in the first four chapters. Chapter 1 gives a tutorial on the basic concept of genetic algorithms, including the encoding issue, adaptation issue, genetic optimizations, and a summary on recent doctoral dissertations. In Chapter 2 we discuss applications of genetic algorithms to combinatorial optimization problems, such as the set-covering problem, bin-packing problem, knapsack problem, and minimum spanning tree problem. In Chapter 3 we give a comprehensive survey on the-state-of-the-art of applying genetic algorithms to solve multiobjective optimization problems. Because many optimization problems in the industrial engineering world are combinatorial optimization problems subject to multiobjective and complex constraints, the book will provide readers with a solid foundation for their genetic algorithms practice. In Chapter 4 we show how to solve the fuzzy optimization problem with genetic algorithms. Handling uncertainty and imprecision is an important issue in engineering design and optimization. Each of the next five chapters presents a specific case study of the adaptation and application of genetic algorithms to the reliability design problem, scheduling problem, transportation problem, network design and routing problem, and manufacturing cell design problem, respectively.

We would like to express our sincere appreciation to Dr. Hamid R. Parsaei, University of Louisville, Editor-in-Chief of John Wiley & Sons' Series *Engineering Design and Automation,* for giving us the chance to include the book as one in this series. The book benefited from numerous discussions with colleagues and friends. We would like to thank the following people for the their valuable comments: Dr. Ronald Wolff and Dr, Shmuel Oren, University of California, Berkeley; Dr. Alice E. Smith, Auburn University; Dr. Gursel A. Süer, University of Puerto Rico; Dr. Sang M. Lee, University of Nebraska; Dr. Frank A. Tillman and Dr. Young-Jou Lai, Kansas State University; Dr. Way Kuo, Texas A&M University; Dr. Chaiho Kim, Sana Clara Univeristy; Dr. Shu-Cherng Fang and Dr. Jeffrey A. Joines, North Carolina State University; Dr. Anup Kumar, University of Louisville; Dr. Chiung Moon, University of Ulsan; Dr. Weixuan Xu, IPM, Chinese Academy of Sciences; Dr. Baoding Liu, Tsinghua University, China; Dr. Dingwei Wang and Dr. Zhihou Yang, Northeastern University, China; Dr. Hawk Hwang and Dr. Jong-Hwan Kim, Korean Advanced Institute of Science & Technology; Dr. Moo Young Jung, Pohang University of Science & Technology; Dr. Xin Yao, University of Birmingham; Dr. Kin K. Lai, City University of Hong Kong; Dr. Andreas Bastian, K-EFE Engineering Volkswagen AG; Dr. Osamu Katai, Kyoto University; Dr. Genji Yamazaki, Tokyo Metrop. Inst. of Technology; Dr. Hideo Tanaka, Dr. Hiroshi Ohta, Dr. Hidetomo Ichihashi, and Dr. Hisao Ishibuchi, Osaka Prefecture University; Dr. Katsuhisa Ohno, Nagoya Institute of Technology; Dr. Hiroaki Ishii, Osaka University; Dr. Tomoharu Nagao and Dr. Shigenobu Kobayashi, Tokyo Institute of Technology; Dr. Zenji Katagata, Kinjo Gakuin University; Dr. Tsutomu Usui, Takasaki Shoka College; Dr. Masao Mukaitono, Meiji University; Dr. Yoshiharu Sato, Hokkaito University; Dr. Mitsuo Yamashiro, Dr. Takao Yokota, Dr. Yasuhiro Tsujimura, Dr. Masato Sasaki, and Dr. Tadahiko Murata, Ashikaga Institute of Technology.

We are also indebted to many researchers who have developed the underlying concepts that permeate this book. Although far too numerous to mention, we have tried to recognize their contributions through bibliographic references at the end of the book.

We would like to give our special thanks to Ph.D. graduates of Ashikaga Institute of Technology: Dr. Gengui Zhou, Dr. D. Gong, and Dr. Yinzhen Li as well as the following graduate students at the Institute for their creative and devoted work with us during the past few years: Takeaki Taguchi, Jong Ryul Kim, Changyoon Lee, Juno Choi, and Yuichiro Mafune.

It was a real pleasure working with John Wiley & Sons professional editorial and production staffs, especially Robert L. Argentieri, Executive Editor of the series, Ms. Millie Torres-Matias, and Ms. Akemi Takada.

This project was supported by the International Scientific Research Program through Grants-in-Aid for Scientific Research (No. 07045032:1995.4–1998.3; No. 10044173:1998.4–2001.3) by the Ministry of Education, Science and Culture, the Japanese government.

We thank our wives, Eiko Gen and Liying Zhang, and children for their love, encouragement, understanding, and support during the preparation of this book.

Sept. 30, 1999

MITSUO GEN
in Berkeley, Visiting Professor
University of California, Berkeley

RUNWEI CHENG
in Tokyo, Visiting Researcher
Ashikaga Institute of Technology

1
FOUNDATIONS OF GENETIC ALGORITHMS

1.1 INTRODUCTION

Since the 1960s there has been increasing interest in imitating living beings to develop powerful algorithms for difficult optimization problems. A term now in common use to refer to such techniques is *evolutionary computation*. The best known algorithms in this class include genetic algorithms, developed by Holland [303]; evolution strategies, developed by Rechenberg [530] and by Schwefel [567]; evolutionary programming, developed by Fogel et al. [200]; and genetic programming, developed by Koza [375]. There are also many hybrid versions that incorporate various features of the foregoing paradigms. State-of-the-art overviews of the field of evolutionary computation have been given by Bäck and Schwefel [33], Michalewicz [454], and Fogel [198].

Genetic algorithms, as powerful and broadly applicable stochastic search and optimization techniques, are perhaps the most widely known types of evolutionary computation methods today. In the past few years the genetic algorithm community has turned much of its attention to optimization problems in industrial engineering, resulting in a fresh body of research and applications [28, 198, 219, 239, 249, 455, 567]. A bibliography on genetic algorithms has been prepared by Alander [11].

In general, a genetic algorithm has five basic components, as summarized by Michalewicz [455]:

1. A genetic representation of solutions to the problem
2. A way to create an initial population of solutions
3. An evaluation function rating solutions in terms of their fitness
4. Genetic operators that alter the genetic composition of children during reproduction
5. Values for the parameters of genetic algorithms

The genetic algorithm maintains a population of individuals, say $P(t)$, for generation t. Each individual represents a potential solution to the problem at hand. Each individual is evaluated to give some measure of its fitness. Some individuals undergo stochastic transformations by means of genetic operations to form new individuals. There are two types of transformation: *mutation,* which creates new individuals by making changes in a single individual, and *crossover,* which creates new individuals by combining parts from two individuals. New individuals, called *offspring* $C(t)$, are then evaluated. A new population is formed by selecting the more fit individuals from the parent population and the offspring population. After several generations, the algorithm converges to the best individual, which hopefully represents an optimal or suboptimal solution to the problem. A general structure of the genetic algorithms is as follows:

```
Procedure: Genetic Algorithms
begin
 t ← 0;
 initialize P(t);
 evaluate P(t);
 while (not termination condition) do
 begin
  recombine P(t) to yield C(t);
  evaluate C(t);
  select P(t + 1) from P(t) and C(t);
  t ← t + 1;
 end
end
```

There are two important issues with respect to search strategies: exploiting the best solution and exploring the search space [71]. The genetic algorithms provide a directed random search in complex landscapes. Genetic operators perform essentially a blind search; selection operators hopefully direct the genetic search toward the desirable area of the solution space. One general principle for developing an implementation of genetic algorithms for a particular real-world problem is to make a good balance between exploration and exploitation of the search space. To achieve this, all the components of the genetic algorithms must be examined carefully. Additional heuristics should be incorporated in the algorithm to enhance the performance.

1.1.1 Encoding Issue

How to encode a solution of the problem into a chromosome is a key issue when using genetic algorithms. The issue has been investigated from many aspects, such as mapping characters from genotype space to phenotype space when individuals are decoded into solutions, and metamorphosis properties when individuals are manipulated by genetic operators.

Classification of Encodings. In Holland's work, encoding is carried out using binary strings [303]. Binary encoding for function optimization problems is known to have severe drawbacks due to the existence of *Hamming cliffs,* pairs of encodings having a large Hamming distance while belonging to points of minimal distance in phenotype space [427]. For example, the pair 01111111111 and 10000000000 belong to neighboring points in phenotype space (points of minimal Euclidean distance) but have maximum Hamming distance in genotype space. To cross the Hamming cliff, all bits have to be changed simultaneously. The probability that crossover and mutation will occur can be very small. In this sense, the binary code does not preserve the locality of points in the phenotype space.

For many problems in the industrial engineering world, it is nearly impossible to represent their solutions with binary encoding. During the last 10 years, various encoding methods have been created for particular problems to provide effective implementation of genetic algorithms. According to what kind of symbol is used as the alleles of a gene, the encoding methods can be classified as follows:

- Binary encoding
- Real-number encoding
- Integer or literal permutation encoding
- General data structure encoding

Real-number encoding is best used for function optimization problems. It has been widely confirmed that real-number encoding performs better than binary or Gray encoding for function optimizations and constrained optimizations [180, 447, 455, 647]. Since the topological structure of the genotype space for real-number encoding is identical to that of the phenotype space, it is easy to form effective genetic operators by borrowing useful techniques from conventional methods. Integer or literal permutation encoding is best used for combinatorial optimization problems. Since the essence of combinatorial optimization problems is the search for a best permutation or combination of items subject to constraints, literal permutation encoding can be the best way to this type of problem. For more complex real-world problems, an appropriate data structure is suggested as the allele of a gene, to capture the nature of the problem. In such cases, a gene may be an n-ary or more complex data structure.

According to the structure of encodings, the encoding methods can also be classified into the following two types: (1) one-dimensional encoding and (2) multidimensional encoding. In most practices, one-dimensional encoding is used. However, many real-world problems require solutions for multidimensional structures. It is natural to use a multidimensional encoding method to represent those solutions. For example, Vignaus and Michalewicz used an allocation matrix as encoding for the transportation problem [641]. Cohoon and Paris used two-dimensional encoding for the VLSI circuit placement problem [127]. Anderson, Jones, and Ryan used a two-dimensional grid type of encoding [13]. Moon and Kim used two-dimensional encoding for graph problems [462]. Ono, Yamamura, and Kobayashi used a job-sequence matrix as encoding for job-shop scheduling problems [493]. A general discussion of multidimensional encoding and crossover was given by Bui and Moon [79], who argued that to fit solutions of multidimensional problems into one-dimensional encoding entails losing a considerable amount of the information contained in the multidimensional structure.

According to the contents encoded, the following encoding methods can also be used: (1) solution only and (2) solution + parameters. In genetic algorithm practice, the first method is widely used to develop suitable encoding for a given problem. The second method is used in the evolution strategies of Rechenberg and Schwefel [567]. An individual consists of two parts: the first is the solution to a given problem, and the second, the strategy parameters, comprise variances and covariances of the normal distribution for mutation. The purpose of incorporating strategy parameters into the representation of individuals is to facilitate the evolutionary self-adaptation of these parameters by applying evolutionary operators to them. The search will then be performed in the space of solutions and strategy parameters together. In this way a suitable adjustment and diversity of mutation parameters should be provided under arbitrary circumstances.

Infeasibility and Illegality. Genetic algorithms work on two types of spaces alternatively: coding space and solution space, or in other words, genotype space and phenotype space. Genetic operators work on genotype space, and evaluation and selection work on phenotype space. Natural selection is the link between chromosomes and the performance of decoded solutions. The mapping from genotype space to phenotype space has a considerable influence on the performance of genetic algorithms. One outstanding problem associated with mapping is that some individuals correspond to infeasible solutions to a given problem. This problem may become very severe for constrained optimization problems and combinatorial optimization problems.

We need to distinguish between two basic concepts: infeasibility and illegality (Figure 1.1). These are often misused in the literature. *Infeasibility* refers to the phenomenon that a solution decoded from chromosome lies outside the feasible region of a given problem; *illegality* refers to the phenomenon that a chromosome does not represent a solution to a given problem.

The infeasibility of chromosomes originates from the nature of the constrained optimization problem. Whichever technique is used, conventional methods or genetic algorithms, it must handle the constraints. For many optimization problems, the feasible region can be represented as a system of

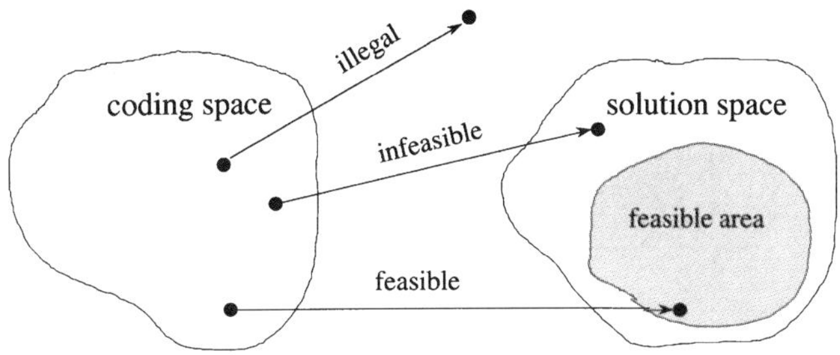

Figure 1.1. Infeasibility and illegality.

equalities or inequalities. For such cases, penalty methods can be used to handle infeasible chromosomes [218, 453]. In constrained optimization problems, the optimum typically occurs at the boundary between the feasible and infeasible areas. The penalty approach will force the genetic search to approach the optimum from both sides of the feasible and infeasible regions.

The illegality of chromosomes originates from the nature of encoding techniques. For many combinatorial optimization problems, problem-specific encodings are used, and such encodings usually yield illegal offspring by a simple one-cutpoint crossover operation. Because an illegal chromosome cannot be decoded to a solution, the penalty techniques are inapplicable to this situation. Repair techniques are usually adopted to convert an illegal chromosome to a legal one. For example, the well-known PMX operator is essentially a two-cutpoint crossover for permutation representation, together with a repair procedure to resolve the illegitimacy caused by simple two-cutpoint crossover. Orvosh and Davis [494] have shown that for many combinatorial optimization problems, it is relatively easy to repair an infeasible or illegal chromosome, and the repair strategy does indeed surpass other strategies, such as the rejecting or the penalizing strategy.

Properties of Encodings. When a new encoding method is given, it is usually necessary to examine whether we can set up an effective genetic search using the encoding. Several principles have been proposed to evaluate an encoding [370, 530]:

- Nonredundancy
- Legality
- Completeness
- Lamarckian property
- Causality

Property 1.1 (Nonredundancy). The mapping between encodings and solutions must be 1-to-1.

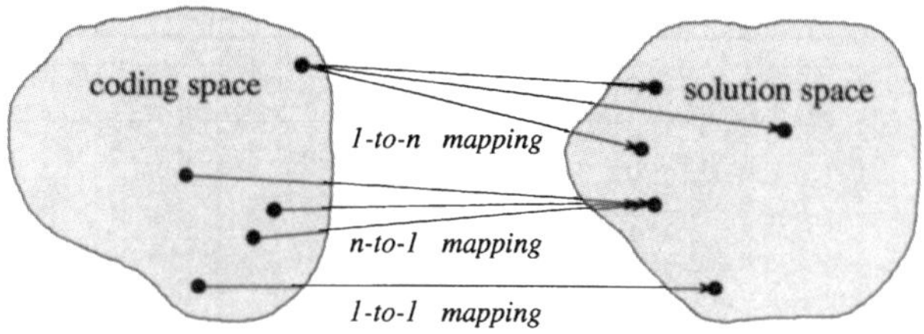

Figure 1.2. Mapping from chromosomes to solutions.

Mapping from encoding to solutions may belong to one of the following three cases (Figure 1.2):

1. 1-to-1 mapping
2. n-to-1 mapping
3. 1-to-n mapping

The most desired case, 1-to-1 mapping, ensures that no trivial operations will occur when creating offspring. If n-to-1 mapping occurs, genetic algorithms will waste time while searching. Because two individuals are duplicated in the phenotype space but not in the genotype space, distance measures in the genotype space cannot treat individuals as identical. It then becomes one reason for genetic algorithms to convergence prematurely. The most undesirable case is 1-to-n mapping, because we need another procedure performed on the phenotype space to determine one solution among many possible solutions.

Property 1.2 (Legality). Any permutation of an encoding corresponds to a solution.

This property guarantees that most existing genetic operators can easily be applied to the encoding.

Property 1.3 (Completeness). Any solution has a corresponding encoding.

This property guarantees that any point in solution space is accessible for a genetic search.

Property 1.4 (Lamarckian Property). The meaning of alleles for a gene is not context dependent.

The Lamarkian property for encoding concerns the issue of whether or not one chromosome can pass on its merits to future populations through common genetic operations [114]. Let us look at an example. The problem of a nine-city tour by a traveling salesman can be encoded by Bean's random key encoding method as follows [47]:

random key: [0.34 0.09 0.88 0.66 0.91 0.01 0.23 0.21 0.44]

A tour is obtained by sorting keys in descending order as follows: (6–2–8–7–1–9–4–3–5). Consider a subtour [8 7 1 9]. For random key representation, the subtour is [0.08 0.66 0.91 0.01]. Usually, it refers to a different subtour in offspring but not the same 8–7–1–9 subtour as that of its parent. The subtour that it refers to will depend on the values of other genes in the offspring. In fact, the offspring receive nothing (about a good subtour) from the parent but mere random chance. Such an encoding property is called *no Lamarckian;* that is, the meaning of alleles for a gene is interpreted in a context-dependent manner. Therefore, offspring usually inherit nothing from parents. Generally, we hope an encoding technique has the Lamarckian property so that offspring can inherit goodness from parents. In the *half-Lamarckian* type, some of the segments inherited from parents refer to the same things as they do in the parents, and some refer to different things.

Property 1.5 (Causality). Small variations on the genotype space due to mutation imply small variations in the phenotype space.

This property is suggested by Rechenberg in relation to evolution strategies [530]. It focuses on the conservation of neighborhood structures; that is, for the successful introduction of new information by mutation, the mutation operator should preserve the neighborhood structure in the corresponding phenotype space. The perspective is common among practitioners of genetic optimizations [179]. Search processes that do not destroy the neighborhood structure are said to exhibit *strong causality*. The opposite extreme is *no causality*. *Weak causality* describes the case where small changes in the genotype space correspond to larger changes in the phenotype space, and vice versa. Sendhoff, Kreuts, and Seelen suggested a condition to measure the causality associated with mapping from genotype to phenotype in combination with a mutation operator [571].

1.1.2 Genetic Operators

Search is one of the more universal problem-solving methods for problems in which one cannot determine a priori the sequence of steps leading to a solution. Typically, there are two types of search behaviors: random search and local search. *Random search* explores the entire solution and is capable of achieving escape from a local optimum. *Local search* exploits the best solution and is capable of climbing upward toward a local optimum. The two types of search abilities form the mutual complementary components of a serach. An ideal search should possess both types simultaneously. It is nearly impossible to design such a search method with conventional techniques. Genetic algorithms are a class of general-purpose search methods combining elements of directed and stochastic searches which can make a good balance between exploration and exploitation of the search space. In genetic algorithms, accumulated information is exploited by the selection mechanism, while new regions of the search space are explored by means of genetic operators.

In conventional genetic algorithms, the crossover operator is used as the principal operator and the performance of a genetic system is heavily dependent on it. The mutation operator which produces spontaneous random changes in various chromosomes, is used as a background operator. In essence, genetic operators perform a random search and cannot guarantee to yield improved offspring. It has been discovered that the speed of convergence of genetic algorithms is rather slow in many large combinatorial optimization problems. There are many empirical studies on a comparison between crossover and mutation. It is confirmed that mutation can sometimes play a more important role than crossover.

There are two hypotheses for the explanation of how genetic algorithms exploit the distributed information to generate good solutions: (1) the building-block hypothesis and (2) the convergence-controlled variation hypothesis. The building-block hypothesis was proposed by Holland [303] and refined by Goldberg [249]. According to the hypothesis, crossover recombines features from two parents to produce offspring. Sometimes crossover will recombine the best features from two parents, resulting in superior offspring. Since the fitness of an individual will often depend on complex patterns of simple features, it is important that the operator be able to propagate to offspring those patterns of features that contribute to the fitness of the parents. The concept of *epistasis* refers to strong interaction among genes in an encoding. In other words, epistasis measures the extent to which the contribution to fitness of one gene depends on the values of other genes. For a given problem, a high degree of epistasis means that building blocks cannot form.

The convergence-controlled variation hypothesis was given by Eshelman, Mathias, and Schaffer [178]. The hypothesis suggests using the convergence of a population to constrain the search. New points are sampled from a distribution that is a function of the population distribution at any point in time. As the population converges, the variation becomes more focused. Whereas the building-block hypothesis stresses recombining, and hence propagating, features that have survived in the parent population, the convergence-controlled variation hypothesis stresses randomly sampling from a distribution that is a function of the current population distribution [179].

How we conceptualize the genetic search will affect how we design genetic operators. From the point of view of search abilities, it is expected that a search provided by a method can possess the abilities of random search and directed search simultaneously. Cheng and Gen suggest the following approach for designing genetic operators [105]. For the two genetic operators, crossover and mutation, one is used to perform a random search to try to explore the area beyond a local optimum, and the other is used to perform a local search to try to find an improved solution. The genetic search then possesses two types of search abilities. With this approach, the mutation operator will play the same important role as that of the crossover operator in a genetic search.

1.1.3 Selection

The principle behind genetic algorithms is essentially Darwinian natural selection. Selection provides the driving force in a genetic algorithm. With too much force, genetic search will terminate prematurely; with too little force, evolutionary progress will be slower than necessary. Typically, a lower selection pressure is indicated at the start of a genetic search in favor of a wide exploration of the search space, while a higher selection pressure is recommended at the end to narrow the search space. The selection directs the genetic search toward promising regions in the search space. During the past two decades, many selection methods have been proposed, examined, and compared. Common types are as follows:

- Roulette wheel selection
- $(\mu + \lambda)$-selection
- Tournament selection
- Steady-state reproduction
- Ranking and scaling
- Sharing

Roulette wheel selection, proposed by Holland, is the best known selection type [303]. The basic idea is to determine selection probability or survival probability for each chromosome proportional to the fitness value. Then a model roulette wheel can be made displaying these probabilities. The selection process is based on spinning the wheel the number of times equal to population size, each time selecting a single chromosome for the new population. The wheel features the selection method as a stochastic sampling procedure. Baker proposed a stochastic universal sampling method [36] that uses a single wheel spin. The wheel is constructed in the same way as is a standard roulette wheel, with a number of equally spaced markers equal to the population size. The basic strategy underlying this approach is to keep the expected number of copies of each chromosome in the next generation.

In contrast with proportional selection, $(\mu + \lambda)$-selection and (μ,λ)-selection as proposed by Bäck are deterministic procedures that select the best chromosomes from parents and offspring [27]. Note that both methods prohibit selection of duplicate chromosomes from the population, so many researchers prefer to use this method to deal with combinatorial optimization problems. Truncation selection and block selection are also deterministic procedures that rank all individuals according to their fitness and select the best as parents [619]. Elitist selection is generally used as supplementary to proportional selection to preserve the best chromosome in the new generation if it is not selected through a proportional selection process.

Generational replacement, replacing an entire set of parents by their offspring, can be viewed as another version of the deterministic approach. The steady-state reproduction of Whitley [662] and Syswerda [603] belongs to this class, in which the n worst parents are replaced by offspring (n is the number of offspring).

Another type of selection procedure contains random and deterministic features simultaneously. A typical example is the tournament selection of Goldberg, Korb, and Deb [251]. This method randomly chooses a set of chromosomes and picks out the best chromosome for reproduction. The number of chromosomes in the set is called the *tournament size*. A common tournament size is 2; this is called a *binary tournament*. Stochastic tournament selection was suggested by Wetzel [659]. In this method, selection probabilities are calculated normally and successive pairs of chromosomes are drawn using roulette wheel selection. After drawing a pair, the chromosome with higher fitness is inserted in the new population. The process continues until the population is full. Remainder stochastic sampling, proposed by Brindle, is a modified version of his deterministic sampling [76]. In this method, each chromosome is allocated samples according to the integer part of the expected number, and then chromosomes compete for the remaining places in the population, according to the fractional parts of the number expected.

In the proportional selection procedure, the selection probability of an individual is proportional to its fitness. This simple scheme exhibits some undesirable properties. For example, in early generations, there is a tendency for a few superchromosomes to dominate the selection process; in later generations, when the population has largely been converged, competition among chromosomes is less strong and random search behavior will emerge.

The scaling and ranking mechanisms are proposed to mitigate these problems. The scaling method maps raw objective function values to positive real values, and the survival probability for each chromosome is determined according to these values. Fitness scaling has a twofold intention: (1) to maintain a reasonable differential between relative fitness ratings of chromosomes, and (2) to prevent too-rapid takeover by some superchromosomes to meet the requirement to limit competition early but to stimulate it later.

Since De Jong's work, use of scaling objective functions has become widely accepted, and several scaling mechanisms have been proposed. According to the type of function used to transform the raw fitness into scaled fitness, scaling methods can be classified as linear scaling [270], sigma truncation [203], power law scaling [245], logarithmic scaling [270], an so on. If the transformation relation between scaled fitness and raw fitness is constant, it is called a *static scaling method;* if the transformation is variable with respect to some factors, it is called a *dynamic scaling method.* The windowing technique introduces a moving baseline into fitness proportionate selection to maintain more constant selection pressure [281]. The normalizing technique is also one type of dynamic scaling proposed by Cheng and Gen [105].

For most scaling methods, scaling parameters are problem dependent. Fitness ranking has an effect similar to that of fitness scaling but avoids the need for extra scaling parameters [531]. Baker introduced the notion of ranking selection with genetic algorithms to overcome the scaling problems of the direct fitness-based approach [35, 250, 281]. The ranking method ignores the actual object function values; instead, it uses a ranking of chromosomes to determine survival probability. The idea is straightforward: Sort the population from best to worst and assign the selection probability of each chromosome according to the ranking but not its raw fitness. Two methods are in common use: linear ranking and exponential ranking.

Sharing techniques, introduced by Goldberg and Richardson [248] for multimodel function optimization, are used to maintain the diversity of population. A sharing function is a way of determining the degradation of an individual's fitness due to a neighbor at some distance. With the degradation, the reproduction probability of individuals in a crowd peak is restrained while other individuals are encouraged to give offspring.

1.1.4 Genetic Local Search

The idea of combining genetic algorithms and local search heuristics for solving optimization problems has been investigated extensively during the past decade, and various methods of hybridization have been proposed. One of most common forms of hybrid genetic algorithm involves incorporating local optimization as an add-on extra to a canonical genetic algorithm. With the hybrid approach, local optimization is applied to each newly generated offspring to move it to a local optimum before injecting it into the population. Genetic search is used to perform global exploration among the population, and local search is used to perform local exploitation around chromosomes. Because of the complementary properties of genetic algorithms and local search methods, the hybrid approach often outperforms either method operating alone.

There are two common forms of genetic local search. One features Lamarckian evolution and the other features the Baldwin effect [663]. Both approaches use the metaphor that an individual learns (hill climbs) during its lifetime (generation). In the Lamarckian case, the resulting individual (after hill climbing) is put back into the population. In the Baldwinian case, only the fitness is changed; the genotype remains unchanged. According to Whitley, Gordon, and Mathias' experiences on some test problems, the Baldwinian search strategy can sometimes converge to a global optimum when the Lamarckian strategy converges to a local optimum using the same form of local search. However, in all the cases they examined, the Baldwinian strategy is much slower than the Lamarckian strategy.

The early work that linked genetic and Lamarkian evolution theory included that of Grefenstette, who introduced Lamarckian operators into genetic algorithms [269]; of Davidor, who defined Lamarkian probability for mutations in order to enable a mutation operator to be more controllable and to introduce some qualities of a local hill-climbing operator [141]; and of Shaefer, who added to a standard genetic algorithm that is Lamarkin in nature, intermediate mapping between the chromosome and solution spaces [572].

Kennedy provided an explanation of hybrid genetic algorithms combined with Lamarckian evolution theory [353, 663]. The canonical genetic algorithms of Holland were inspired by Darwin's theory of natural selection. In the nineteenth century, Darwin's theory was challenged by Lamarck, who proposed that environmental changes throughout an organism's life cause structural changes that are transmitted to offspring. This theory lets organisms pass along the knowledge and experience they acquire in their lifetime. Although no biologist today believes that traits acquired in the natural world can be inherited, the power of Lamarckian theory is illustrated by societal evolution. Ideas and knowledge are passed from generation to generation through structured language and culture. Genetic algorithms, as artificial organisms, can benefit from the advantages of Lamarckian theory. By letting some individuals' experiences be passed along to future individuals, we can improve the ability of GAs to focus on the most promising areas. Following a more Lamarckian approach, a traditional hill-climbing routine could use offspring as a starting point and perform rapid, localized optimization. After they have learned to climb the local landscape, we can put the offspring through the evaluation and selection phases. Offspring have a chance to pass on experience to future offspring through common crossover.

Let $P(t)$ and $C(t)$ be parents and offspring in current generation t. The general structure of hybrid genetic algorithms is outlines as follows:

```
Procedure: Hybrid Genetic Algorithms
begin
 t ← 0;
 initialize P(t)
 evaluate P(t)
 while (not termination condition) do
 begin
  recombine P(t) to yield C(t);
  climb C(t) locally;
  evaluate C(t);
  select P(t + 1) from P(t) and C(t);
  t ← t + 1;
 end
end
```

In the hybrid approach, artificial organisms first pass through Darwin's biological evolution and then pass through Lamarckian's intelligence evolu-

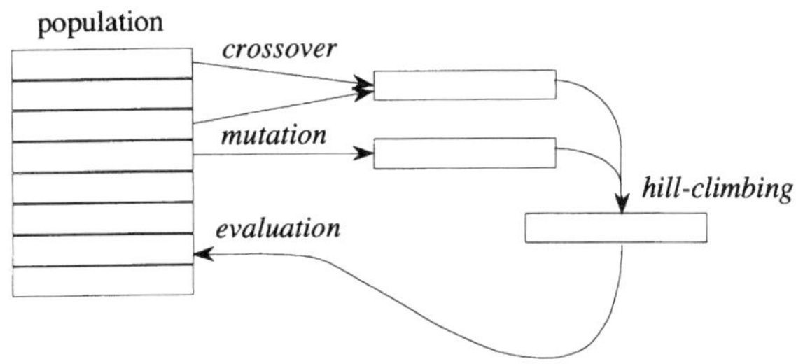

Figure 1.3. General structure off hybrid genetic algorithms.

tion (Figure 1.3). A traditional hill-climbing routine is used as Lamarckian evolution to try to inject some "smarts" into the offspring organism before returning it to be evaluated.

Moscato and Norman have introduced the term *memetic algorithm* to describe genetic algorithms in which local search plays a significant part [468]. The term is motivated by Dawkin's notion of a *meme* as a unit of information that reproduces itself as people exchange ideas [149]. A key difference exists between genes and memes. Before a meme is passed on, it is typically adapted by the person who transmits it as that person thinks, understands, and processes the meme, whereas genes get passed on whole. Moscato and Norman linked this thinking to local refinement and therefore promoted the term *memetic algorithm* to describe genetic algorithms that use local search heavily.

Radcliffe and Surry gave a formal description of *memetic algorithms* [520], which provided a homogeneous formal framework for considering memetic and genetic algorithms. According to Radcliffe and Surry, if a local optimizer is added to a genetic algorithm and applied to every child before it is inserted into the population, a memetic algorithm can be though of simply as a special kind of genetic search over the subspace of local optima. Recombination and mutation will usually produce solutions that are outside this space of local optima, but a local optimizer can then repair such solutions to produce final children that lie within this subspace, yielding the memetic algorithm shown in Figure 1.4.

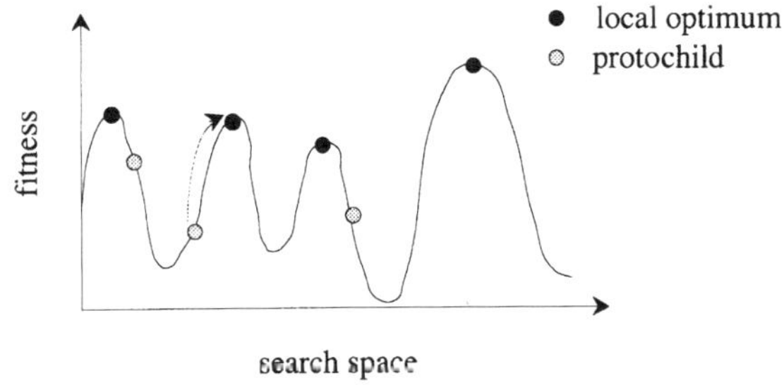

Figure 1.4. Memetic algorithm and local optimizer. (Adapted from Radcliffe and Surry [520].)

The role of local search in the context of genetic algorithms has been receiving serious consideration, and many successful applications strongly favor such a hybrid approach. Both memetic algorithms and Lamarkian evolution try to give a reasonable explanation of the hybrid approach based on different natural phenomena.

1.2 ADAPTATION OF GENETIC ALGORITHMS

Since genetic algorithms are inspired from the idea of evolution, it is natural to expect that the adaptation is used not only for finding solutions to a given problem, but also for tuning genetic algorithms to the particular problem. During the past few years, many adaptation techniques have been suggested and tested to obtain an effective implementation of genetic algorithms to real-world problems. In general, there are two types of adaptations:

1. Adaptation to problems
2. Adaptation to evolutionary processes

The difference between these two adaptations is that the first advocates modifying some components of genetic algorithms, such as representation, crossover, mutation, and selection, in order to choose an appropriate form of the algorithm to meet the nature of a given problem. The second adaptation suggests a way to tune the parameters of the changing configurations of genetic algorithms while solving the problem. According to Herrera and Lozano, the latter type of adaption can be further divided into the following classes [294]:

- Adaptive parameter settings
- Adaptive genetic operators
- Adaptive selection
- Adaptive representation
- Adaptive fitness function

Among these classes, parameter adaptation has been studied extensively in the past decade because strategy parameters such as mutation ratio, crossover ratio, and population size are key factors in determination of the exploitation versus exploration trade-off. It has long been acknowledged that these strategy parameters have a significant impact on the performance of genetic algorithms [146, 268].

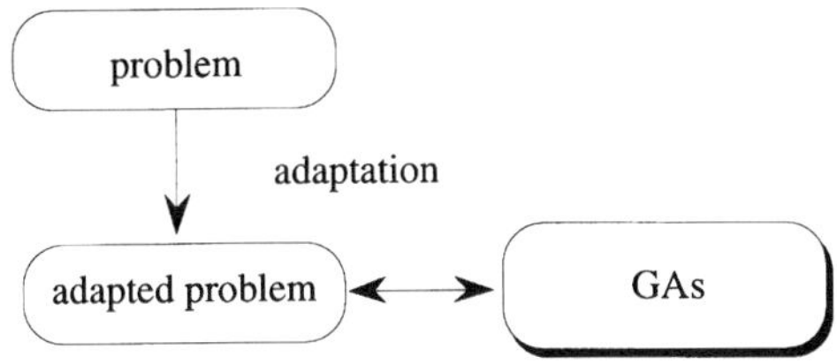

Figure 1.5. Adapting a problem to genetic algorithms.

1.2.1 Structure Adaptation

The genetic algorithms were created as a generic and weak method featuring binary encoding and binary genetic operators. This approach requires modification of an original problem into a form suitable for genetic algorithms, as shown in Figure 1.5. The approach includes mapping between potential solutions and binary representation, care of decoders and repair procedures, and so on. For complex problems such an approach usually fails to provide successful applications.

To overcome such problems, various nonstandard implementations of genetic algorithms have been created for particular problems. As shown in Figure 1.6, this approach leaves the problem unchanged and adapts genetic algorithms by modifying a chromosome representation of a potential solution and applying appropriate genetic operators.

But in general, it is not a good choice to use the whole original solution of a given problem as the chromosome because many real problems are too complex to permit suitable implementation of genetic algorithms with the entire solution representation. Generally, encoding methods can be either direct or indirect. In *direct encoding,* the entire solution for a given problem is used as a chromosome. For a complex problem, however, such a method will make almost all conventional genetic operators unusable because a vast number of offspring will be infeasible or illegal. By contrast, in *indirect encoding,* just the necessary part of a solution is used as a chromosome. Solutions can then be generated by a decoder. A *decoder* is a problem-specific procedure used to generate a solution according to the permutation and/or combination

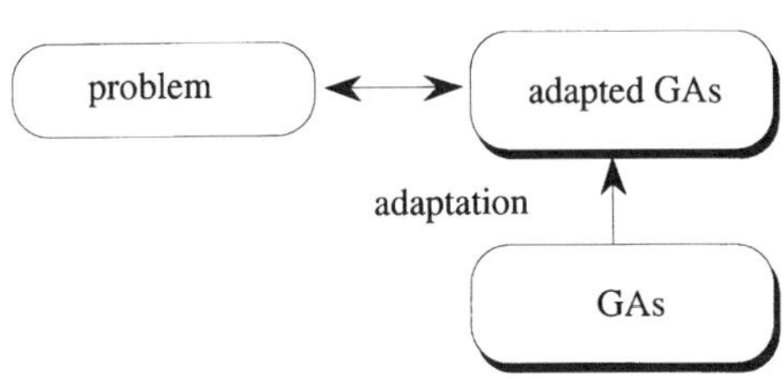

Figure 1.6. Adapting genetic algorithms to a problem.

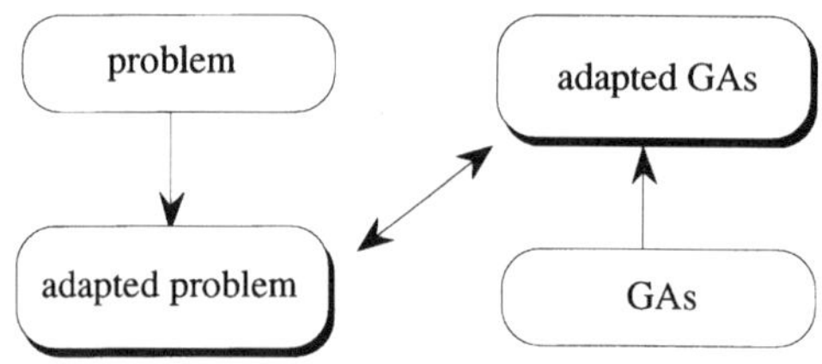

Figure 1.7. Third approach: adapt both the genetic algorithms and the problem.

of items produced by genetic algorithms. When used in this way, genetic algorithms will focus their search solely on the interesting part of solution space.

A third approach is to adapt both the genetic algorithms and the problem, as shown in Figure 1.7. A common feature of combinatorial optimization problems is to find a permutation and/or combination of items associated with side constraints. If the permutation and/or combination can be determined, a solution can then be derived using a problem-specific procedure. With this third approach, genetic algorithms are used to evolve an appropriate permutation and/or combination of items under consideration, and a heuristic method is subsequently used to construct a solution according to the permutation and combination. This approach has been applied successfully in the area of industrial engineering and has recently become the main approach for the practical use of genetic algorithms [219, 455].

1.2.2 Parameter Adaptation

The behavior of genetic algorithms is characterized by a balance between exploitation and exploration in the search space. The balance is strongly affected by strategy parameters such as population size, maximum generation, crossover ratio, and mutation ratio. How to choose a value for each parameter and how to find the values efficiently are very important, promising areas of research into genetic algorithms. Recent surveys on adaptation techniques are those of Herrera and Lozano [294] and Hinterding, Michalewicz, and Eiben [299].

Fixed parameters are used in most applications of genetic algorithms. The parameter values are determined using a set-and-test approach. Since a genetic algorithm is an intrinsically dynamic and adaptive process, the use of constant parameters is thus in contrast to the general evolutionary spirit. Therefore, it is natural to try to modify the values of strategy parameters during the run of the algorithm. It is possible to do this in various ways: (1) by using a rule, (2) by taking feedback information from the current state of search, or (3) by employing a self-adaptive mechanism. According to Hinterding, Michalewicz, and Eiben's classification of adaptation, there are three principal categories: (1) deterministic, (2) adaptive and (3) self-adaptive.

Deterministic Adaptation. Deterministic adaptation takes place if the value of a strategy parameter is altered by a deterministic rule. Generally, a time-varying approach is used, measured by the number of generations. For example, the mutation ratio is decreased gradually along with the elapse of generations using the following equation:

$$p_m = 0.5 - 0.3\frac{t}{G}$$

where t is the current generation number and G is the maximum generation. Hence the mutation ratio will decrease from 0.5 to 0.2 as the number of generations increases to G. A time dependency of the mutation ratio was first suggested by Holland, although he did not give a detailed choice of parameter [303]. Early examples of this method are those of Fogarty [193] and Hesser and Männer [296].

Adaptive Adaptation. Adaptive adaptation takes place if there is some form of feedback from the evolutionary process, which is used to determine the direction and/or magnitude of the change in the strategy parameter. Early examples of this type of adaptation include Rechenberg's $\frac{1}{5}$ *success rule* in evolution strategies, which was used to vary the step size of mutation [530]. The rule states that the ratio of successful mutations to all mutations should be $\frac{1}{5}$; hence, if the ratio is greater than $\frac{1}{5}$, the step size should be increased, and if the ratio is less than $\frac{1}{5}$, the step size should be decreased. Davis's *adaptive operator fitness* utilizes feedback on the success of a larger number of reproduction operators to adjust the ratio being used [147]. Julstrom's adaptive mechanism regulates the ratio between crossovers and mutations based on their performance [341]. An extensive study of these types of learning-rule mechanisms has been undertaken by Tuson and Ross [635].

Self-adaptive Adaptation. Self-adaptive adaptation enables strategy parameters to evolve within the evolutionary process. The parameters are encoded onto the individuals' chromosomes and undergo mutation and recombination. The encoded parameters do not affect the fitness of individuals directly, but better values will lead to better individuals, which will be more likely to survive and produce offspring, hence propagating better parameter values. The parameters for self-adaption can be those that control the operation of genetic algorithms, the operation of reproduction or other operators, or the probabilities of using alternative processes. Schwefel developed this method for self-adaption of mutation step size and mutation rotation angles in evolution strategies [567]. Hinterding used a multichromosome to implement self-adaptation with contiguity int he cutting stock problem, where self-adaptation is used to adapt the probability of using one of the two available mutation operators and the strength of the group mutation operator.

1.2.3 Fuzzy Logic Controller

Around the late 1980s, fuzzy control emerged as one of the most active and fruitful areas for research in the application of fuzzy set theory. A comprehensive survey of this research was given by Lee [395, 396]. Fuzzy logic is much closer in spirit to human thinking and natural language than are traditional logical systems. Basically, it provides an effective means of capturing the approximate, inexact nature of the real world. Viewed in this perspective, the essential part of the fuzzy logic controller is a set of linguistic control rules related by the dual concepts of fuzzy implication and the compositional rule of inference. In essence, the fuzzy logic controller provides an algorithm that can convert linguistic control strategy based on expert knowledge into an automatic control strategy. In particular, this methodology appears very useful when the processes are too complex for analysis using conventional techniques or when the available sources of information are interpreted qualitatively, inexactly, or with uncertainty. Thus fuzzy logic control may be viewed as a step toward a rapprochement between conventional precise mathematical control and human or humanlike decision making.

The pioneering work in extending the fuzzy logical technique to dynamic adjustment of the strategy parameters of genetic algorithms were those of Lee and Takagi [402] and Xu and Vukovich [673]. The main idea is to use a fuzzy logic controller to compute new strategy parameter values that will be used by the genetic algorithms. The inputs to the controller are any combination of the performance measures and current parameters of the genetic algorithms. The outputs are the parameters.

According to Lee, a fuzzy logic controller is comprised of four principal components [395]:

1. A knowledge base
2. A fuzzification interface
3. An inference system
4. A defuzzification interface

The experts' knowledge is stored in the knowledge base in the form of linguistic control rules. The fuzzification interface is used to transform crisp data into fuzzy data. The inference system, the heart of the controller, provides approximate reasoning based on the knowledge base. The defuzzification interface translates fuzzy control action to nonfuzzy control action. The generic structure of a fuzzy logic controller is shown in Figure 1.8.

The fuzzy logic controller is a type of rule-based system built on fuzzy logic and fuzzy set theory. Parametric models are used to realize an input–output mapping that can be specified using linguistic values. The difference between fuzzy and nonfuzzy systems is in how they partition the space and resolve conflicts. In a fuzzy system, the partitions can be overlaid and a point in the input space may belong to more than one partition. When action has

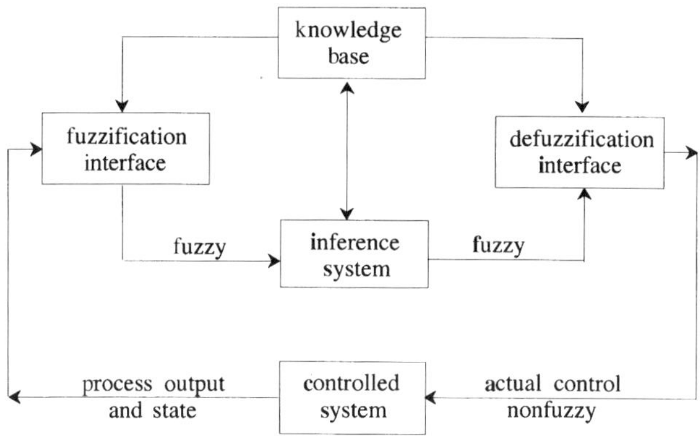

Figure 1.8. Generic structure of fuzzy logic controller. (From Lee [395].)

to be taken, each partition to which a point belongs contributes to the decision to the extent that the point belongs to each partition. This mixing mechanism gives fuzzy rule-based systems their smooth interpolative behavior.

The dynamic behavior of a fuzzy system is characterized by a set of linguistic descriptive rules based on expert knowledge. The expert knowledge is usually of the form

> **if** *a set of conditions are satisfied,* **then** *a set of consequences can be inferred*

Since the antecedents and consequents of these if–then rules are associated with fuzzy concepts (linguistic terms), they are often called *fuzzy conditional statements.* Furthermore, several linguistic variables might be involved in the antecedents and conclusions of these rules. When this is the case, the system is referred to as a *multi-input/multi-output fuzzy system.* In fuzzy logic controllers, a fuzzy control rule is a fuzzy conditional statement in which the antecedent is a condition in its application domain and the consequent is a control action for the system under control. Basically, fuzzy control rules provide a convenient way of expressing control policy and domain knowledge.

The knowledge base is composed of two components: a database and a rule base. The database contains the definitions of linguistic variables used in the antecedents and consequents of the if–then rules. The definitions are made up of membership functions for the fuzzy sets. The rule base is the collection of fuzzy control rules representing expert knowledge.

When utilizing a fuzzy logic controller to adjust the strategy parameters of genetic algorithms, diversity measure, fitness values, and current parameters are taken as inputs of if–then rules. Outputs indicate values of strategy parameters or changes in these parameters, such as crossover ratio, mutation ratio, population size, and selective pressure. Each input and output should have an associated set of linguistic labels. The meaning of these labels is specified through membership functions of fuzzy sets. Therefore, it is necessary that every input and output have a bounded range of values within which to define these membership functions. After selecting the inputs and outputs and defining the database, the fuzzy rules describing the relation between them should be defined. There are two ways to do this: (1) using the experience and knowledge of experts or (2) using an automatic learning technique.

Generally, the behavior of genetic algorithms depends on many uncertain factors, and only incomplete knowledge and imprecise information are available for identification of the relationship between the strategy parameters and the behavior of genetic algorithms. Therefore, it is acceptable for fuzzy logic controllers to adjust these parameters dynamically.

Zeng and Rabenasolo's Approach. The basic structure of the method given by Zeng and Rabenasolo consists of three fuzzy logic controllers: one for control crossover ratio, one for mutation ratio, and one for crossover position [692]. Fuzzy logic controllers are designed to approximate the relationship between the strategy parameters of genetic algorithms and several measures in populations. These measures, which characterize the behavior of genetic algorithms, constitute the input variables of the controller; and the strategy parameters p_c, p_m, and p_{os} are taken as the output variables of the controller. The fuzzy rules of the controller are built based on the knowledge extracted from their experiments for optimizing some sampling functions.

Some heuristic principles for optimize the behavior of genetic algorithms are adjusting crossover ratio so as to:

- To maintain the diversity of each population, two distant samples have more chance to be selected for crossover.
- To enhance searching in optimal regions, two near samples with high fitness values have more change to be selected for crossover.
- To avoid convergence to local optima, crossover operations have to be enhanced if the variance of fitness values is very small.
- To stabilize optimal populations, crossover operations have to be reduced if the specific fitness values of the samples to be selected are close enough to the maximal fitness value of the current population.

There are some apparent conflicts in these principles. However, good behavior of a genetic algorithm can be obtained by proper adjustment of these conflicts.

Let $P = \{x_1,x_2,\ldots,x_n\}$ be the current population, where n is the size of P. The corresponding fitness function is denoted by F. The objective of the genetic algorithm is to find the optimal x so that $F(x) = \max\{F(x_i)|i = 1,2,\ldots,n\}$. According to the previous principles, the crossover probability p_c for the pair (x_i,x_k) is designed to be function of the following parameters:

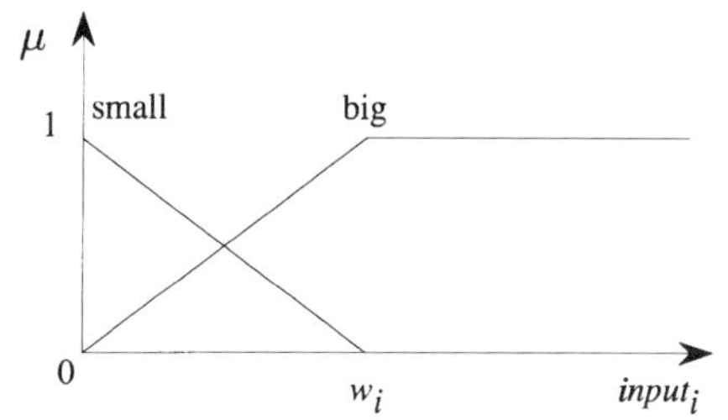

Figure 1.9. Definition of membership function for input variables: Var, F_1, F_2 and D.

- The variance of fitness values Var $= (F_{max} - \overline{F})/(F_{max} - F_{min})$
- The distance between the fitness value of x_i and F_{max}: $G = [F_{max} - F(x_i)]/(F_{max} - F_{min})$
- The distance between x_j and x_k: $D = d(x_j, x_k)$
- The fitness values $F_1 = F(x_j)/F_{max}$ and $F_2 = F(x_k)/F_{max}$

The parameters Var, F_1, F_2, and D are included in the interval [0,1]. Of these parameters, Var is determined by the entire population, whereas the others are defined by specific individuals. In a fuzzy logic controller, these parameters are taken as input variables and the values of p_c are generated as part of its output. For each input variable, the membership functions are defined in Figure 1.9, where $input_i \in \{\text{Var}, G, F_{,1}, F_2\}$ is any input variable of the controller and w_i is the corresponding parameter obtained from its experience on genetic algorithms. The membership function of p_c is defined in Figure 1.10.

Using the definitions, the principles above can be transformed into the following fuzzy rules:

1. **If** G is large, **then** p_c is large.
2. **If** Var is small **and** G is small, **then** p_c is small.
3. **If** D is small **and** F_1 is large **and** F_2 is large **then** p_c is large.
4. **If** D is small **and** (F_1 **or** F_2) is small, **then** p_c is small.

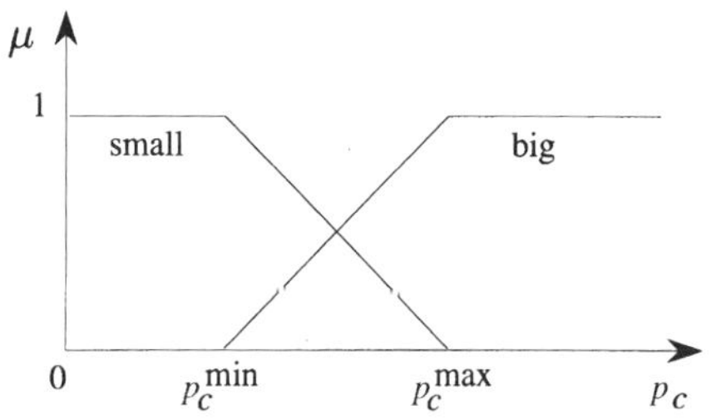

Figure 1.10. Definition of membership function for output variable p_c.

TABLE 1.1. Fuzzy Decision Table for Crossover

$\Delta c(t)$		$\Delta f(t-1)$								
		NR	NL	NM	NS	ZE	PS	PM	PL	PR
	NR	NR	NL	NL	NM	NM	NS	NS	ZE	ZE
	NL	NL	NL	NM	NM	NS	NS	ZE	ZE	PS
	NM	NL	NM	NM	NS	NS	ZE	ZE	PS	PS
	NS	NM	NM	NS	NS	ZE	ZE	PS	PS	PM
$\Delta f(t)$	ZE	NM	NS	NS	ZE	ZE	PS	PS	PM	PM
	PS	NS	NS	ZE	ZE	PS	PS	PM	PM	PL
	PM	NS	ZE	ZE	PS	PS	PM	PM	PL	PL
	PL	ZE	ZE	PS	PS	PM	PM	PL	PL	PR
	PR	ZE	PS	PS	PM	PM	PL	PL	PR	PR

5. **If** D is large, **then** p_c is large.

The parameter p_m is designed so that the value of the mutation ratio is increased when the genetic algorithm tends to a local optimum, and the value is decreased when the current population varies greatly or a globally optimal population is obtained. The corresponding fuzzy rules include:

1. **If** Var is small **and** G is large, **then** p_m is large.
2. **If** Var is small **and** G is small, **then** p_m is small.
3. **If** Var is large, **then** p_m is small.

Wang, Wang, and Hu's approach. The basic structure of this method consists of two fuzzy logic controllers: one for the crossover ratio and one for the mutation ratio [655]. The heuristic updating principle of the crossover ratio is to consider changes in the average fitness of the population, denoted as $\Delta f(t)$. Assume that the following three conditions are satisfied:

1. The change is sufficiently small, that is, $|\Delta f(t)| < \epsilon$, where ϵ is a given small real number.
2. The change is greater than zero; that is, $\Delta f(t) > 0$.
3. The changes keep the same sign in consecutive generations.

Then the crossover ratio is increased; otherwise, the crossover ratio is decreased. If the change is in the proximity of zero, the crossover ratio should be increased rapidly. The input variables of the fuzzy logic controller are the change in average fitness at instant t, $\Delta f(t)$, and that at the instant $t-1$, $\Delta f(t-1)$. These permit computation of $\Delta^2 f(t)$ $[\Delta^2 f(t) = \Delta f(t) - \Delta f(t-1)]$. The output is the change in crossover ratio, $\Delta c(t)$ (Table 1.1).

The membership functions of fuzzy imports and outputs variables are illustrated in Figure 1.11. The nomenclature is: NR, negative larger; NL, neg-

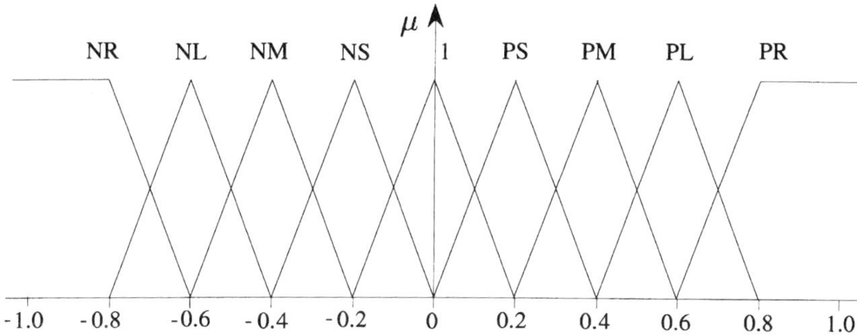

Figure 1.11. Membership function of $\delta f(t-1)$, $\delta f(t)$, and $\delta c(t)$.

ative large; NM, negative medium; NS, negative small; ZE, zero; PS, positive small; PM, positive medium; PL, positive large; PR, positive larger.

Table 1.2 is a look-up table for actions of a fuzzy logic controller. In this table, z is a minimum integer not greater than $\alpha x + (1\alpha)_y$, where α is an adaptive coefficient that varies with changes in fitness. The output of the fuzzy logic controller is given by the following function:

$$\Delta c(t) = z[x][y] \times 0.02 \times \beta \tag{1.1}$$

where β is another adaptive coefficient which is less than 1.0 when the changes in fitness of whole populations are less than 0.02. Therefore, the crossover ratio is calculated by the following equation:

$$p_c(t) = p_c(t-1) + \Delta c(t) \tag{1.2}$$

Similarly, adjustment of the mutation ratio is given by the following equations:

TABLE 1.2. Fuzzy Membership Value for Control Action of Crossover

z		x								
		−4	−3	−2	−1	0	1	2	3	4
	−4	−4	−3	−3	−2	−2	−1	−1	0	0
	−3	−3	−3	−2	−2	−1	−1	0	0	1
	−2	−3	−2	−2	−1	0	0	1	1	2
	−1	−2	−2	−1	−1	2	1	1	2	2
y	0	−2	−1	−1	0	1	1	2	2	3
	1	−1	−1	0	0	1	1	?	2	3
	2	−1	0	0	1	1	2	2	3	3
	3	0	0	1	1	2	2	3	3	4
	4	0	1	1	2	2	3	3	4	4

$$\Delta m(t) = z[x][y] \times 0.002 \times \beta \tag{1.3}$$

$$p_m(t) = p_m(t-1) + \Delta m(t) \tag{1.4}$$

Lee and Takagi's Approach. In Lee and Takagi's *dynamic parametric genetic algorithms* [402], the inputs to the fuzzy logic controller are any combination of performance measures of genetic algorithms or current parameter settings, and outputs may be any of the parameters of genetic algorithms. An automatic fuzzy design technique is proposed for obtaining both the database and the rule base for the case where knowledge or experitese is not available. The phenotypic diversity measure is used to monitor the performance of genetic algorithms [294]. Let $\overline{f}$ denote the average fitness, f_{best} the best fitness, and f_{worst} the worst fitness. Two phenotypic diversity measures given by Lee and Tagaki are:

$$\text{PDM}_1 = \frac{f_{\text{best}}}{\overline{f}} \tag{1.5}$$

$$\text{PDM}_2 = \frac{\overline{f}}{f_{\text{worst}}} \tag{1.6}$$

PDM_1 and PDM_2 belong to the interval [0,1]. If they are near 1, convergence has been reached, whereas if they are near 0, the population shows a high level of diversity.

The heuristic principles for adjusting strategy parameters are given as follows:

1. **If** PDM_1 is big, **then** *pop_size* should increase.
2. **If** PDM_2 is small, **then** *pop_size* should decrease.
3. **If** mutation is small **and** population is small, **then** *pop_size* should increase.

Xu and Vukovich's Approach. In Xu and Vukovich's *fuzzy evolutionary algorithms* [673], the input variables of the fuzzy logic controller are two parameters: generation and population size. The outputs are strategy parameters: p_c and p_m. The rules for adjustment of the strategy parameters are given in Table 1.3.

Herrera, Viedma, Lozano, and Verdegay's Approach. In their fuzzy logic controller, dispersion statistical measures are used as the input variables [295]. Let L denote the length of a chromosome and N the population size. Let S_{ij} denote the gene in position j for chromosome i. Define the following statistical values:

TABLE 1.3. Fuzzy Decision Table for Crossover and Mutation

	Population Size		
Generation	Small	Medium	Big
Short	Large	Medium	Small
Medium	Medium	Small	Very small
Long	Small	Very small	Very small

$$\overline{S}_i = \frac{\sum_{j=1}^{L} S_{ij}}{L} \tag{1.7}$$

$$\overline{S} = \frac{\sum_{i=1}^{L} \sum_{j=1}^{L} S_{ij}}{LN} \tag{1.8}$$

$$\overline{S}_j = \frac{\sum_{i=1}^{L} S_{ij}}{L} \tag{1.9}$$

Then the average variance chromosome AVC is defined as follows:

$$\text{AVC} = \frac{\sum_{i=1}^{N} (\overline{S}_i - \overline{S})^2}{N} \tag{1.10}$$

The average variance allele AVA is defined as follows:

$$\text{AVA} = \frac{\sum_{j=1}^{L} \sum_{i=1}^{N} (S_i - \overline{S}_j)^2}{N} \tag{1.11}$$

The fuzzy rules are given as follows:

1. **If** VAC is low, **then** p_c should be adjusted upward slightly.
2. **If** VAC is high, **then** p_c should be forced downward slightly.
3. **If** AVA is low, **then** p_m should be adjusted upward slightly.
4. **If** AVA is high, **then** p_m should be forced downward slightly.

Herrera and Lozano's Appoach. Herrera and Lozano proposed adaptive real-coded genetic algorithms based on the fuzzy logic controller, called ARGAF [294]. The major features of ARGAF are:

1. It applies two different crossover operators: one with exploitation properties and another with exploration properties.
2. It uses the linear ranking selection mechanism [35].

TABLE 1.4. Fuzzy Decision Table for Parameter p_e

	PDM_1		
ED	Low	Medium	High
Low	Small	Small	Medium
Medium	Big	Big	Medium
High	Big	Big	Medium

3. Two parameters, p_e and η_{min}, are adjusted using the fuzzy logic controller.

The parameters p_e defines the frequency of application of each crossover type: exploitative and explorative. The parameter η_{min} determines the selective pressure. If η_{min} is low, high pressure is achieved, whereas if it is high, the pressure is low.

Two diversity measures are considered as inputs. One is the genotypic diversity measure ED (Euclidean distance) and the other is the phenotypic diversity measure PDM_1. The first measure represents the quantity of the genetic material in the population, and the second, the quality of the diversity.

Let P denote the set of population, $\boldsymbol{v}_i$ the ith individual, $\boldsymbol{v}^*$ the best individual in the population, and $d(\boldsymbol{v}_i,\boldsymbol{v}_j)$ the Euclidean distance between two individuals. Define:

$$\bar{d} = \frac{1}{N}\sum_{i=1}^{N} d(\boldsymbol{v}^*,\boldsymbol{v}_i) \tag{1.12}$$

$$d_{max} = \max\{d(\boldsymbol{v}^*,\boldsymbol{v}_i)|\boldsymbol{v}_i \in P\} \tag{1.13}$$

$$d_{min} = \min\{d(\boldsymbol{v}^*,\boldsymbol{v}_i)|\boldsymbol{v}_i \in P\} \tag{1.14}$$

Then ED is defined as follows:

$$\text{ED} = \frac{\bar{d} - d_{min}}{d_{max} - d_{min}} \tag{1.15}$$

The range of ED is [0,1]. If ED is low, most individuals in the population are concentrated around the best individuals, so convergence is achieved. If ED is high, most individuals are not biased toward the current best individual.

The rules describing the relation between the inputs ED and PDM_1 and the output p_e are given in Table 1.4 The rules descrbe the relation between the inputs ED and PDM_1 and the output η_{min} are given in Table 1.5.

TABLE 1.5. Fuzzy Decision Table for Parameter η_{min}

	PDM$_1$		
ED	Small	Medium	Large
Low	Small	Medium	Big
Medium	Small	Big	Big
High	Small	Small	Big

1.3 GENETIC OPTIMIZATIONS

Optimization deals with problems of minimizing or maximizing a function with several variables usually subject to equality and/or inequality constraints. It plays a central role in operations research, management science, and engineering design. Many industrial engineering design problems are very complex and difficult to solve using conventional optimization techniques. In recent years, genetic algorithms have received considerable attention regarding their potential as a novel optimization technique. Because of their simplicity, ease of operation, minimal requirements, and parallel and global perspective, genetic algorithms have been applied successfully in a wide variety of problem domains [147, 219, 455]. A brief introduction to genetic optimization techniques is described in this section, including major fields of optimization, such as global, constrained, combinatorial, and multiobjective optimizations.

1.3.1 Global Optimizations

Global optimizations utilize techniques that can distinguish between the global optimum and numerous local optima within a region of interest. Global optimization problems usually take the form of unconstrained optimization, that is, the problem is one of minimizing or maximizing a function in the absence of restrictions [45]. In general, an unconstrained optimization problem can be represented mathematically as follows [428]:

$$\begin{aligned} \min \quad & f(\boldsymbol{x}) \\ \text{s.t.} \quad & \boldsymbol{x} \in \Omega \end{aligned}$$

where f is a real-valued function and Ω, the feasible set, is a subset of E^n. When attention is restricted to the case where $\Omega = E^n$, it corresponds to the completely unconstrained case. In many applications we just need to consider the case where Ω is a particular subset of E^n. A point $\boldsymbol{x}^* \in \Omega$ is said to be a *local minimum* of f over Ω if there is an $\epsilon > 0$ such that $f(\boldsymbol{x}) \geq f(\boldsymbol{x}^*)$ for all $\boldsymbol{x} \in \Omega$ within a distance ϵ of $\boldsymbol{x}^*$. A point $\boldsymbol{x}^* \in \Omega$ is said to be a *global minimum* of f over Ω if $f(\boldsymbol{x}) \geq f(\boldsymbol{x}^*)$ for all $\boldsymbol{x} \in \Omega$. Even though most practical optimization problems have side restrictions that must be satisfied, the study of techniques for unconstrained optimization provides a basis for further study.

Conventional global optimization methods can roughly be categorized into two classes: (1) deterministic methods and (2) stochastic methods [623]. Genetic algorithms have been fairly successful at solving problems of the type that are too ill-behaved, nondifferentiable, and discontinuous for conventional hill-climbing and derivative-based techniques. Examples of such problems are multimodal, nondifferentiable, and discontinuous problems. Since the emergence of genetic algorithms in the early 1970s, global optimization has been one of their major targets, and a lot of effort has been devoted to developing powerful algorithms for global optimization problems.

The usual way of applying genetic algorithms to solving global optimization problems is to encode each decision variable as a bit using either standard binary coding or Gray coding. For a problem having n variables, $\boldsymbol{x} = (x_1, x_2 \ldots, x_n)$, a chromosome contains n segments of a bit string as follows:

$$\underbrace{1001...01}_{x_1}\ \underbrace{1001...01}_{x_2}\ \cdots\ \underbrace{1001...01}_{x_n}$$

Thus, if each variable x_i is coded in l_i bits, a complete chromosome has length $\Sigma_{i=1}^{n} l_i$. The strong preference for using binary representations of solutions in the genetic algorithms is typically derived from the *schema theory* of genetic algorithms, which tries to analyze genetic algorithms in terms of their expected schema sampling behavior. However, practical experience as well as some theoretical hints regarding the binary encoding of continuous variables indicate strongly that the binary representation has serious disadvantages. It might introduce an additional multimodality into the objective function, thus making the combined objective function more complex than the original problem [33]. Therefore, a major trend for solving function optimization is to adopt a real-number representation, which leads to a very active research field, sometimes called *real-coded genetic algorithms* [147, 180, 647].

In the real-number representation, each chromosome is encoded as a vector of real numbers. For a problem with n variables, the real-number vector will be $\boldsymbol{x} = (x_1, x_2, \ldots, x_n)$. During the past two decades, several genetic operators have been proposed for real-number codings, which can roughly be put into four classes:

1. Conventional operators
2. Arithmetical operators
3. Direction-based operators

4. Stochastic operators

Conventional operators are made by extending the operators for binary representation into the real-coding case. Arithmetic operators are constructed by borrowing the concept of linear combination of vectors from the area of convex set theory. The direction-based operators are formed by introducing the approximate gradient (subgradient) direction into genetic operators. The stochastic operators give offspring by altering parents by random numbers with some distribution, typically a Guassian distribution. A brief introduction to some well-used genetic operators is given below.

Arithmetical Crossover. The basic concept of this type of operator is borrowed from convex set theory [44]. Generally, the weighted average of two vectors $\boldsymbol{x}_1$ and $\boldsymbol{x}_2$ is calculated as follows:

$$\lambda_1 \boldsymbol{x}_1 + \lambda_2 \boldsymbol{x}_2 \tag{1.16}$$

If the multipliers are restricted as

$$\lambda_1 + \lambda_2 = 1, \qquad \lambda_1 > 0, \quad \lambda_2 > 0 \tag{1.17}$$

the weighted form (1.16) is known as *convex combination.* If the nonnegative condition on the multipliers is dropped, the combination is known as an *affine combination.* If the multipliers are simply required to be in real space E_1, the combination is known as a *linear combination.*

Similarly, arithmetic crossover is defined as the combination of two vectors (chromosomes) as follows:

$$\boldsymbol{x}'_1 = \lambda_1 \boldsymbol{x}_1 + \lambda_2 \boldsymbol{x}_2 \tag{1.18}$$

$$\boldsymbol{x}'_2 = \lambda_1 \boldsymbol{x}_2 + \lambda_2 \boldsymbol{x}_1 \tag{1.19}$$

According to the restriction on multipliers, it yields to three types of crossovers: *convex crossover, affine crossover,* and *linear crossover.* Let us look at a geometric explanation of arithmetic operators. For two parents $\boldsymbol{x}_1$ and $\boldsymbol{x}_2$, the collection of all convex combinations is called a *convex hull.* Similarly, we can define a *affine hull* as the collection of all affine combinations, and a *linear hull* as the collection of all linear combinations. Figure 1.12 shows them in a simple case in two-dimensional space. The offspring generated with convex crossovers lie on the solid line, the offspring from affine crossovers lie on the dotted line, and the offspring from linear crossovers lie within the entire space.

A special case is that of $\lambda_1 = \lambda_2 = 0.5$, called *average crossover* by Davis [147] and *intermediate crossover* by Schwefel [566]. Wright restricted the

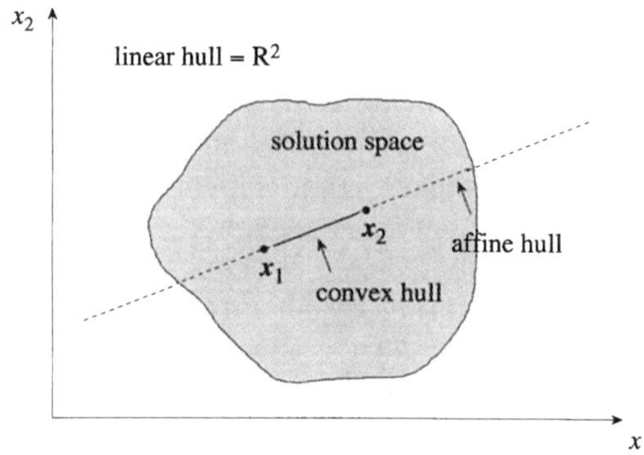

Figure 1.12. Convex hull, affine hull, and linear hull.

multipliers as follows: $\lambda_1 = 1.5$ and $\lambda_2 = -0.5$, which is a special case of affine crossover [667]. In another case of affine crossover, considered by Mühlenbein and Schlierkamp-Voosen [709], called *extended intermediate crossover,* one multiplier was given as a random real number in the interval $[-d, 1 + d]$. Cheng and Gen restricted the multipliers as follows:

$$\lambda_1 + \lambda_2 \leq 2, \qquad \lambda_1 > 0, \quad \lambda_2 > 0$$

which is a special case of linear crossover [113].

Blend Crossover. Essentially, blend crossover creates offspring randomly within a hyper-rectangular defined by the parent points [180]. Consider a one-dimensional case; that is, the problem has just one variable. Suppose that the first parent has the value p_1, the second parent p_2, and that $p_2 > p_1$. Let $I = |p_1 - p_2|$ and $0 < \alpha, \beta < 1$; then an offspring is generated by randomly choosing a point within the interval

$$[p_1 - \alpha I, p_2 + \beta I] \tag{1.20}$$

It is usually specified as BLX–α–β. Figure 1.13 illustrates the blend crossover. If the problem has two variables, a new offspring will be a point that lies within the rectangle defined by the two intervals, and if the problem has n variables, an offspring will be a point within the hyper-rectangular defined

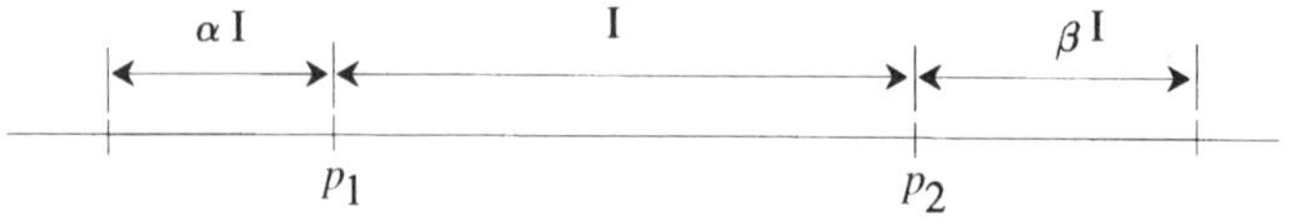

Figure 1.13. Blend crossover in a one-dimensional case.

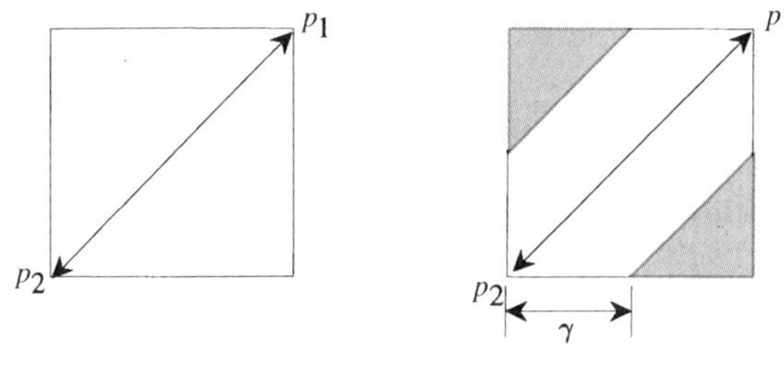

Figure 1.14. Box-BLX and directional-BLX in a two-dimensional case.

by the n intervals. This version of blend crossover is referred to as *box-BLX* to distinguish it from another type of blend crossover given by Eshelman, Mathias, and Schaffer, called *directional-BLX* [179]. Box-BLX samples the hyper-rectangular defined by the parents uniformly, whereas directional-BLX has a sampling bias toward a region that parallels the diagonal defined by the parents. There is an assumption that if the population distribution lies along a region that forms a diagonal in the problem space, then instead of sampling points formed by a box defined by two parents, it makes more sense to sample in a diagonal region defined by the parents. Directional-BLX is specified as BLX–α–β–γ. The easiest way of describing directional-BLX is to think of it as a form of box-BLX in which the corners opposite the diagonal connecting the two parents have been trimmed. The parameter γ indicates the amount that the box is to be trimmed. Figure 1.14 illustrates box-BLX and directional-BLX. In the case BLX–0.0–0.0, the sampling region is constrained within the box specified by two parents; in the case BLX–0.0–0.0–0.5, two corners that are opposite the diagonal are trimmed of the sampling region.

Unimodal Normal Distribution Crossover. The UNDX crossover method proposed by Ono and Kobayshi [492], based on an idea similar to Eshleman's directional-BLX, gives the crossover a sampling bias toward the diagonal region defined by parents. The UNDX generates two children from a region of normal distribution defined by three parents. Figure 1.15 illustrates UNDX in a two-dimensional case. In one dimension defined by two parents, p_1 and p_2, the standard deviation of the normal distribution is proportional to the distance between parents p_1 and p_2. In the dimension orthogonal to the first, the standard deviation of the normal distribution is proportional to the distance of the third parent, p_3 from the line. The distance is divided by $\sqrt{n}$ to reduce the influence of the third parent.

Let P_1 and P_2 denote the parent vectors, C_1 and C_2 the child vectors, n the number of variables, d_1 the distance between parents p_1 and p_2, d_2 the distance of parent p_3 from the axis connecting parents p_1 and p_2, z_1 a random number with the normal distribution $N(0,\sigma_1^2)$, z_k a random number with the normal

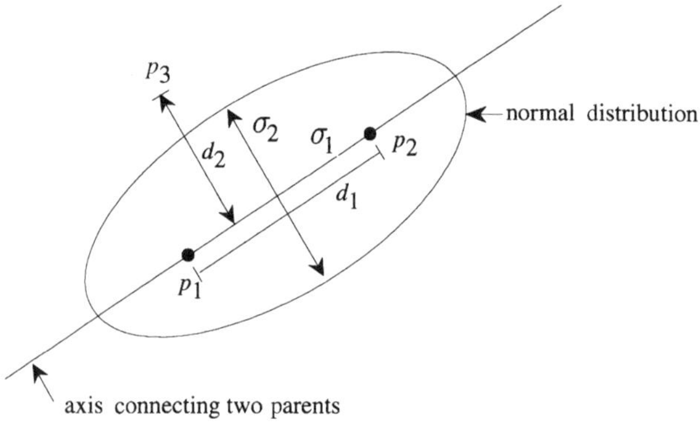

Figure 1.15. Unimodal normal distribution crossover.

distribution $N(0,\sigma_k^2)$, $k = 2,3...,n$, and α and β constants. The children are generated as follows:

$$C_1 = m + z_1 e_1 + \sum_{k=2}^{n} z_k e_k \tag{1.21}$$

$$C_2 = m - z_1 e_1 - \sum_{k=2}^{n} z_k e_k \tag{1.22}$$

$$m = \frac{P_1 + P_2}{2} \tag{1.23}$$

$$z_k \sim N(0,\sigma_k^2), \qquad k = 1,2,...,n \tag{1.24}$$

$$\sigma_1 = \alpha d_1, \qquad \sigma_2 = \frac{\beta d_2}{\sqrt{n}} \tag{1.25}$$

$$e_1 = \frac{P_2 - P_1}{|P_2 - P_1|} \tag{1.26}$$

$$e_i \perp e_j, \qquad i,j = 1,2,....,n, \text{ and } i \neq j \tag{1.27}$$

The UNDX is relatively independent of coordinate systems since it produces children based not on the axes of the coordinate system, but on the line connecting the parents.

Boundary Operators. This method was proposed by Schoenauer and Michalewicz [565]. The significance of this line of research is based on the observation that the global solution for many optimization problems usually lies on the boundary of the feasible region. Thus, for may constrained optimization problems, it may be beneficial to search just the boundary of the search space defined by a set of constraints. The *sphere crossover* is an example of such an approach. It produces one offspring $(z_2, z_2, \ldots, z_n)$ from two parents $(x_1, x_2, \ldots, x_n)$ and $(y_1, y_2, \ldots, y_n)$ as follows:

$$z_i = \sqrt{\alpha x_i^2 + (1 - \alpha) y_i^2}, \qquad i = 1,2,\ldots,n \tag{1.28}$$

where α is a random number chosen in $[1,n]$.

Direction-Based Crossover. This method uses the values of the objective function in determining the direction of genetic search [456]. The operator generates a single offspring $\boldsymbol{x}'$ from two parents $\boldsymbol{x}_1$ and $\boldsymbol{x}_2$ according to the following rule:

$$\boldsymbol{x}' = r(\boldsymbol{x}_2 - \boldsymbol{x}_1) + \boldsymbol{x}_2 \tag{1.29}$$

where r is a random number between 0 and 1. It also assumes that the parent $\boldsymbol{x}_2$ is not worse than $\boldsymbol{x}_1$; that is, $f(\boldsymbol{x}_2) \geq f(\boldsymbol{x}_1)$ for maximization problems and $f(\boldsymbol{x}_2) \leq f(\boldsymbol{x}_1)$ for minimization problems.

Nonuniform Mutation. This method is given by Janilow and Michalewicz [455]. It is designed for fine-tuning capabilities aimed at achieving high precision. For a given parent $\boldsymbol{x}$, if the element x_k of it is selected for mutation, the resultant offspring is $\boldsymbol{x}' = (x_1, \ldots, x_k', \ldots, x_n)$, where x_k' is randomly selected from the following two possible choices:

$$x_k' = x_k + \Delta(t, x_k^U - x_k) \tag{1.30}$$

$$x_k' = x_k - \Delta(t, x_k - x_k^L) \tag{1.31}$$

The function $\Delta(t,y)$ returns a value in the range $[0,y]$ such that the value of $\Delta(t,y)$ approaches 0 as t increases (t is the generation number). This property causes the operator to search the space uniformly initially (when t is small), and very locally at later stages. The function $\Delta(t,y)$ is given as follows:

$$\Delta(t,y) = yr\left(1 - \frac{t}{T}\right)^b \tag{1.32}$$

where r is a random number from $[0,1]$, T the maximum generation number, and b a parameter determining the degree of nonuniformity. It is possible for the operator to generate an offspring that is not feasible. In such a case, we can reduce the value of random number r.

Directional Mutation. This method is given by Gen, Liu, and Ida [237, 238]. Let $\boldsymbol{d}$ be the approximate direction of the gradient. The offspring after mutation would be

$$\boldsymbol{x}' = \boldsymbol{x} + r \cdot \boldsymbol{d}$$

where r is a random nonnegative real number, which is selected in such a way that the offspring is a feasible solution. The ith component of the approximate gradient d is calculated as follows:

$$\frac{f(x_1,\ldots,x_i + \Delta x_i,\ldots,x_n) - f(x_1,\ldots,x_i,\ldots,x_n)}{\Delta x_i} \tag{1.33}$$

where Δx_i is a small real number.

The direction can also be given randomly as a free direction to avoid the chromosomes jamming into a *corner*. If a chromosome is near the boundary, the mutation direction given by some criteria might point toward the close boundary, and jamming would occur. In such a case we can choose a random direction instead of the given direction.

Gaussian Mutation. This method was originally developed for evolution strategies [31, 32]. Generally, an individual in evolution strategies consists of two components $(\boldsymbol{x},\boldsymbol{\sigma})$, where the first vector $\boldsymbol{x}$ represents a point in the search space, and the second vector $\boldsymbol{\sigma}$ represents stanndard deviation. An offsprig $(\boldsymbol{x}',\boldsymbol{\sigma}')$ is generated as follows:

$$\boldsymbol{\sigma}' = \boldsymbol{\sigma} e^{N(0,\Delta\boldsymbol{\sigma})} \tag{1.34}$$

$$\boldsymbol{x}' = \boldsymbol{x} + N(0,\Delta\boldsymbol{\sigma}') \tag{1.35}$$

where $N(0,\Delta\boldsymbol{\sigma}')$ is a vector of independent random Gaussion numbers with a mean of zero and standard deviations $\boldsymbol{\sigma}$.

1.3.2 Constrained Optimizations

Nonlinear programming (or constrained optimization) deals with the problem of optimizing an objective function in the presence of equality and/or inequality constraints. Nonlinear programming is an extremely important tool used in almost every area of engineering, operations research, and mathematics because many practical problems cannot be modeled successfully as a linear program. The general nonlinear programming may be written as follows:

$$\max \quad f(\boldsymbol{x}) \tag{1.36}$$

$$\text{s.t.} \quad g_i(\boldsymbol{x}) \leq 0, \qquad i = 1,2,\ldots,m_1 \tag{1.37}$$

$$h_i(\boldsymbol{x}) = 0, \qquad i = m_1 + 1,\ldots,m \ (= m_1 + m_2) \tag{1.38}$$

$$\boldsymbol{x} \in X \tag{1.39}$$

where f, $g_1,g_2,\ldots,g_{m_1}$, $h_{m_1+1},h_{m_1+2},\ldots,h_m$ are real-valued functions defined on

E_n, X a subset of E_n, and $\boldsymbol{x}$ an n-dimensional real vector with components $x_1,x_2,\ldots,x_n$. The problem above must be solved for the values of the variables $x_1,x_2,\ldots,x_n$ that satisfy the restrictions and meanwhile minimize the function f. The function f is usually called the *objective function* or *criterion function*. Each of the constraints $g_i(\boldsymbol{x}) \leq 0$ is called an *inequality constraint*, and each of the constraints $h_i(\boldsymbol{x}) = 0$ is called an *equality constraint*. The set X, which might typically include lower and upper bounds on the variables, is usually called the *domain constraint*. A vector $\boldsymbol{x} \in X$ satisfying all the constraints is called a *feasible solution* to the problem. The collection of all such solutions forms the *feasible region*. The nonlinear programming problem then is to find a feasible point $\bar{\boldsymbol{x}}$ such that $f(\boldsymbol{x}) \leq f(\bar{\boldsymbol{x}})$ for each feasible point $\boldsymbol{x}$. Such a point is called an *optimal solution*. Unlike linear programming problems, the conventional solution methods for nonlinear programming are very complex and not very efficient. In the last few years, there has been a growing effort to apply genetic algorithms to nonlinear programming problems [537]. The section shows how to solve nonlinear programming problems with genetic algorithms in general.

The central difficulty in applying genetic algorithms to these types of problems is how to handle constraints, because genetic operators used to manipulate the chromosomes often yield infeasible offspring. Recently, several techniques have been proposed to handle constraints with genetic algorithms [452, 494]. The existing techniques can be roughly classified as follows: rejecting methods, repairing methods, and penalty methods. A rejecting method discards all infeasible chromosomes created throughout evolutionary process. It is the simplest but least effective way to handle the problem. Repairing methods involve taking an infeasible chromosome and generating a feasible one through a repair procedure. For many combinatorial optimization problems, it is relatively easy to create a repair procedure.

The penalty technique is perhaps the most common technique used in genetic algorithms for constrained optimization problems. In essence, this technique transforms a constrained problem into an unconstrained problem by penalizing infeasible solutions, in which a penalty term is added to the objective function for any violation of the constraints.

The basic idea of the penalty technique is borrowed from conventional optimizations. It is a natural to ask: Is there any difference when we use the penalty method in conventional optimization and in genetic algorithms? In conventional optimization, the penalty technique is used to generate a sequence of infeasible points whose limit is an optimal solution to the original problem. The major concern is how to choose a proper penalty value so as to hasten convergence and avoid premature termination. In genetic algorithms, the penalty technique is used to keep a certain number of infeasible solutions in each generation so as to enforce the genetic search toward an optimal solution from both sides of feasible and infeasible regions. We do not simply reject infeasible solutions in each generation because some may provide much more useful information about optimal solutions than do some feasible solutions. The major concerns are how to determine the penalty term so as to

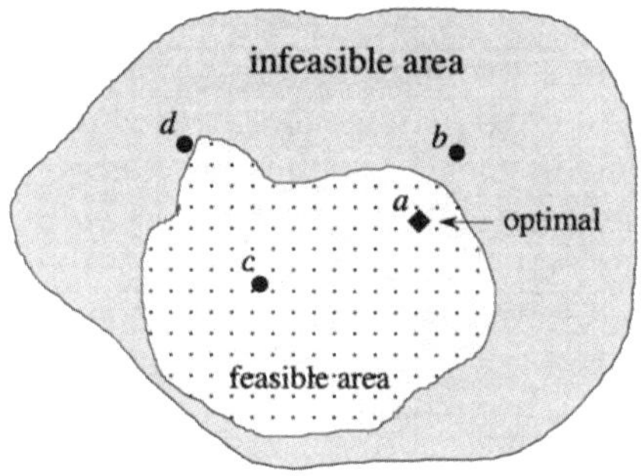

Figure 1.16. Solution space: feasible area and infeasible area.

strike a balance between information preservation (keeping some infeasible solutions) and selective pressure (rejecting some infeasible solutions) and avoiding both underpenalty and overpenalty.

In general, the solution space contains two parts: feasible and infeasible areas. We do not make any assumptions about these subspaces; in particular, they need not be convex or connected as shown in Figure 1.16. Handling infeasible chromosomes is far from trivial. From the figure we can see that infeasible solution b is much nearer to optimal solution a than infeasible solution d and feasible solution c. We may hope to give a lower penalty to b than to d even though it is a little farther than d from the feasible area. We also can believe that b contains much more information about the optimal solution than c does even though it is infeasible. However, we have no a priori knowledge about the optimal solution, so generally it is very hard for us to judge which solution is best. The principal issue of the penalty strategy is how to design penalty function $p(\boldsymbol{x})$ so as to effectively guide the genetic search toward a promising area of solution space. The relationship between infeasible chromosomes and the feasible part of the search space plays a significant role in penalizing infeasible chromosomes. The penalty value corresponds to the "amount" of its infeasibility according to some measurement. There are no general guidelines for designing a penalty function, and constructing an efficient penalty function is quite problem dependent.

Evaluation Function with Penalty Term. Penalty techniques transform the constrained problem into an unconstrained problem by penalizing infeasible solutions. In general, there are two possible ways to construct an evaluation function with a penalty term. One is to take the addition form, expressed as follows:

$$\text{eval}(\boldsymbol{x}) = f(\boldsymbol{x}) + p(\boldsymbol{x}) \tag{1.40}$$

where $\boldsymbol{x}$ represents a chromosome, $f(\boldsymbol{x})$ the objective function of problem, and $p(\boldsymbol{x})$ the penalty term. For maximization problems, we usually require that

$$\begin{aligned} p(\boldsymbol{x}) &= 0, \quad && \text{if } \boldsymbol{x} \text{ is feasible} \\ p(\boldsymbol{x}) &< 0, \quad && \text{otherwise} \end{aligned} \tag{1.41}$$

Let $|p(\boldsymbol{x})|_{\max}$ and $|f(\boldsymbol{x})|_{\min}$ be the maximum of $|p(\boldsymbol{x})|$ and minimum of $|f(\boldsymbol{x})|$ among the current population, respectively. We also require that

$$|p(\boldsymbol{x})|_{\max} \le |f(\boldsymbol{x})|_{\min} \tag{1.42}$$

to avoid negative fitness value. For minimization problems, we usually require that

$$\begin{aligned} p(\boldsymbol{x}) &= 0, \quad && \text{if } \boldsymbol{x} \text{ is feasible} \\ p(\boldsymbol{x}) &> 0, \quad && \text{otherwise} \end{aligned} \tag{1.43}$$

The second way is to take the multiplication form expressed as follows:

$$\text{eval}(\boldsymbol{x}) = f(\boldsymbol{x})p(\boldsymbol{x}) \tag{1.44}$$

In this case, for maximization problems we require that

$$\begin{aligned} p(\boldsymbol{x}) &= 1, \quad && \text{if } \boldsymbol{x} \text{ is feasible} \\ 0 \le p(\boldsymbol{x}) &< 1, \quad && \text{otherwise} \end{aligned} \tag{1.45}$$

and for minimization problems, we require that

$$\begin{aligned} p(\boldsymbol{x}) &= 1, \quad && \text{if } \boldsymbol{x} \text{ is feasible} \\ p(\boldsymbol{x}) &> 1, \quad && \text{otherwise} \end{aligned} \tag{1.46}$$

Note that for minimization problems, the fitter chromosome has the lower value of eval($\boldsymbol{x}$). For some selection methods it is required to transform the objective values into fitness values such way that the fitter one has the larger fitness value.

Classification of the Penalty Function. Several techniques for handling infeasibility have been proposed in the area of genetic algorithms. In general, we can classify them into two classes: (1) constant penalty and (2) variable penalty. The constant penalty approach is known to be less effective for complex problems, and most recent research work focuses on the variable penalty.

In general, the variable penalty approach contains two components: (1) a variable penalty ratio and (2) a penalty amount for the violation of constraints. The variable penalty ratio can be adjusted according to (1) the degree of violation of constraints and (2) the iteration number of genetic algorithms. The first approach increases the penalty pressure as the violation becomes severe, which leads to the class of *static penalty,* and the second approach increases the penalty pressure along with growth of the evolutionary process, which leads to the class of *dynamic penalty* as discussed by Michalewicz [453].

Essentially, the penalty is a function of the distance from the feasible area. This can be given in the following three possible ways:

1. As a function of absolute distance of a single infeasible solution
2. As a function of relative distance of all infeasible solutions in the current population
3. As a function of the adaptive penalty term

Most methods take the first approach. For highly constrained problems, the ratio of infeasible to feasible solutions is relatively high in each generation. In such cases, the second and third approaches may be used to make a good balance between the preservation of information and the pressure an infeasible solutions.

The penalty approaches can be further distinguished as problem dependent or problem independent. Most penalty techniques belong to the problem-dependent approach.

The penalty approaches can also be distinguished as with parameter or without parameter. Most penalty techniques follow a parameterized approach. Parameterized penalty methods tend to be problem dependent.

1.3.3 Combinatorial Optimizations

Combinatorial optimization problems are characterized by a finite number of feasible solutions. Although, in principle, the optimal solution to such a finite problem can be found by a simple enumeration, in practice it is frequently impossible, especially for practical problems of realistic size where the number of feasible solutions can be extremely high. A major trend in solving such hard problems is to utilize a heuristic search. In the past decade, genetic algorithms have experienced increased interest from the combinatorial optimization community and show great promise both in experimentation and in practice in many industrial engineering areas.

Combinatorial optimizations contain a huge body of problems with different features and properties. Although these problems are quite different from each other, the essence of the problems can be characterized as one of the following types:

- To determine a permutation of some items associated with the problem
- To determine a combination of some items
- To determine both permutation and combination of some items
- Any one of the above subject to some constraints

For example, the essence of resource-constrained project scheduling problems and vehicle routing and scheduling problems is to determine a permutation of some items subject to some constraints, the essence of set-covering problems and grouping problems is to determine a combination of items, and the essence of the parallel machine scheduling problem is to determine both a permutation and a combination of items subject to certain constraints.

A common feature of the problems is that if the permutation and/or combination can be determined, a solution can then easily be derived with a problem-specific procedure. So the general approach for applying genetic algorithms to these problems is as follows:

1. Use genetic algorithms to evolve an appropriate permutation and/or combination of items under consideration.
2. Then use a heuristic method to construct a solution according to the permutation and combination.

How to adapt genetic algorithms to combinatorial optimization problems is both challenging and frustrating. A key step is how to encode a solution of the problem into a chromosome. Because of the existence of complex constraints inherent in combinatorial optimization problems, a simple binary string does not work at all, since it indubitably yields infeasible or even illegal solutions. The virgin efforts of most researchers have been devoted to the invention of a new and efficient encoding method to the problem, resulting in a very active research field called *order-based genetic algorithms* [249]. In Chapter 2 we explain how to solve some typical combinatorial optimization problems with genetic algorithms, including the set-covering problem, the bin packing problem, the knapsack problem, and the minimum spanning tree problem.

1.3.4 Multiobjective Optimizations

Multiple-objective optimization problems have received increased interest from researchers with various backgrounds since early 1960. In a multiobjective optimization problem, multiple objective functions need to be optimized simultaneously. In the case of multiple objectives, there does not necessarily exist a solution that is best with respect to all objectives because of incommensurability and confliction among objectives. A solution may be best in one objective but worst in another. Therefore, there usually exist a set of solutions for the multiple-objective case which cannot simply be compared with each other. For such solutions, called *nondominated solutions* or *Pareto optimal solutions,* no improvement is possible in any objective function without sacrificing at least one of the other objective functions.

In the past few years, there has been a boom in applying genetic algorithms to solving the multiobjective optimization problem, known as *evolutionary multiobjective optimization* or *genetic multiobjective optimization*. The basic feature of genetic algorithms is the multiple directional and global search, in which a population of potential solutions is maintained from generation to generation. The population-to-population approach is beneficial in the exploration of Pareto solutions.

One of the special issues arising in solving multiobjective optimization problems by use of genetic algorithms is how to determine the fitness value of individuals according to multiple objectives. The issue of fitness assignment mechanisms has been studied extensively during the past decade, and several methods have been suggested and tested. The details of this topic are introduced in Chapter 3.

1.4 RECENT GENETIC ALGORITHM DISSERTATIONS

During the past 10 years, more then 200 Ph.D. dissertations have been published on studies of genetic algorithms and applications in diversity disciplinary ranging from biology to engineering optimizations. The incomplete stochastic data show the ongoing trend of genetic algorithm–related studies in universities around the world. The genetic algorithm was developed originally as a technique of function optimization derived from the principles of evolutionary theory. In recent years it has been found very useful for solving many problems in the fields of biology. For example, genetic algorithms were used to predict protein tertiary structure and interpret the electrical spiking behavior of neurons.

Biology

Year	Author	Title of Dissertation
1993	Grand	Application of the genetic algorithm to protein tertiary structure prediction [263]
1994	Xiao	Computer-assisted drug design: genetic algorithms and structures of molecular clusters of aromatic hydrocarbons and actinomycin D-deoxyguanosine [670]
1995	Besson	Evaluation of the genetic algorithm as a computational tool in protein NMR [60]
1996	Bangalore	Data analysis strategies for qualitative and quantitative determination of organic compounds by Fourier transform infrared spectroscopy [42]
1996	Eichler	On the development and interpretation of parameter manifolds for biophysically robust compartmental models of CA3 hippocampal neurons [173]
1996	Hase	Simulation on mutual adaptation of a body form and biped walking by nerve muscle skelation model and genetic algorithms [288]

Biology ***(Continued)***

Year	Author	Title of Dissertation
1996	Lohn	Automated discovery of self-replicating structures in cellular space automata models [423]
1996	Parrill	Applications of artificial intelligence in drug design [504]
1997	Ho	Simulation of condensed-phase physical, chemical, and biological systems on massively parallel computers [301]
1997	Rocha	Evidence sets and contextual genetic algorithms: exploring uncertainty, context, and embodiment in cognitive and biological systems [538]
1998	Niesse	Structural optimization of atomic and molecular microclusters using a genetic algorithm in real-valued space-fixed coordinates [484]
1997	Palazolo	Use of genetic algorithms in bounded search for design of biological nitrification/denitrification waste treatment systems [497]

Combinatorial optimization problems are a major research target in the genetic algorithm community. Genetic algorithms have demonstrated great power with very promising results for many difficult-to-solve problems.

Combinatorial Optimization

Year	Author	Title of Dissertation
1994	Levine	Parallel genetic algorithm for the set partitioning problem [406]
1995	Lin	High-quality tour hybrid genetic schemes for TSP optimization problems [417]
1995	Nakamura	Study on protocol design and combinatorial optimization in resource assignment problems on the basis of net theory and genetic algorithms [480]
1995	Tagami	Study on application method of genetic algorithms for combinatorial optimization problems [606]
1998	Ikonen	Genetic algorithm for a three-dimensional non-convex bin packing problem [317]
1999	Zhou	Study on constrained spanning tree problems with genetic algorithms [695]

Genetic algorithms have also been applied to solving many problems in the fields of computer science, including parallelizing and scheduling programs on multicomputers, distributed database design, and others.

Computer Science

Year	Author	Title of Dissertation
1994	Adar	Allocation and scheduling on multi-computers using genetic algorithms [5]
1994	Brown	Optimal rate concept acquisition using version spaces and genetic algorithms [78]
1995	Rho	Distributed database design: allocation of data and operations to nodes in distributed database systems [536]

Control of dynamic behavior of engineering systems is a difficult and challenging problem. Recently, genetic algorithms have been applied to control design: the design problem is formulated as a parameter search problem, and genetic algorithms are used to search for appropriate values of control parameters and the nominal trajectory to yield good control performance.

Engineering Control

Year	Author	Title of Dissertation
1994	Jeon	Genetic algorithm–based non-linear modeling and its application to control and mechanical diagnostics [329]
1995	Hachino	Study on identification of continuous time-delay systems and lower dimensionizing of models by genetic algorithms [275]
1995	Memon	Optimization methods for real-time traffic control [449]
1995	Munemoto	Study on disperse control–type dynamic load balance using genetic algorithms [473]
1995	Yang	Study on signature verification using a genetic algorithm method [678]
1996	Cheng	Control design and robustness measurement for biped locomotion [102]
1997	Abido	Intelligent techniques approach to power system identification and control [1]
1997	Hajda	Genetic algorithms for control of wastewater conveyance systems [276]
1997	Marchelya	Enforcing pollution control laws: Stackelberg–Markov game models for improving efficiency and effectiveness [437]
1997	Masters	Evolutionary design of controlled structures [443]

Engineering design occupies a major body of research and applications of genetic algorithms: for example, topological structural design, network design, VLSI layout, dynamic system design and integration, nonlinear filtering, facility location, manufacturing cell design, pollution control, kinematic design of turbine blade fixtures, and so on.

Engineering Design

Year	Author	Title of Dissertation
1992	Bowden	Genetic algorithm–based machine learning applied to the dynamic routing of discrete parts [72]
1992	Jensen	Topological structural design using genetic algorithms [327]
1992	Ros	Learning Boolean functions with genetic algorithms: a PAC analysis [541]
1992	Yao	Parameter estimation for nonlinear systems [679]
1993	Lee	Three new algorithms for exact D-optimal design problems [394]
1994	Fluerent	Algorithmes génétiques hybrides pour l'optimisation combinatoire [192]
1994	Palmer	Approach to a problem in network design using genetic algorithms [499]
1994	Shahookar	Application of the genetic algorithm for computer-aided design of VLSI layout [573]
1995	Abuali	Using determinant and cycle basis schemes in genetic algorithms for graph and network applications [2]
1995	Crossley	Using genetic algorithms as an automated methodology for conceptual design of rotorcraft [138]
1995	Huang	Data-driven description of reservoir petrophysical properties [311]
1995	Nims	Contingency ranking for voltage stability using a genetic algorithm [485]
1995	Noga	Performance analysis and simulation of a class of numerical demodulation techniques for large-deviation FM signals [486]
1995	Pheatt	Construction of D-optimal experimental designs using genetic algorithms [508]
1995	Sun	Evolving population-based search algorithms through thermodynamic operation: dynamic system design and integration [600]
1995	Tay	Automated generation and analysis of dynamic system designs [617]
1995	Tomasz	Nonlinear adaptive filtering: the genetic algorithm approach [624]
1995	Yu	Crystal deformation due to a dipping fault in an elastic gravitational layer overlying a viscoelastic gravitational layer overlying a visoelastic gravitational half-space [687]

Engineering Design (*Continued*)

Year	Author	Title of Dissertation
1996	Brown	Using the facility location problem to explore operator policies and constraint-handling methods for genetic algorithms [77]
1996	Chaer	Mixture-of-experts approach to adaptive estimation [90]
1996	Joines	Hybrid genetic search for manufacturing cell design [332]
1996	Larson	Vibration testing by design: excitation and sensor placement using genetic algorithms [391]
1996	Lee	Genetic algorithms in multidisciplinary design of low-vibration rotors [400]
1996	Li	Genetic algorithm–based design of multiplierless two-dimensional state-space digital filters [415]
1996	Liu	Application of multiobjective genetic algorithms to control systems design [422]
1996	Lu	Study of genetic algorithms on an application to industrial design [426]
1996	Mock	Intelligent information filtering via hybrid techniques: hill climbing, case-based reasoning, index patterns, and genetic algorithms [458]
1996	Monem	Performance evaluation and optimization of irrigation canal system using genetic algorithm [460]
1996	Patel	Genetic algorithms: a tool for quantitative structure-activity relationship analyses [505]
1996	Starns	Optimal synthesis of a planar four-bar mechanism with prescribed timing using generalized reduced gradient, simulated annealing, and genetic algorithms [585]
1996	Streifel	Application of computational intelligence to electromechanical systems [591]
1996	Houck	Metaheuristics for manufacturing problems [309]
1997	Lazio	Genetic algorithms, pulsar planets, and ionized interstellar microturbulence [392]
1997	Maifeld	Genetic-based unit commitment algorithm [433]
1997	Mantawy	Unit commitment by artificial intelligence techniques [436]
1997	Marchelya	Enforcing pollution control laws: Stackelberg–Markov game models for improving efficiency and effectiveness [437]
1997	Tang	Genetic algorithms for optimal operation of soil aquifer treatment systems [615]
1998	Gold	Application of the genetic algorithm to the kinematic design of turbine blade fixtures [247]
1998	Guan	Applications of genetic algorithms in groundwater quality management [271]
1998	Jallas	Improved model-based decision support by modeling cotton varability and using evolutionary algorithms [323]
1999	Sasaki	Studies on solution methods for fuzzy reliability design problems by hybrid genetic algorithms [560]

The need to solve optimization problems arises in almost every field and, in particular, is a dominant theme in the engineering world. Many optimization problems from the engineering world are very complex in nature and quite difficult to solve by conventional optimization techniques. Genetic algorithms have received considerable attention regarding their potential as an optimization technique for complex problems and have been applied successfully in the area of industrial engineering.

Engineering Optimization

Year	Author	Title of Dissertation
1991	Annaiyappa	Critical analysis of genetic algorithms for global optimization [16]
1992	Lee	Tolerance optimization using genetic algorithm and approximated simulation [399]
1993	Yunker	Optimization of simulation models by genetic algorithms: a comparative study [689]
1994	Hart	Adaptive global optimization with local search [287]
1994	Nakanishi	Optimization of a structure phase by homology theory and genetic algorithms [481]
1995	Areibi	Toward optimal circuit layout using advanced search techniques [19]
1995	DeChaine	Stochastic fuel management optimization using genetic algorithms and heuristic rules [152]
1995	Dozier	Constraint processing using adaptive microevolutionary/systematic hill climbing [166]
1995	Golden	Adaptive approximation and optimization of transform functions [253]
1995	Maria	Genetic algorithm for multimodel continuous optimization problems [438]
1995	Nolte	Global optimization algorithms for seismic inversion [487]
1995	Pinon	Investigation of the applicability of genetic algorithms to spacecraft trajectory optimization [511]
1995	Tompkins	Optimization of qualitative variables in discrete event simulation models [625]
1996	Yokota	Study on solving system reliability optimization problems with interval data by genetic algorithms [683]
1996	Liu	Application of multiobjective genetic algorithms to control systems design [422]
1996	Michael	Form-finding analysis of vibrating towered shells of revolution as an inverse eigenproblem and as a genetic algorithm optimization problem [451]
1996	Moon	High-performance simulation-based optimization environment for large-scale systems [466]

Engineering Optimization (*Continued*)

Year	Author	Title of Dissertation
1996	Murata	Genetic algorithms for multiobjective optimization [475]
1997	Canpolat	Optimization of seasonal irrigation scheduling by genetic algorithms [84]
1997	Filho	Optimizing hydrocarbon field development using a genetic algorithm–based approach [188]
1998	Matthew	Applications and limitations of genetic algorithms for the optimization of multivariate calibration techniques [445]

Since the early 1970s when Holland first specified the genetic algorithms, enormous effort has been devoted to theoretical investigation of genetic algorithms to explain why and how genetic algorithms work. The foundational issues include coding and representation, variation and recombination, fitness landscapes and genetic operators, selection and convergence, parallelization, deception, genetic diversity, parameter adaptation, and others.

Foundation of Genetic Algorithms

Year	Author	Title of Dissertation
1992	Collins	Studies in artificial evolution [131]
1992	Shu	Impact of data structures on the performance of genetic algorithm–based learning [575]
1993	Kommu	Enhanced genetic algorithms in constrained search spaces with emphasis on parallel environments [373]
1993	Rankin	Consideration of rapidly converging genetic algorithms designed for application to problems with expensive evaluation functions [543]
1994	Corcoran	Techniques for reducing the disruption of superior building blocks in genetic algorithms [133]
1994	Giddens	Determination of invariant parameters and operators of the traditional genetic algorithm using an adaptive genetic algorithm generator [244]
1994	Gordon	Exploitable parallelism in genetic algorithms [258]
1995	Cavaretta	Cultural algorithms and real-valued function optimization [87]
1995	Cedeño	The multiniche crowding genetic algorithm: analysis and applications [88]
1995	Chung	Analysis of the effects of different representation schemes on genetic algorithms [124]

Foundation of Genetic Algorithms (*Continued*)

Year	Author	Title of Dissertation
1995	Jones	The Hereford Ranch algorithm: an improvement in genetic algorithms using selective breeding [340]
1995	Mahfoud	Niching methods for genetic algorithms [432]
1995	Meddin	Genetic algorithms: a Markov chain and detail balance approach [448]
1995	Takashi	Study of evolutional mechanisms of cooperative problem solving: a hybrid genetic algorithm for multialgorithm problems [610]
1995	Wu	Noncoding DNA and floating building blocks for the genetic algorithm [669]
1996	Crawford	Role of recombination in genetic algorithms for fixed-length subset problems [137]
1996	Eberlein	GA heuristic generically has hyperbolic fixed points [169]
1996	Hightower	Computational aspects of antibody gene families [297]
1996	Kargupta	Search, polynomial complexity, and the fast messy genetic algorithm [346]
1996	Merkle	Analysis of linkage-friendly genetic algorithms [450]
1996	Picardo	Framework for the analysis of evolutionary algorithms [509]
1996	Sato	TA proposal and analysis on the alternation of generation models of genetic algorithms [561]
1996	Stumpf	Studies on enhanced operator-oriented genetic algorithms [592]
1996	Wang	Matrix genome encoding for genetic algorithms [654]
1996	Wong	Performance analysis for genetic algorithms [665]
1996	Xie	Properties and characteristics of genetic algorithms as a problem-solving method [671]
1997	Harik	Learning gene linkage to efficiently solve problems of bounded difficulty using genetic algorithms [286]
1997	Kaiser	Dynamic load distributions for adaptive computations on MIMD machines using hybrid genetic algorithms [342]
1997	Potter	Design and analysis of a computational model of cooperative coevolution [513]
1997	Rosin	Coevolutionary search among adversaries [542]
1998	Rasheed	GADO: a genetic algorithm for continuous design optimization [527]
1998	Sarma	Analysis of decentralized and spatially distributed genetic algorithms [559]

Foundation of Genetic Algorithms (*Continued*)

Year	Author	Title of Dissertation
1998	Spears	Role of imitation and recombination in evolutionary algorithms [580]
1998	Voicu	Multiresolution aspects in genetic algorithms [643]

Recently, there has been increasing interest in combining fuzzy concepts and genetic algorithms, known as *fuzzy genetic algorithms*. Two basic approaches to the use of fuzzy logic–based techniques have been investigated: to model different components of genetic algorithms and to manage problems in an imprecise environment.

Genetic Algorithms and Fuzzy Logic

Year	Author	Title of Dissertation
1994	Lee	Automatic design and adaptation of fuzzy systems and genetic algorithms using soft computing techniques [401]
1994	Takagi	Study on an automatic structure method using a fuzzy reasoning controller with neural networks and genetic algorithms [609]
1995	Ye	Fuzzy modeling a nonlinear systems and eugenicofs-based genetic algorithms [681]
1996	Chbat	Direct-learning neural and fuzzy control of two unknown dynamic systems using Powell optimization and genetic algorithms [99]
1996	Velthuizen	Feature extraction with genetic algorithms for fuzzy clustering [637]
1996	Xu	Genetic algorithms' searching capability and their application for optimization of a model reference fuzzy adaptive controller for linear systems with time delay [674]
1997	Egan	Validity-guided robust fuzzy clustering methods for characterizing clusters in noisy data [172]
1997	Inoue	Study on an automatic generation method of fuzzy rules by genetic algorithms [318]
1997	Li	Oil tanker market modeling, analysis, and forecasting using neural networks, fuzzy logic, and genetic algorithms [409]
1998	Li	Multiobjective optimum design of static and seismic-resistant structures with genetic algorithms, fuzzy logic, and game theory [407]
1998	Rajyabaquse	Fuzzy logic application of genetic algorithm techniques for adaptive control of nonlinear systems [524]
1998	Wozniak	Knowledge evolution in active database systems via fuzzy constraint management [666]

Genetic algorithms and neural networks are both inspired by computation in biological systems. A good deal of biological neural architecture is determined genetically. It is therefore not surprising that many researchers explored how to evolve neural architectures with genetic algorithms.

Genetic Algorithms and Neural Networks

Year	Author	Title of Dissertation
1992	Guo	Nuclear power plant fault diagnostics and thermal performance studies using neural networks and genetic algorithms [272]
1992	Rogers	Optimal groundwater remediation using artificial neural networks and a genetic algorithm [539]
1993	Caskey	Genetic algorithms and neural networks applied to manufacturing scheduling [86]
1993	White	GANNet: a genetic algorithm for searching topology and weight spaces in neural network design—the first step in finding a neural network solution [661]
1995	Hanagandi	Process control, optimization, and modeling of chemical systems using genetic algorithms and neural networks [280]
1995	Hashimoto	Study of neural network and genetic algorithms on an application to production planning problems [289]
1995	Igata	Study on estimation of short-time rainy areas using optimized neural networks by genetic algorithms [314]
1996	Arguelles	Hybrid artificial neural network/genetic algorithm approach to the on-line optimization of electrical power systems [20]
1996	Han	Modeling and optimization of plasma-enhanced chemical vapor deposition using neural networks and genetic algorithms [279]
1996	Jin	Study on optimal neural network design and its application using genetic algorithms [330]
1996	Kermani	On using artificial neural networks and genetic algorithms to optimize performance of an electric noise [355]
1996	Oh	On-line optimization of substrate feeding in a hybrid cell culture using an artificial neural network and a genetic algorithm [489]
1997	Chen	Hybrid intelligent system for process modeling control using a neural network and a genetic algorithm [710]

Genetic Algorithms and Neural Networks (*Continued*)

Year	Author	Title of Dissertation
1997	Kang	Learning evasive maneuvers using evolutionary algorithms and neural networks [344]
1997	Zhou	Study on the convergence of genetic algorithms and its application to structural determination of feedforward neural networks [694]
1993	Fujimoto	Study on parallel processing architecture on neural networks and genetic algorithms [207]
1998	Gong	Evolutionary computation and artificial neural networks for optimization problems and their applications [255]

Image processing and pattern recognition are additional research targets for the genetic algorithm community.

Image Processing and Pattern Recognition

Year	Author	Title of Dissertation
1993	Tadikonda	Automated image segmentation and interpretation using genetic algorithms and semantic nets [605]
1995	Chintrakulchai	High-performance fractal image compression [118]
1996	Bakalis	Encoding the invariant measure of iterated affine transforms with a genetic algorithm [34]
1996	Chunduru	Global and hybrid optimization in geophysical inversion [123]
1996	Matsui	Expression of genetic algorithms to macroadaptation degree and its application to picture processing systems [444]
1996	Swets	Self-organizing heirarchical optimal subspace learning and inference framework for view-based recognition and image retrieval [602]
1998	Huang	Detection strategies for face recognition using learning and evolution [310]

Scheduling is an important practical problem in industry that is hard to tackle using traditional optimization methods, due to conflicting requirements, interrelated constraints, and high computational complexity. Genetic planning and genetic scheduling have been demonstrated as very promising in numerous research studies and in practice.

Planning and Scheduling

Year	Author	Title of Dissertation
1993	Timothy	Optimization of sequencing problems using genetic algorithms [621]
1994	Awadth	Manufacturing process planning model using genetic algorithms [24]
1994	Davern	Architecture for job shop scheduling with genetic algorithms [140]
1994	Wright	Genetic algorithm approach to scheduling resources for a space power system [668]
1995	Hoschei	Tabu search/genetic algorithm hybrid heuristic for solving a master production scheduling problem with sequence-dependent changeover times [307]
1995	Kou	Decision support for irrigated project planning using a genetic algorithm [374]
1995	Wang	Scheduling in flowshops with reentrant flows and pallet requirements [653]
1996	Edirisinghe	Optimization of radiotherapy planning using genetic algorithms [170]
1996	Morikawa	Study on scheduling using genetic algorithms [467]
1996	Wall	Genetic algorithm for resource-constrained scheduling [646]
1997	Al-harkan	On merging sequencing and scheduling theory with genetic algorithms to solve stochastic job shop problems [10]
1997	Cheng	Study on genetic algorithm–based optimal scheduling techniques [103]
1997	Hocaogle	Multipath planning using evolutionary computation with specification [302]
1997	Lin	Genetic algorithm–based scheduling system for dynamic job shop scheduling problem [416]
1997	Wang	Genetic algorithm–based approach to subtask matching and scheduling in heterogeneous computing environments and a comparative study of parallel genetic algorithms [652]
1999	Li	Study on hybridized genetic algorithms for production distribution planning problems [411]

Other Applications

Year	Author	Title of Dissertation
1993	Cowgill	Monte Carlo validation of two genetic clustering algorithms [135]
1993	Elketroussi	Relapse from tobacco smoking cessation: mathematical and computer microsimulation modeling including parameter optimization with genetic algorithms [176]

Other Applications (*Continued*)

Year	Author	Title of Dissertation
1995	Zhuang	Ergonomic evaluation of assistive devices and manual methods for resident-handling tasks in nursing homes [704]
1996	Dighe	Human pattern nesting strategies in a genetic algorithm framework [163]
1997	Hughell	Simulated adaptive management for timber and wildlife under uncertainty [312]
1997	Xie	Genetic algorithms: the method of regularization and their application to hypocenter location [672]
1998	Freeman	Genetic algorithms: a new technique for solving the neutron spectrum unfolding problem [206]
1998	Graybeal	Pest and resistance management simulation model of the Colorado potato beetle on potatoes [264]

2

COMBINATORIAL OPTIMIZATION PROBLEMS

2.1 INTRODUCTION

Combinatorial optimization study problems, which are characterized by a finite number of feasible solutions, abound in everyday life, particularly in engineering design. An important and widespread area of applications concerns the efficient use of scarce resources to increase productivity. Typical engineering design problems relate to set covering, bin packing, knapsack packing, quadratic assignment, minimum spanning tree determination, machine scheduling, sequencing and balancing, cellular manufacturing design, vehicle routing, network densing, facility location and layout, traveling salesman assignment, and so on.

Although, in principle, the optimal solution to such problems can be found by simple enumeration, in practice it is frequently impossible, especially for practical problems of realistic size, where the number of feasible solutions can be extremely high. One of the most challenging problems in combinatorial optimization is to deal effectively with the *combinatorial explosion.* A major trend in solving such difficult problems is to the use of genetic algorithms. It is expected that significant strides will be made in this area in the next few years.

In this chapter we explain how to solve set-covering, bin-packing, knapsack, and minimum spanning tree problems using genetic algorithms. All techniques most recently developed for such problems and applicable to other combinatorial problems, such as job-shop scheduling, machine scheduling, sequencing and balancing, cellular manufacturing design, vehicle routing, facility layout, network design, and so on, are discussed in subsequent chapters.

2.2 SET-COVERING PROBLEM

The set-covering problem is a typical problem in combinatorial optimization. The problem is to cover the rows of an m-row/n-column zero–one matrix

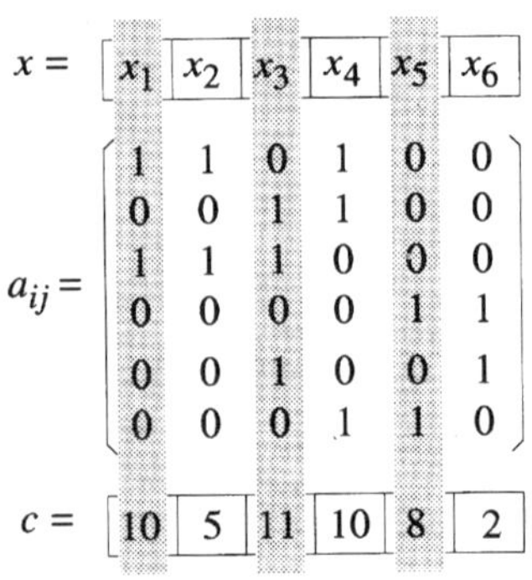

Figure 2.1. Simple example of the set-covering problem.

with a subset of columns at minimal cost. Consider a vector $\boldsymbol{x}$ such that $x_j = 1$ if column j (with a cost $c_j > 0$) is in the solution and $x_j = 0$ otherwise ($j = 1,2,...,n$). The set-covering problem is then formulated as

$$\min \quad z(\boldsymbol{x}) = \sum_{j=1}^{n} c_j x_j \tag{2.1}$$

$$\text{s.t.} \quad \sum_{j=1}^{n} a_{ij} x_j \geq 1, \qquad i = 1,2,...,m \tag{2.2}$$

$$x_j \in \{0,1\}, \qquad j = 1,2,...,n \tag{2.3}$$

Equation (2.2) ensures that each row is covered by at least one column and equation (2.3) is the integrality constraint. If all the cost coefficients c_j are equal, the problem is called a *unicost set-covering problem.* If equation (2.2) is an equality function, the problem above is referred to as a *set partitioning problem.*

Consider the following example, with six rows and six columns:

$$C = [10 \;\; 15 \;\; 11 \;\; 10 \;\; 8 \;\; 2]$$

$$A = \begin{bmatrix} 1 & 1 & 0 & 1 & 0 & 0 \\ 0 & 0 & 1 & 1 & 0 & 0 \\ 1 & 1 & 1 & 0 & 0 & 0 \\ 0 & 0 & 0 & 0 & 1 & 1 \\ 0 & 0 & 1 & 0 & 0 & 1 \\ 0 & 0 & 0 & 1 & 1 & 0 \end{bmatrix}$$

A feasible solution to the instance is $\boldsymbol{x} = [1, 0, 1, 0, 1, 0]$ with cost 29, as shown in Figure 2.1.

Several network problems can be modeled as a set-covering problem with a particular $A = (a_{ij})$ matrix, such as the node-covering problem of Salkin

TABLE 2.1. Applications of the Set-Covering Problem

Authors	Applications
Arabeyrre et al. (1969) [18]	Airline crew scheduling
Balinski (1965) [39]	Switching circuit scheduling
Balinski and Quandt (1964) [40]	Trucking dispatching
Bellmore (1971) [53]	Other instances
Bellmore, Greenberg, and Jarvis (1970) [54]	Network attack and defence
Bellmore (1971) [53]	Network attack and defense
Busacker and Saaty (1965) [82]	Map coloring
Charnes and Miller (1956) [96]	Railroad crew scheduling
Cobham, Fridshal, and North (1961) [126]	PERT/CPM analysis
	Switching circuit scheduling
	Symbolic logic
Day (1965) [150]	Information retrieval
Garey and Johnson (1979) [208]	Resource allocation
Garfinkel (1968) [209]	Political redistricting
Labordere (1969) [389]	Circuit design
Levin (1969) [405]	Vehicle routing
Revelle, Marks, and Leibman (1970) [535]	Facilities location
Root (1964) [540]	Symbolic logic
Rubin (1973) [544]	Assignment problem
Salverson (1955) [557]	Assembly line balancing
Steinmann and Schwinn (1969) [587]	Assembly line balancing
Toregas (1971) [626]	Facilities location
Valenta (1969) [636]	Capital investment
Walker (1974) [645]	Assignment problem

and Saha [555], the matching problem of Balinski [39], and the maximum flow problem of Sallkin and Mathur [556]. Besides, many applications of real-world problems have been reported as listed in Table 2.1.

The set-covering problem has been proven to be NP (nondetermined polynomial)-complete [208]. Both optimal and heuristic solutions to this problem have been reported in the literature. Balas and Ho proposed a cutting plane method [38]. Fisher and Kedia presented a dual heuristic algorithm for solving problems up to 200 rows and 2000 columns [189]. Beasley and Jornsten combined a Lagrangian heuristic, feasible solution exclusion constraints, Gomory f-cuts, and an improved branching strategy to solve problems up to 400 rows and 4000 columns [50]. Harche and Thompson developed an exact method, called the *column subtraction algorithm,* capable of solving large sparse instances of set-covering problems [284].

As with other sizable combinatorial problems, there has recently been an increasing interest in evolutionary computation solution techniques. Jacobs and Brusco developed a simulated annealing algorithm and reported considerable success on problems with up to 1000 rows and 10,000 columns [321]. Sen investigated the performances of a simulated annealing algorithm and a simple genetic algorithm [570]. Beasley and Chu [49] and González Hernándex and Corne [257] both proposed genetic algorithm approaches to set-covering problems of a larger scale.

2.2.1 Airline Crew Scheduling Problems

The airline crew scheduling problem of Salkin and Mathur [556] is a typical application. Suppose we have to assign a set of m flights that must be flown over a given time period to a set of n crews to minimize operation cost. Subject to time, geography, crew ability (e.g., not all pilots are licensed to fly Boeing 767 aircraft), and other limitations, the combinations of flights that each crew may take are listed. Let k be the index of crews and t the index of flights (or combinations of flights). A decision variable $x_{k(t)} = 1$ denotes that the kth crew is assigned the tth combination. A coefficient $a_{ii(t)} = 1$ means that the ith "flight leg" (e.g., Flight 347, Cleveland to Chicago) can be assigned to at least (or exactly) one crew, we have the following constraints:

$$\sum_k \sum_t a_{ik(t)} x_{k(t)} \geq 1, \qquad i = 1,2,\ldots,m \tag{2.4}$$

Let $c_{k(t)}$ be the cost of the tth combination of the kth crew. The problem is to find the best assignment of crew to flights to minimize total cost.

Table 2.2 shows a simple case with four crews and five flight legs. Crew 1 can take flights 1, 2, 4, and 5, or 1, 3, and 4, or 1 and 3. Crew 2 can be assigned to flights 1, 2, and 3, or 1 and 2, and so on. Provided with the cost of the tth combination of the kth crew, it is going to determine the airline crew scheduling with minimal cost.

2.2.2 Genetic Representation

Two representation schemes have been proposed for the set-covering problem: (1) column-based representation and (2) row-based representation.

Column-Based Representation. Column-based representation is an obvious choice for the set-covering problem since it represents the underlying 0–1 integer variables i.e., for an n-column problem, an n-bit binary string can be used as the chromosome structure. A value of 1 for the ith bit implies that column i is in the solution. An example is shown in Figure 2.2.

The fitness of an individual $f(\boldsymbol{x})$ is calculated simply by

$$f(\boldsymbol{x}) = \sum_{j=1}^{6} c_j x_j \tag{2.5}$$

The initial population can be generated randomly. Note that individuals in the column-based representation are not guaranteed to be feasible. It means that

TABLE 2.2. Example of Airline Crew Scheduling

Crew, k		1		2		3		4		
Combination, t	1	2	3	1	2	1	2	1	2	
Variable, $x_{k(t)}$	$x_{1(1)}$	$x_{1(2)}$	$x_{1(3)}$	$x_{2(1)}$	$x_{2(2)}$	$x_{3(1)}$	$x_{3(2)}$	$x_{4(1)}$	$x_{4(2)}$	
Flight leg, i	$a_{i1(1)}$	$a_{i1(2)}$	$a_{i1(3)}$	$a_{i2(1)}$	$a_{i2(2)}$	$a_{i3(1)}$	$a_{i3(3)}$	$a_{i4(1)}$	$a_{i4(2)}$	
1	1	1	1	1	1	0	0	0	0	≥ 1
2	1	0	0	1	1	1	1	1	1	≥ 1
3	0	1	1	1	0	0	1	1	0	≥ 1
4	1	1	0	0	0	1	0	0	0	≥ 1
5	1	0	0	0	0	0	1	0	0	≥ 1

column (gene)	x_1	x_2	x_3	x_4	x_5	x_6
bit string	1	0	1	0	1	0

Figure 2.2. Column-based representation.

not all rows will be covered. This needs some correction mechanisms for converting an infeasible representation to a feasible representation. Generally, there are three ways to deal with infeasible solutions.

One way is to reject all infeasible solutions generated in the reproductive process, which leads to the possibility that no feasible solution will be obtained. Another way is to apply a penalty function to penalize the fitness of infeasible solutions. Bäck carried out a comparative study using a penalty function and a repair heuristic on the set-covering problem [28]. A third way is to repair the chromosomes in infeasible solutions throughout the reproductive process. Beasley and Chu preferred a repair strategy using additional heuristic operators to transform infeasible solutions into feasible solutions [49].

Row-Based Representation. The problem of maintaining feasibility (all rows are covered) may be resolved by using a row-based representation. One possible representation is to have an individual's size equal to the number of rows for a given problem. In this representation, the location of each gene corresponds to a *row* and the encoded value of each gene is a *column* that covers that row.

An example is illustrated in Figure 2.3, where row 1 is covered by column 2, row 2 is covered by column 3, and so on. With this representation, feasibility can generally be maintained throughout the crossover and mutation procedures. But evaluation of fitness may become ambiguous because the same solution can be represented in different forms, and each form may give different fitness, depending on how the string is represented. Because the same column may be represented in more than one gene location, modified decoding method for fitness evaluation has to be used by extracting only the unique set of columns that the chromosome represents (i.e., repeated columns are only counted once). González Hernández and Corne adopted this representation scheme [257].

2.2.3 Genetic Operators

Beasley and Chu proposed a generalized fitness-based crossover operator called the *fusion operator* [49]. This operator takes into account both the

row (gene)	1	2	3	4	5	6
string	2	3	3	6	3	4

Figure 2.3. Row-based representation of an individual.

structure and relative fitness of the parent solutions. The fusion operator produces just a single child, whereas the conventional crossover operators usually produce two children.

Let P_1 and P_2 be the parent strings. Let f_{P_1} and f_{P_2} be the fitnesses of parent strings P_1 and P_2, respectively. Let C be the child string. The fusion operator works as follows:

Procedure: Fusion Operator

Step 1. $i = 1$.
Step 2. If $P_1[i] = P_2[i]$, then $C[i]: = P_1[i] = P_2[i]$.
Step 3. If $P_1[i] \neq P_2[i]$, then
 (3.1) $C[i] := P_1[i]$ with probability $p = f_{P_2}/(f_{P_1} + f_{P_2})$.
 (3.2) $C[i] := P_2[i]$ with probability $1 - p$.
Step 4. If $i = n$, stop; otherwise, set $i = i + 1$ and go to step 1.

Mutation usually works by inverting each bit of individuals with a small probability. Bäck suggested a rate of $1/n$ as a lower bound on an optimal mutation rate, where n is the length of the individual [26]. Beasley and Chu suggested a variable mutation schedule [49]. The number of bits to be mutated is determined by

$$\left\lceil \frac{m_f}{1 + \exp[-4m_g(c - m_c)/m_f]} \right\rceil \tag{2.6}$$

where c is the number of child individuals that have been generated, m_f specifies the final stable mutation rate, m_c specifies the number of child individuals generated at which a mutation rate of $m_f/2$ is reached, and m_g specifies the gradient at $c = m_c$. The value of m_g is specified by the user, and the values of m_c and m_g are determined based on the rate at which the genetic algorithm converges for a particular problem.

Beasley and Chu adopted the column-based representation [49]. To avoid the infeasibility of individuals, a heuristic operator is given, which includes two major components: repairing the infeasibility of a given individual, and performing an additional local optimization in an attempt to make the genetic algorithm more effective.

The steps required to make each individual feasible involve (1) identifying all uncovered rows and (2) adding some necessary columns such that all rows are covered. Once columns are added and an individual becomes feasible, a local optimization step is applied to remove any redundant columns in the individual. A redundant column is one such that by removing it from the individual, the individual remains feasible.

Without loss of generality, suppose that the columns are sorted in increasing order of cost. Columns with equal cost are sorted in decreasing order of the number of rows that they cover. In stating the procedure, we use the following notations:

I = set of all rows

J = set of all columns

α_i = set of columns that cover row i, $i \in I$

β_j = set of rows covered by column j, $j \in J$

S = set of columns in a solution

U = set of uncovered rows

w_i = number of columns that cover row i, $i \in I$ in S

The procedure for repairing infeasible solutions is outlined as follows:

Procedure: Heuristic Operator

Step 1. Initialize $w_i := |S \cap \alpha_i|$, $\forall i \in I$.
Step 2. Initialize $U := \{i | w_i = 0, \forall i \in I\}$.
Step 3. For each row i in U (in increasing order of i):
 (3.1) Find the first column j (in increasing order of j) in α_i that minimizes $c_j / |U \cap \beta_j|$.
 (3.2) Add j to S and set $w_i := w_i + 1$, $\forall i \in \beta_j$; set $U := U - \beta_j$.
Step 4. For each column j in S (in decreasing order of j), if $w_i \leq 2$, $\forall i \in \beta_j$, set $S := S - j$ and set $w_i := w_i - 1$, $\forall i \in \beta_j$.
Step 5. S is now a feasible solution containing no redundant columns.

Steps 1 and 2 identify uncovered rows; steps 3 and 4 are "greedy" heuristics in the sense that in step 3, columns with low cost ratios are considered first, and in step 4, columns with high cost are dropped first whenever possible.

2.2.4 Genetic Algorithm

The genetic algorithm approach to the set-covering problem proposed by Beasley and Chu is summarized as follows:

Procedure: Genetic Algorithm for the Set-Covering Problem

Step 1. Generate an initial population of N random solutions. Set $t := 0$.

Step 2. Select two solutions, P_1 and P_2, from the population using binary tournament selection.

Step 3. Combine P_1 and P_2 to form a new solution, C, using the fusion crossover operator.

Step 4. Mutate k randomly selected columns in C, where k is determined by the variable mutation schedule.

Step 5. Make C feasible and remove redundant columns in C by applying the heuristic operator.

Step 6. If C is identical to any of the solutions in the population, go to step 2; otherwise, set $t := t + 1$ and go to step 7.

Step 7. Replace a randomly chosen solution with above-average fitness in the population by C (steady-state replacement method).

Step 8. Repeat steps 2 to 7 until $t = M$ (i.e., M nonduplicate) solutions have been generated. The best solution found is the one with the smallest fitness in the population.

2.2.5 Computational Experience

The test problems by Beasley and Chu range from 200×1000 to $1000 \times 10{,}000$ [49], which can be obtained electronically from OR-Library (e-mail address *scpinfo* to *or.library@ic.ac.uk*) [48]. Among those test problems, some have known optimal solution values, but some have only the previous best-known solutions. According to Beasley and Chu's computational results, the genetic algorithm approach can get much better near-optimal solutions than those for previous best-known solutions.

In their experiments, 10 trials, each with a different random seed, were made for each test problem. Each trial terminated with $M = 100{,}000$ nonduplicate solutions. The population size N was set to 100 for all the problems. Table 2.3 provides some of their computational results (the best case and the worst case for each problem of the same size).

The optimal solution values are given in [50] and the previous best-known solution values in [32]. The average percentage deviation (σ) is calculated as $\Sigma_{i=1}^{10}(S_{Ti} - S_o)/10S_o \times 100\%$, where S_{Ti} is the ith trial minimum (best) solution value and S_o is the optimal solution value. A negative value means that the genetic algorithm approach gets a better result than that of the previous best-known solution. The solution time is measured in CPU seconds and is the time that the genetic algorithm takes to reach the final best solution. The execution time is the time (CPU seconds) that the genetic algorithm takes to generate 100,000 nonduplicate child solutions.

2.3 BIN-PACKING PROBLEM

The bin-packing problem consists of placing n objects into a number of bins (at most n bins). Each object has a weight ($w_i > 0$) and each bin has a limited bin capacity ($c_i > 0$). The problem is to find a best assignment of objects to bins such that the total weight of the objects in each bin does not exceed its capacity and the number of bins used is minimized. The problem can be illustrated as follows:

TABLE 2.3. Computational Results for the Set-Covering Problem Using a Genetic Algorithm Approach

Problem Size, $m \times n$	Density (%)	Optimal or PBS[a]	BS[b]	Average, σ (%)
200 × 1000	2	512	512	0.00
		641	641	0.33
200 × 2000	2	279	279	0.00
		226	228	0.88
200 × 1000	5	131	131	0.00
		146	146	0.14
300 × 3000	2	236	236	0.00
		253	253	0.79
300 × 3000	5	72	72	0.00
		80	80	0.00
400 × 4000	2	215	215	0.05
		243	243	1.40
400 × 4000	5	60	60	0.00
		72	72	0.28
500 × 5000	10	28	28	0.00
		27	27	2.60
500 × 5000	20	14	13	−2.14
		14	14	0.00
1000 × 10,000	2	158	155	−1.08
		168	168	0.83
1000 × 10,000	5	60	59	−1.50
		55	55	0.18

[a]PBS, previous best-known solution value.
[b]BS, best in each of the 10 trials.

Object	1	2	···	n
Weight	w_i	w_2	···	w_n
Capacity	c_i	c_2	···	c_n

where for n bins, the weights w_i and capacity c_i are positive real numbers for i 1,2,...,n.

Usually, all bins have the same capacity ($c > 0$). Obviously, placing each object into each bin is a feasible solution to the problem but is not an optimal solution. A mathematical formulation for the bin-packing problem is given as follows [441]:

$$\min \quad z(\mathbf{y}) = \sum_{i=1}^{n} y_i \tag{2.7}$$

$$\text{s.t.} \quad \sum_{j=1}^{n} w_j x_{ij} \leq c y_i, \qquad i \in N = \{1,2,\ldots,n\} \tag{2.8}$$

$$\sum_{i=1}^{n} x_{ij} = 1, \qquad j \in N \tag{2.9}$$

$$y_i = 0 \text{ or } 1, \qquad i \in N \tag{2.10}$$

$$x_{ij} = 0 \text{ or } 1, \qquad i, \quad j \in N \tag{2.11}$$

where

$$y_i = \begin{cases} 1, & \text{if bin } i \text{ is used} \\ 0, & \text{otherwise} \end{cases}$$

$$x_{ij} = \begin{cases} 1, & \text{if object } j \text{ is assigned to bin } i \\ 0, & \text{otherwise} \end{cases}$$

The bin-packing problem underlies many optimization problems of practical importance [182]. In construction, tubes and rods must often be cut from stock of constant length. Wires for electric wiring come in coils of constant length. Wallpaper sheets must be cut from packs of a given length. Stones of various lengths (but constant width) have to be transported to the construction site on pallets. In the metalworking industry, steel sheets must be cut from master sheets. The dimensions of the objects do not have to be geometrical. For instance, in the electronics industry, microcode routines of various sizes (in bytes) have to be stored in microprocessor memory divided into banks of a given size. In the transportation industry, it is often the case that only the weight of the various loads counts in distributing them over trucks of a given maximum load. In the entertainment industry, song collections are released on media (compact disks, cassettes) of a given maximum capacity. In operations management, tasks of various durations must be allocated to workers under the constraint of a given time period for completing all the tasks. The important problem of line balancing (allocation of tasks to workstations along a production line) is a bin-packing problem with additional precedence constraints. The bin-packing model is one of the special cases of capacitated plant location models as shown in Section 7.5.

2.3.1 Heuristic Algorithms

The bin-packing problem belongs to the class of NP hard problems [208]. There is no known optimal algorithm for running it in polynomial time. Only a few heuristic algorithms were proposed to deal with it. But most heuristics are greedy methods characterized by adopting some simple rules, such as next fit, first fit, or best fit. Garey and Johnson stated that simple heuristics can be shown to be no worse (but also no better) than a rather small factor multiplied by the optimum [208].

Next-Fit Heuristic. The simplest heuristic approach to the bin-packing problem is the *next-fit* (NF) *algorithm.* The first object is assigned to bin 1, and objects 2,...,n are then considered by their increasing indices. Each object is assigned to the current bin if it fits; otherwise, it is assigned to a new bin. The new bin then becomes the current bin. The complexity of the algorithm is clearly $O(n)$. For a given problem I, it is easy to prove that the solution value NF(I) provided by the algorithm satisfies the following bound:

$$\mathrm{NF(I)} \leq 2z(\mathrm{I}) \tag{2.12}$$

where z(I) denotes the optimal solution value.

First-Fit Heuristic. The *first-fit* (FF) *algorithm* also considers the objects according to their increasing indices. But each object is assigned to the lowest-indexed initialized bin that it fits. When the current object cannot fit into any initialized bin, a new bin is introduced. It has been proved by Johnson et al. [331] that

$$\mathrm{FF(I)} \leq \frac{17}{10}\, z(\mathrm{I}) + 2 \tag{2.13}$$

for all instances I of the bin-packing problem.

Best-Fit Heuristic. The *best-fit* (BF) *algorithm* is modified from the FF algorithm by assigning the current object to the feasible bin having the smallest residual capacity. Ties are broken in favor of the lowest-indexed bin. Johnson et al. have proved that BF satisfies the same worst-case bounds as FF [331]. The time complexity of both FF and BF is $O(n \log n)$.

Other Heuristics. If the objects are sorted in descending order,

$$w_1 \geq w_2 \geq \cdots \geq w_n$$

and then NF, FF, or BF is used, the resulting algorithms, of time complexity $O(n \log n)$, are called *next-fit decreasing* (NFD), *first-fit decreasing* (FFD), and *best-fit decreasing* (BFD), respectively. These heuristics are generally used to testify to the effectiveness of newly proposed algorithms on the bin-packing problem. Often, they are embedded in a genetic algorithm approach

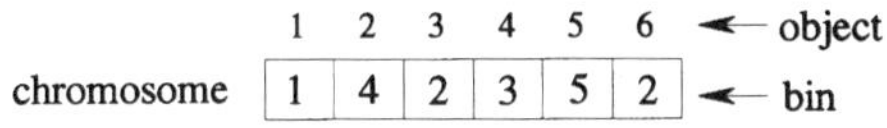

Figure 2.4. Representation of membership of objects.

to enhance the genetic search to find a better nearly optimal solution for the bin-packing problem.

2.3.2 Genetic Representation

Three representation schemes have been proposed for bin-packing problems: (1) bin-based representation, (2) object-based representation, and (3) group-based representation.

Bin-Based Representation. The most straightforward approach is to encode the *membership* of objects in the solution. In bin-based representation, the position of a gene is used to represent an object, and the value of the gene is used to represent a bin in which the corresponding object is put. For instance, the chromosome 1 4 2 3 5 2 would encode a solution where the first object is in bin 1, the second in bin 4, the third in bin 2, the fourth in bin 3, the fifth in bin 5, and the sixth in bin 2. The representation of the bin-packing problem is illustrated in Figure 2.4.

This representation has the advantage that the chromosomes are of constant length (equal to the number of objects), which allows for the application of standard genetic operators. However, as Falkenauer [181] points out, this encoding suffers several drawbacks. First, the encoding is highly redundant. Indeed, the cost function depends only on the number of bins rather than on the numbering of the objects to bins. For instance, 1 2 3 1 4 4 and 3 1 4 3 2 2 both encode the same solution of the problem; namely, the first and fourth objects are assigned one bin, the fifth and sixth are assigned one bin, and the second and third are assigned another two bins.

The degree of redundancy (i.e., the number of distinct chromosomes encoding the same solution of the original problem) of this scheme grows *exponentially* with the number of bins, that is, indirectly, with the size of the problem. Thus the size of the space the genetic algorithm has to search is much larger than the original space of solutions. Consequently, the power of the genetic algorithm is seriously impaired.

Second, this encoding leads to the highly undesirable effect of casting context-dependent information out of context under the standard crossover. For instance, let us apply the standard two-point crossover to two chromosomes and their offspring as shown in Figure 2.5. Generally, a recombination of two identical individuals must produce progeny that are again identical to the parents. The two parents in Figure 2.5 are identical because they both encode the same solution of the problem. Hence a recombination operator

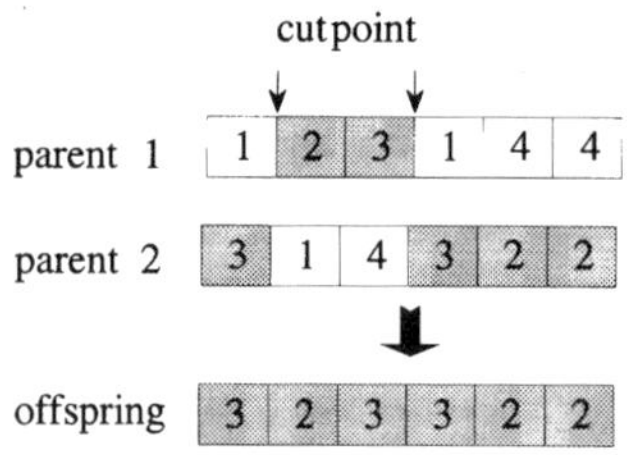

Figure 2.5. Crossover operation.

should produce an individual that encodes the same solution. However, the resulting child encodes a solution that has nothing in common with the solution its parents encode, since there are two bins instead of four. In other words, while the schemata are well transmitted with respect to the chromosomes under the standard encoding/crossover, their meaning with respect to the problem to solve (i.e., the cost function to optimize) is lost in the process of recombination.

The third drawback is that the representation may yield infeasible solutions in the sense that a bin may be assigned too many objects, exceeding its capacity.

Object-Based Representation. Another representation for the bin-packing problem is to encode the permutations of objects and then apply a decoder to retrieve the corresponding solutions. Suppose that there are six objects to pack, numbered 1 through 6. A chromosome can be represented as the permutation of objects from 1 to 6 as follows:

$$1\ 2\ 3\ 4\ 5\ 6$$

This encoding also suffers various drawbacks. First, this encoding is also highly redundant. Indeed, suppose that the objects in the chromosome above are partitioned as follows:

$$1\ 2\ 3\ |\ 4\ 5\ |\ 6$$

that is, there are three bins, one containing objects 1 through 3, the second containing objects 4 and 5, and the third containing object 6. Now any permutation of the objects having the same bin contents, such as

$$3\ 2\ 1\ |\ 4\ 5\ |\ 6$$

or

$$4\ 5\ |\ 2\ 1\ 3\ |\ 6$$

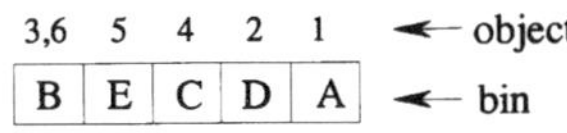

Figure 2.6. Representation of grouping of objects.

encodes the same solution to the bin-packing problem. In the same way, the degree of redundancy of this encoding grows exponentially with the size of the instance, and the power of the genetic algorithm is seriously diminished.

Second, given the fact that the decoder retrieves the numbering of objects to bins corresponding to a given permutation by considering the objects from left to right, the position of a given object depends strongly on the "head" of the chromosome, the object on the leftmost position. However, the position of a given gene provides nothing useful. As a result, it cannot maintain the heritage from parents in the evolutionary process. In contrast with bin-based representation, object-based representation never produces infeasible solutions.

Group-Based representation. The two representations above are not suitable for grouping problems such as the bin-packing problem because those chromosomes are *item* oriented rather than *group* oriented. In short, the encodings above are not adapted to the cost function optimization, as the cost function of a bin-packing problem depends on grouping of objects in bins. For the bin-packing problem, a better representation would consist of two parts. The first part provides information as to which objects belong to which bin (group). The second part encodes the bins that are used. Based on this consideration, Falkenauer [181] proposed a representation scheme, together with a grouping of objects, which encode the objects on a one-gene-for-one-bin basis.

More concretely, let us consider the chromosome suggested in Section 2.1. Numbering the objects from 1 through 6, the object part of the chromosome can be written explicitly as 1 4 2 3 5 2. This means that object 1 is in bin 1, object 2 in bin 4, objects 3 and 6 in bin 2, object 4 in bin 3, and object 5 in bin 5. The group part of the chromosome represents only the bins. here we adopt letters instead of integers to represent bins (i.e., the chromosome above can be represented as ADBCEB). By a look-up in the object part, we can establish what the group names stand for, namely

$$B = \{3,6\},\quad E = \{5\},\quad C = \{4\},\quad D = \{2\},\quad \text{and}\quad A = \{1\}$$

Furthermore, integration of the chromosome with the grouping of objects can be illustrated as shown in Figure 2.6. This encoding scheme makes the genes represent both objects and groups (bins). The rationale is that for the

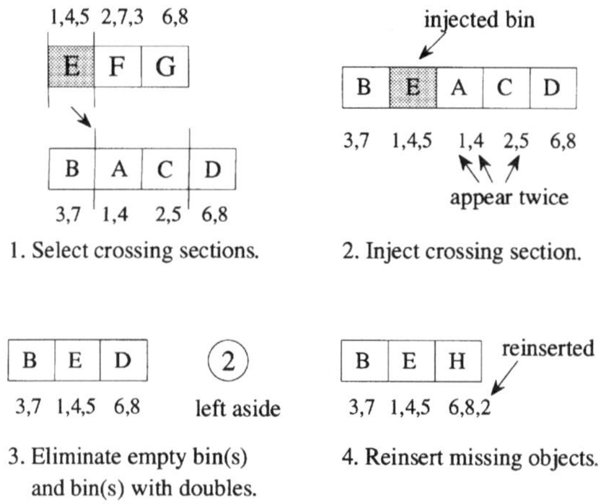

Figure 2.7. Crossover.

bin-packing problem the groups are the meaningful building blocks (i.e., the smallest piece of a solution that can convey information on the expected quality of the solution of which they are a part). The important point is that the genetic operators will work with the group part of the chromosomes, the object part of the chromosomes serving merely to identify which objects actually form which group. Note in particular that this implies that the operators will have to handle chromosomes of variable length.

2.3.3 Genetic Operators

In this section we focus on the genetic operators for the group-based representation given by Falkenauer [182].

Crossover. As pointed out in the preceding section, group-based representation consists of two parts: bins and groups of genes. Therefore, crossover will have to work with variable-length chromosomes, with genes representing the bins. The crossover procedure can be illustrated as follows and as shown in Figure 2.7.

Procedure: Crossover

Step 1. Select at random two crossing sites, delimiting the crossing section, in each of the two parents.

Step 2. Inject the contents of the crossing section of the first parent at the first crossing site of the second parent. Recall that crossover works with the group part of the chromosome, so this means injecting some of the groups (bins) from the first parent in the second.

Step 3. Eliminate all objects now occurring twice from the bins they were members of in the second parent, so that the "old" membership of these objects gives way to the membership specified by the "new" injected bins. Consequently, some of the old groups coming from the second parent are altered: They no longer contain all the same objects, because some of the objects had to be eliminated.

Step 4. If necessary, adapt the resulting bins, according to the hard constraints and the cost function to be optimized. At this stage, local problem-dependent heuristics such as FFD can be applied.

Step 5. Apply steps 2 to 4 to the two parents with their roles permuted to generate the second child.

Mutation. A mutation operator for the bin-packing problem must work with groups (bins) rather than objects. As for the mutation, the implementation details of the operator depend on the particular grouping problem on hand. Nevertheless, two general strategies can be outlined: Either create a new bin or eliminate an existing one. When the objects that composed those bins are thus missing from the solution, we can use the FF or FFD heuristics to insert them back in a random order.

2.3.4 Fitness Function

There are two ways to evaluate the objective of the bin-packing problem in a genetic algorithm approach: (1) minimize the number of bins used and (2) minimize the number of bins used and fill all used bins. Falkenauer and Delchambre [184] proposed the following evaluation function for the bin-packing problem, which belong to the second method listed above:

$$f_{\text{BPP}} = \frac{\Sigma_{i=1}^{N} (F_i/C)^k}{N} \tag{2.14}$$

where N is the number of bins used in the solution, F_i the sum of sizes of the objects in bin i(the fill of the bin), C the bin capacity, and k a constant, $k > 1$. The constant k expresses concentration on the helite bins compared to the less-filled bins. The larger k is, the more a few well-filled bins are preferred to a collection of about equally filled bins. According to their limited computational experience, it is found that $k = 2$ gives good results.

2.3.5 Initial Population

For the bin-packing problem, there are two ways to produce the initial population: randomly and heuristically. In the first, all initial individuals are generated randomly. However, it is also relatively easy to generate all individuals by applying the FF heuristic to the bin-packing problem. This means that we can pack all objects into bins according to the FF heuristic rule. This may yield a population of individuals containing a large variety of bins (i.e., genes).

TABLE 2.4. Computational Results for the Bin-Packing Problem

Problem Size	Optimal Solution	FFD	GA
30	9	11	9
60	18	22	18
90	27	33	27
120	36	44	36
150	45	55	45

2.3.6 Computational Experience

Some results have been reported by Bilchev [711] and Falkenauer [181]. Bilchev adopted object-based representation to deal with the bin-packing problem. Table 2.4 shows the effectiveness of the genetic algorithm approach compared with the FFD heuristic.

Falkenauer [181] adopted group-based representation to cope with the bin-packing problem. The test data were constructed as follows: First generate objects demonstrating perfect packing (i.e., $f_{\text{BPP}} = 1$), and then subtract from randomly chosen objects a total of p percent of the size of one bin. For example, with $p = 10\%$ and bin capacity $C = 255$, the total size of the objects was 25.5 less than the total capacity of the bins in perfect packing. Thus the test reflected the ability of the algorithm to get as close as possible to the optimum, rather than its eventual ability to find the optimally perfect packing. Indeed, the latter test would be a test of optimality, which would be equivalent to asking whether we can solve an NP-complete decision problem in a reasonable period of time—something the algorithm wasn't and couldn't be designed to do.

The results are summarized in Figure 2.8 for instances of 64 objects. The figure shows as a function of p the proportion of test cases successfully optimized by the FFD heuristic and genetic algorithm with $k = 2$. To compute

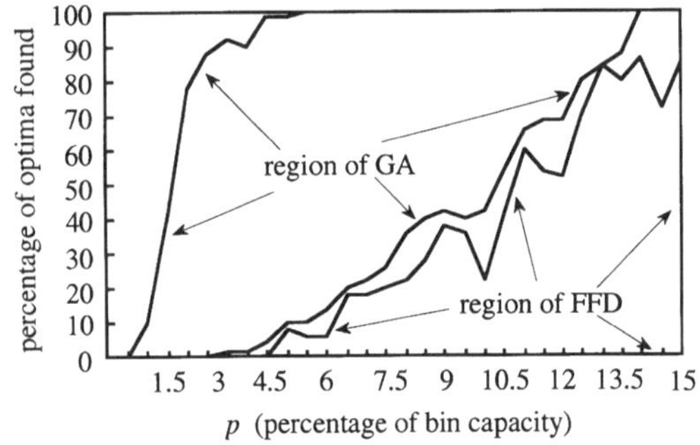

Figure 2.8. Relative performance of the FFD heuristic and genetic algorithm.

TABLE 2.5. Weights and Profits in the Knapsack Problem

Item Number	1	2	3	4	5	6	7	8
Weight	35	50	30	15	10	35	25	40
Cost	40	60	25	20	5	60	40	25

the proportions, 50 test cases were generated by the procedure described above for each selected value of p (1.5 through 15% of C).

2.4 KNAPSACK PROBLEM

Knapsack problems have been studied intensively in past decades, attracting both theorists and practicioners. The theoretical interest arises mainly from their simple structure, which both allows exploitation of a number of combinatorial properties and permits more complex optimization problems to be solved through a series of knapsack-type subproblems. From a practical point of view, these problems can model many industrial situations: the most classical applications being capital budgeting, cargo loading, and cutting stock [441,478].

Suppose that we want to fill up a knapsack by selecting some objects among various objects (generally called *items*). There are n different items available and each item j has a *weight* of w_j and a *cost* of c_j. The knapsack can hold a weight of at most W. The problem is to find an optimal subset of items so as to minimize the total cost subject to the knapsack's weight capacity. The costs, weights, and capacity are positive integers.

Consider an eight-item knapsack problem. Each weight and cost is defined in Table 2.5. Weight capacity is defined as $W = 140$. The problem is illustrated clearly in Figure 2.9.

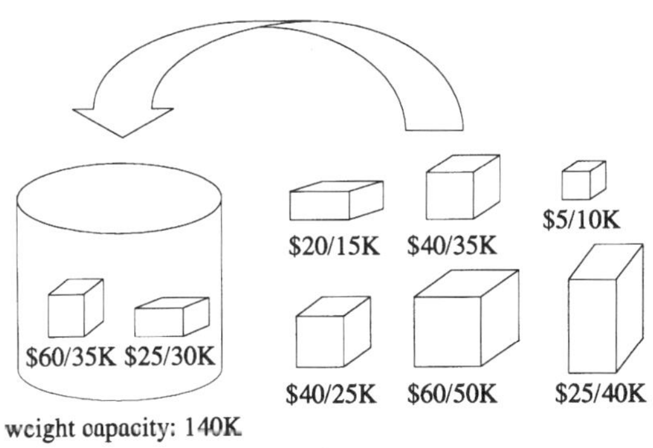

Figure 2.9. Knapsack problem.

Let x_j be binary variables given as follows:

$$x_j = \begin{cases} 1, & \text{if item } j \text{ is selected} \\ 0, & \text{otherwise} \end{cases}$$

The knapsack problem can be formulated mathematically as follows:

$$\min \quad \sum_{j=1}^{n} c_j x_j \tag{2.15}$$

$$\text{s.t.} \quad \sum_{j=1}^{n} w_j x_j \leq W \tag{2.16}$$

$$x_j = 0 \text{ or } 1, \qquad j = 1,2,\ldots,n \tag{2.17}$$

This is known as the *zero–one knapsack problem,* pure integer programming with a single constraint which forms a very important class of integer programming. There are many variations of knapsack problems, such as the multiple-choice, bounded, unbounded, and multiconstrained knapsack problems [*441*].

Knapsack problems have been proven NP-hard [208]. Salkin and De Kluyver presented a number of industrial applications and results in transforming integer linear programs to knapsack problems (an approach which appeared very promising at that time) [554]. Martello and Toth considered exact algorithms for the zero–one knapsack problem and their average computational performance [439]. The study was extended to other linear knapsack problems and to approximate algorithms by Martello and Toth [440]. Dudzinski and Walukiewicz analyzed dual methods of solving Lagrangian and linear programming relaxation [168].

Like other combinatorial optimization problems, many researchers have tried to solve knapsack problems using genetic algorithms (GAs). Among them are the genetic algorithm approaches by Olsen [491], by Kubota and Fukuda [377] using binary representation, by Hinterding [298] using order representation, and by Gordon and Whitley [259] using variable-length representation. In this section we introduce the genetic algorithm approach to the multiple-choice knapsack problem by Gen et al. [220] and the genetic algorithm approach to the multiconstrained knapsack problem by Raidl [522].

2.4.1 Multiple-Choice Knapsack Problem

Problem Description. The multiple-choice knapsack problem is defined as a knapsack problem with the addition of disjoint multiple-choice constraints [441]. The general description of the problem is given as follows: There is one knapsack with limited capacity. Objects to be packed in the knapsack are

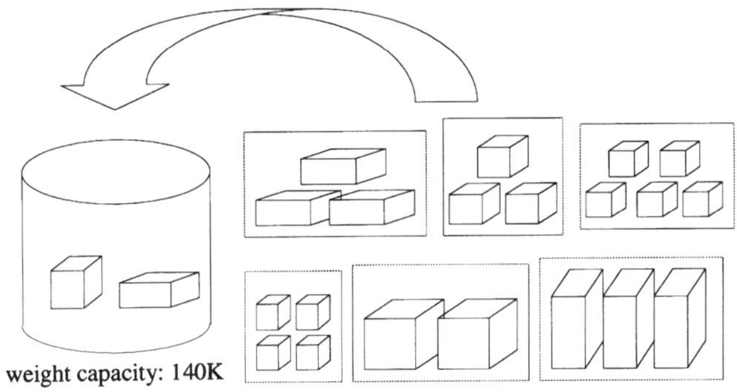

Figure 2.10. Multiple-choice knapsack problem.

classified into multiple mutually exclusive classes. Within each class there are several different items. The problem is to select some items from a class so as to minimize the total cost while the total size of the items selected does not exceed the limited capacity of the knapsack. The problem is illustrated in Figure 2.10.

The multiple-choice knapsack problem can be formulated mathematically as follows:

$$\min \quad \sum_{i=1}^{m} \sum_{j=1}^{n_i} c_{ij} x_{ij} \tag{2.18}$$

$$\text{s.t.} \quad \sum_{i=1}^{m} \sum_{j=1}^{n_i} w_{ij} x_{ij} \leq W \tag{2.19}$$

$$\sum_{j=1}^{n_i} x_{ij} = 1, \qquad i = 1,2,\ldots,m \tag{2.20}$$

$$x_{ij} \in \{0,1\}, \qquad \forall i,j$$

where i is the index of class, j the index of item within each class, n_i the number of total items within the ith class, c_{ij} the cost for item j within class i, m the number of total classes. w_{ij} the weight of item j within class i, and W the limited capacity of the knapsack. The use of double subscripts makes m multiple-choice sets of variables mutually exclusive. Constraints (2.20) are called *multiple-choice constraints* or *generalized upper bounds* (GUBs).

The problem is a kind of the generalized assignment problem and is NP-hard. The most successful techniques for solving the problem are branch-and-bound algorithms that use either linear programming [291], Lagrangian relaxation [243, 601], or variations of Lagrangian relaxation [46] for bounding purposes.

y_1	y_2	y_3	y_4	...	y_m

Figure 2.11. Permutation enccoding.

Genetic Representation. Several researchers have solved the knapsack problem by using genetic algorithms, which can be classified roughly into three groups [219]: (1) the binary representation approach, (2) the order representation approach, and (3) the variable-length representation approach.

But for the multiple-choice knapsack problem, a kind of permutation is more natural and effective for encoding candidate individuals [220]. A chromosome can be defined to have m genes corresponding to m classes of items. The ith gene takes an integer number from a mutually exclusive set n_i, where N_i is a set $\{1,2,\ldots,n_i\}$ at each class i. Thus the position of a gene is used to represent one class of object, and the value of the gene is used to represent the item selected from the class. By defining an indicator variable, y_i is as follows:

$$y_i = j \qquad \text{if } x_{ij} = 1, \quad j \in N_i, \quad i = 1,2,\ldots,m \tag{2.21}$$

The genetic representation can be defined formally as shown in Figure 2.11.

Consider the following example:

$$\begin{aligned}
\min \quad & x_{11} + 4x_{12} + 5x_{13} + 7x_{14} + 7x_{21} + 9x_{22} + 10x_{23} \\
& + 5x_{31} + 6x_{32} + 11x_{33} \\
\text{s.t.} \quad & 10x_{11} + 8x_{12} + 5x_{13} + 4x_{14} + 6x_{21} + 5x_{22} + 3x_{23} \\
& +8x_{31} + 6x_{32} + 4x_{33} \le 16 \\
& x_{11} + x_{12} + x_{13} + x_{14} = 1 \\
& x_{21} + x_{22} + x_{23} = 1 \\
& x_{31} + x_{32} + x_{33} = 1
\end{aligned}$$

Figure 2.12 shows an example of this encoding method, which stands for a solution of $x_{12} = 1$, $x_{23} = 1$, $x_{33} = 1$. It is easy to verify that it is a feasible solution.

2	3	3

Figure 2.12. Example of permutation encoding.

Crossover and Mutation. Simply, uniform crossover and random perturbation can be adopted for genetic operations. In random perturbation, the gene selected, y_i, will be replaced by a random integer only from the set N_i.

Evaluation and Selection. A penalty term is added to a fitness function to force the genetic search to approach the optimal solution from both the feasible and infeasible sides of the solution space. The evaluation function has two terms: total cost and penalty to illegality. The total cost is calculated using the objective function. Let v_k be the kth chromosome in the current population and x_{ij}^k be the corresponding decision variables; then the cost for the kth chromosome can be calculated as follows:

$$f_k(v_k) = \sum_{i=1}^{m} \sum_{j \in N_i} c_{ij} x_{ij}^k$$

In general, an encoding may correspond to an infeasible solution due to the fact that the total weight of the items selected exceeds the limited capacity of the knapsack. The penalty coefficient p_k is calculated proportionally to the extra weight necessary for the solution as follows:

$$p_k = \begin{cases} 0, & \text{if constraint (2.19) is satisfied} \\ \alpha_0 \left(\sum_{i=1}^{m} \sum_{j \in N_i} w_{ij} x_{ij}^k - W \right), & \text{otherwise} \end{cases}$$

where α_0 is the large positive penalty value. Therefore, the evaluation function can be expressed as follows:

$$\text{eval}(v_k) = \frac{1}{f_k(v_k) + p_k}, \qquad k = 1,2,\ldots,pop\_size$$

As to selection, the *elitist* approach was embedded in *roulette wheel* selection to enforce preserving the best chromosome in the next generation and overcoming the stochastic errors involved in sampling. With elitist selection, if the best individual in the preceding generation is not reproduced in the new generation through the process of roulette wheel selection, remove one chromosome randomly from the new population and add the best chromosome of the preceding population to the new population.

Experimental Results. In Gen et al.'s genetic algorithm approach to this problem, all the test problems were generated randomly [220]. A uniform distribution was used to generate the c_{ij} and w_{ij} values, ensuring that the same value for c_{ij} and w_{ij} was not repeated within a class. The following numerical example is a multiple-choice knapsack problem with eight class items, and each class has different numbers of total items.

$$\begin{aligned}
\min \quad & 3x_{11} + 4x_{12} + 5x_{13} + 4x_{14} + 8x_{15} + 5x_{16} + 4x_{17} + 4x_{18} \\
& + 7x_{21} + 9x_{22} + 8x_{23} + 4x_{24} + 5x_{25} \\
& + 5x_{31} + 6x_{32} + 9x_{33} + 8x_{34} + 4x_{35} + 6x_{36} \\
& + 2x_{41} + 8x_{42} + 7x_{43} + 5x_{44} + 4x_{45} + 3x_{46} + 7x_{47} + 8x_{48} \\
& + 8x_{51} + 5x_{52} + 7x_{53} + 9x_{54} + 7x_{55} + 4x_{56} + 7x_{57} \\
& + 6x_{61} + 8x_{62} + 7x_{63} + 9x_{64} + 5x_{65} \\
& + 6x_{71} + 9x_{72} + 7x_{73} + 4x_{74} + 3x_{75} + 7x_{76} + 9x_{77} + 8x_{78} \\
& + 9x_{81} + 2x_{82} + 5x_{83} + 5x_{84} + 8x_{85} + 6x_{86} \\
\text{s.t.} \quad & 6x_{11} + 8x_{12} + 5x_{13} + 4x_{14} + 9x_{15} + 5x_{16} + 9x_{17} + 3x_{18} \\
& + 6x_{21} + 5x_{22} + 3x_{23} + 5x_{24} + 9x_{25} \\
& + 8x_{31} + 6x_{32} + 4x_{33} + 8x_{34} + 4x_{35} + 3x_{36} \\
& + 9x_{41} + 4x_{42} + 8x_{43} + 9x_{44} + 4x_{45} + 9x_{46} + 4x_{47} + 5x_{48} \\
& + 3x_{51} + 7x_{52} + 4x_{53} + 2x_{54} + 3x_{55} + 8x_{56} + 5x_{57} \\
& + 8x_{61} + 3x_{62} + 2x_{63} + 8x_{64} + 6x_{65} \\
& + 5x_{71} + 4x_{72} + 3x_{73} + 8x_{74} + 9x_{75} + 2x_{76} + 4x_{77} + 4x_{78} \\
& + 2x_{81} + 9x_{82} + 2x_{83} + 6x_{84} + 4x_{85} + 3x_{86} \le 40
\end{aligned}$$

$$\begin{aligned}
& x_{11} + x_{12} + x_{13} + x_{14} + x_{15} + x_{16} + x_{17} + x_{18} = 1 \\
& x_{21} + x_{22} + x_{23} + x_{24} + x_{25} = 1 \\
& x_{31} + x_{32} + x_{33} + x_{34} + x_{35} + x_{36} = 1 \\
& x_{41} + x_{42} + x_{43} + x_{44} + x_{45} + x_{46} + x_{47} + x_{48} = 1 \\
& x_{51} + x_{52} + x_{53} + x_{54} + x_{55} + x_{56} + x_{57} = 1 \\
& x_{61} + x_{62} + x_{63} + x_{64} + x_{65} = 1
\end{aligned}$$

TABLE 2.6. Computation Results in Example

Problem	Problem Size: Total Items	Problem Size: Total Classes	GA Parameter Setting: *pop_size*	GA Parameter Setting: *max_gen*	p_c	p_m	Time (s)	Frequency
1	10	3	20	20	0.2	0.1	0.04	1
2	15	4	20	20	0.2	0.1	0.04	1
3	31	6	40	100	0.2	0.1	0.07	0.87
4	53	8	40	100	0.2	0.1	0.15	0.67

The optimal value is 34 and three optimal solutions found by the enumeration method are

$$x_{11} = 1,\quad x_{24} = 1,\quad x_{35} = 1,\quad x_{45} = 1$$

$$x_{56} = 1,\quad x_{63} = 1,\quad x_{75} = 1,\quad x_{83} = 1$$

$$x_{18} = 1,\quad x_{24} = 1,\quad x_{35} = 1,\quad x_{45} = 1,$$

$$x_{52} = 1,\quad x_{65} = 1,\quad x_{75} = 1,\quad x_{83} = 1$$

$$x_{18} = 1,\quad x_{24} = 1,\quad x_{35} = 1,\quad x_{45} = 1$$

$$x_{56} = 1,\quad x_{65} = 1,\quad x_{74} = 1,\quad x_{83} = 1$$

The genetic system environment for the instance was set as follows: Population size was 40, maximum iteration was 100, crossover ratio was 0.2, and mutation ratio was 0.1. A total of 40 runs of the algorithm were made with different random number seeds, and the optimal solution was found with a probability of 0.675.

More numerical experiments, with comparisons of on problem size, genetic algorithm parameter setting, execution times, and frequency of obtaining optimal solutions, are given in Table 2.6.

2.4.2 Multiconstraint Knapsack Problem

Problem Description. The multiconstraint knapsack problem is defined as a knapsack problem with a set of constraints, such as *weight, size, reliability,* and so on. Also called the multiple-knapsack problem or multidimensional knapsack problem [441], it can be described as follows: There are m knapsacks of capacities W_1, W_2,...,W_m, and n objects, each of which has a cost c_j, $j = 1,2,...,n$. Unlike the simple knapsack problem, in which the weights of the objects are fixed, the weight of the jth object in the ith constraint takes the jth object weight w_{ij} when it is considered for possible inclusion in the jth knapsack of capacity W_i. Once more, we are interested in putting all

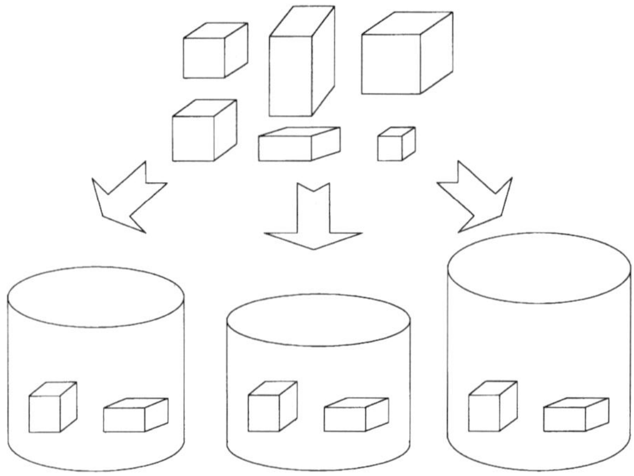

Figure 2.13. Multiconstraint knapsack problem.

objects into a knapsack, with no knapsack overfilled, and obtaining the minimum total cost. The problem is illustrated in Figure 2.13 using only three constraints.

The multiconstraint knapsack problem can be formulated mathematically as follows:

$$\min \sum_{j=1}^{n_i} c_j x_j \tag{2.22}$$

$$\text{s.t.} \quad \sum_{j=1}^{n_i} w_{ij} x_{ij} \leq W_i, \qquad i = 1,2,\ldots,m \tag{2.23}$$

$$x_{ij} \in \{0,1\}, \qquad \forall i,j$$

We refer to any vector or string $\boldsymbol{x} = (x_1,x_2,\ldots,x_n)$ satisfying the binary constraints as a solution. If $\boldsymbol{x}$ satisfies constraint (2.23), it is called feasible. The multiconstraint knapsack problem is a spacial case of zero–one linear programming with a great variety of applications in such areas as resource allocation and capital budgeting. It also appears as a subproblem in models for allocating processors and databases in a distributed computer system. Various algorithms for solving this NP hard problem have been proposed in the literature [155, 167, 212].

Genetic Algorithm Approach. Like many other combinatorial optimization problems, the multiconstraint knapsack problem has been approached using genetic algorithms. Khuri, Bäck, and Heitkötter directly adopted binary string encoding to represent a possible solution to the problem [360]. If the jth position has a value of 1 (i.e., $x_j = 1$), the jth object is in all knapsacks; otherwise, it is not. Encoding in this way, an infeasible solution will be generated. Rather than discarding infeasible strings and ignoring infeasible regions of the search space, they allowed infeasibly bred strings to join the population. Based on the observation that the farther away from feasibility the string is, the higher its penalty term should be, they maximized the following fitness function (their problem is to maximize the total profits p_j, $j = 1,2,\ldots,n$):

$$f(\boldsymbol{x}) = \sum_{j=1}^{n} p_j x_j - s \max\{p_j\}$$

where $s = |\{i | \Sigma_{j=1}^{n} w_{ij} x_j > W_i\}|$. In other words, s $(0 \leq s \leq m)$ denotes the number of overfilled knapsacks. The fitness function uses a graded penalty term, $\max\{p_j\}$. The number of times this term contributes to the weakening of the infeasible string's fitness is equal to the number of overfilled knapsacks it produces. With the fitness function above and several test problems from [48], they used the genetic algorithm software package GENEsYs [25] (available via anonymous ftp to *lupti.informatik.uni-dortmund.de* as file *GENEs Ys-1.o.tar.Z* in */pub/GA/src*) to deal with the multiconstraint knapsack problem.

Recently, Raidl developed an improved genetic algorithm to deal with the multiconstraint knapsack problem [522]. Bit string representation is also used, but linear programming (LP) is used to generate a meaningful starting solution for the problem. Based on the solution of $\boldsymbol{x}^{\text{LP}} = (x_1^{\text{LP}}, x_2^{\text{LP}}, \ldots, x_n^{\text{LP}})$ of the LP-relaxed multiconstraint knapsack problem, the algorithm proceeds as follows:

```
Procedure: Initialize x
 begin
 x ← 0;
 P ← random permutation of (1,2,...,n);
 for j ← 1 to n do
  if x^LP_P[j] > R_j then
   x_P[j] ← 1;
   if any W_i is violated then
    x_P[j] ← 0;
   end
  end
 end
end
```

All variables x_j are set to 0 initially. Then a random permutation P of indices 1 to n is generated to process all the variables x_j in random order. In the following loop, each variable $x_{P[j]}$ is first set to 1 with a probability equal to the LP solution's value $x_{P[j]}^{\text{LP}}$. Pseudo-random numbers $R_j (0 \leq R_j \leq 1)$ are used for this proposal. If setting $x_{P[j]}$ to 1 violates any constraint, $x_{P[j]}$ is reset to 0. Thus only feasible solutions are generated. Due to the random elements of the permutation, the algorithm generates different starting solutions and maintains population diversity.

Instead of penalizing infeasible solutions created by standard uniform crossover and bitwise mutation, a repair operator is used and illustrated as follows:

```
Procedure: Repair x
 begin
 P ← random permutation of (1,2,...,n) with
 x^LP_P[j] ≤ x^LP_P[j+1] (j = 1,...,n − 1);
 for j ← 1 to n do
  if x_P[j] = 1 and any W_i is violated then
   x_P[j] ← 0;
  end
 end
end
```

This operator is applied to each infeasible solution immediately before its fitness evaluation. Again, a permutation P of indices 1 to n is first determined to check all the variables x_j in a specific order. But now the indices are sorted according to increasing LP-solution values x_j^{LP}. Only indices with the same values are shuffled randomly. The aim is therefore to process variables with low x_j^{LP} first via variables with high x_j^{LP}, which are assumed to be more valuable when set to 1 later. Inside the loop, all variables $x_{P[j]}$ set to 1 one after the other reset to 0 as long as any constraint W_i is violated. Thus, in the worst case, all variables are reset to 0 and $\boldsymbol{x}$ is guaranteed to be a feasible solution.

Furthermore, a local improvement technique is adopted to improve each newly generated solution after making it feasible. The local improvement is illustrated in the following procedure:

```
Procedure: Locally Improve x
 begin
 P ← random permutation of (1,2,...,n) with
 x^LP_P[j] ≤ x^LP_P[j+1] (j = 1,...,n − 1);
 for j ← 1 to n do
  if x_P[j] = 0 then
   x_P[j] ← 1;
   if any W_i is violated then
    x_P[j] ← 0;
   end
  end
 end
end
```

In contrast to the repair operator, the variables x_j are now processed in decreasing order of LP solution values x_j^{LP}. As in the initialization process, each variable $x_{P[j]} = 0$ will be set to 1 if this does not violate a constraint. Therefore, solutions have the chance to become better while feasibility is retained. The major advantage of this improved genetic algorithm seems to be its much faster convergence rate, especially for larger-scale problems.

2.5 MINIMUM SPANNING TREE PROBLEM

The minimum spanning tree problem has a venerable history in combinatorial optimization. The problem was first formulated by Boruvka in 1926 when he developed a solution to finding the most economical layout of a power-line network [262]. Since then the minimum spanning tree formulation has been widely applied to many combinatorial optimization problems, such as transportation problems, telecommunication network design, distribution systems, and so on [357].

Consider a connected graph $G = (V,E)$, where $V = \{v_1,v_2,\ldots,v_n\}$ is a finite set of *vertices* and $E = \{e_1,e_2,\ldots,e_m\}$ is a finite set of *edges* representing connections between these vertices. Each edge has a positive real number denoted by $W = \{w_1,w_2,\ldots,w_m\}$, representing distance or cost. The minimum spanning tree problem is to seek a least-weight subgraph connecting all vertices on graph G.

Let $\boldsymbol{x}$ be a binary decision variable defined as:

$$x_i = \begin{cases} 1, & \text{if edge } e_i \text{ is selected} \\ 0, & \text{otherwise} \end{cases}$$

and let T denote the set of all spanning trees corresponding to the graph G. The minimum spanning tree problem can then be formulated as follows:

$$\min\left\{ z(\boldsymbol{x}) = \sum_{i=1}^{m} w_i x_i \mid \boldsymbol{x} \in T \right\} \tag{2.24}$$

Minimum spanning tree algorithms have been studied extensively and a variety of fast algorithms for the problem exist which attain nearly linear time complexity in the number of edges or number of vertices [164, 376, 515, 579].

Much research work shows that a spanning tree structure is the best topology for telecommunication network designs [357]. Spanning trees play a pivotal role in a majority of such network design and analysis problems. However, in a real-life network optimization situation, the problem often requires satisfying additional constraints. Therefore, we have constrained spanning tree problems [160], some of which are listed in Table 2.7.

TABLE 2.7. Examples of Constrained Minimum Spanning Tree Problems

Authors	Problem
Bertsimas (1990) [59]	Probabilistic minimum spanning tree problem
Fernandes and Gouveia (1998) [186]	Leaf-constrained minimum spanning tree problem
Ishii, Shiode, and Nishida (1981) [320]	Stochastic minimum spanning tree problem
Kershenbaum (1974) [356]	Capacitated minimum spanning tree problem
Narula and Ho (1980) [482]	Degree-constrained minimum spanning tree problem
Xu (1984) [675]	Quadratic minimum spanning tree problem
Zhou and Gen (1996) [698]	

Most constrained spanning tree problems are NP-hard and no polynomial-time solutions for them exist. Because of their complexity, some researchers use genetic algorithms to deal with them (see Table 2.8).

2.5.1 Quadratic Minimum Spanning Tree Problem

The quadratic minimum spanning tree problem takes two types of costs into consideration: direct cost and interactive cost. *Direct cost* is the cost associated with each edge. *Interactive cost* was introduced by Xu [675] to capture the fact that a cost occurs due to a pair of edges selected simultaneously within a tree. The objective function in equation (2.24) is no longer linear, so interactive costs have to be considered. Let c_{ik} denote the interactive cost

TABLE 2.8. Genetic Algorithm Approach to Constrained Minimum Spanning Tree Problems

Authors	Problem
Abuali, Wainwright, and Schoenfeld (1995) [3]	Probabilistic minimum spanning tree problem
Palmer and Kershenbaum (1995) [500]	Probabilistic minimum spanning tree problem
Zhou and Gen (1997) [699, 701]	Degree-constrained minimum spanning tree problem
Zhou and Gen (1998) [697]	Degree-constrained minimum spanning tree problem
Zhou and Gen (1997) [699]	Capacitated minimum spanning tree problem
Zhou and Gen (1998) [702]	Quadratic minimum spanning tree problem
Zhou and Gen (1998) [703]	Multicriteria minimum spanning tree problem
Zhou and Gen (1998) [697]	Leaf-constrained minimum spanning tree problem

due to a pair of edges (e_i,e_k). Thus the problem is formulated in the following form:

$$\min\left\{ z(\boldsymbol{x}) = \sum_{i=1}^{m} \sum_{k=1,k\neq i}^{m} c_{ik}x_ix_k + \sum_{i=1}^{m} w_ix_i \mid \boldsymbol{x} \in T \right\} \tag{2.25}$$

The minimum spanning tree problem with quadratic objective function above is denoted as the quadratic minimum spanning tree problem [675].

Heuristic Algorithms. There are no effective polynomial-time algorithms for the quadratic minimum spanning tree problem; only two heuristic algorithms were given by Xu [675].

Heuristic Algorithm H1 (Average Contribution Method). If the edge e_k is singly taken into consideration and all terms are separated from containing index k, the objective $z(\boldsymbol{x})$ in equation (2.25) can be rewritten as follows:

$$z(\boldsymbol{x}) = \left[w_k + \sum_{j\neq k} (c_{jk} + c_{kj})x_j \right] x_k + z_2(\boldsymbol{x})$$

where the term $z_2(\boldsymbol{x})$ no longer involves x_k. In the summation inside the brackets there is a total of $(m - 1)$ terms, all but $(n - 1)$ terms are bound to be zero. Suppose that e_k is the edge selected in the solution tree. The average contribution of e_k to the objective can be estimated according to the formula

$$p_k = w_k + \frac{n-1}{m-1} \sum_{j\neq k} (c_{jk} + c_{kj}), \qquad 1 \leq k \leq m \tag{2.26}$$

After evaluating the average contribution p_k for all edges e_k $(k = 1,2,...,m)$, the q-MST solution can be obtained by solving the following MST problem:

$$P = \min\left\{ \sum_{k=1}^{m} p_kx_k \mid \boldsymbol{x} \in T \right\} \tag{2.27}$$

Heuristic Algorithm H2 (Sequential Fixing Method). In heuristic algorithm H1, the average contributions of all edges are estimated independently. Suppose that fixing certain edges in the tree would affect the estimated contributions of the remaining edges. The average contributions of the remaining edges are then reassessed prior to fixing another edge. In this way the edges in the spanning tree are determined sequentially.

Let U be the index set of the edges that have been spanned into the tree, S that of the excluded edges whose inclusion would result in a cycle in the

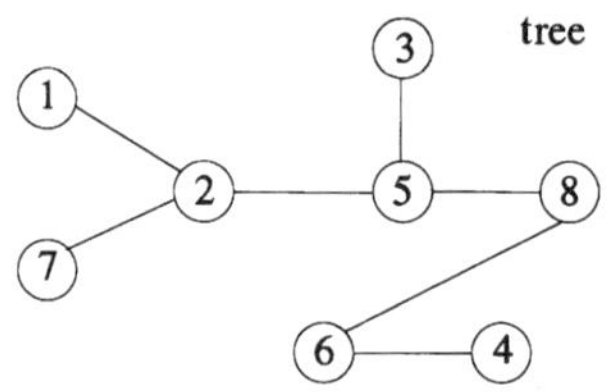

2	5	6	8	2	5

Figure 2.14. Tree and its Prüfer number.

tree, and F that of the free edges which are the candidates for the tree. The average contribution of e_k to the objective in each iteration can be estimated sequentially as follows:

$$q_k = \hat{w}_k + \frac{n_1}{m_1} \sum_{j \in F, j \neq k} (c_{jk} + c_{kj}) \tag{2.28}$$

where $m_1 = |F| - 1$ and $n_1 = n - 1 - |U|$, which means that there are $|F| - 1$ terms of which all but $n - 1 - |U|$ must vanish, and

$$\hat{w}_k = w_k + \sum_{j \in U} (c_{kj} + c_{jk})$$

The heuristic algorithm is illustrated in the following procedure.

Procedure: Heuristic H2

Step 1. Let $U \leftarrow \{\emptyset\}$, $F \leftarrow \{1,2,\ldots,m\}$, $n_1 \leftarrow n - 1$, and $m_1 \leftarrow m - 1$.

Step 2. Calculate q_k for all $k \in F$.

Step 3. Select the e_l such that $q_l = \min\{q_k | k \in F\}$. Let $x_l \leftarrow 1$, $U \leftarrow U \cup \{l\}$, $F \leftarrow F \backslash \{l\}$, and $n_1 \leftarrow n_1 - 1$. If $n_1 = 0$, stop.

Step 4. If for any k in F, adding edge e_k results in a cycle with the edges in U, set $x_k \leftarrow 0$ and $F \leftarrow F \backslash \{k\}$. Return to step 2.

Genetic Algorithms Implementation. Zhou and Gen proposed a genetic algorithm method to deal with the quadratic minimum spanning tree problem [702]. They adopted the *Prüfer number* to represent all candidate solutions of the problem, as it is capable of representing all possible trees equally. Figure 2.14 illustrates this encoding. For a detailed description, the reader may refer to Gen and Cheng [219].

Crossover and Mutation. Because Prüfer number encoding can always represent a tree after any crossover or mutation operations, the uniform crossover operator is adopted. Mutation is performed as a random perturbation within the range 1 to n.

TABLE 2.9. Comparison of Heuristic Algorithms and the Genetic Algorithm Approach[a]

Problem Size (Nodes)	H1	H2	GA	Improvement[b] (%)
10	925	921*	834	9.45
20	3,554*	3,749	3,233	9.03
30	8,500*	9,402	7,742	8.92
40	15,539	14,805*	14,275	3.58
50	25,182	23,487*	22,980	2.16

[a]H1 and H2, heuristic algorithms of Xu; GA, genetic algorithm of Zhou and Gen. An asterisk indicates the best-known solution by heuristic algorithms.

[b]Improvement = $(H_{\text{best}} - \text{GA})/H_{\text{best}} \times 100\%$, where H_{best} is the best-known solution using heuristics.

Evaluation and Selection. Evaluation consists of the following two steps:

1. Convert a chromosome into a tree, which includes transformation of the Prüfer number into a tree in the form of an edge set.
2. Calculate the total costs of the tree, in which the fitness of each chromosome can be calculated directly according to the objective in equation (2.25).

The $(\mu + \lambda)$-selection method is adopted [30]. It selects the μ best chromosomes from μ parents and λ offspring. If μ distinct chromosomes are not available, the vacant population pool is filled up with newly produced individuals.

Numerical Examples. The test problems have from 10 to 50 nodes [702]. All graphs are complete graphs with n nodes and $m = n(n - 1)/2$ edges. The diagonal elements in the intercost matrix for each quadrant minimum spanning tree problem are integers that are generated randomly over the range (0,100)]. The off-diagonal elements in the intercost matrix are randomly generated integers over the range (0,20)].

The parameters for the genetic algorithms are set as follows: *pop_size* = 300, $p_c = 0.2$, $p_m = 0.2$, *max_gen* = 500, and run length = 20.

Table 2.9 compares heuristic algorithms of Xu with the genetic algorithm approach of Zhou. Apparently, the genetic algorithm approach can produce much better results. The best heuristic results can be improved by an average of 9.6% (at most by 12.83%) using the genetic algorithm approach.

2.5.2 Degree-Constrained Minimum Spanning Tree Problem

In the minimum spanning tree problem, if we assume that there is a degree constraint on each vertex such that at each vertex v_j, the degree value d_j is at most a given value b_j, the number of edges incident to each vertex is constrained. Then the problem is denoted as a degree-constrained minimum spanning tree and can be formulated as follows [482]:

$$\min\left\{z(\boldsymbol{x}) = \sum_{i=1}^{m} w_i x_i \mid d_j \leq b_j, j \in V, \boldsymbol{x} \in T\right\} \tag{2.29}$$

One intuitive approach to dealing with this problem is the use of heuristic algorithms: (1) finding a minimum spanning tree without a degree constraint, and (2) modifying the minimum spanning tree combined with a degree constraint. But it is difficult to operate effectively as the problem scale gets larger. Several heuristic algorithms to solve this problem were suggested by Narula and Ho [482], Savelsbergh [562], and Volgenant [644]. Zhou and Gen once proposed a genetic algorithm approach to dealing with this problem [219]. More recently, Zhou and Gen developed a new tree encoding based on the degree value of a tree for solving the degree-constrained minimum spanning tree problem [697].

Genetic Algorithm Implementation

Degree-Based Permutation. For a degree-constrained minimum spanning tree problem, two factors should be encoded in a chromosome: (1) the connectivity among vertices and (2) the degree value of each vertex. Therefore, an intuitive idea is to use a two-dimensional structure to encode a spanning tree with a degree constraint: one dimension for the connection among vertices and the other dimension for the degree value of each vertex. A $2 \times n$ matrix is needed to represent a chromosome for a n-vertex tree. Genes in the vertex dimension take integers from 1 to n exclusively; genes in the degree dimension take integers from 1 to b, the constrained degree value of all vertices. This representation was termed degree-based permutation by Zhou and Gen [697].

For an undirected tree, we can take any vertex as the root vertex. All other vertices are then regarded as being connected to it hierarchically. For a given vertex (current vertex), the vertex incident to it on the upper hierarchy is termed its predecessor vertex, and the vertex incident to it on the lower hierarchy, its successor vertex. Obviously, the root vertex has no predecessor vertex and the leaf vertex has no successor vertex. Based on this observation, degree-based permutation of a tree can be encoded as in the following procedure:

Procedure: Degree-Based Permutation Encoding

Step 1. Select any vertex (root vertex) in a labeled tree T, set it as the first digit in the vertex dimension of the permutation and its degree value as the first digit in the degree dimension, and let that vertex be the current vertex.

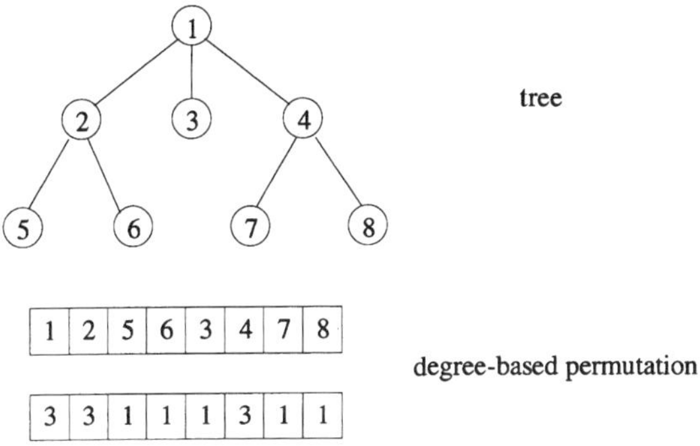

Figure 2.15. Tree and its degree-based permutation.

Step 2. Check for a successor vertex to the current vertex from the left branch to the right branch. If there is a successor vertex, put it into the vertex dimension and its degree value into the degree dimension (here we build the permutation by appending digits to the right), and then go to step 3. If there is no such vertex, let the predecessor vertex be the current vertex, and return to step 2.

Step 3. If the successor vertex is not a leaf vertex, let the successor vertex be the current vertex, then go to step 2. If the successor vertex turns to be a leaf vertex, delete it and go to step 4.

Step 4. If all vertices have been checked, stop; otherwise, go to step 2.

Figure 2.15 illustrates a degree-based permutation.

It is easy to obtain a tree from degree-based encoding. Let $P_1(k)$ denote the vertex dimension for a given individual and $P_2(k)$ denote the degree dimension for a given individual. The decoding procedure works as follows:

Procedure: Degree-Based Permutation Decoding

Step 1. Set $k \leftarrow 1$ and $j \leftarrow 2$.

Step 2. Select a vertex in $P_1(k)$, say v_r, where $r = P_1(k)$; select a vertex in $P_1(j)$, say v_s, where $s = P_1(j)$; and add the first edge from r to s into a tree.

Step 3. Let $P_1(j) \leftarrow P_2(j) - 1$.

Step 4. If $P_2(j) \geq 1$, let $k \leftarrow j$ and $j \leftarrow j + 1$.

Step 5. If $k < 1$, stop.

Step 6. If $P_2(k) \geq 1$, select a vertex in $P_1(k)$, say v_r, where $r = P_1(k)$, and let $P_2(k) \leftarrow P_2(k) - 1$; otherwise, let $k \leftarrow k - 1$ and go to step 5.

Step 7. If $P_2(j) \geq 1$, select a vertex in $P_1(j)$, say v_s, where $s = P_1(j)$, and let $P_2(j) \leftarrow P_2(j) - 1$; otherwise, let $j \leftarrow j + 1$ and go to step 6.

Step 8. Add the edge from r to s into the tree and go to step 4.

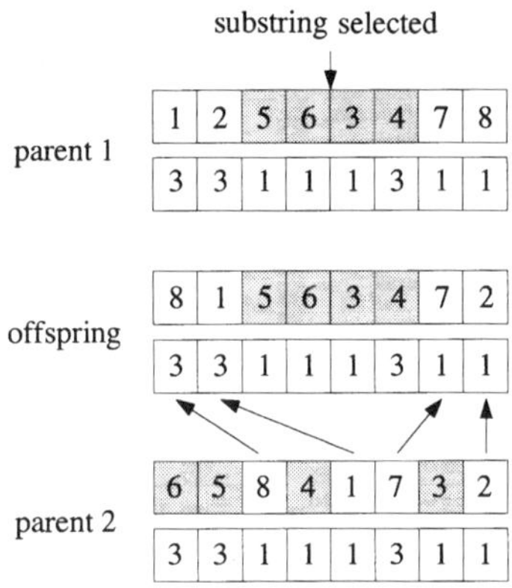

Figure 2.16. Order crossover on vertices.

Obviously, any spanning tree can be encoded by the representation scheme and any encoding represents a spanning tree. However, the relation between the encoding and its spanning tree may not be 1-to-1 mapping on an undirected graph, and different encodings may represent the same spanning tree. The encoding possesses the locality in the sense that small changes in the genotype make small changes in the phenotype, the tree.

Initial Population. To keep the degree constraint and connectivity between vertices, the genes in the degree dimension need to satisfy the following conditions: For an n-vertex tree, the degree value for any vertex is at least 1 and the total degree value for all vertices is $2(n - 1)$. Suppose that d_{rest} is the total lower bound of the degree values for all those vertices whose degree value in the degree dimension have not been assigned and that d_{used} is the total degree value of the vertices whose degree value in the degree dimension have been assigned. Then the degree value for the current vertex in the degree dimension should hold: no less than 1 or no greater than the given degree value. The degree value for the current vertex together with the number of the rest vertex should hold: no less than d_{rest} and no greater than $2(n - 1) - d_{\text{used}}$.

Order Crossover on Vertices. Order crossover was proposed by Davis [145]. To avoid illegal connections among vertices, crossover is operated only on the vertex dimension; the degree dimension remains unchanged. This is illustrated in Figure 2.16. Crossover will dramatically change the tree structure among generations. This is useful for exploration on shole solution space.

Swap Mutation on Vertices. Swap mutation selects two genes (vertices) at random and then swaps them. This is illustrated in Figure 2.17.

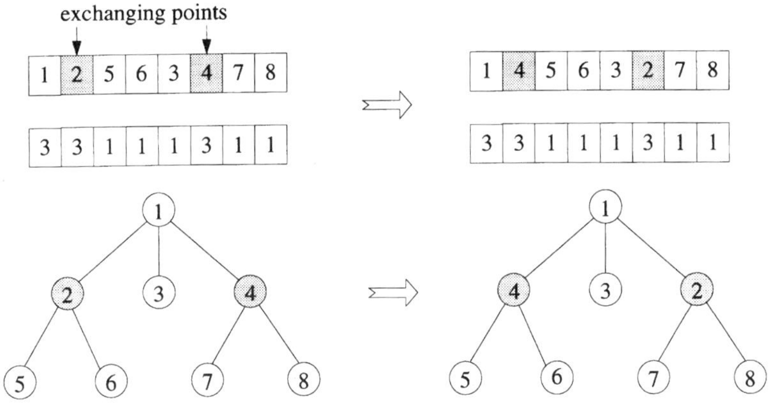

Figure 2.17. Swap mutation on vertices.

Insertion Mutation. Insertion mutation selects a string of genes (branch) at random and inserts it in a random gene (vertex). In both removing and inserting a gene to change the degree value of the vertex, it is necessary to modify the gene (degree) in the corresponding vertex, as illustrated in Figure 2.18. This operator can maintain a good heritage from generation to generation. The result is exploitation in the evolutionary process.

Numerical Examples. Table 2.10 is a simple example of a nine-vertex complete graph where the constrained degree value of all vertices is 3. It was given by Savelsbergh and Volgenant, who solved it using a heuristic algorithm denoted as *edge exchange,* and the optimal solution is 2256 [562].

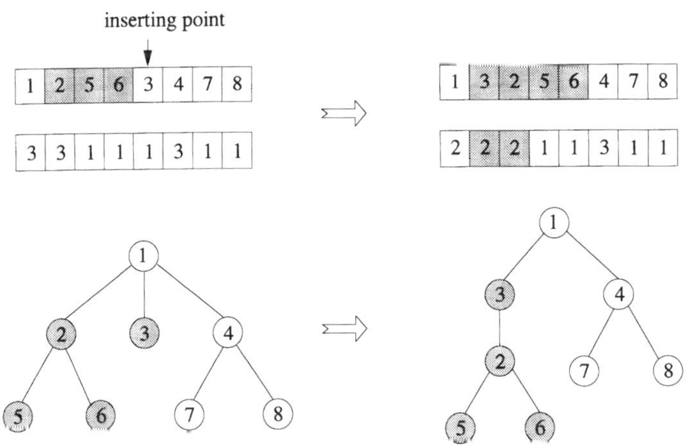

Figure 2.18. Insertion mutation.

TABLE 2.10. Edge Weights of the Nine-Vertex Problem

i	1	2	3	4	5	6	7	8	9
1	—	224	224	361	671	300	539	800	943
2		—	200	200	447	283	400	728	762
3			—	400	566	447	600	922	949
4				—	400	200	200	539	583
5					—	600	447	781	510
6						—	283	500	707
7							—	361	424
8								—	500
9									—

The parameter for the proposed genetic algorithm approach with degree-based permutation are set as *pop_size* = 100, crossover rate = 0.4, mutation rate for exchange = 0.2 and for insertion = 0.6, and *max_gen* = 500.

The optimal solution, 2256, can be obtained easily using the genetic algorithm. Table 2.11 shows the results on five randomly generated instances. The weights on each edge are generated randomly and distributed uniformly over [10,100]. The lower bounds in each instance are calculated using Prim's algorithm [515] by relaxing the degree constraint. The results using the genetic algorithm approach are within less than 6% of those of their lower bounds.

2.5.3 Bicriteria Minimum Spanning Tree Problem

In the real world there are usually cases in which in determining a minimum spanning tree, one has to consider multiple criteria simultaneuosly, because multiple attributes are defined on each edge. When designing a layout for a telecommunication system, for example, in addition to the cost of connection

TABLE 2.11. Comparison of Results Using Genetic Algorithms and Their Lower Bounds[a]

Problem Scale (Vertices)	LBs		dpGAs	
	min_val	Degree	*min_val*	%
10	117	5(2)	123	5.13
20	233	4(2)	237	1.72
30	316	4(4)	321	1.58
40	419	7(3)	428	2.15
50	513	6(4)	531	3.51

[a]dpGAs, GAs with degree-based permutation; *min_val*, minimal value; %, percentage of minimum of GAs over LBs.

between cities or terminals, factors such as the time for communication or construction, the complexity of construction, and even the reliability are important and have to be taken into consideration. In all these cases, the minimum spanning tree with multicriteria is a more realistic representation of the practical problem [698].

Consider the minimum spanning tree problem in problem (2.24), where each edge has two associated positive real numbers, representing two attributes defined on it and denoted by $w_i = (w_{1i}, w_{2i})$ $(i = 1,2,...,m)$. Then the bicriteria minimum spanning tree problem can be formulated as

$$\min \quad z_1(\boldsymbol{x}) = \sum_{i=1}^{m} w_{1i} x_i$$

$$\min \quad z_2(\boldsymbol{x}) = \sum_{i=1}^{m} w_{2i} x_i \qquad (2.30)$$

$$\text{s.t.} \quad \boldsymbol{x} \in T$$

Enumeration Method. In general, we cannot obtain an optimal solution of the problem because the two objectives usually conflict with each other in practice. The solutions to this problem are a set of Pareto optimal solutions [97]. A possible way to approach this problem is to scale two objectives into a single objective such as is done in the common method when dealing with multicriteria optimization problems. But only one single-point Pareto optimal solution can be obtained, not to mention the difficulty in properly transforming the two objectives into a single objective. In practice, however, decision makers may prefer one (Pareto optimal) point over the others, depending on their preferences on different objectives. Consequently, it may be useful to have (all) other possible Pareto optimal solutions. Zhou and Gen proposed a enumeration method by enumerating all Pareto optimal solution for this problem to verify the effectiveness and efficiency of their genetic algorithm approach.

Simulating the process of edge growth in obtaining a spanning tree, we can enumerate all spanning trees to find the nondominated solutions for the problem. In the process of edge growth, we cannot determine which edge has the least weight, but we can determine whether or not a particular subtree is dominated if one edge is added to that subtree. Therefore, once an edge is spanned into a subtree, the set of nondominated subtrees is updated. Finally, we can obtain the set of all nondominated solutions until all subtrees have $n - 1$ edges.

Two principal factors should be taken into consideration in enumerating all Pareto optimal solutions:

1. For any subtree in the set of nondominated subtrees, all adjacent edges are checked but only one edge is spanned into the subtree in each step.

2. Because each subtree will result in more than one nondominated subtree, all the nondominated subtrees generated from different subtrees with the same number of edges have to be checked again regarding Pareto optimality and the set of nondominated subtree updated.

Suppose that $T_r^{(t)}$ is the rth subtree containing t edges, $r \in I^{(t)}$, which is the index set representing the number of such subtrees; $W_k(T_r^{(t)})$ is the total weight of the subtree $T_r^{(t)}$ in the kth objective; and $E(T_r^{(t)})$ is the set of all those edges adjacent to the subtree $T_r^{(t)}$.

Procedure: Enumeration

Step 1. Set $t \leftarrow 0$, $T_0^{(0)} \leftarrow \emptyset$, $I^{(0)} \leftarrow \emptyset$. and $W_k(T_0^{(0)}) \leftarrow 0$, $k = 1{,}2$.

Step 2. Select an arbitary vertex in V, say v_1. Let $E(T_0^{(0)})$ contain all edges adjacent to the vertex v_1. Set $r \leftarrow 1$ and $t \leftarrow t + 1$. Go to step 6.

Step 3. Set $r \leftarrow 1$ and $t \leftarrow t + 1$. If $t \geq n - 2$, stop; otherwise, continue.

Step 4. If $r \notin I^{(t-1)}$, go back to step 3; otherwise, continue.

Step 5. If $E(T_r^{(t-1)})$ is empty, then let $r \leftarrow r + 1$ and go back to step 4; otherwise, continue.

Step 6. Select an edge e_l from the set $E(T_r^{(t-1)})$ randomly. Update the edge set with $E(T_r^{(t-1)}) \leftarrow E(T_r^{(t-1)}) \backslash \{e_l\}$.

Step 7. If the edge $\{e_l\}$ forms a cycle with all other edges in $T_r^{(t-1)}$, go back to step 5; otherwise, continue.

Step 8. Make a temporary subtree with $T^{(t)} \leftarrow T_r^{(t-1)} \cup \{e_l\}$ and calculate its two objective function values with $W_k(T^{(t)}) \leftarrow W_k(T_r^{(t-1)}) + w_{kl}$, $k = 1{,}2$.

Step 9. Compare the subtree $T^{(t)}$ with all subtrees $T_s^{(t)}$, $s \in I^{(t)}$ in the sense of Pareto optimality. If it is a nondominated one, update the nondominated set $\{T_s^{(t)}, s \in I^{(t)}\}$ and the index set of $I^{(t)}$. Go back to step 5.

When all those subtrees $T_s^{(t)}$, $s \in I^{(t)}$, contain $n - 1$ edges, we get the final Pareto optimal solutions for the problem. Obviously, this method can be extended to the minimum spanning tree problem with multicriteria.

Genetic Algorithm Approach. Zhou and Gen proposed a genetic algorithm approach to deal with the bicriteria minimum spanning tree problem [703]. The Prüfer number was adopted as the tree encoding [517], as the encoding is capable for representing all possible spanning trees for a graph equally and uniquely and is easily found through a genetic algorithm approach. Together with the Prüfer number, uniform crossover [603] and perturbation mutation were used for the genetic operations.

To evaluate the fitness of each individual, they adopted two strategies for the genetic algorithm approach to the bicriteria minimum spanning tree problem. The first strategy uses an adaptive evaluation function based on the method discussed in Section 3.6. The evaluation procedure can be outlined as follows:

Procedure: Evaluation for Strategy I

Step 1. Decode all chromosomes and calculate their objective values in each objective function.

Step 2. Determine the fitness value eval(T) of all individuals according to the following formula:

$$\text{eval}(T) = \sum_{k=1}^{2} \lambda_k z_k$$

where λ_k ($k = 1,2$) are the weighting coefficients.

Only $(\mu + \lambda)$-selection was adopted.

The second strategy uses the nondominated sorting technique discussed in Section 3.6. The entire procedure can be formulated concisely as follows:

Procedure: Evaluation for Strategy II

Step 1. Determine all nondominated individuals P_c from the current population, and assign a large dummy fitness value to them.

Step 2. Calculate each individual's *niche count* m_j:

$$m_j = \sum_{k \in P_c} \text{sh}(d_{jk})$$

where

$$\text{sh}(d_{jk}) = \begin{cases} 1 - \left(\dfrac{d_{jk}}{\sigma_{\text{share}}} \right)^2, & \text{if } d_{jk} < \sigma_{\text{share}} \\ 0, & \text{otherwise} \end{cases}$$

d_{jk} is the phenotypic distance between two individuals j and k in the current nondominated solution set, and σ_{share} is the maximum phenotypic distance allowed between any two individuals to become members of a niche.

Step 3. Calculate the shared fitness value of each individual by dividing its dummy fitness value by its niche count.

Step 4. Ignore all sorted nondominated individuals, go to step 1, and continue the process until the entire population is sorted.

The overall pseudocode procedure for the bicriteria minimum spanning tree problem is outlined as follows:

TABLE 2.12. Results by the Genetic Algorithm and Enumeration Approaches[a]

Problem Size (Vertices)	Enumeration		GA with Strategy I		GA with Strategy II	
	TPO	Run Time (min)	Percent	Run Time (min)	Percent	Run Time (min)
10	56	20	100	3.02	100	4.20
20	123	56	100	4.31	100	6.50
30	262	132	90	6.18	94	9.32
40	417	336	86	8.05	91	12.12
50	635	744	80	10.96	85	15.58

[a]TPO is the total number of Pareto optimal solutions; percent is the percentage of the Pareto optimal solutions.

```
Procedure: Genetic Algorithm
begin
 t ← 0;
 initialize the population of parents P(0);
 determine the set of nondominated solutions E(0);
 while (not termination condition) do
 begin
  recombine P(t) to yield the population of offspring
  C(t);
  update E(t);
  if (preference on Strategy I) then
   evaluate P(t) and C(t) by strategy I;
   select P(t + 1) from P(t) and C(t) by strategy I;
  else
   evaluate P(t) and C(t) by strategy II;
   select P(t + 1) from P(t) and C(t) by strategy II;
  end
  t ← t + 1;
 end
end
```

Numerical Example. The performance of the multicriteria minimum spanning tree problem by a genetic algorithm approach is tested on five numerical examples from a 10- to 50-vertex complete graph generated randomly. Each edge has two weights. The weights are generated randomly and distributed uniformly over [10, 100] and [10, 50], respectively.

The parameters for the genetic algorithm are set as follows: $pop\_size = 200$, crossover probability $p_c = 0.2$, mutation probability $p_m = 0.05$, and $max\_gen = 500$.

Table 2.12 shows that genetic algorithm approaches are much more efficient than the method of Pareto optimal enumeration. All the results obtained

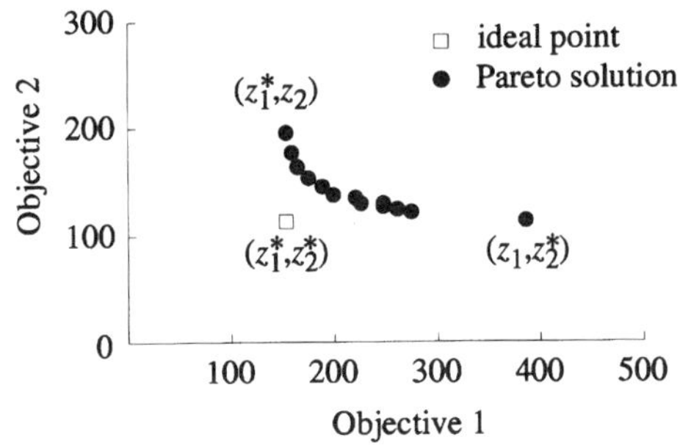

Figure 2.19. Genetic algorithm on mc-MST with 10-vertex by strategy I.

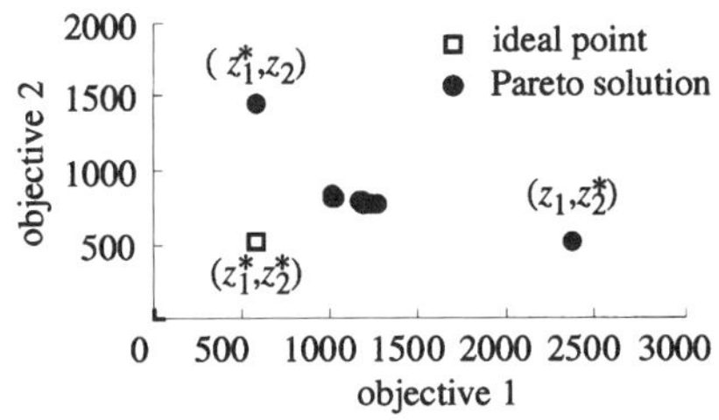

Figure 2.20. Genetic algorithm on mc-MST with 50-vertex by strategy I.

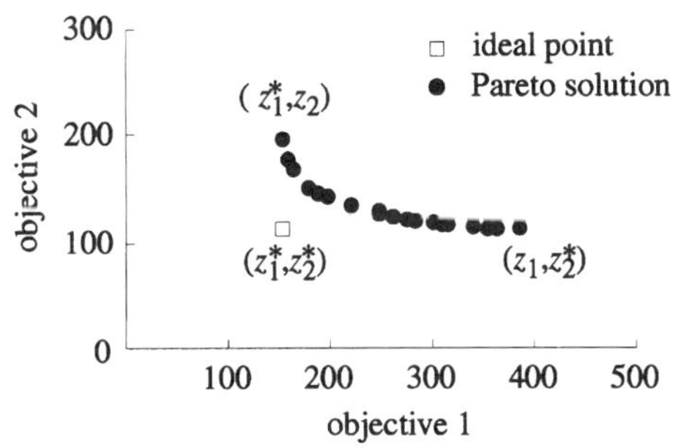

Figure 2.21. Genetic algorithm on mc-MST with 10-vertex strategy II.

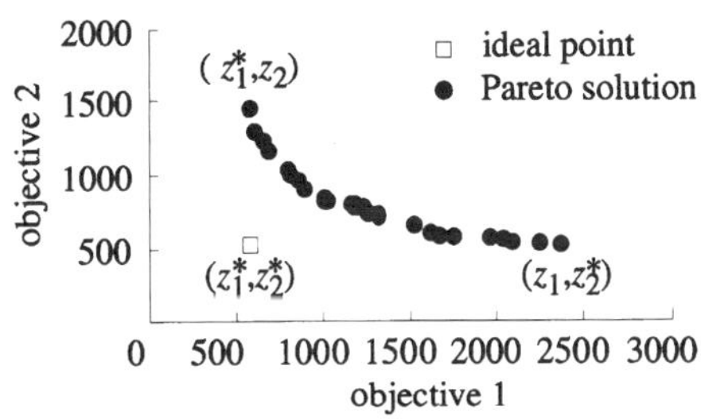

Figure 2.22. Genetic algorithm on mc-MST with 50-vertex strategy II.

by genetic algorithm approaches are Pareto optimal solutions when the problem scale is small and most of them are Pareto optimal solutions when the problem scale is larger.

The average results in 10 trials by the genetic algorithm approach with strategy I are illustrated in Figures 2.19 and 2.20. They give out both the ideal points and all Pareto optimal solutions. They clearly show that all Pareto optimal solutions are close to the ideal point under the action of the adaptive evaluation function. In the case of 50 vertices shown in Figure 2.20, it is particularly obvious that the adaptive evaluation function forces individuals to evolve toward the ideal point and focus on the Pareto frontier near the ideal point. This reflects such a decision maker's preference; that is, all objectives have the same importance, and the closer all objectives are to the ideal point, the better.

Figures 2.21 and 2.22 present the results using strategy II. They show that all solutions are distributed along the Pareto frontier instead of in specific regions. It provides more solutions by allowing decision makers to choose among possible alternatives.

3

MULTIOBJECTIVE OPTIMIZATION PROBLEMS

3.1 INTRODUCTION

Optimization deals with the problem of seeking solutions over a set of possible choices to optimize certain criteria. If there is only one criterion to consider, it becomes a single-objective optimization problem, a type studied extensively for the past 50 years. If there is more than one criterion and they must be treated simultaneously, we have a multiple-objective optimization problem [161, 588]. Multiple-objective problems arise in the design, modeling and planning of many complex real systems in the areas of industrial production, urban transportation, capital budgeting, forest management, reservoir management, layout and landscaping of new cities, and energy distribution, for example. Almost every important real-world decision problem involves multiple and conflicting objectives that need to be tackled while respecting various constraints, leading to overwhelming problem complexity. Multiple-objective optimization problems have been of increasing interest to researchers of various backgrounds since the early 1960s [313]. A number of scholars have made significant contributions to the problem. Among them, Pareto is perhaps one of the most recognized pioneers in the field [503]. Interested readers are referred to [583] and [584]. Genetic algorithms have received considerable attention as a novel approach to multiobjective optimization problems, resulting in a fresh body of research and applications known as genetic multiobjective optimizations.

3.2 BASIC CONCEPTS OF MULTIOBJECTIVE OPTIMIZATIONS

A single-objective optimization problem is usually given in the following form:

$$\max \quad z = f(\boldsymbol{x}) \tag{3.1}$$

$$\text{s.t.} \quad g_i(\boldsymbol{x}) \leq 0, \qquad i = 1,2,\ldots,m \tag{3.2}$$

where $\boldsymbol{x} \in R^n$ is a vector of n decision variables, $f(\boldsymbol{x})$ an objective function, and $g_i(\boldsymbol{x})$ inequality constraint m functions which form an area of feasible solutions. We usually denote the feasible area in decision space by the set S, as follows:

$$S = \{\boldsymbol{x} \in R^n | g_i(\boldsymbol{x}) \leq 0,\ i = 1,2,\ldots,m,\ \boldsymbol{x} \geq 0\} \tag{3.3}$$

Without loss of generality, a multiple-objective optimization problem can be represented formally as follows:

$$\max \quad \{z_1 = f_1(\boldsymbol{x}),\ z_2 = f_2(\boldsymbol{x}), \ldots,\ z_q = f_q(\boldsymbol{x})\} \tag{3.4}$$

$$\text{s.t.} \quad g_i(\boldsymbol{x}) \leq 0. \qquad i = 1,2,\ldots,m \tag{3.5}$$

We sometimes graph the multiple-objective problem in both decision space and criterion space. S is used to denote the feasible region in the *decision space* and Z is used to denote the feasible region in the *criterion space*.

$$Z = \{z \in R^q | z_1 = f_1(\boldsymbol{x}),\ z_2 = f_2(\boldsymbol{x}), \ldots, z_q = f_q(\boldsymbol{x}), \quad \boldsymbol{x} \in S\} \tag{3.6}$$

where $z \in R^q$ is a vector of values of q objective functions. In other words, Z is the set of images of all points in S. Although S is confined to the nonnegative orthant of R^n, Z is not necessarily confined to the nonnegative orthant of R^q.

3.2.1 Nondominated Solutions

In principle, multiple-objective optimization problems are very different from single-objective optimization problems. In the single-objective case, one attempts to obtain the best solution, which is absolutely superior to all other alternatives. In the case of multiple objectives, there does not necessarily exist a solution that is best with respect to all objectives because of incommensurability and conflict among objectives. A solution may be best in one objective but worst in other objectives. Therefore, there usually exist a set of solutions for the multiple-objective case which cannot simply be compared with each other. For such solutions, called *nondominated* solutions or *Pareto optimal* solutions, no improvement in any objective function is possible without sacrificing at least one of the other objective functions. For a given nondominated point in the criterion space Z, its image point in the decision space S is called *efficient* or *noninferior*. A point in S is efficient if and only if its image in Z is nondominated.

Definition 3.1. For a given point $z^0 \in Z$, it is *nondominated* if and only if there does not exist another point $z \in Z$ such that for the maximization case,

$$z_k > z_k^0, \qquad \text{for some } k \in \{1,2,\ldots,q\}$$

$$z_l \geq z_l^0, \qquad \text{for all } l \neq k$$

where z^0 is a *dominated* point in the criterion space Z.

Definition 3.2. For a given point $\boldsymbol{x}^0 \in S$, it is *efficient* if and only if there does not exist another point $\boldsymbol{x} \in S$ such that for the maximization case,

$$f_k(\boldsymbol{x}) > f_k(\boldsymbol{x}^0), \qquad \text{for some } k \in \{1,2,\ldots,q\}$$
$$f_l(\boldsymbol{x}) \geq f_l(\boldsymbol{x}^0), \qquad \text{for all } l \neq k$$

where $\boldsymbol{x}^0$ is *inefficient.*

A point in the decision space is efficient if and only if its image is a nondominated point in the criterion space Z.

To illustrate the definitions above, consider the following linear programming problem:

$$\min \quad z_1(x_1,x_2) = x_1 - 3x_2 \tag{3.7}$$

$$\min \quad z_2(x_1,x_2) = 3x_1 + x_2 \tag{3.8}$$

$$\text{s.t.} \quad g_1(x_1,x_2) = x_1 + 2x_2 - 2 \leq 0 \tag{3.9}$$

$$g_2(x_1,x_2) - 2x_1 + x_2 - 2 \leq 0 \tag{3.10}$$

$$x_1,x_2 \geq 0 \tag{3.11}$$

The feasible region S in the decision space is shown in Figure 3.1. The extreme points in the feasible region are $\boldsymbol{x}^1$ (0,0), $\boldsymbol{x}^2$ (1,0), $\boldsymbol{x}^3$ (⅔, ⅔), and $\boldsymbol{x}^4$ (0,1). The feasible region Z in the criterion space is obtained by mapping set S by using two objectives, (3.7) and (3.8), as shown in Figure 3.2. The corresponding extreme points are z^1 (0,0), z^2 (1,3), z^3 (−4⁄3, 8⁄3), and z^4 (−3,1). From the figures we observe that both regions are convex, and the extreme points of Z are the images of extreme points of S. Having portions of Z outside the nonnegative orthant of R^2 is a likely occurrence when the problem has objective functions with negative coefficients. The slashed border of the feasible region Z is identified as the set of nondominated solutions. This set can be found by simple visual inspection on the feasible region Z. Looking at the points between z^1 and z^4, it is noted that as $f_2(x_1,x_2)$ increases from 0 to 1, $f_1(x_1,x_2)$ decreases from 0 to -3, and accordingly, all points between z^1 and

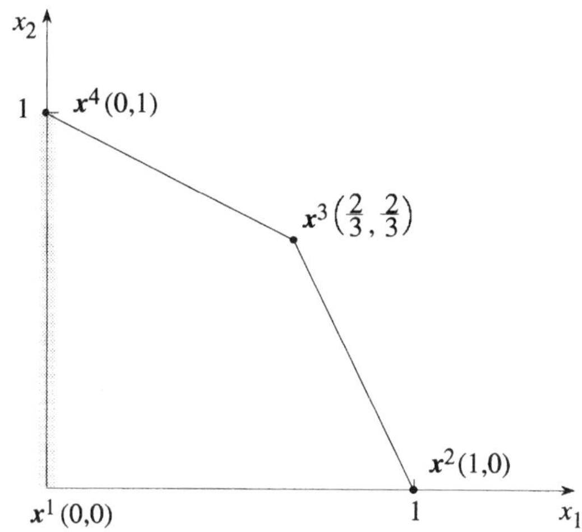

Figure 3.1. Feasible region and efficient solutions in decision space.

z^4 are nondominated points. The corresponding efficient points in the decision space are the segment between points $\boldsymbol{x}^1$ and $\boldsymbol{x}^4$.

There is a special point in the criterion space Z called an *ideal point* or a *positive ideal solution,* denoted by $z^* = (z_1^*, z_2^*, ..., z_q^*)$, where $z_k^* = \sup\{ f_k(\boldsymbol{x}) | \boldsymbol{x} \in S\}$, $k = 1,2,...,q$. The point $\mathbf{z}^*$ is called an ideal point because usually it is not attainable. Note that, individually, z_k^* may be attainable. But to find a point z^* that can maximize each $f_k(\cdot)$, $k = 1,2,...,q$, simultaneously is usually very difficult.

A *negative ideal solution* is also defined to represent the pessimistic status in the criterion space, denoted by $z^- = (z_1^-, z_2^-, ..., z_q^-)$, where $z_k^- =$

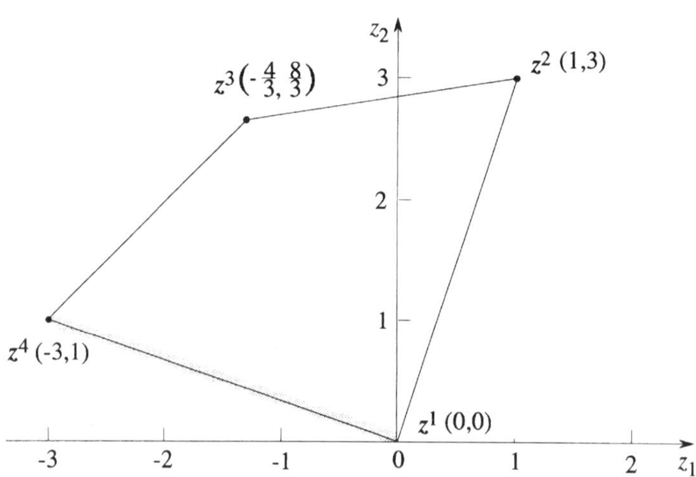

Figure 3.2. Feasible region and nondominated solutions in criterion space.

$\inf\{ f_k(\boldsymbol{x}) | \boldsymbol{x} \in S \}$, $k = 1,2,\ldots,q$. Similar to the positive ideal point, the negative ideal point is usually not attainable. Individually, z_k^- may be attainable.

3.2.2 Preference Structures

We know that a basic feature of solutions for multiple-objective problems is that there is a set of *efficient solutions* that cannot simply be compared with each other. In real decision-making cases, we are usually asked to select one of the nondominated solutions as a final solution to a given problem. It is most likely, however, that we will be unable to settle for one of those solutions without providing additional preferences regarding various objectives. Therefore, how to make a final choice from those alternative solutions essentially depends on one's subjective preference. Conceptually, the *preference* intends to give an *order* to the incomparable solutions within the efficient set by using one's value judgments on objectives. The preference reflects either one's trade-offs among objectives or an emphasis on particular objectives according to prior knowledge of the problem. With a given preference we can order the alternative solutions in the nondominated set, and then we can obtain a final solution, which is the usual outcome of a decision-making process. Such a final solution is called the *best-compromised solution.*

A preference is usually represented by a binary relation. A *binary relation* is a set of ordered pairs. For a given pair, say u and v, one and only one of the following relations can occur:

1. u is better than or preferred to v, denoted by $u > v$.
2. u is worse than or less preferred to v, denoted by $u < v$.
3. u is equivalent to or equally preferred to v, denoted by $u \sim v$.
4. The preference relation between u and v is indefinite, denoted by $u?v$.

The symbol $>$, $<$, $\sim$, and ? are *operators* defining the comparison and relations. The preference relations between any pair of points in the decision space Z can be specified by one of the four operators above. Any revealed preference information can then be represented by a subset of the *Cartesian product* $Z \times Z$ as follows:

1. $\{>\}$ is the set of preferred relations.
2. $\{<\}$ is the set of less preferred relations.
3. $\{\sim\}$ is the set of equally preferred relations.
4. $\{?\}$ is the set of indefinite preference relations.

For example, $(u,v) \in \{>\}$ means that $u > v$. The set $\{>\}$ and $\{<\}$ are *symmetric,* that is, $(u,v) \in \{>\}$ if and only if $(v,u) \in \{<\}$. Thus if $\{>\}$ is known, $\{<\}$ is also known, and vice versa. Therefore, a preference structure can be determined uniquely by $\{>\}$, $\{\sim\}$, and $\{?\}$.

3.2.3 Basic Solution Approaches

The general expectation for a decision-making process can be either to obtain a compromised or preferred solution or to identify all nondominated solutions. Therefore, there are basically two kinds of techniques for constituting a solution method to multiple-objective optimization problems: (1) generating approaches and (2) preference-based approaches. The generating approaches have been developed to identify an entire set of Pareto solutions or an approximation. Preference-based approaches attempt to obtain a compromised or preferred solution. If we have no prior knowledge of preference structure over objectives, we have to adopt the generating approach to examine all nondominated alternatives. If we have some idea of the relative importance of objectives, we can quantify the preference. With the preference information, a compromised or preferred solution can be identified.

Both generating and preference-based methods exhibit strengths and weaknesses. Preference-based techniques require decision makers to articulate their preferences in a formal and structured way, while generating techniques need decision makers to make a necessary value judgment by selecting from among entire pareto solutions. There are, however, additional problems with generating techniques that are not observed with most preference-based techniques. Problems with two or three criteria permit the clear presentation of choices through graphical means. But displaying results and making a choice become very complicated in higher-dimensional problems, increasing in difficulty approximately exponentially with the number of criteria. Computational costs also increase rapidly with the number of criteria.

From the viewpoint of solution techniques, most traditional methods reduce multiple objectives to a single objective, then solve the problem using mathematical programming tools. To utilize mathematical programming tools to solve multiple-objective problems, we first need to express our preference in terms of numbers, so that the larger the number, the stronger the preference. The needs foster the development of various scalarization techniques, by whose use multiple-objective optimization problems are usually transformed into a single objective or a sequence of single-objective optimization problems; then traditional techniques such as the utility function approach, weighted-sums approach, and compromise approach can be adapted to solve the altered problems.

Weighted-Sum Approach. The basic idea of assigning weights to each objective function and combining them into a single-objective function was first proposed by Zadeh [690]. The weighted-sum method can be represented as follows:

$$\max \quad z(\boldsymbol{x}) = \sum_{k=1}^{q} w_k f_k(\boldsymbol{x}) \tag{3.12}$$

$$\text{s.t.} \quad \boldsymbol{x} \in S \tag{3.13}$$

The weight w_k can be interpreted as the relative emphasis or worth of that objective compared to other objectives. In other words, the weight can be interpreted as representing our preference over objectives. Therefore, an optimal solution to problem (3.12)–(3.13) relates to a particular preference structure. Moreover, the optimal solution to the problem is a nondominated solution provided that all the weights are positive.

For two given points $\boldsymbol{x}^1$ and $\boldsymbol{x}^2$ in the decision space S, $z(\boldsymbol{x}^1) > z(\boldsymbol{x}^2)$ if and only if $\boldsymbol{x}^1 > \boldsymbol{x}^2$, and $z(\boldsymbol{x}^1) = z(\boldsymbol{x}^2)$ if and only if $\boldsymbol{x}^1 \sim \boldsymbol{x}^2$. Then we have

$$\{>\} = \{(\boldsymbol{x}^1,\boldsymbol{x}^2) \in S \times S | z(\boldsymbol{x}^1) > z(\boldsymbol{x}^2)\}$$
$$\{\sim\} = \{(\boldsymbol{x}^1,\boldsymbol{x}^2) \in S \times S | z(\boldsymbol{x}^1) = z(\boldsymbol{x}^2)\}$$
$$\{?\} = \emptyset$$

Because of the numerical ordering in the weighted-sum function, there is no ambiguity in preference comparison. For any two points, either is better, worse, or equivalent to the other. Only one of the three cases must ensure. There is no such thing as an indefinite set in a preference structure.

Utility Function Approach. Utility function is a mathematical representation of a preference structure which maps points in the criterion space into real numbers so that the greater the number, the more preferred the point in criterion space. Let a real-valued function $U(\cdot) : Z \rightarrow R$ be a utility function with a preference on Z such that for any two points z^1 and z^2 in the criterion space Z, $U(z^1) > U(z^2)$ if and only if $z^1 > z^2$, and $U(z^1) = U(z^2)$ if and only if $z^1 \sim z^2$. Then we have

$$\{>\} = \{(z^1,z^2) \in Z \times Z | U(z^1) > U(z^2)\}$$
$$\{\sim\} = \{(z^1,z^2) \in Z \times Z | U(z^1) = U(z^2)\}$$
$$\{?\} = \emptyset$$

As in weighted-sum methods, because of numerical ordering by the utility function, there is no ambiguity in preference comparison. Regarding the preference relation for any two points in the criterion space, only one of the three cases—preferred, less preferred, or equally preferred—must result.

For a given preference structure, if we are able to construct a utility function $U(\cdot)$, the ideal way to solve a multiple-objective optimization problem would be to solve the *utility function program*

$$\max\{U(z_1,\ldots,z_q) | z_1 = f_1(\boldsymbol{x}), \ldots, z_q = f_q(\boldsymbol{x}), \boldsymbol{x} \in S\} \tag{3.14}$$

In practice we may never know the mathematical representation of a particular utility function over a preference structure. Because of the difficulty in obtaining function U and the almost certain nonlinearity of the utility function program, this method is somewhat limited for solving a multiple-objective optimization problem. However, the utility approach is useful for conceptual reasons. The literature on utility theory is abundant. An excellent representation of the method with applications to a variety of real-world problems appears in a book by Keeney and Raiffa [352].

Compromise Approach. The compromise approach can be regarded as a kind of mathematical formulation of goal-seeking behavior in terms of a distance function. Because it is simple to understand and easy to compute, the concept has much general appeal. The compromise approach identifies solutions closest to the *ideal point.* Recall the definition of an ideal point z^* given earlier. For each component of a criterion vector, we cannot do better than the ideal points. As z^* is usually not attainable, a compromise is needed if no other alternative is available to dissolve the objectives conflict. Now given $z \in Z$, the *regret* of using z instead of obtaining the ideal point may be approximated by the following distance function:

$$r(z) = \|z - z^*\| \tag{3.15}$$

where $\|z - z^*\|$ is the distance from z to z^* according to a specified norm. Typically, the L_p-norm will be used because of its clarity. For a given positive number $p \geq 1$, we have

$$r(z;p) = \|z - z^*\|_p = \left[\sum_{j=1}^{q} |z_j - z_j^*|^p \right]^{1/p} \tag{3.16}$$

The compromise solution with respect to the L_p-norm is a point $z \in Z$ that minimizes $r(z;p)$ over Z, or is a point $\boldsymbol{x} \in S$ that minimizes $r(z(\boldsymbol{x});p)$ over S.

The regret function $r(z;p)$ threatens to give each value of $|z_j - z_j^*|$ the same importance. If the criteria have different degrees of importance, a weight vector $\boldsymbol{w} = (w_1, w_2, \ldots, w_q)$ may be assigned to signal the different degrees of importance. In this case we have the following weighted L_p-norm:

$$r(z;p,\boldsymbol{w}) = \|z - z^*\|_{p,w} = \left[\sum_{j=1}^{q} w_j^p |z_j - z_j^*|^p \right]^{1/p} \tag{3.17}$$

In terms of parameter p, the sum of regrets of all objectives is most strongly emphasized when $p = 1$:

$$r(z;1,\boldsymbol{w}) = \sum_{j=1}^{q} w_j |z_j - z_j^*| \tag{3.18}$$

The individual regret function of objectives is emphasized most strongly when $p = \infty$:

$$r(z;\alpha,w) = \max\{w_j|z_j - z_j^*|, j = 1,2,\ldots,q\} \tag{3.19}$$

In general, the choice of p reflects concerns with respect to the maximum regret among objectives. The larger the value of p, the greater the concern.

The regret function $r(\cdot) : Z \rightarrow R$ associated with a parameter p is a real-valued function for the preference of compromise on Z such that for any two points z^1 and z^2 in the criterion space Z, $r(z^1;p) > r(z^2;p)$ if and only if $z^1 > z^2$ in the sense of preference of compromise, and $r(z^1;p) = r(z^2;p)$ if and only if $z^1 \sim z^2$. Then we have

$$\{>\} = \{(z^1,z^2) \in Z \times Z | r(z^1;p) > r(z^2;p)\}$$
$$\{\sim\} = \{(z^1,z^2) \in Z \times Z | r(z^1;p) = r(z^2;p)\}$$
$$\{?\} = \emptyset$$

As in the weighted-sum and utility function methods, because of the numerical ordering in the regret function, there is no ambiguity in preference comparison. Regarding the preference relation for any two points in the criterion space, exactly one of the three cases—preferred, less preferred or equally preferred—must happen.

Lexicographic Ordering Approach. In this type of preference, order is imposed on objectives. Let $z = \{z_1,z_2,\ldots,z_q\}$ be indexed so that the kth component is overwhelmingly more important than the $(k + 1)$th component for $k = 1,2,\ldots,q - 1$. A lexicographic ordering preference is defined as follows: Point z^1 is preferred to z^2 if and only if $z_1^1 > z_1^2$ or there is somc $r \in \{1,2,\ldots,q\}$ so that $z_r^1 > z_r^2$ and $z_i^1 = z_i^2$ for $i = 1,2,\ldots,r - 1$. No two distinct points can be equal lexicographically. We then have

$$\{>\} = \{(z^1,z^2) \in Z \times Z | z^1 \text{ is lexicographically preferred to } z^2\}$$
$$\{\sim\} = \{(z,z) \in Z \times Z | z \in Z\}$$
$$\{?\} = \emptyset$$

Pareto Approach. The Pareto approach assumes that no information on the preference among objectives is available and that all we know is that for each component z_j, the greater value is preferred. The Pareto preference is defined as follows: For any two points z^1 and z^2 in the criterion space Z, point z^1 is preferred to point z^2 if and only if $z^1 \geq z^2$, and that, there is at least one component, say the rth component for which $z_r^1 > z_r^2$ and that for the others $z_k^1 \geq z_k^2$, $k = 1,2,\ldots,q$, $k \neq r$. Then we have

$$\{>\} = \{(z^1,z^2) \in Z \times Z | z^1 \geq z^2\}$$
$$\{\sim\} = \{(z,z) \in Z \times Z | z \in Z\}$$
$$\{?\} = \{(z^1,z^2) \in Z \times Z | \text{ neither } z^1 \geq z^2 \text{ nor } z^1 \leq z^2\}$$

Compared with other methods, once the proper regret function, or value function, or weight coefficients are determined, the set {?} is empty and the problem becomes a one-dimensional comparison or a mathematical programming problem. With the Pareto approach, we tackle problems with the set {?} not empty.

3.2.4 Structures and Properties of Problems

In general, multiple-objective optimization problems are too complex to be solved easily. Two types of difficulty associated with problem solving need to be distinguished: (1) difficulties inherent in problems and (2) difficulties related to solution techniques. The most profound drawback of many existing methods is that they are very sensitive to the value of weights, or the prescribed order of objectives, or the shape of utility functions. In essence, such difficulties are caused by solution techniques, not by the problem itself.

In general, a multiple-objective optimization problem is characterized by the structure and property of the following factors: (1) objective functions and criterion space, (2) constraint functions and solution space, and (3) size of the problem. The functions of objectives and constraints can be linear or nonlinear; convex, concave, or neither; differentiable or nondifferentiable; continuous or semicontinuous; unimodal or multimodal. Accordingly, the criterion and solution spaces can be convex or nonconvex, compact or noncompact, connected or disjoined. The size of the problem is determined by the numbers of decision variables, constraints, and objectives. Major conventional approaches have been restricted to the simple case of linear functions and convex solution space. Such multiple-objective optimization problems are generally transformed into a single objective or a sequence of single-objective optimization problems; then gradient- or simplex-based methods are extended to solve the transformed problems.

3.3 GENETIC MULTIOBJECTIVE OPTIMIZATION

During the last two decades, genetic algorithms have received considerable attention regarding their potential as a novel approach to multiobjective optimization problems, known as evolutionary or genetic multiobjective optimization. This topic has been surveyed by Fonseca and Fleming [202], Horn [305], and Tamaki, Kita, and Kobayashi [612]. In this section we describe briefly some basic ideas behind various approaches to genetic algorithm implementations. The details of some representative methods are given in subsequent sections.

3.3.1 Features of Genetic Search

The inherent characteristics of genetic algorithms demonstrate why genetic search may be well suited to multiple-objective optimization problems. The basic feature of genetic algorithms is multiple directional and global search through maintaining a population of potential solutions from generation to generation. The population-to-population approach is useful when exploring Pareto solutions.

Genetic algorithms do not have many mathematical requirements and can handle all types of objective functions and constraints. Because of their evolutionary nature, genetic algorithms can be used to search for solutions without regard to the specific inner workings of the problem. Therefore, it is hoped that many more complex problems can be solved using genetic algorithms than using conventional methods.

Because genetic algorithms, as a kind of metaheuristics, provide us with great flexibility to incorporate conventional methods into the main framework, we can exploit the advantages of both genetic algorithms and conventional methods to establish much more efficient implementations to problems. The growing research on applying genetic algorithms to multiple-objective optimization problems presents a formidable theoretical and practical challenge to the mathematical community [65].

3.3.2 Fitness Assignment Mechanism

The genetic algorithms are essentially a type of metastrategy of solutions. When applying genetic algorithms to solving a given problem, it is necessary to refine each of their major components, such as encoding methods, recombination operators, fitness assignment, selections, constraint handling, and so on, to obtain effective implementations to the given problem. Because multiple-objective optimization problems are natural extensions of constrained and combinatorial optimization problems, many useful methods and techniques employing genetic algorithms developed for constrained optimization problems and combinatorial optimization problems during the past two decades are readily applicable. Therefore, when considering how to adapt genetic algorithms to multiple-objective optimization problems, we just need to examine some special issues concerning the problems.

One special issue arising in solving multiple-objective optimization problems by use of genetic algorithms is how to determine the fitness value of individuals according to multiple objectives. The fitness assignment mechanism has been studied extensively during the past decade and several methods have been suggested and tested. Roughly, these methods can be classified as follows:

1. Vector evaluation approach
2. Weighted-sum approach
3. Pareto-based approach
4. Compromise approach
5. Goal programming approach

According to how much preference information is incorporated into the fitness function, these approaches range from complete preference information given, as when combining objective functions directly or prioritizing them, to no preference information given, as with Pareto-based approaches. In addition, a notable feature is that given a rough preference, progressive refinement of the preference can be carried out by evolutionary search. The progressive refinement of preferences is like an interactive procedure often used in multiobjective optimization, where preferences are modified at each iteration by the decision maker. What makes it unique is the refinement mechanism: The preference is refined gradually through evolutionary search by some adaptive refinement mechanism, not by intervention by the decision maker at each iteration. Of course, an interactive procedure can also be embedded in a genetic search to guide preference refining.

Perhaps the first notable work in extending simple genetic algorithms to solve multiple-objective optimization problems is the vector evaluation approach [563]. Instead of using a scalar fitness measure to evaluate each chromosome, it uses a vector fitness measure to create the next population. For a given problem with q objectives, the selection step in each generation will become a loop: It is repeated exactly q times, and one objective will be used in only one repetition in turn. At each repetition through the loop, a portion of the next generation is selected on the basis of one of the objectives.

There are two types of Pareto-based approach: Pareto ranking and Pareto tournament. The Pareto ranking–based fitness assignment method was first suggested by Goldberg as a means of achieving equal reproductive potential for all Pareto individuals [249]. It includes two major steps:

1. Sort the population based on Pareto ranking.
2. Assign selection probabilities to individuals according to the ranking.

The ranking procedure consists of assigning rank 1 to nondominated individuals and removing them from contention; finding the nondominated individuals among the remaining ones, ranked 2; and so on. In this way, the Pareto ranking approach assigns all nondominated solutions an identical fitness value in order to give them an equal probability of reproduction. Note that most other fitness assignment methods give nondominated solutions with different values, which is in contrast to what the definition of nondominance would suggest.

The Pareto tournament method was proposed by Horn, Nafpliotis, and Goldberg [306]. Instead of nondominated sorting and ranking selection, a niched Pareto concept is used in the tournament method, where a pareto solution with the least number of individuals in its neighbor wins the competition.

The weighted-sum approach takes its basic premise from conventional multiobjective optimizations. It assigns weights to each objective function and combines the weighted objectives into a single objective function. Conceptually, it is simple to understand and easy to compute. To make it work requires only a proper weighting vector. Therein, however, lies the difficulty. Once embedded in a genetic algorithm, its weakness can be compensated for by the powers of population-based and evolutionary search. Recently, several weight-adjusted methods have been proposed to fully utilize the power of genetic search: (1) fixed-weight approach, (2) random-weights approach, and (3) adaptive weight approach.

In the fixed-weight approach, weights will not be changed during an entire evolutionary process. Weights can be determined by prior knowledge of objectives, or randomly if there is no way to get such knowledge in advance. This approach gives selective pressure to a fixed direction determined by the weights.

Murata, Ishibuchi, and Tanaka suggested a random-weight approach [476] in which weights are randomly reset at each step in the selection procedure to give an even chance to all possible combinations. Statistically, this approach gives a uniform selective force to the Pareto frontier. This method ignores the information available on the Pareto solutions at each generation.

Gen and Cheng have proposed an adaptive weight approach [107, 693] in which weights are adjusted adaptively based on the current generation to obtain search pressure toward the positive ideal point. Because useful information from the current population is used to readjust the weights at each generation, the approach provides a selection pressure between those of the fixed approach and the random approach.

A compromise-based fitness assignment method has been suggested by Chang and Gen for solving bicriteria shortest path problems [109]. Its basic idea and techniques are borrowed from conventional multiobjective optimizations. The compromise approach identifies solutions closest to the ideal solution as determined by some measure of distance. As we know, the ideal solution is usually not attainable. However, it can serve as a standard for evaluation of the attainable nondominated solutions. Since all would prefer the ideal point if it were attainable, it can be argued that finding a solution that is as close to ideal as possible is a reasonable substitute. One frequently used measure of closeness is a family of weighted L_p metrics. For many complex problems, finding an ideal point is also difficult. To overcome the difficulty, the concept of a proxy ideal point is suggested to replace the ideal point. The proxy ideal point is the point corresponding to the current generation but not to a given problem. In other words, it is calculated based on the partial solution space explored, not on the entire solution space. The proxy ideal point is easy to obtain for each generation. Along with evolutionary progress, the proxy ideal point will gradually approximate the real ideal point.

Goal programming is one powerful technique for solving multicriteria optimization problems. Gen and Liu investigated the application of genetic algorithms to solving a nonlinear goal programming problem [237]. A rank-based fitness assignment method is used to assess the merit of each individual. Because lexicographic ordering among objectives is preferred in goal programming, individuals are sorted on the value of objectives in a lexicographic manner as follows: Sort the population on the first priority objective; if some individuals have the same objective value, sort them on the second priority objective, and so on. A tie is broken randomly. Individual fitness values are then assigned by interpolating from best to worst according to an exponential function.

From the viewpoint of methodology, there are two basic approaches to multiobjective optimization: the generating and preference-based approaches. Generating approaches are used to identify an entire set of Pareto solutions or an approximation, whereas preference-based approaches attempt to obtain a compromised or preferred solution. Conceptually, the vector evaluation approach, Pareto ranking–based approach, and random-weighting approach are designed as generating methods; the compromise approach, adaptive weighting approach, and goal programming approach are designed as preference-based approaches.

In multiobjective optimizations, generating and preference-based methods both exhibit strengths and weaknesses. Generating techniques require decision makers to make judgments by selecting from among entire Pareto solutions. For problems with more than three criteria, making a choice becomes very complicated, increasing in difficulty approximately exponentially with the number of criteria. Computational costs also increase rapidly with more criteria. In genetic multiobjective optimizations, the situation essentially does not change. In contrast, preference-based techniques seem not to put as great a burden on decision makers when using multiobjective optimization. Because preferences can be refined gradually within the evolutionary process, a rough preference can be made to work by evolutionary search.

In multiobjective optimizations, a solution method can be designed as either a generating method for obtaining an entire set of Pareto solutions or as a preference-based method for obtaining a preferred or compromised solution. We cannot have a solution method that implements the two distinct ideas into a single solution procedure, but in genetic multiobjective optimization, we can.

How to handle infeasible solutions is a very important issue in genetic optimizations because genetic operators usually produce infeasible, even illegal, solutions. This issue receives less discussion in studies of genetic multiobjective optimization because major focus has been put on how to generate Pareto solutions. Most test problems are small artificial instances without complex constraints. For many real-world applications, it becomes a relatively severe problem because in case of complex constraints, infeasible solutions will affect a relatively large portion of the entire population [218]. Repairing and penalizing strategies are two major techniques used extensively in genetic optimization practices. Essentially, penalizing techniques can be used together with any scaling approaches, such as the compromise and weighted-sum approaches, while repairing techniques can be used together with Pareto ranking–based approaches if necessary.

3.3.3 Fitness Sharing and Population Diversity

Fitness sharing is a technique used to maintain population diversity. First, we need to distinguish between the following two types of sharing techniques: (1) sharing in multimodal functions and (2) sharing along the Pareto frontier. Fitness sharing was introduced by Goldberg and Richardson [248] for multimodal function optimization. Afterward, sharing techniques are extended in genetic multiobjective optimization to maintain individuals all along the nondominated frontier.

One of the problems of genetic algorithms for solving multimodal functions is that the finite populations will eventually converge to only one optimum, due to stochastic errors in the selection process. This phenomenon is known as *genetic drift.* The goal of the sharing technique is to prevent genetic drift and to promote uniform sampling, in other words, to distribute populations over a number of different peaks in the search space, with each peak receiving a fraction of the population in proportion to the height of that peak. To achieve this, a concept of fitness degradation is introduced. If a peak has too many individuals, the fitness value for all individuals in the peak will be degraded to reduce their reproduction abilities.

A sharing function is a way of determining the degradation of an individual's fitness due to crowding by its neighbor. The metric of distance between two individuals can be defined over two different spaces, yielding two types of fitness sharing: genotypic sharing and phenotypic sharing. In genotypic sharing, the distance between two individuals is measured in terms of encoding space; in phenotypic sharing, the distance is measured on the basis of decoded space. Deb's work indicated that, in general, phenotypic sharing is superior to genotypic sharing [151].

Let s_i denote a string or an individual in encoded form. We may have a metric over strings directly:

$$d_{ij} = d(s_i, s_j) \tag{3.20}$$

By use of the metrics over strings, fitness sharing is called genotypic sharing. Let x_i denote a solution or an individual in decoded form. We may have a metric over decision variable space (i.e., solution space):

$$d_{ij} = d(x_i, x_j) \tag{3.21}$$

By use of metrics over the solution space, fitness sharing is called phenotypic sharing. Then we define a sharing function Sh(·) as a function of the metric value d_{ij}. Typically, the following *power law function* is used as the sharing function:

$$\mathrm{Sh}(d_{ij}) = \begin{cases} 1 - \left(\dfrac{d_{ij}}{\sigma_{\text{share}}}\right)^{\alpha}, & \text{if } d_{ij} < \sigma_{\text{share}} \\ 0, & \text{otherwise} \end{cases} \tag{3.22}$$

where α is a constant and σ_{share} is the *niche radius,* fixed by the user at some estimate of the minimal separation desired or expected between individuals. Individuals within the distance of a shared radius of each other degrade each other's fitness value.

Once a metric and a sharing function are determined, the *shared fitness* f_i' for an individual is defined by dividing its fitness f_i by its *niche count* m_i as follows:

$$f_i' = \frac{f_i}{m_i} \tag{3.23}$$

The niche count m_i for a given individual i is taken as the sum of all sharing function values over the entire population:

$$m_i = \sum_{j=1}^{pop\_size} \mathrm{Sh}(d_{ij}) \tag{3.24}$$

The sum includes the individual itself. Thus, if an individual is all by itself in its own niche ($m_i = 1$), it receives its full potential fitness value. Otherwise, the sharing function derates fitness according to the number and closeness of neighboring points.

Fitness sharing was originally combined with the fitness proportionate selection. When sharing is combined with tournament selection, the genetic algorithms exhibit chaotic behavior. Wild fluctuations in niche subpopulations can be avoided by *continuously updated sharing,* suggested by Oei, Goldberg, and Chang [488], in which niche counts are calculated using not only the current population but also the partially filled next-generation population.

In genetic multiobjective optimization, the sharing techniques can be divided into the following two categories: (1) sharing on the objective space and (2) sharing on the solution space. Sharing on the objective space attempts to achieve uniform sampling of the entire Pareto set; sharing on the solution space attempts to achieve uniform sampling of the admissible solution set. Sharing on the objective space was suggested by Fonseca and Fleming [201]. In multimodal function optimization, sharing is conducted on the solution space. It usually requires a priori knowledge of a finite number of peaks and uniform niche placement in the solution space. Upon convergence, local optima are occupied by a number of individuals proportional to their fitness values. On the contrary, the situation is quite different for the multiobjective optimization. There are no longer any local peaks. All solutions in the Pareto frontier are equally good. We are more concerned with obtaining a set of globally nondominated solutions, possibly uniformly spaced and illustrative of the global trade-off surface. The use of ranking already forces the search to concentrate exclusively on global optima. By implementing fitness sharing in the objective value domain rather than the decision variable domain, and only between pairwise nondominated individuals, we can expect to be able to evolve a uniformly distributed representation of the global trade-off surface.

Sharing on the solution space was suggested by Srinivas and Deb [581]. A major concern is that diversity along the Pareto frontier may not correspond to diversity over the solution space, a matter of importance for a decision maker. Sharing is achieved by calculating the following sharing function value between two individuals:

$$\mathrm{Sh}(d_{ij}) = \begin{cases} 1 - \left(\dfrac{d_{ij}}{\sigma_{\mathrm{share}}}\right)^2, & \text{if } d_{ij} < \sigma_{\mathrm{share}} \\ 0, & \text{otherwise} \end{cases} \tag{3.25}$$

where d_{ij} is the phenotypic distance between two individuals and σ_{share} is the maximum phenotypic distance allowed between any two individuals to become members of a niche.

Mating restriction is another aspect that becomes relevant as the population distributes itself around multiple regions of optimality. Mating restriction assumes that neighboring fit individuals are genotypically similar, so that they can form stable niches. In the multimodal function optimization it is easy to understand that different regions of local peaks may generally have very different genetic representations. Therefore, mating restriction may help crossover operators to perform fine tuning around local optima. For the multiobjective optimization case, it is not so apparent that two individuals within relatively small regions of the Pareto set have similar genotypic representations. Fonseca and Fleming implemented the mating restriction based on the distance between individuals in the objective domain. Nevertheless, the use of mating restrictions in genetic multiobjective optimization does not appear to be widespread [202].

3.3.4 The Concept of Pareto Solution

In a strict sense, the term *Pareto solution* as used in genetic algorithms has a different meaning than as used in a conventional way. In the original definition, a point is said to be a Pareto solution if and only if it is a nondominated point with respect to all points in the criterion space Z for a given problem. In some genetic algorithms (not in all), Pareto solutions are identified in each generation. Because a population in each generation contains only partial solutions of the original problem, a Pareto solution has meaning only with respect to all solutions currently examined. A nondominated solution in one generation may become dominated by a new solution emerging in a later generation. Therefore, for a given generation of genetic algorithms, a Pareto solution obtained in that generation may be a true Pareto solution to the problem or may not be. There is no guarantee that a genetic algorithm will produce Pareto solutions to a given problem. But genetic algorithms will provide better approximations of Pareto solutions.

How to maintain a set of nondominated individuals during the evolutionary process is a special issue for multiple-objective optimization problems. Basically, there are two different ways to handle Pareto solutions, which lead to two different overall structures of genetic algorithm implementations: (1) preserving Pareto solutions separately from the population pool and (2) without preserving mechanisms.

In most existing methods, Pareto solutions are identified at each generation and used only to calculate fitness values or ranks for each chromosome. No mechanism is provided to guarantee that Pareto solutions generated during the evolutionary process enter the next generation. In other words, some Pareto solutions may get lost during the evolutionary process. To avoid such sampling errors, a preserving mechanism for Pareto solutions has been suggested by many researchers [214, 319, 476]. A special pool for preserving Pareto solutions is added to the basic structure of genetic algorithms. In each generation, the set of Pareto solutions is updated by deleting all dominated solutions and adding all newly generated Pareto solutions. The overall structure of the approach is given as follows:

```
Procedure: Pareto Genetic Algorithms
begin
  t ← 0;
  initialize P(t);
  objectives P(t);
  Pareto E(t);
  fitness P(t);
  while (not termination condition) do
  begin
    crossover P(t);
    mutation P(t);
    objectives P(t);
    update Pareto E(t);
    fitness P(t);
    select P(t + 1) from P(t);
    t ← t + 1;
  end
end
```

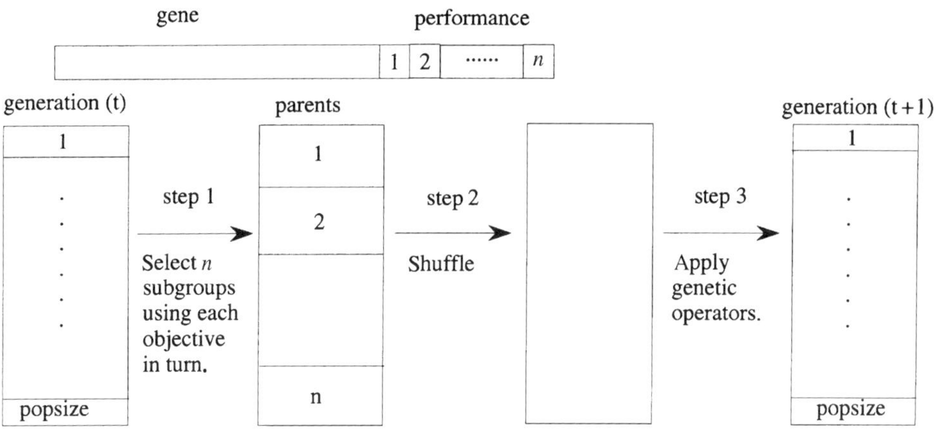

Figure 3.3. VEGA selection.

Without a preserving mechanism, the Pareto solutions can be gathered only from the last generation. If the method used has a tendency to *speciation,* as mentioned by Schaffer [563], the entire population will converge toward individual optimum regions after a large number of generations. The preserving mechanism is, to a certain extent, helpful in minimizing speciation through a Pareto-preserving procedure at each generation.

3.4 VECTOR-EVALUATED GENETIC ALGORITHMS

Being aware of the potential of genetic algorithms in solving multiple-objective optimization, Schaffer proposed an extension of a simple genetic algorithm to accommodate incommensurable multiple objectives in 1985, known as the vector-evaluated genetic algorithm (VEGA) [563]. In the VEGA approach, the selection step in each generation becomes a loop. Each time through the loop the appropriate fraction of the next generation, or subpopulations, is selected on the basis of each objective. Then the entire population is shuffled thoroughly to apply crossover and mutation operators. This is performed to achieve the mating of individuals of different subpopulations. Nondominated individuals are identified by monitoring the population as it evolves, but this information is not used by VEGA itself. This approach, illustrated in Figure 3.3, protects the survival of the best individuals on one of the objectives and, simultaneously, provides appropriate probabilities for multiple selection of individuals which are better than average on more than one objective.

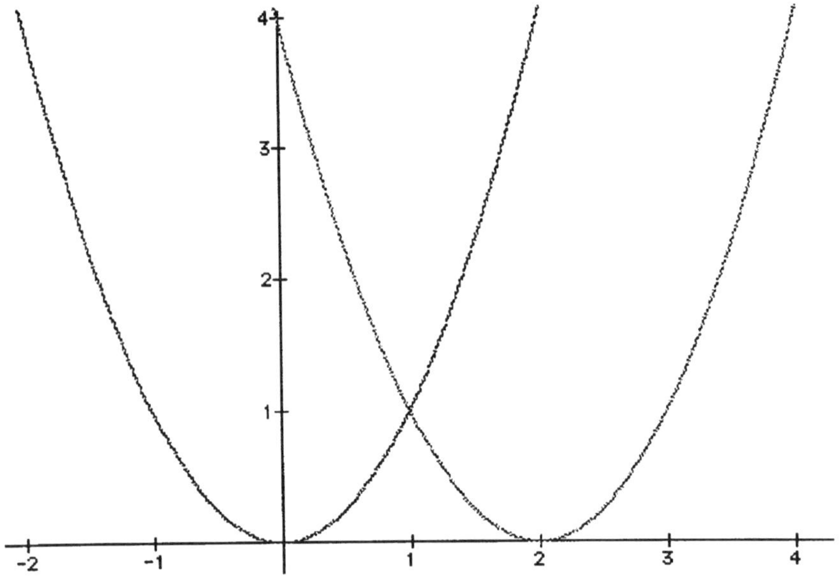

Figure 3.4. Objectives versus x.

Schaffer implemented VEGA by modifying Grefenstette's GENESIS program [265], where a loop-around selection procedure is created so that selection is repeated for each objective to fill up a portion of the generation. The expected total number of offspring produced by each parent becomes the sum of the expected numbers of offspring produced by that parent according to each objective. Because Schaffer used a proportional fitness assignment, these were, in turn, proportional to the objectives themselves. The resulting overall fitness for a given individual therefore corresponded to a linear function of the objectives, where the weights depended on the distribution of the population in each generation. As a consequence, different nondominated individuals are generally assigned different fitness values, like all other weighted-sum approaches. Note that an individual has no explicit form in overall fitness because VEGA does not use a scalar fitness function. We use the phrase to refer just to the equivalent effect that a certain number of offspring would be produced by an individual with corresponding fitness value in a normal case.

A simple two-objective problem with one variable is used by Schaffer to test the properties of VEGA:

$$\begin{aligned} \min \quad & f_1 = x^2 \\ \min \quad & f_2 = (x-2)^2 \\ \text{s.t.} \quad & x \in R^1 \end{aligned}$$

Figure 3.4 plots the problem by restricting variable x within the region

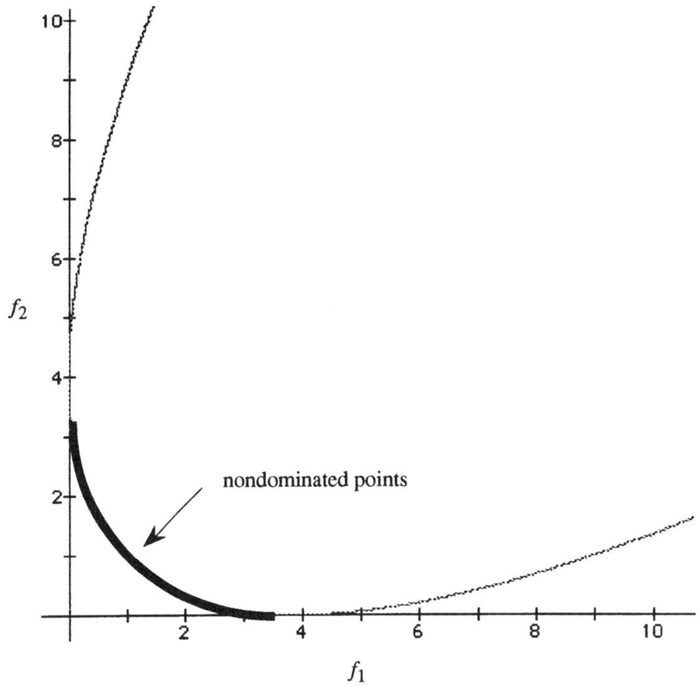

Figure 3.5. Objectives in criterion space.

from −2 to 4. It is clear that the Pareto solutions constitute all x values varying from 0 to 2. Figure 3.5 plots the problem in the criterion space.

In the simulation results of VEGA given by Srinivas and Deb [581], the parameters of genetic algorithms are set as follows: maximum generation, 500; population size, 100; string length (binary code), 32; probability of crossover, 1.0; and probability of mutation, 0. The initial range for the variable x is (-10.10). The initial population is shown in Figure 3.6. In generation 10, the population converged toward the nondominated region, as shown in Figure 3.7. In generation 100, almost the entire population fell in the region of nondominated solutions, as shown in Figure 3.8. Observe that in generation 500, the population converged to only three subregions, as shown in Figure 3.9. This is the phenomenon of *speciation.*

Schaffer's works are commonly regarded as the first effort in opening the domain of multiple-objective optimizations to genetic algorithms. Even though VEGA cannot give a satisfactory solution to the multiple-objective optimization problem, it provides some useful hints for developing other new implementations of genetic algorithms. Most subsequent work refers to VEGA as a paradigm for comparing performances.

Fourman also addressed multiobjective optimization problems with a non-aggregating approach [205]. Two versions of implementations were given. In

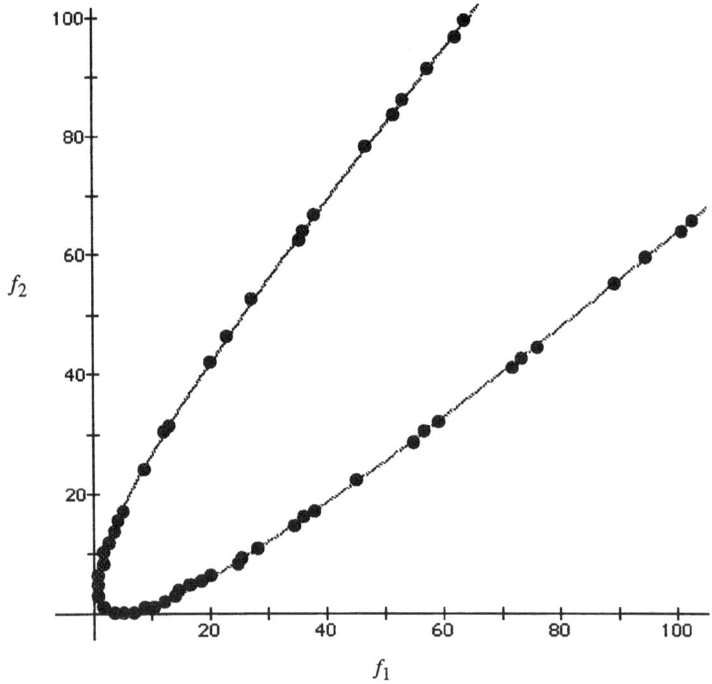

Figure 3.6. Population in generation 0 obtained using VEGA.

the first version, objectives are assigned different priorities in advance, and selection is performed by comparing pairs of individuals according to the objective, with the highest priority first. If this results in a tie, the objective with the second-highest priority is used, and so on. As mentioned earlier, this is known as *lexicographic ordering.* In the second version, an objective is randomly selected in each comparison. It was reported to work surprisingly well. The resulting populations showed greater variability with a random selection of objectives than with a deterministic selection of objectives. Kursawa implements a random selection of objects almost identical to that of Fourman [383].

3.5 PARETO RANKING AND TOURNAMENT METHODS

3.5.1 Pareto Ranking Method

A Pareto ranking–based fitness assignment method was first suggested by Goldberg [249], as a means of achieving equal reproductive potential for all Pareto individuals. The procedure is similar to Baker's ranking selection procedure [36] but the population is ranked on the basis of nondominated indi-

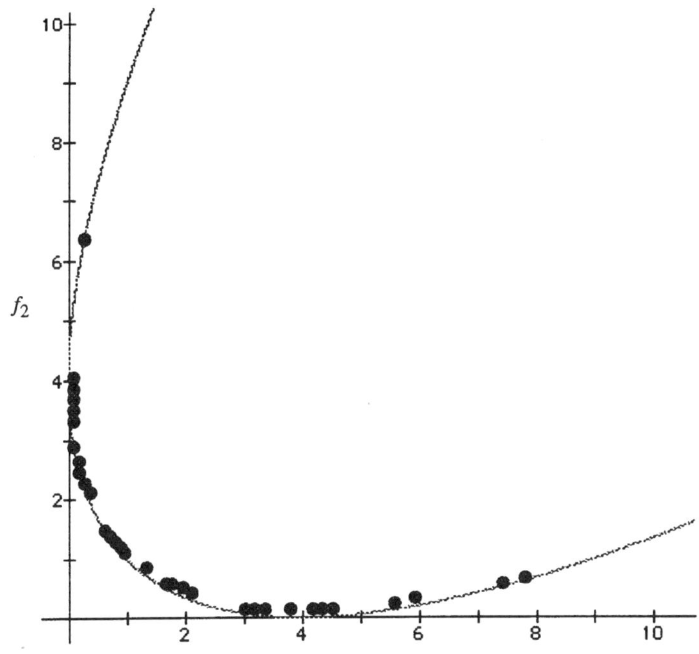

Figure 3.7. Population in generation 10 obtained using VEGA.

viduals. The ranking procedure is as follows: Rank 1 is given to the nondominated individuals and they are removed from contention; then the next set of nondominated individuals is found and rank 2 is assigned to them. The process continues until the entire population is ranked. This approach is illustrated in Figure 3.10 for a simple case with two objectives to be maximized simultaneously.

The idea of Baker's *ranking selection* is straightforward:

1. Sort the population from best to worst.
2. Assign selection probabilities to individuals according to the ranking.

If we let p_k denote the selection probability for the kth individual in the ranked population, the linear ranking method of Baker takes the form

$$p_k = q - (k - 1) \times r$$

where parameter q is the probability for the best individual. Let q_0 be the probability for the worst individual and K be the number of the last rank. Then parameter r can be determined as follows:

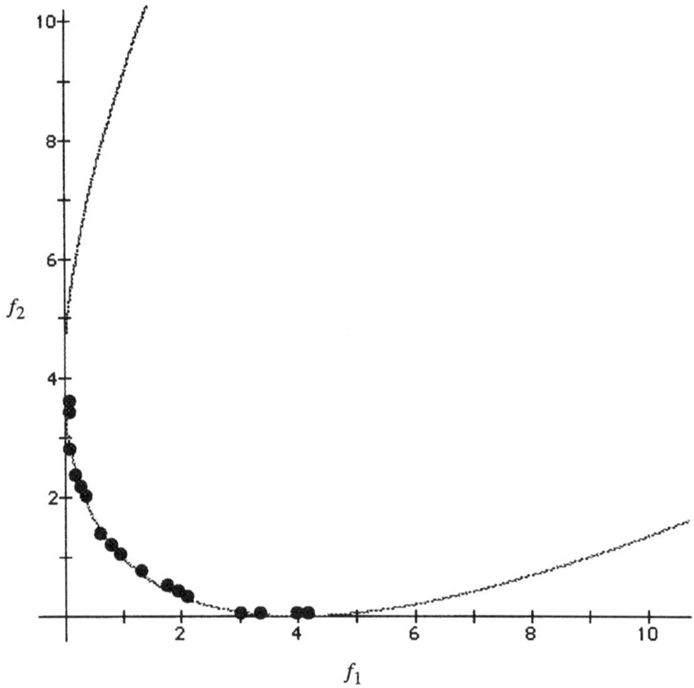

Figure 3.8. Population in generation 100 obtained using VEGA.

$$r = \frac{q - q_0}{K - 1}$$

Intermediate individuals' fitness values are decreased from q to q_0 in proportion to their rank. Letting q_0 be 0 provides maximum selective pressure.

Fonseca and Fleming proposed a slightly different method of Pareto ranking [201]. All nondominated individuals are assigned rank 1, and any other indiviudal is assigned a rank equal to the number of its dominating individuals plus one. The method is illustrated in Figure 3.11. The method proceeds by sorting the population according to the ranks, and ties may be broken by random choice. Fitness is assigned by interpolating, linearly or nonlinearly, from the best to the worst individuals in the population and then averaging between individuals with the same rank. Selection is a *stochastic universal sampling* method proposed by Baker [36].

Srinivas and Deb implemented a similar sorting and fitness asignment procedure [581]. The population is ranked based on Goldberg's Pareto ranking method. The nondominated individuals are first identified and assigned a large dummy fitness value. To maintain diversity in the population, these individuals are shared with their dummy fitness values. After sharing, these nondominated individuals are ignored temporarily and the second nondominated front

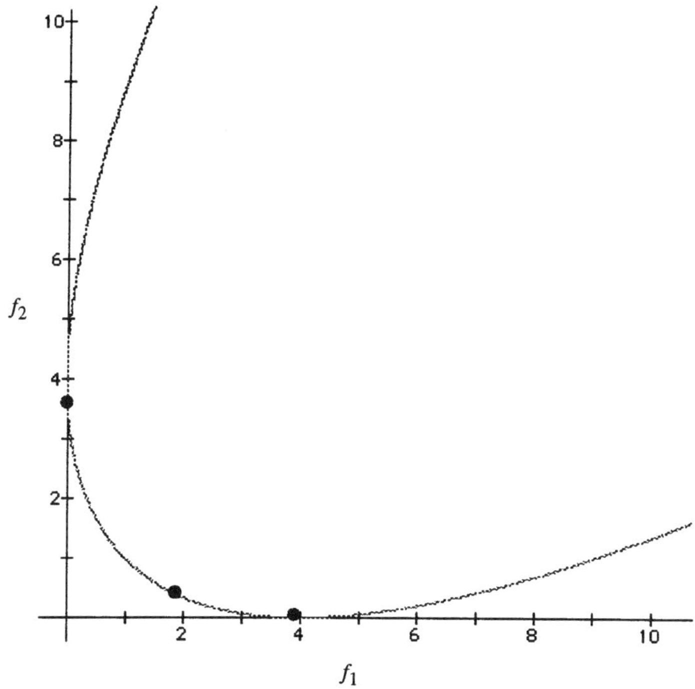

Figure 3.9. Population in generation 500 obtained using VEGA.

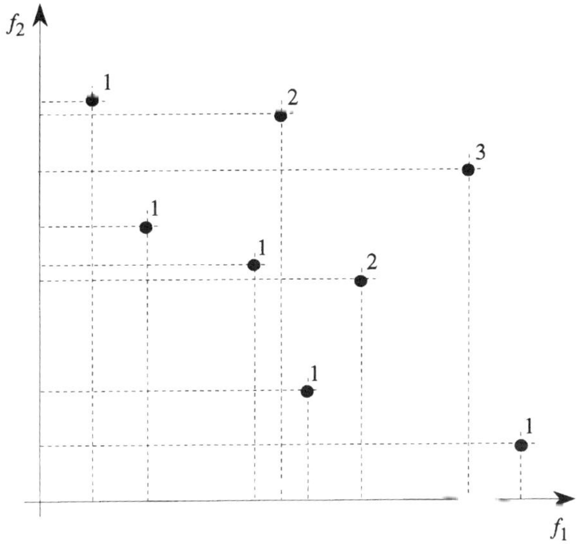

Figure 3.10. Goldberg's ranking.

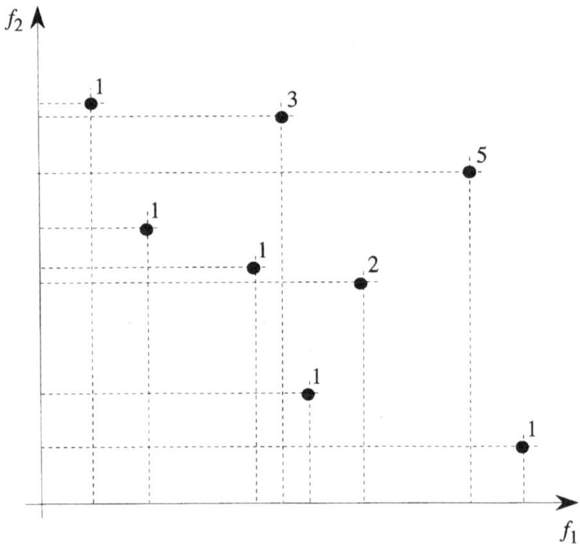

Figure 3.11. Fonseca and Fleming's ranking.

in the rest of the population is identified and assigned a dummy fitness value that is kept smaller than the minimum shared dummy fitness of the previous front. This process is continued until the entire population is classified into several fronts. A *stochastic remainder proportionate selection* is used to reproduce a new generation [76]. The flowchart of this algorithm is shown in Figure 3.12.

3.5.2 Pareto Tournament Method

The Pareto tournament method was proposed by Horn, Nafpliotis, and Goldberg [306]. In this method, two candidates for selection are picked at random from a population. A comparison set of indivdiuals is also picked randomly from the population. Each candidate is then compared against each individual in the comparison set. Two types of outcomes may occur:

1. If one candidate is dominated by the comparison set but the other is not, the nondominated candidate is selected for reproduction.
2. If both candidates are either nondominated or dominated, a sharing method is used to choose the winner.

As we know, the sharing method is a way of determining the degradation of an individual's fitness due to its niche count. Horn et al. proposed a new sharing technique that did not implement any form of fitness degradation

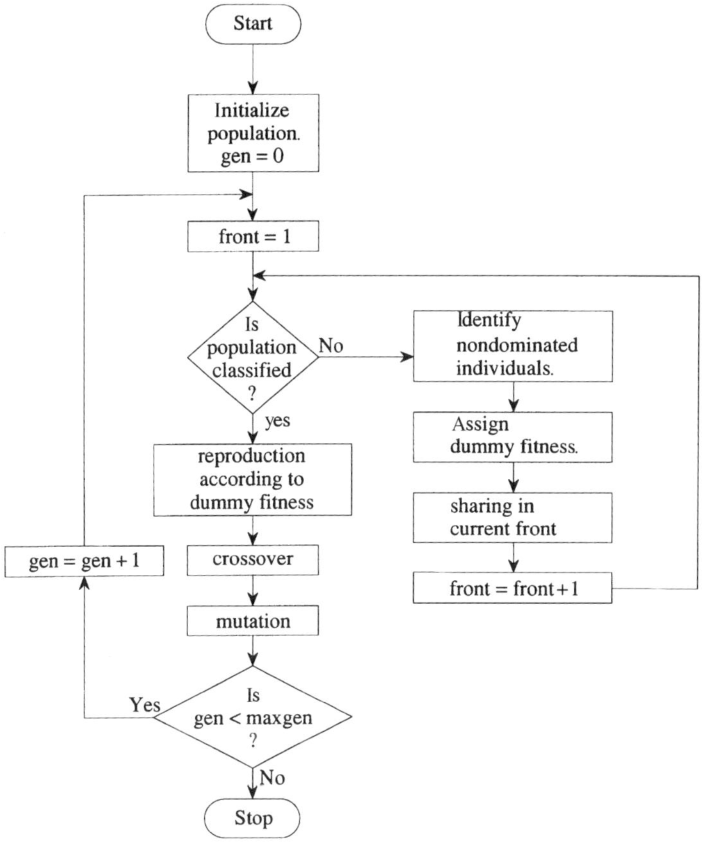

Figure 3.12. Flowchart of Srinivas and Deb's nondominated sorting method.

according to the niche count. Instead, the candidate with least niche count is selected as the winner. The niche count is calculated by counting the number of individuals in the population within a certain distance of the candidate.

Figure 3.13 illustrates how this sharing method works between two nondominated individuals. It is a case of maximizing the first objective and minimizing the second objective. Two candidates for selection are in the Pareto solution set and are not dominated by the comparison set. From the point of view of Pareto solution, neither candidate is preferred. If we want to maintain the diversity of population, it is apparent that it would be best to choose the candidate that has the smaller niche count. In this case that is candidate 2.

Since the nondominance is computed by comparing a candidate with a randomly chosen comparison set of population, the success of the algorithm depends on the parameter of the size of the comparison set. If it is too small, it may result in a few nondominated individuals in the population. If it is too large, premature convergence may occur.

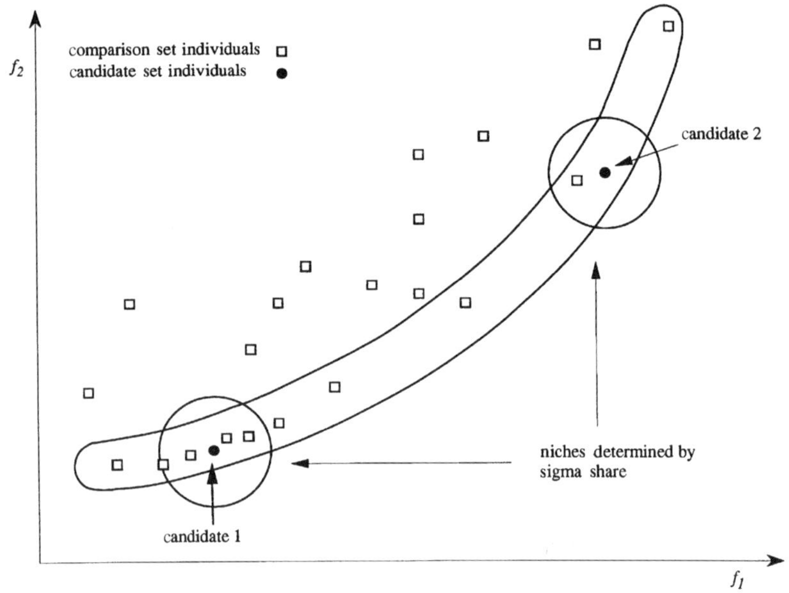

Figure 3.13. Horn's sharing.

3.6 WEIGHTED-SUM APPROACH

Conceptually, the weighted-sum approach can be viewed as an extension of methods used in multiobjective optimization to genetic algorithms. It assigns weights to each objective function and combines the weighted objectives into a single objective function. In fact, the weighted-sum approaches used in the genetic algorithms are very different in nature from that in conventional multiobjective optimizations. In the multiobjective optimization, the weighted-sum approach is used to obtain a compromise solution. To make the method work, all that is needed is a good weighting vector. It is usually very difficult to determine a set of appropriate weights for a given problem. In the genetic algorithms, the weighted-sum approach is used primarily to adjust genetic search toward the Pareto frontier. Weights are readjusted adaptively along with the evolutionary process. Therefore, a good weighting vector is not a mandatory precondition to making genetic algorithms work. In addition, the drawbacks exhibited in the multiobjective optimization can be compensated by the powers of population-based search and evolutionary search. Recently, three weight-setting mechanisms have been proposed: the fixed-weight approach, random-weight approach, and adaptive weight approach [112]. The fixed-weight approach can be viewed as analogous to conventional scalarization techniques, while the random-weight and adaptive weight approaches are designed for genetic algorithms to fully utilize the power of genetic search, which can only work due to the nature of population-based evolutionary search of genetic algorithms.

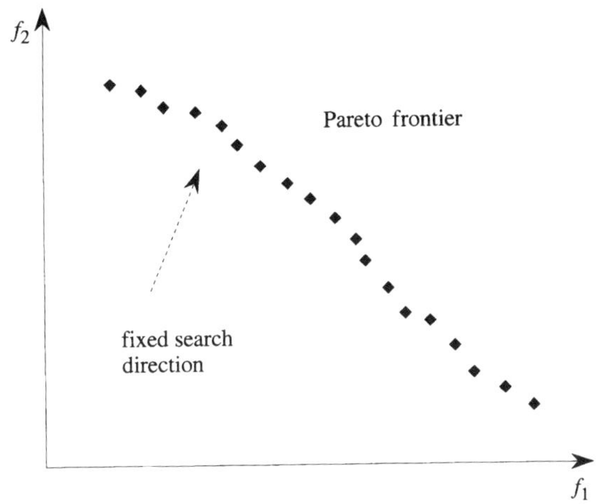

Figure 3.14. Search in a fixed direction in criterion space.

3.6.1 Random-Weight Approach

Murata, Ishibuchi, and Tanaka proposed a random-weight approach to obtaining a variable search direction toward the Pareto frontier [319, 476]. Typically, there are two types of search behavior in the objective space: fixed-direction search and multiple-direction search, as demonstrated in Figures 3.14 and 3.15. The fixed-weight approach gives the genetic algorithms a tendency to sample the area toward a fixed point in the criterion space, while the random-weight approach gives the genetic algorithms a tendency to demonstrate a variable search direction, therefore, the ability to sample the area uniformly over the entire frontier.

Suppose that we are going to maximize q objective functions. The weighted-sum objective is given as follows:

$$z = \sum_{k=1}^{q} w_k f_k(\boldsymbol{x}) \tag{3.26}$$

The random weights w_k are calculated by the equation

$$w_k = \frac{r_k}{\sum_{j=1}^{q} r_j}, \qquad k = 1,2,\ldots,q \tag{3.27}$$

where r_i are nonnegative random numbers.

Before selecting a pair of parents for crossover operation, a new set of random weights is specified by equation (3.27), and fitness values for each individual are calculated by equation (3.26). The selection probability p_i for individual i is then defined by the following linear scaling function:

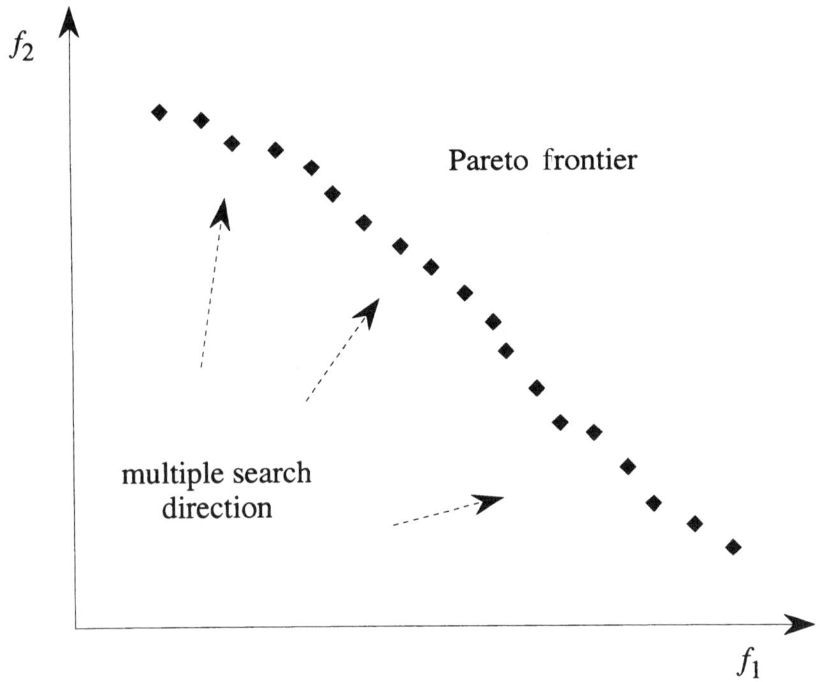

Figure 3.15. Search in multiple directions in criterion space.

$$p_i = \frac{z_i - z_{\min}}{\sum_{j=1}^{pop\_size}(z_j - z_{\min})} \tag{3.28}$$

where $z_{\min}$ is the worst fitness value in the current population.

A tentative set of Pareto solutions is stored and updated at each generation. For a problem with q objectives, there are q extreme points in the Pareto solutions, each of which maximizes one objective. An *elite preserving strategy* is suggested for putting the n extreme points plus some randomly selected Pareto solutions into the next population. Let N_{pop} denote the population size and N_{elite} denote the number of elite solutions to preserve. The overall structure of their implementation of genetic algorithms is given as follows:

Step 1. Initialization. Randomly generate an initial population containing N_{pop} individuals.

Step 2. Evaluation. Calculate the values of q objective functions for each individual. Update a tentative set of Pareto solutions.

Step 3. Selection. Repeat the following steps to select $(N_{\text{pop}} - N_{\text{elite}})$ pairs of parent: Specify random weights by equation (3.27), calculate fitness value by equation (3.26), calculate selection probability by equation (3.28), and select a pair of parent individuals for a crossover operation.

Step 4. Crossover. For each pair selected, apply a crossover operation to generate offspring.

Step 5. Mutation. Apply a mutation operation to each offspring generated by the crossover operation.

Step 6. Elitist strategy. Randomly select N_{elite} individuals from the tentative set of Pareto solutions. Add the selected solutions N_{elite} to $(N_{\text{pop}} - N_{\text{elite}})$ individuals generated in the foregoing steps to construct a population of N_{pop} of individuals.

Step 7. Termination test. If a prespecified stopping condition is satisfied, stop the run; otherwise, return to step 1.

3.6.2 Adaptive Weight Approach

Gen and Cheng proposed an adaptive weight approach which utilizes some useful information from the current population to readjust weights to obtain a search pressure toward a positive ideal point [107, 214–216, 677, 693]. Without loss of generality, consider the following maximization problem with q objectives:

$$\max \quad \{z_1 = f_1(\boldsymbol{x}),\ z_2 = f_2(\boldsymbol{x}),\ \ldots,\ z_q = f_q(\boldsymbol{x})\} \tag{3.29}$$

$$\text{s.t.} \quad g_i(\boldsymbol{x}) \leq 0, \qquad i = 1,2,\ldots,m \tag{3.30}$$

For the solutions examined in each generation, we define two extreme points: the maximum extreme point z^+ and the minimum extreme point z^- in criteria space as follows:

$$z^+ = \{z_1^{\max}, z_2^{\max}, \ldots, z_q^{\max}\} \tag{3.31}$$

$$z^- = \{z_1^{\min}, z_2^{\min}, \ldots, z_q^{\min}\} \tag{3.32}$$

where $z_k^{\min}$ and $z_k^{\max}$ are the maximal value and minimal value for objective k in the current population. Let P denote the set of the current population. For a given individual $\boldsymbol{x}$, the maximal value and minimal value for each objective are defined as follows:

$$z_k^{\max} = \max\{\, f_k(\boldsymbol{x}) | \boldsymbol{x} \in P\}, \qquad k = 1,2,\ldots,q \tag{3.33}$$

$$z_k^{\min} = \min\{\, f_k(\boldsymbol{x}) | \boldsymbol{x} \in P\}, \qquad k = 1,2,\ldots,q \tag{3.34}$$

The hyperparallelogram defined by the two extreme points is a minimal hyperparallelogram containing all current solutions. The two extreme points are renewed at each generation. The maximum extreme point will gradually approximate the positive ideal point. The adaptive weight for objective k is calculated by the following equation:

$$w_k = \frac{1}{z_k^{\max} - z_k^{\min}}, \qquad k = 1,2,...,q \tag{3.35}$$

For a given individual $\boldsymbol{x}$, the weighted-sum objective function is given by the following equation:

$$z(\boldsymbol{x}) = \sum_{k=1}^{q} w_k(z_k - z_k^{\min}) \tag{3.36}$$

$$= \sum_{k=1}^{q} \frac{z_k - z_k^{\min}}{z_k^{\max} - z_k^{\min}} \tag{3.37}$$

$$= \sum_{k=1}^{q} \frac{f_k(\boldsymbol{x}) - z_k^{\min}}{z_k^{\max} - z_k^{\min}} \tag{3.38}$$

As the extreme points are renewed at each generation, the weights are renewed accordingly. Equation (3.38) is a hyperplane defined by the following extreme points in current solutions:

$$(z_1^{\max}, z_2^{\min},..., z_k^{\min},..., z_q^{\min}) \tag{3.39}$$

$$(z_1^{\min}, z_2^{\max},..., z_k^{\min},..., z_q^{\min}) \tag{3.40}$$

...

$$(z_1^{\min}, z_2^{\min},..., z_k^{\max},..., z_q^{\min}) \tag{3.41}$$

...

$$(z_1^{\min}, z_2^{\min},..., z_k^{\min},..., z_q^{\max}) \tag{3.42}$$

The hyperplane divides the criteria space Z into two half-spaces: One half-space contains the positive ideal point, denoted as Z^+, and the other half-space contains the negative ideal point, denoted as Z^-. All Pareto solutions examined lie in the space Z^+, and all points lying in Z^+ have larger fitness values than those in the points in the space Z^-. As the maximum extreme point approximates the positive ideal point along with the evolutionary progress, the hyperplane will gradually approach the positive ideal point. Therefore, the adaptive weight method can readjust its weights according to the current population to obtain a search pressure toward the positive ideal point.

For the minimization case, we just need to transform the original problem into its equivalent maximization problem and then apply equation (3.38). For a maximization problem, equation (3.38) can be simplified as follows:

$$z(\boldsymbol{x}) = \sum_{k=1}^{q} w_k z_k \tag{3.43}$$

$$= \sum_{k=1}^{q} \frac{z_k}{z_k^{\max} - z_k^{\min}} \tag{3.44}$$

$$= \sum_{k=1}^{q} \frac{f_k(\boldsymbol{x})}{z_k^{\max} - z_k^{\min}} \tag{3.45}$$

Let us look at an example of a bicriteria maximization problem:

$$\max \quad \{z_1 = f_1(\boldsymbol{x}), z_2 = f_2(\boldsymbol{x})\} \tag{3.46}$$

$$\text{s.t.} \quad g_i(\boldsymbol{x}) \leq 0, \qquad i = 1,2,\ldots,m \tag{3.47}$$

For a given generation, two extreme points are identified as

$$z_1^{\max} = \max\{z_1(\boldsymbol{x}^j), j = 1,2,\ldots,pop\_size\} \tag{3.48}$$

$$z_2^{\max} = \max\{z_2(\boldsymbol{x}^j), j = 1,2,\ldots,pop\_size\} \tag{3.49}$$

$$z_1^{\min} = \min\{z_1(\boldsymbol{x}^j), j = 1,2,\ldots,pop\_size\} \tag{3.50}$$

$$z_2^{\min} = \min\{z_2(\boldsymbol{x}^j), j = 1,2,\ldots,pop\_size\} \tag{3.51}$$

and the adaptive weights are calculated as

$$w_1 = \frac{1}{z_1^{\max} - z_1^{\min}} \tag{3.52}$$

$$w_2 = \frac{1}{z_2^{\max} - z_2^{\min}} \tag{3.53}$$

The weighted-sum objective function is then given by

$$z(\boldsymbol{x}) = w_1 z_1 + w_2 z_2 \tag{3.54}$$

$$= w_1 f_1(\boldsymbol{x}) + w_2 f_2(\boldsymbol{x}) \tag{3.55}$$

It is an adaptive moving line defined by the extreme points $(z_1^{\max}, z_2^{\min})$ and $(z_1^{\min}, z_2^{\max})$, as shown in Figure 3.16. The rectangle defined by the extreme points $(z_1^{\max}, z_2^{\max})$ and $(z_1^{\min}, z_2^{\max})$ is the minimal rectangle containing all current solutions.

One of important issues in the genetic multiobjective optimization is how to handle constraints because genetic operators used to manipulate chromo-

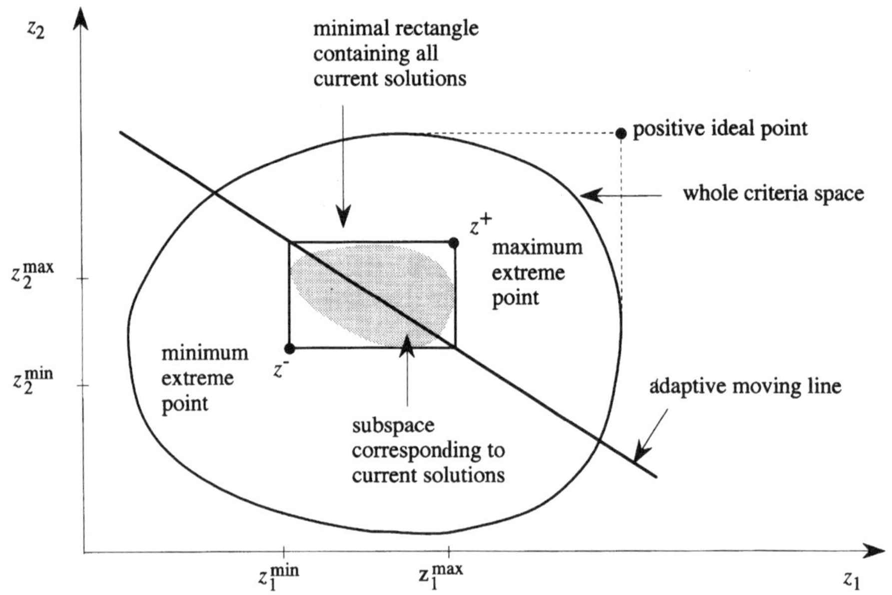

Figure 3.16. Adaptive weights and adaptive hyperplane.

somes often yield infeasible offspring. Gen and Cheng suggested an adaptive penalty method to deal with the infeasible individuals [216, 218]. Given an individual $\boldsymbol{x}$ in current population $P(t)$, the adaptive penalty function is constructed as follows:

$$p(\boldsymbol{x}) = 1 - \frac{1}{m}\sum_{i=1}^{m}\left(\frac{\Delta b_i(\boldsymbol{x})}{\Delta b_i^{\max}}\right)^{\alpha} \tag{3.56}$$

where

$$\Delta b_i(\boldsymbol{x}) = \max\{0, g_i(\boldsymbol{x}) - b_i\} \tag{3.57}$$

$$\Delta b_i^{\max} = \max\{\epsilon, \Delta b_i(\boldsymbol{x}) | x \in P(t)\} \tag{3.58}$$

where $\Delta b_i(\boldsymbol{x})$ is the value of violation for constraint i for the ith chromosome, $\Delta b_i^{\max}$ the maximum of violation for constraint i among current population, and ϵ a small positive number used to have penalty avoid from zero division. For highly constrained optimization problems, the infeasible solutions take a relative big portion among population at each generation. The penalty approach adjusts the ratio of penalties adaptively at each generation to make a balance between the preservation of information and the pressure for infeasibility and avoid overpenalty. With the penalty function, the fitness function then takes the following form:

$$\text{eval}(\boldsymbol{x}) = z(\boldsymbol{x})p(\boldsymbol{x}) \tag{3.59}$$

The overall procedure of adaptive weight approach is summarized as follows:

Step 1. Initialization. Generate the initial population randomly.

Step 2. Evaluation. Calculate the objective value, penalty value, and fitness value for each individual.

Step 3. Pareto set. Update the set of Pareto solutions.

Step 4. Selection. Select the next generation using the roulette wheel method.

Step 5. Production. Produce offspring with crossover and mutation.

Step 6. Termination. If the maximal generation is reached, stop; otherwise, go to step 2.

3.7 DISTANCE METHOD

Osyczka and Kundu proposed a distance method to calculate the fitness values [495, 496]. The basic idea of this method is to assign fitness values to each individual according to a distance measure with reference to the Pareto solutions obtained in the preceding generation.

3.7.1 General Idea of the Distance Method

A concept of *potential value* is used in the method. It is a scalar value only assigned to each Pareto solution, which is different from the fitness value of a given Pareto solution. After each updating of a Pareto set, an identical value is assigned to all Pareto solutions so that each new solution can be assigned with a reasonable fitness value.

As shown in Figure 3.17, existing Pareto solutions may have different potential values, denoted as $(p_1,p_2,\ldots,p_5)$. For a newly generated solution, the distances to all existing Pareto solutions are denoted as $(d_1,d_2,\ldots,d_5)$. Among them, d_3 is the minimum distance, which is used to calculate the fitness value for the new solution.

A newly generated solution may fall into any of the following three types:

1. A Pareto solution dominates some or all Pareto solutions
2. A Pareto solution that does not dominate any existing Pareto solutions
3. A solution dominated by at least one existing Pareto solution

In the first case, the fitness value of the new solution is calculated as the sum of the maximum potential value and the minimum distance. Then the set of Pareto solutions is updated by removing dominated solutions from the set and adding the new solution to the set. The potential value for the new so-

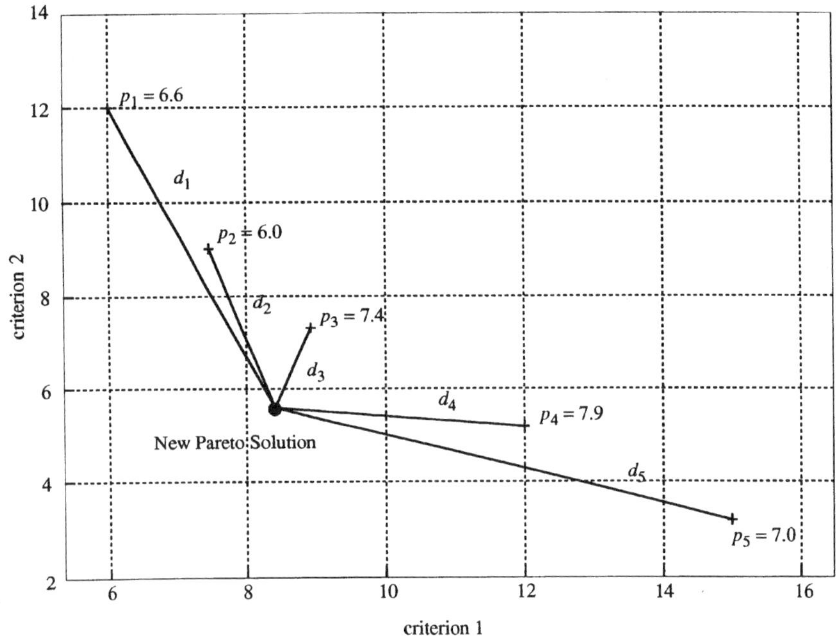

Figure 3.17. Potential value and distance measure in criteria space for the first case.

lution equals its fitness value. Let f denote the fitness value of the new solution and p denote the total number of existing Pareto solutions. We have

$$f = p_{\max} + d_{\min}$$

where

$$p_{\max} = \max\{p_i \mid i = 1,2,\ldots,p\}$$

$$d_{\min} = \min\{d_i \mid i = 1,2,\ldots,p\}$$

As shown in Figure 3.17, the new solution dominates the solution with potential value p_3. Its fitness value is $f = p_{\max} + d_{\min} = p_4 + d_3$. The dominated one is then removed from the Pareto solution set.

In the second case, the fitness value of the new solution is calculated as the sum of the potential value of its nearest Pareto solution and the minimum distance. The solution is added to the existing set of Pareto solutions with the potential value equal to its fitness value. As shown in Figure 3.18, the new solution is a new Pareto solution. Its fitness value is $f = p_4 + d_4$. The solution has the potential value $p_6 = f$.

In the third case, the fitness value of the new solution is calculated by subtracting the minimum distance from the potential value of its nearest Pa-

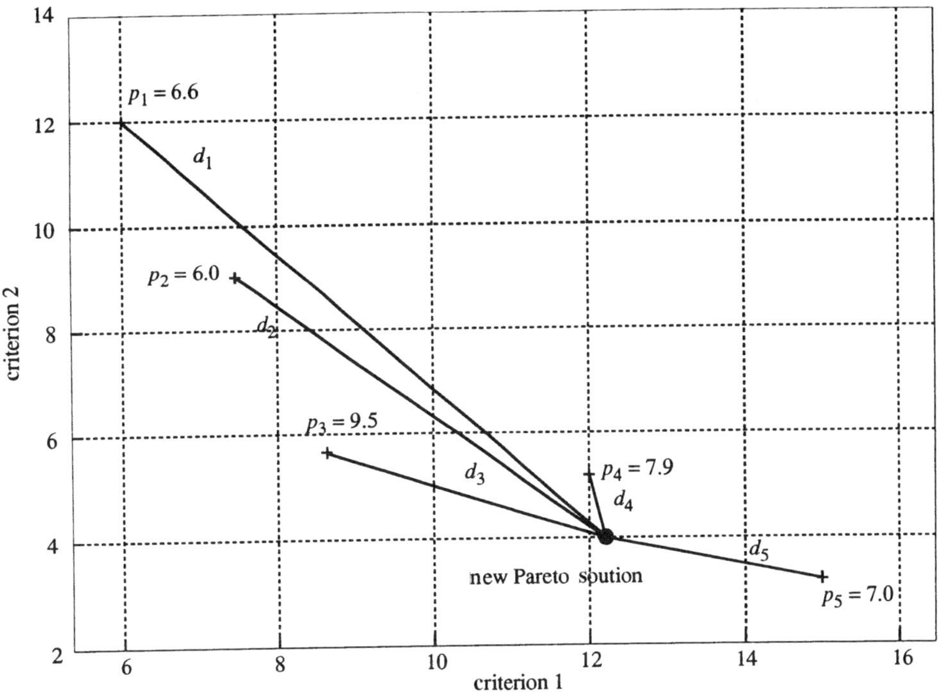

Figure 3.18. Potential value and distance measure in criteria space for the second case.

reto solution. As shown in Figure 3.19, the fitness value for the new solution is $f = p_4 \quad d_4$.

3.7.2 Calculation of Distance Measure

Consider the following general nonlinear multicriteria optimization problem:

$$\min \quad \{ f_1(\boldsymbol{x}), f_2(\boldsymbol{x}),\ldots, f_q(\boldsymbol{x})\} \tag{3.60}$$

$$\text{s.t.} \quad g_i(\boldsymbol{x}) \geq 0 \qquad i = 1,2,\ldots,m_0 \tag{3.61}$$

$$h_i(\boldsymbol{x}) = 0 \qquad i = 1,2,\ldots,m_1 \tag{3.62}$$

Osyczka and Kundu transformed the problem into an unconstrained problem using an exterior penalty function as follows [495]:

$$\phi_k(\boldsymbol{x}) = f_k(\boldsymbol{x}) + r \sum_{i=1}^{m_0} [h_i(\boldsymbol{x})]^2 + r \sum_{i=1}^{m_1} G_i[g_i(\boldsymbol{x})]^2, \qquad k = 1,2,\ldots,q \tag{3.63}$$

where G_i is a Heaviside operator such that $G_i = 0$ for $g_i(\boldsymbol{x}) \geq 0$ and $G_i = 1$

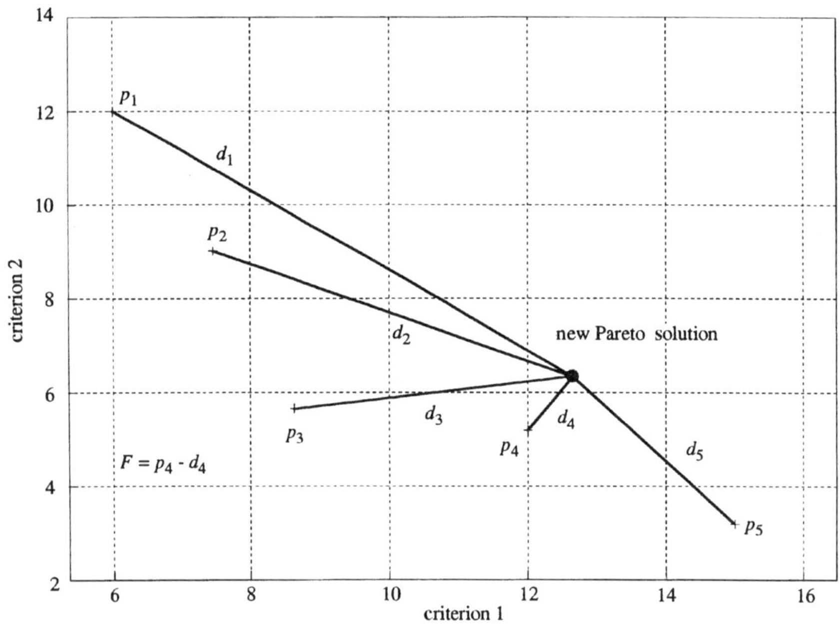

Figure 3.19. Potential value and distance measure in criteria space for the second case.

for $g_i(\boldsymbol{x}) \leq 0$, and r is a positive multiplier that controls the magnitude of the penalty terms.

Suppose that there are p Pareto solutions obtained by genetic search. Let f_{kl} denote the value of the kth objective function for the lth Pareto solution. For a new solution $\boldsymbol{x}$ generated by genetic operators, its distances to all existing Pareto solutions are calculated by the following formula:

$$d_l(\boldsymbol{x}) = \sqrt{\sum_{k=1}^{q} \left(\frac{f_{kl} - \phi_k(\boldsymbol{x})}{f_{kl}} \right)^2}, \qquad l = 1,2,\ldots,p \tag{3.64}$$

and the minimum distance, which is used to calculate fitness values, is given as

$$d_{l^*} = \min\{d_l(\boldsymbol{x}) \mid l = 1,2,\ldots,p\} \tag{3.65}$$

where l^* indicates the nearest existing Pareto solution to the new solution $\boldsymbol{x}$.

Let t be the index of the current generation, J the total number of solutions generated within the tth generation, j the index of a solution generated within the tth generation, and T the last generation. The overall procedure for the distance method is summarized below.

Step 1. Let $t = 1$ and $j = 1$. The first solution generated is taken as a Pareto solution with the potential value p_1, which is an arbitrarily chosen value called the starting potential value.

Step 2. Substitute $j = j + 1$ and check if $j \leq J$. If this is true, go to step 3; otherwise, go to step 5.

Step 3. Next, generate a new solution $\boldsymbol{x}$ and calculate the relative distances to all existing Pareto solutions using equation (3.64). Then find the minimum distance from then using equation (3.65).

Step 4. Compare the new solution $\boldsymbol{x}$ with all existing Pareto solutions in the following way:

(4.1) If the solution is a new Pareto solution and it dominates at least one of the existing Pareto solutions, calculate its fitness value using the following formula:

$$F = p_{\max} + d_{l^*}(\boldsymbol{x}) \tag{3.66}$$

Then substitute $p_{\max} = F$. Update the set of Pareto solutions. Let the potential value of the new solution be F. Then go to step 2.

(4.2) If the solution is a new Pareto solution, calculate its fitness value using the following formula:

$$F = p_{l^*} + d_{l^*}(\boldsymbol{x}) \tag{3.67}$$

Add it to the Pareto solution set with a potential value of F. If $F > p_{\max}$, substitute $p_{\max} = F$. Then go to step 2.

(4.3) If the solution is not a new Pareto solution, calculate its fitness value using the formula

$$F = p_{l^*} - d_{l^*}(\boldsymbol{x}) \tag{3.68}$$

If $F < 0$, substitute $F = 0$ to avoid a negative fitness value. Then go to step 2.

Step 5. Substitute all existing Pareto solutions with $p_{\max}$; that is,

$$p_l = p_{\max} \qquad \text{for } l = 1,2,\ldots,p \tag{3.69}$$

Step 6. Substitute $t = t + 1$. If $t > T$, terminate the run. If $t \leq T$, check if there is an improvement through the last several generations of a prescribed number. If there is any improvement, substitute $j = 0$ and go to step 2; otherwise, terminate the run.

Note that the distance method is sensitive to the values of p_1 and r. For an infeasible solution, the higher the value of r, the greater the distance of

$d_l(\boldsymbol{x})$. Therefore, its fitness value will approach zero. Genetic search will fail to work when too many solutions have its zero fitness values. Besides, equation (3.68) will tend to a zero fitness value if the initial potential value p_1 is too small. On the other hand, if the initial potential value is too large, the difference between the fitness of different solutions will be too small, which will cause the selective pressure to be very small or evenly distributed; therefore, the convergence of the genetic algorithms will be very slow.

3.7.3 Application of the Distance Method

Osyczka and Kundu used the following problem to test the distance method [496].

$$f_1(\boldsymbol{x}) = -25(x_1 - 2)^2 - (x_2 - 2)^2 - (x_3 - 1)^2 - (x_4 - 4)^2 - (x_5 - 1)^2$$

$$f_2(\boldsymbol{x}) = (x_1 - 1)^2 + (x_2 - 1)^2 + (x_3 - 1)^2 + (x_4 - 1)^2 + (x_5 - 1)^2$$

$$\begin{aligned}
\text{s.t.}\quad & g_1(\boldsymbol{x}) = x_1 + x_2 - 2 \geq 0 \\
& g_2(\boldsymbol{x}) = 6 - x_1 - x_2 \geq 0 \\
& g_3(\boldsymbol{x}) = 2 + x_1 - x_2 \geq 0 \\
& g_4(\boldsymbol{x}) = 2 - x_1 + 3x_2 \geq 0 \\
& g_5(\boldsymbol{x}) = 4 - (x_3 - 3)^2 - x_4 \geq 0 \\
& g_6(\boldsymbol{x}) = (x_5 - 3)^2 + x_6 - 4 \geq 0 \\
& 0 \leq x_i \leq 10, \qquad i = 1,2,3,4,5,6
\end{aligned}$$

A binary string was used to encode each decision variable. The penalty multiplier r was chosen as 1000 and the initial potential value p_1 was chosen as 10. The maximum generation was set as 50 and the population was set as 400. A comparison of the Pareto solutions obtained using the distance and stochastic multicriteria optimization methods was given in Figure 3.20, which showed that genetic algorithms can produce much better solutions. Figure 3.21 shows Pareto frontier movement along with the evolutionary progress.

3.8 COMPROMISE APPROACH

Compromise solution–based fitness assignment has been proposed by Cheng and Gen as a means to obtain a compromised solution instead of generating all Pareto solutions [109]. For many real-world problems, the set of Pareto solutions may be very large, possibly exponential in size. Thus the computational effort required to solve it can increase exponentially with problem size in the worst case. Having to evaluate a large set of Pareto solutions in

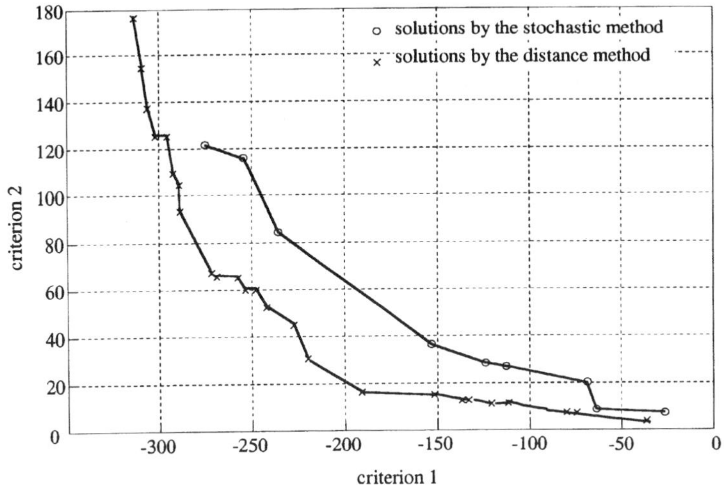

Figure 3.20. Comparison of Pareto solutions obtained.

order to select the best one poses a considerable cognitive burden on the decision maker. Therefore, in such cases, obtaining the entire set of Pareto solutions is of little interest to decision makers. In contrast to generating methods, the compromise approach searches for compromised solutions to overcome such difficulty.

The compromise approach identifies solutions that are closest to the ideal solution as determined by some measure of distance. The weighted L_p-norm is useful as a distance measure.

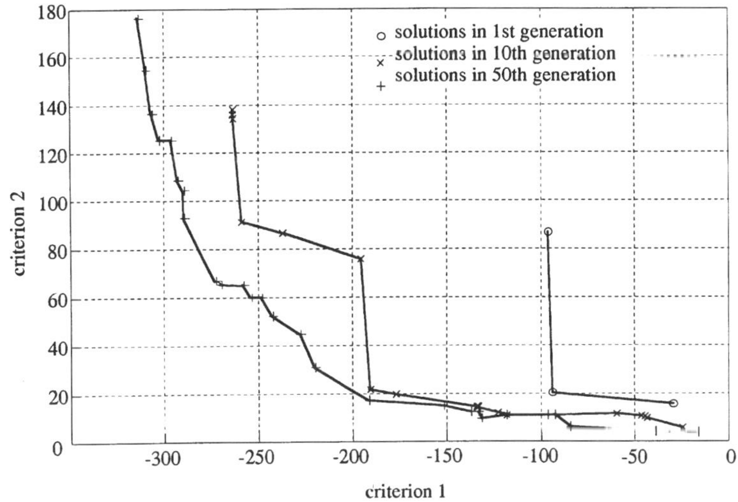

Figure 3.21. Pareto frontier movement showing evolutionary progress.

$$r(z;p,w) = \|z - z^*\|_{p,w} = \left(\sum_{j=1}^{q} w_j^p |z_j - z_j^*|^p \right)^{1/p} \tag{3.70}$$

where, $(z_1^*, z_2^*, \ldots, z_q^*)$ is the positive ideal point in the criterion space Z and weights $(w_1, w_2, \ldots, w_q)$ are assigned to objectives to emphasize different degrees of importance. As we know, the ideal solution is not usually attainable. However, it can serve as a good standard for evaluation of the nondominated solutions attainable. For many complex problems, to find an ideal point is also a difficult task. To overcome the difficulty, the concept of a proxy ideal point is suggested to replace the ideal point. The proxy ideal point is the ideal point corresponding to the current generation but not to a given problem. Let P denote the set of the current population; then the proxy ideal point $(z_1^{\text{mini}}, z_2^{\min}, \ldots, z_q^{\min})$ is calculated as follows:

$$z_1^{\min} = \min\{z_1(\boldsymbol{x}) \mid \boldsymbol{x} \in P\} \tag{3.71}$$

$$z_2^{\min} = \min\{z_2(\boldsymbol{x}) \mid \boldsymbol{x} \in P\} \tag{3.72}$$

$$\cdots$$

$$z_q^{\min} = \min\{z_q(\boldsymbol{x}) \mid \boldsymbol{x} \in P\} \tag{3.73}$$

The proxy ideal point is easy to obtain at each generation. Along with evolutionary progress, the proxy ideal point will gradually approximate the real ideal point.

Consider a minimization problem. Since the smaller the regret value, the better the individual, we have to convert the regret value into a fitness value to ensure that a fitter individual has a larger fitness value. Let $r(\boldsymbol{x})$ denote the regret value of individual $\boldsymbol{x}$, $r_{\max}$ the biggest regret value, and $r_{\min}$ the smallest regret value in the current generation. The transformation is given as follows:

$$\text{eval}(\boldsymbol{x}) = \frac{r_{\max} - r(\boldsymbol{x}) + \gamma}{r_{\max} - r_{\min} + \gamma} \tag{3.74}$$

where γ is a positive real number usually restricted within the open interval (0,1). Its purpose is twofold: (1) to prevent equation (3.74) from zero division and (2) to make it possible to adjust the selection behavior from fitness-proportional selection to pure random selection.

3.9 GOAL PROGRAMMING APPROACH

The basic idea of goal programming is to establish a specific numeric goal for each objective and then seek a solution that minimizes the weighted sum of deviations of objective functions from respective goals [98, 222, 315, 403].

Essentially, there are two kinds of goal programming problems. In one, *non-preemptive goal programming,* all goals are of roughly comparable importance. The other is *preemptive goal programming,* where there is a hierarchy of priority levels for the goals so that the goals of primary importance receive first-priority attention, those of secondary importance receive second-priority attention, and so on. The term *goal programming* today usually refers to the preemptive approach. It has become an essential and widely recognized approach for solving multiple-criteria decision-making problems.

Nonlinear goal programming is one of the mathematical programming techniques developed to solve problems with conflicting nonlinear objectives and constraints wherein the user provides levels or targets of achievement for each objective and prioritizes the order in which the goals have to be achieved. It finds an optimal solution that satisfies as many of the goals as possible in the order specified. The generalized formulation of nonlinear goal programming is given as follows:

$$\min \quad z_0 = \sum_{k=1}^{q} \sum_{i=1}^{m} p_k(w_{ki}^{+} d_i^{+} + w_{ki}^{-} d_i^{-}) \tag{3.75}$$

$$\text{s.t.} \quad f_i(x) + d_i^{-} - d_i^{+} = b_i, \qquad i = 1,2,\ldots,m_0 \tag{3.76}$$

$$g_i(x) \le 0, \qquad i = m_0 + 1,\ldots,\overline{m}_1(= m_0 + m_1) \tag{3.77}$$

$$h_i(x) = 0, \qquad i = \overline{m}_1 + 1,\ldots,m(= \overline{m}_1 + m_2) \tag{3.78}$$

$$d_i^{-}, d_i^{+} \ge 0, \qquad i = 1,2,\ldots,m_0 \tag{3.79}$$

where

p_k = kth preemptive priority ($p_k >> p_{k+1}$ for all k)
d_i^{+} = positive deviation variable representing overachievement of goal i
d_i^{-} = negative deviation variable representing underachievement of goal i
w_{ki}^{+} = positive weight assigned to d_i^{+} at priority p_k
w_{ki}^{-} = negative weight assigned to d_i^{-} at priority p_k
x = n-dimensional decision vector
f_i = function: $R^n \to R^1$ in goal constraints
g_i = function: $R^n \to R^1$ in real inequality constraints
h_i = function: $R^n \to R^1$ in real equality constraints
b_i = target value or aspiration level of goal i
q = number of priorities
m_0 = number of goal constraints
m_1 = number of real inequality constraints
m_2 = number of real equality constraints

Sometimes, the objective function is expressed as follows:

$$\text{lexmin}\left\{z_1 = \sum_{i=1}^{m}(w_{1i}^{+}d_i^{+} + w_{1i}^{-}d_i^{-}), \ldots, z_q = \sum_{i=1}^{m}(w_{qi}^{+}d_i^{+} + w_{qi}^{-}d_i^{-})\right\} \tag{3.80}$$

where lexmin means lexicographically minimizing the objective.

Gen and Liu [236] proposed a genetic algorithm for solving nonlinear goal programming problems. A rank-based fitness assignment method is used to assess the merit of each chromosome. Because the lexicographic ordering among objectives is preferred in goal programming, individuals are sorted in lexicographic objective order. A fitness value is then assigned to each individual by interpolating from best to worst according to an exponential function. The evaluation procedure consists of three steps:

Step 1. Calculate the objective values according to equation (3.80). There are q objective values associated with each individual, that is,

$$\left\{\sum_{i=1}^{m}(w_{1i}^{+}d_i^{+} + w_{1i}^{-}d_i^{-}), \sum_{i=1}^{m}(w_{2i}^{+}d_i^{+} + w_{2i}^{-}d_i^{-}), \ldots, \sum_{i=1}^{m}(w_{qi}^{+}d_i^{+} + w_{qi}^{-}d_i^{-})\right\}$$

Step 2. Sort individuals on the first priority objective $\Sigma_{i=1}^{m}(w_{1i}^{+}d_i^{+} + w_{1i}^{-}d_i^{-})$. If some individuals have the same objective value, sort on the second priority objective $\Sigma_{i=1}^{m}(w_{2i}^{+}d_i^{+} + w_{2i}^{-}d_i^{-})$, and so on. A tie is broken randomly. Thus individuals are arranged in ascending order according to the objective values.

Step 3. Assign each individual a rank-based fitness value. Let r_k be the rank of individual $\boldsymbol{x}_k$. For a parameter $a \in (0,1)$ specified by the user, the rank-based fitness function is defined as follows:

$$\text{eval}(\boldsymbol{x}_k) = a(1 - a)^{r_k - 1}$$

where $r_k = 1$ means the best individual and $r_k = pop\_size$, the worst individual. We have that

$$\sum_{k=1}^{pop\_size} \text{eval}(\boldsymbol{x}_k) \approx 1$$

Taguchi, Ida, and Gen proposed an implementation of the genetic algorithms to solve the nonlinear goal programming problem with interval coefficients [607]. For a given individual $\boldsymbol{x}$, the fitness function proposed by them takes the following weighted-sum form:

$$\text{eval}(\boldsymbol{x}) = \sum_{k=1}^{q}\sum_{i=1}^{m_0} p_k(w_{ki}^{+}d_i^{+} + w_{ki}^{-}d_i^{-}) \tag{3.81}$$

where p_k is the preemptive priority calculated as follows:

$$p_k = 10^{2*(q-k)}, \qquad k = 1,2,\ldots,q \tag{3.82}$$

Note that equation (3.81) has the property that the smaller the value of fitness, the fitter the individual. Therefore, individuals are sorted on the fitness values in ascending order and the next generation is selected by the first *pop_size* distinct individuals.

Recently, Gen and Liu developed an evolution algorithm for solving the optimal capacity expansion problem [235], Taguchi, Gen, and Ida reported a genetic algorithm for solving the multiple nonlinear integer programming problem [726], Kim et al. proposed a genetic algorithm for solving the multiobjective assembly line balancing problem [718], Dev and Goyal developed a flexible optimization procedure for the machanical component design based on genetic adaptive search [711].

4
FUZZY OPTIMIZATION PROBLEMS

4.1 INTRODUCTION

Most optimization problems can be described as mathematical programming problems. Mathematical programming is used for programming and decision making at various levels of management, production, and so on, but in these cases clear objective functions and constraints are used. In real problems there is leeway in both the functions that express things such as profit or loss, and the constraints that express things such as possible investment money. A better representation of such considerations is necessary to handle such ambiguities. Various forms of fuzzy mathematical programming have been proposed to attend to this situation, including fuzzy linear programming, fuzzy nonlinear programming, fuzzy integer programming, fuzzy mixed-integer programming, fuzzy multiobjective programming, and so on [349, 547, 550, 618. 706]. Several papers (see, e.g., Luhandjula [430, 431] and Yazenin [680]) have considered fuzzy linear programming or fuzzy multiobjective linear programming problems and proposed a series of ideas for translating the original constraints into crisp equivalents via the possibility theory provided by Zadeh [691]. A generally theoretical framework of chance-constrained programming, multiobjective programming, and goal programming, without the linearity assumption, has been investigated [418, 419].

There are many successful applications of genetic algorithms for a variety of mathematical programming problems. Unfortunately, the same implementations of genetic algorithms for crisp mathematical programming problems cannot be directly adopted to solve fuzzy ones. We must employ some special tricks, which can treat fuzziness in evolution processes to solve fuzzy mathematical programming problems using genetic algorithms.

In this chapter, some successful applications of genetic algorithms for typical fuzzy optimization problems, such as fuzzy linear programming problem, fuzzy nonlinear programming, the fuzzy nonlinear mixed-integer goal programming problem, and fuzzy multiobjective integer programming, are introduced. Some numerical examples are presented for deeper understanding.

4.2 FUZZY LINEAR PROGRAMMING

Since Bellman and Zadeh proposed the concept of fuzzy decisions, research on fuzzy mathematical programming has been an active area [52, 390, 408, 429, 616, 655, 706, 707]. According to Bellman and Zadeh's definition, a *fuzzy decision* is the confluence of fuzzy objectives and constraints, defined by a max-min operator. Based on this definition, Zimmermann developed a tolerance approach for symmetric model of fuzzy linear programming. It is noted as the first practical method to solve linear programming having fuzzy constraints and objectives. It had the potential to be applied to actual problems such as production planning, resource allocation, and so on. Since then, fuzzy linear programming has been developed in a number of directions with many successful applications. Fuzzy programming is considered to be an important part of multiobjective optimization under fuzziness. The bicriteria transportation problem with fuzzy coefficients in Section 7.6 is a typical real-world fuzzy multiobjective optimization problem.

In this section, after an overview of the basic concepts of linear programming via a simple numerical example of fuzzy linear programming as proposed in Zimmermann's approach, an inexact method is introduced to solve objective and resource types of fuzzy linear programming problems. Instead of finding an exact optimal solution, this approach used a genetic algorithm with mutation along the weighted gradient direction to find a family of inexact solutions with acceptable membership degree. Then, by means of human–computer interaction, the solution preferred by the decision maker can be sought by a convex combination of the solutions selected from the family. Finally, numerical experiment has shown that more satisfactory results can be achieved using this new approach.

4.2.1 Fuzzy Linear Programming Model

First, consider the following simple production planning problem as an example of a problem that can be solved using the linear programming.

Example 4.1. A plant is going to produce two types of products, A and B, in the coming month. The net profits of A and B are \$2 million and \$3 million. Production of A and B requires labor time and a principal material. The labor and material requirements of each unit of A and 4 people-months and 2 units, respectively; those of B, 2 people-months and 4 units. Eight people-months of labor is available and eight units of material is available (see Table 4.1). Given this limited labor time and material, management is trying to figure out how many units of products A and B should be produced to maximize

TABLE 4.1. Production Conditions and Profit

	A	B	Amount Available
Labor time	4	2	8
Material	2	4	8
Profit (million dollars)	2	3	

total profit. This production planning problem can be formulated as the following linear programming problem: Maximize the linear profit function,

$$\max \quad z' = 2x_1 + 3x_2$$

subject to the linear constraints

$$4x_1 + 2x_2 \leq 8$$
$$2x_1 + 4x_2 \leq 8$$
$$x_1, x_2 \geq 0$$

Convert the problem to that of minimizing z under the foregoing constraint:

$$\begin{aligned} \min \quad & z = -2x_1 - 3x_2 \\ \text{s.t.} \quad & 4x_1 + 2x_2 \leq 8 \\ & 2x_1 + 3x_2 \leq 8 \\ & x_1, x_2 \geq 0 \end{aligned} \tag{4.1}$$

It is easy to see that in the x_1–x_2 plane, the linearly constrained set of points (x_1,x_2) satisfying all the constraints of problem (4.1) become the boundary lines and interior points in Figure 4.1.

The set of points satisfying $z = -2x_1 - 3x_2$ for a fixed z constitute a line. As z is varied, the line is moved parallel to itself. The optimal value of this example is the smallest value of z for which the corresponding line has at least on point in common with the linearly constrained set. As can be seen from Figure 4.1, the optimal solution to this problem is

$$x_1 = \tfrac{4}{3}, \qquad x_2 = \tfrac{4}{3}, \qquad z = \tfrac{20}{3}$$

A linear programming problem with fuzzy resources and objective can be described as follows:

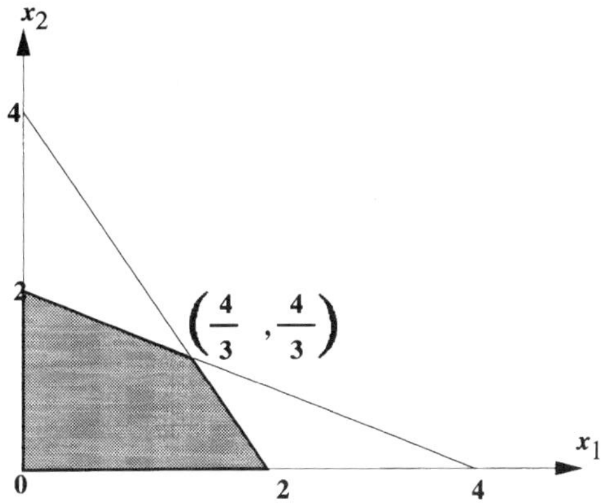

Figure 4.1. Feasible region for Example 4.1.

$$\begin{aligned} \widetilde{\max} \quad & z = \boldsymbol{c}^T \boldsymbol{x} \\ \text{s.t.} \quad & \boldsymbol{a}_i^T \boldsymbol{x} \leq \tilde{b}_i, \qquad i = 1,2,\ldots,m \\ & \boldsymbol{x} \geq \boldsymbol{0} \end{aligned} \tag{4.2}$$

where $\boldsymbol{x} \in R^n$ is the decision vector; $A \in R^{m \times n}$ and $\boldsymbol{c} \in R^n$ are the crisp constraint matrix and objective vector, respectively; $\tilde{\boldsymbol{b}} = (\tilde{b}_1, \ldots, \tilde{b}_m)^T$ is an m-dimensional column vector; and $A = [a_{ij}]$ is an $m \times n$ matrix. The constraint matrix A can be rewritten as

$$A = \begin{bmatrix} \boldsymbol{a}_1^T \\ \boldsymbol{a}_2^T \\ \vdots \\ \boldsymbol{a}_m^T \end{bmatrix} = \begin{bmatrix} a_{11} & a_{12} & \cdots & a_{1n} \\ a_{21} & a_{22} & \cdots & a_{2n} \\ & \cdots & \cdots & \\ a_{m1} & a_{m2} & \cdots & a_{mn} \end{bmatrix}$$

The goal z_0 and its corresponding tolerance p_0 of the fuzzy objective, the fuzzy constraints b_i, and it corresponding tolerances p_i are given initially. The fuzzy objective and the fuzzy constraints are then considered without difference, and their corresponding regions can be described in the intervals $[b_i, b_i + p_i]$, $\forall_i$. Thus equation (4.2) can be considered as follows:

$$\begin{aligned} \text{find} \quad & \boldsymbol{x} \\ \text{s.t.} \quad & \boldsymbol{c}^T \boldsymbol{x} \geq z_0 \\ & \boldsymbol{a}_i^T \boldsymbol{x} \lesssim b_i, \qquad i = 1,2,\ldots,m \\ & \boldsymbol{x} \geq \boldsymbol{0} \end{aligned} \tag{4.3}$$

In fuzzy set theory, the fuzzy objective function and the fuzzy constraints

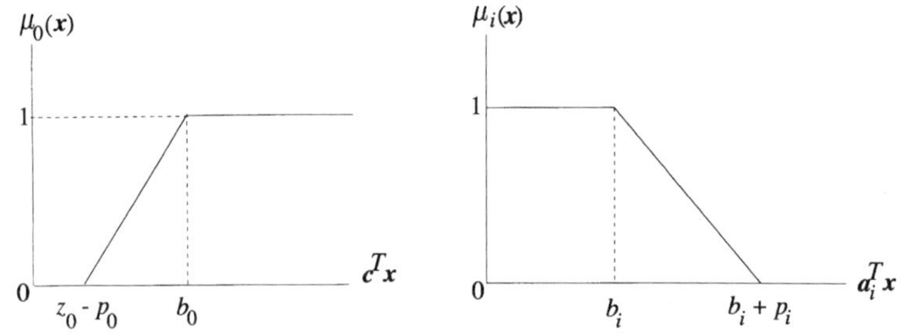

Figure 4.2. Membership functions for the fuzzy objective constraint $\boldsymbol{c}^T\boldsymbol{x} \geq \tilde{z}_0$ and the ith fuzzy constraint $(\boldsymbol{a}_i^T\boldsymbol{x}) \leq \tilde{b}_i$.

are defined by their corresponding membership functions. For simplicity, let us assume that the membership function μ_0 of the fuzzy objective is a nondecreasing continuous linear function, and the membership functions μ_i, $\forall i$, of the fuzzy constraints are nonincreasing continuous linear membership functions and constraints as follows (Figure 4.2): For the objective function,

$$\mu_o(\boldsymbol{x}) = \begin{cases} 1, & \text{if } \boldsymbol{c}^T\boldsymbol{x} > z_0 \\ 1 - \dfrac{z_0 - \boldsymbol{c}^T\boldsymbol{x}}{p_0}, & \text{if } z_0 - p_0 \leq \boldsymbol{c}^T\boldsymbol{x} \leq z_0 \\ 0, & \text{if } \boldsymbol{c}^T\boldsymbol{x} < z_0 - p_0 \end{cases} \tag{4.4}$$

For the ith resource constraint, $i = 1,2,\ldots,m$,

$$\mu_i(\boldsymbol{x}) = \begin{cases} 1, & \text{if } \boldsymbol{a}_i^T\boldsymbol{x} < b_i \\ 1 - \dfrac{\boldsymbol{a}_i^T\boldsymbol{x} - b_i}{p_i}, & \text{if } b_i \leq \boldsymbol{a}_i^T\boldsymbol{x} \leq b_i + p_i \\ 0, & \text{if } \boldsymbol{a}_i^T\boldsymbol{x} > b_i + p_i \end{cases} \tag{4.5}$$

Zimmermann then used Bellman and Zadeh's max-min operator to solve equations (4.3) to (4.5). Thus the optimal solution can be obtained by

$$\max_{\boldsymbol{x}\in(R^n)^+} \mu_{\tilde{S}}(\boldsymbol{x}) = \max\{\min\{\mu_o(\boldsymbol{x}),\mu_i(\boldsymbol{x})|^i = 1,2,\ldots,m\}\} \tag{4.6}$$

The optimization problem (4.6) can be rewritten as

$$\max_{\boldsymbol{x}\in(R^n)^+} \mu_{\tilde{S}}(\boldsymbol{x}) = \max\left\{0,\min\left\{1,1 - \frac{z_0 - \boldsymbol{c}^T\boldsymbol{x}}{p_0}, 1 - \frac{\boldsymbol{a}_i^T\boldsymbol{x} - b_i}{p_i}\right| \right. \right. \\ \left. \left. i = 1,2,\ldots,m\right\}\right\} \tag{4.7}$$

where $\mu_{\tilde{S}}$ is the membership function of the decision space $\tilde{S}$, and $\mu_{\tilde{S}} = \min\{\mu_o,\mu_i|i = 1,2,\ldots,m\}$. If $\alpha = \mu_{\tilde{S}}$, the program will be equivalent to

TABLE 4.2. Production Conditions and Profit

	P_1	P_2	Amount Available
M_1	2.5	5	150
M_2	5	2	120
Profit (million dollars)	1.5	2	

$$\max \quad \alpha \tag{4.8}$$

$$\text{s.t.} \quad \boldsymbol{c}^T\boldsymbol{x} \geq z_0 - (1 - \alpha)p_0 \tag{4.9}$$

$$\boldsymbol{a}_i^T\boldsymbol{x} \leq b_i + (1 + \alpha)p_i, \qquad i = 1,2,\ldots,m \tag{4.10}$$

$$\boldsymbol{x} \geq 0, \quad \alpha \in [0,1] \tag{4.11}$$

Generally, a unique optimal solution $(\boldsymbol{x}^*,\alpha^*)$ can be found from crisp linear programming by the simplex procedure. Let $\boldsymbol{x}^*$ be the solution with the highest membership degree of the fuzzy programming [equation (4.2)]. Its meaning is that the best balance of objective and constraints has been achieved by the optimal solution $\boldsymbol{x}^*$. However, the membership function usually is not the preference function of the decision maker. It is possible that the unique $\boldsymbol{x}^*$ is not desired by the decision maker. On the other hand, a unique exact optimal solution is meaningless in fuzzy programming, because the original data to be used for calculation are imprecise or vague. Wang and Tang propose the concept of a fuzzy optimal instead of a unique optimal solution [616, 649, 650].

Example 4.2. Company A would like to maximize the total profit producing two products, P_1 and P_2, utilizing two different materials, M_1 and M_2. The total amounts of available materials and profit figures are given in Table 4.2. Given these limited materials, the company is trying to figure out how many units of products P_1 and P_2 should be produced to maximize the total profit.

The production planning problem is

$$\begin{aligned} \min \quad & z = -1.5x_1 - 2x_2 \\ \text{s.t.} \quad & 2.5x_1 + 5x_2 \leq 150 \\ & 5x_1 + 2x_2 \leq 120 \\ & x_1,x_2 \geq 0 \end{aligned}$$

The optimal solution to this problem is

$$x_1 = 15, \qquad x_2 = 22.5, \qquad z = -45$$

Instead of giving rigid numerical values for the right-hand-side constraints as in conventional linear programming problems, assume that the decision maker has the fuzzy goal and fuzzy constraints shown in Table 4.3. The membership function and membership constraints can be described as follows: For the objective function,

$$\mu_o(\boldsymbol{x}) = \begin{cases} 1, & -1.5x_1 - 2x_2 \geq -42 \\ 1 - \dfrac{-1.5x_1 - 2x_2 + 47}{5}, & -47 < -1.5x_1 - 2x_2 \leq -42 \\ 0, & -1.5x_1 - 2x_2 \leq -47 \end{cases} \tag{4.12}$$

For the first and second constraints,

$$\mu_{c1}(\boldsymbol{x}) = \begin{cases} 1, & 2.5x_1 - 5x_2 \leq 150 \\ 1 - \dfrac{2.5x_1 + 5x_2 - 150}{10}, & 150 \leq 2.5x_1 + 5x_2 \leq 160 \\ 0, & 2.5x_1 + 2x_2 \geq 160 \end{cases} \tag{4.13}$$

$$\mu_{c2}(\boldsymbol{x}) = \begin{cases} 1, & 5x_1 + 2x_2 \leq 120 \\ 1 - \dfrac{5x_1 + 2x_2 - 120}{5}, & 125 \geq 5x_1 + 2x_2 \geq 120 \\ 0, & 5x_1 + 2x_2 \geq 125 \end{cases} \tag{4.14}$$

Assuming that the linear membership functions are from $\mu = 0$ to $\mu = 1$ for these fuzzy equalities, the flexible formulation of the original problem in the form of a fuzzy linear programming problem can be transformed into the following equivalent conventional linear programming problem:

$$\begin{aligned} \min \quad & \alpha \\ \text{s.t.} \quad & 0.3x_1 + 0.4x_2 + \alpha \geq 8.4 \\ & 0.25x_1 + 0.25x_2 + \alpha \leq 16 \\ & x_1 + 0.4x_2 + \alpha \leq 25 \\ & x_1, x_2 \geq 0 \end{aligned} \tag{4.15}$$

Solving this problem by the simplex of linear programming yields the optimal solution

$$x_1 = 15.069, \qquad x_2 = 23.017, \qquad \alpha = 0.72414.$$

In Table 4.4 the optimal solutions corresponding to the fuzzy linear programming problem as well as the nonfuzzy linear programming problem are

TABLE 4.3. Nonfuzzy and Fuzzy Constraints

	Nonfuzzy	Fuzzy	
		$\mu = 0$	$\mu = 1$
Objective function	−45	−42	−47
First constraint	150	160	150
Second constraint	120	125	120

shown. As a result, in this example it may be natural that the decision maker has about 1.38% additional total profit at the expense of some additional material, as shown in the Table 4.4. Thus, in fuzzy linear programming, the decision maker is not longer forced to formulate the problems in precise and rigid form as in linear programming, which seems to be one of the major advantages of fuzzy linear programming.

4.2.2 Genetic Algorithm Approach

The interactive decision procedure for determining preferred solutions is described as follows:

Step 1. Design several preliminary decision schemes.

Step 2. Check whether the resource constraints can be met by these schemes. If not, adjust the schemes to meet all constraints.

Step 3. Compare all decision schemes and select the best decision scheme with the highest objective value. A genetic algorithm with the mutation along the gradient direction is used to imitate the human decision procedure.

The individuals in the genetic algorithm are produced randomly in $(R^n)^+$; each individual reproduces its children with the selection probability depending on the objective function value; the children are produced along the

TABLE 4.4. Solutions of Nonfuzzy and Fuzzy Linear Programming Problems

Nonfuzzy	Fuzzy
$x_1 = 15$	$x_1 = 15.069$
$x_2 = 22.5$	$x_2 = 23.017$
$z = -45$	$z = -45.621$
Constraints:	Constraints:
1:150	1:152.76
2:120	2:121.38

weighted gradient direction of their parents. After a number of generations, most individuals will approach the unique optimal solution $\boldsymbol{x}^*$.

The optimal problem is an unconstrained max-min optimization problem, but its objective function is not continuously derivable. Thus it cannot be solved by traditional optimization methods. But it is fine for genetic algorithms. Wang and Tang take the membership function of the fuzzy optimal solution, $\mu_{\tilde{S}}(\boldsymbol{x})$, as the fitness function of the genetic algorithm. For a genetic algorithm, the larger the fitness, the higher the probability to produce children. Thus the individuals with higher membership degree have more chance to reproduce children around them.

Wang and Tang have designed a mutation operator that moves along a weighted gradient direction. Individuals will be led by this mutation operator to reach the fuzzy optimal solution set quickly. Given w_o and $w_i(i = 1,2,...,m)$, the weights of the objective function and the ith constraint, respectively, a weighted gradient of $\mu_{\tilde{S}}(\boldsymbol{x})$ can be defined as follows:

$$G(\boldsymbol{x}) \equiv \nabla \mu_{\tilde{S}}(\boldsymbol{x}) = \frac{w_0 c}{p_0} - \sum_{i=1}^{m} \frac{w_i \boldsymbol{a}_i}{p_i} \tag{4.16}$$

Let $\mu_{\min} = \min\{\mu_o(x), \mu_i(x) | i = 1,2,...,m\}$; then

$$w_i = \begin{cases} 1, & \mu_i = \mu_{\min} \\ \dfrac{\sigma}{\mu_i}, & \mu_{\min} < \mu_i < 1 \\ 0, & \mu_i = 1 \end{cases} \tag{4.17}$$

where σ is a sufficiently small positive number (usually, $\sigma = 10^{-6}$). If $\boldsymbol{x}^{k+1}$ is the child of individual $\boldsymbol{x}^k$, the mutation along the weighted gradient direction can be described as follows [199]:

$$\boldsymbol{x}^{k+1} = \boldsymbol{x}^k + \theta G(\boldsymbol{x}) \tag{4.18}$$

where θ is a random step length of the Erlang distribution, which is generated by a random number generator.

From the formulas of weight and mutation, equations (4.16) and (4.17), we can see that constraints that cannot meet the minimum membership function, 0; will get the full weight, 1. The mutation will lead the individual to the feasible area. When $\mu_{\tilde{S}}(\boldsymbol{x})$ is greater than 0, the objective or constraint with minimum membership function will get the full weight. The weighted gradient direction will improve the minimum; thus $\mu_{\tilde{S}}(\boldsymbol{x})$ will be improved immediately. Therefore, mutation along the weighted gradient direction will lead most individuals close to the exact optimal solution $\boldsymbol{x}^*$. It is noted that the $\boldsymbol{x}^{k+1}$ may jump outside the feasible domain. But it is usually led back by the weighted gradient. The random step length, θ, can be controlled to be

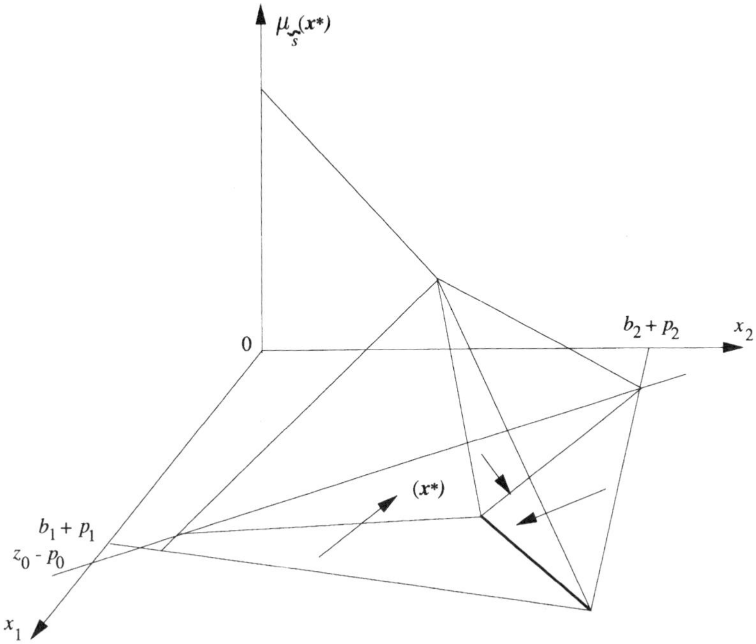

Figure 4.3. Weighted gradient directions in different areas.

smaller as the iteration progresses by decreasing the parameters of the Erlang distribution step by step. Thus the solutions can be kept in the feasible domain when the algorithm terminates. However, it is noted that the weighted gradient direction does not generally point to the exact optimal solution $\boldsymbol{x}^*$. As shown in Figure 4.3, for a problem with $n - 2$ and $m = 2$, the weighted directions in different areas are marked as the vectors. They do not point to $\boldsymbol{x}^*$. Therefore, the genetic algorithm can fast lead individuals to reach the fuzzy optimal solution, but it is very slow to reach the exact optimal solution. Fortunately, the exact optimal solution $\boldsymbol{x}^*$ is not always desired by the decision maker in practice. The remaining problem is how to select the preferred solution from the fuzzy optimal solution.

The real decision vector is used as the gene representation. The "real number" presentation scheme has been very successful for function optimization because it needs no transformation of number systems. The genes consist of n decision variables. Let *pop_size* be the total number in the population. Then for individual j, the real vector

$$\boldsymbol{x}(j) = [x_1(j) \quad x_2(j) \quad \cdots \quad x_n(j)]^T, \qquad j = 1,2,\ldots,pop\_size \tag{4.19}$$

represents the genes. The population size will be kept constant in each generation. We denote individual j in the kth generation by $\boldsymbol{x}^k(j)$, $j = 1,2,\ldots,pop\_size$.

4.2.3 Interactive Approach

For objective and resource problems, the objective used to be the profit or production output and the constraints used to be the different types of manufacturing resources. In practice, the decision maker is usually interested in the objective value or some decision variables, or is concerned with the consumptions of resources. To find the preferred solution from the fuzzy optimal solution, Wang designed a human–computer interactive approach [616, 649, 650]. It can be described as follows. At first, the approach asks the decision maker the acceptable membership degree level of the fuzzy optimal solution, α. Second, it elicits the decision maker to find his or her preference structure by the human–computer interaction. It asks the decision maker to point out which criteria are of most concern to him or her. The criteria may be the objective, resources, or decision variables. Then the solutions in the α-cut of the fuzzy optimal solution $\tilde{S}_\alpha$ with the highest and lowest values of these criteria are stored in memory and updated in each interaction. If the generation number and population size of the genetic algorithm are *max_gen* and *pop_size*, the visited points should be *max_gen* × *pop_size*. It is a large number. Therefore, the solutions preferred by the decision maker will be found with high probability. Since in general there is more than one criterion of concern to the decision maker, there is more than one solution to be found. When the iteration terminates, these solutions with their criteria values will be shown to the decision maker by the human–computer interface. The decision maker can select a couple of solutions preferred by him or her each time. The computer shows the curve representing the criterion of most concern between two solutions. Then the decision maker can determine the best combination of solutions. Repeating this procedure, the decision maker can find the preferred solution. It is evident that the final solution is a convex combination of the solutions with membership degree greater than α. Thus it is in the cut set $\tilde{S}_\alpha$. Combining the genetic algorithm featured by a mutation along the weighted gradient direction with human–computer interactions to select the preferred solution, Wang, Tang, and Fung have proposed a new approach to linear objective or resource optimization problems called an inexact approach. The step-by-step procedure of this approach can be described as follows:

Step 1. Input the acceptable membership degree level of the fuzzy optimal solution α.

Step 2. Input the concerned criterion index set C_I, $C_I \subseteq \{0,1,2,\ldots,n,n+1,n+2,\ldots n+m\}$, where 0 stands for the objective function, $j = 1,2,\ldots,n$ for the decision variables, and $j = n+1, n+2,\ldots,n+m$ for the constraints. Initialize the values and solutions of the minimum and maximum of criterion r, $r \in C_I$, by

$$\min(r) = M, \qquad \max(r) = 0$$

$$z_1(r) = x(1), \qquad z_2(r) = x(1)$$

where M is a sufficiently large positive number (usually, $M = 10^6$), $\min(r)$ and $\max(r)$ are the initial values of the minimum and maximum of criterion r, and $z_1(r)$ and $z_2(r)$ are the solutions to reach $\min(r)$ and $\max(r)$.

Step 3. Find the upper bound of the decision variable x_i, denoted by x_i^{up} $(i = 1,2,\ldots,n)$. For $A \geq 0$ (it is common to our problem) we have

$$x_i^{\mathrm{up}} = \max\left\{\frac{b_j + p_j}{a_{ji}} \middle| j = 1,2,\ldots,m\right\}, \qquad i = 1,2,\ldots,n \tag{4.20}$$

Step 4. Produce the initial population and calculate the membership function by

$$x_i(j) = \xi x_i^{\mathrm{up}}$$

$$\xi \in U(0,1), \qquad i = 1,2,\ldots n, \quad j = 1,2,\ldots,pop\_size$$

$$\mu_{\tilde{S}}(\boldsymbol{x}(j)) = \min\{\mu_o(\boldsymbol{x}), \ \mu_i(\boldsymbol{x})|i{=}1,2,\ldots,m\}$$

where $U(0,1)$ means the random numbers in the uniform distribution from 0 to 1.

Step 5. Set the interaction index $k = 1$.

Step 6. For individual j $(j = 1,2,\ldots,pop\_size)$, calculate its fitness function $F(j)$ and selected probability $P(j)$ by

$$F(j) = \mu_{\tilde{S}}(\boldsymbol{x}(j)) + \epsilon, \qquad P(j) = \frac{F(j)}{\sum_{i=1}^{pop\_size} F(i)}$$

where ϵ is a small positive number (usually $\epsilon = 10^{-8}$). It is used to guarantee a nonzero denominator.

Step 7. By the proportional selection strategy, for child j $(j = 1,2,\ldots,pop\_size)$, select individual i as its parent; then

$$\boldsymbol{x}^k(j) = \boldsymbol{x}^{k-1}(i) + \theta G(\boldsymbol{x}(i))$$

where $G(\boldsymbol{x}(i))$ is the weighted gradient direction and θ is the step length generated by a random number generator of Erlang distribution.

Step 8. For new individual j $(j = 1,2,\ldots,pop\_size)$, calculate $\mu_{\tilde{S}}(\boldsymbol{x}(j))$. Let

$$\boldsymbol{x}^* = \operatorname{argmax}\{\mu_{\tilde{S}}(\boldsymbol{x}(j))|j = 1,2,\ldots,pop\_size\}$$

If $\mu_{\tilde{S}}(\boldsymbol{x}(j)) \geq \alpha$, check if criterion value $V_r(\boldsymbol{x}(j))$, $r \in C_I$, is less than min(r) or greater than max(r); then update min (r), $z_1(r)$ or max(r), $z_2(r)$ by $\boldsymbol{x}(j)$ and $V_r(\boldsymbol{x}(j))$.

Step 9. $k \leftarrow k + 1$. If $k < max\_gen$, go to step 5, where max_gen is a preselected maximum generation number that depends on the required precision. Otherwise, go to step 10.

Step 10. Show the solutions $\boldsymbol{x}^*, z_1(r), z_2(r), r \in C_I$ by the human–computer interface. If the decision maker thinks that z_1 and $z_2 \in \{\boldsymbol{x}^* | z_1(r), z_2(r), r \in C_I\}$ are the preferred solution couple and criterion r is the criterion of concerns show the curve of criterion r between z_1 and z_2. Then the decision maker can determine the preferred combination of z_1 and z_2. Continue this combination determination process until a preferred solution is found.

4.2.4 Numerical Example

Wang and Tang have used the following instances to test their method [616, 649, 650].

Example 4.3. A plant is going to produce two types of products, A and B, in the coming month. The net profits of each A and B are \$6000 and \$4000, respectively. The production of A and B requires labor time and a principal material. The labor and material requirements of each unit of A are 2 people-months and 4 units, respectively; those of B, 3 people-months and 2 units. Seventy people-months of labor is available, and 100 units of material is available. There are also 20 units of material in storage, which is controlled by the general manager of the plant. The decision maker hopes that the objective profit reaches \$200,000 but at least is not less than \$160,000 and that overtime work and consumption of the safety stock are not too high. This is a typical fuzzy objective or resource optimization problem. It can be described by Zimmermann's symmetric model as follows [706, 707]:

$$
\begin{aligned}
\max \quad & \alpha \\
\text{s.t.} \quad & f(\boldsymbol{x}) = 6x_1 + 4x_2 \geq 200 - (1 - \alpha)40 \\
& g_1(\boldsymbol{x}) = 2x_1 + 3x_2 \leq 70 + (1 - \alpha)30 \\
& g_2(\boldsymbol{x}) = 4x_1 + 2x_2 \leq 100 + (1 - \alpha)20 \\
& x_1, x_2 \geq 0, \qquad \alpha \in [0,1]
\end{aligned}
\tag{4.21}
$$

By Zimmermann's tolerance approach, it is easy to determine that the exact optimal solution is $\boldsymbol{x}^* = [20,15]^T$, $\mu_{\tilde{S}}(\boldsymbol{x}^*) = 0.5$ and the optimal value is 180. First, let us analyze the behavior of the mutation along the weighted gradient direction in the genetic algorithm. When the population size $pop\_size = 1$, the moving route of a random point during the first eight iterations is shown in Table 4.5 and Figure 4.4.

TABLE 4.5. Route of Mutation Along the Weighted Gradient Direction

	Iteration							
	1	2	3	4	5	6	7	8
x_1	5.885	14.128	17.315	24.004	22.939	21.058	18.207	19.873
x_2	5.901	11.396	13.521	17.980	17.181	16.241	14.815	15.926

From Table 4.5 and Figure 4.4 we see that this random point can quickly reach $\tilde{S}$, the fuzzy optimal solution, by mutations along the weighted gradient direction. Note that not every point can reach $\tilde{S}$—for example, the points in area D, as shown in Figure 4.4. They will die quickly and be buried in the other area by the genetic algorithm as long as the population size is large enough. For above objective or resource problem, the criteria 0,1, and 2 are of concern to the decision maker. He cares about the net profit and the production quantities of products A and B. The population size in 100, and the acceptable membership degree $\alpha = 0.3$. After 228 generations, the result of the genetic algorithm is as shown in Table 4.5.

From Table 4.6 we see that the polygon taking these solutions as vertices almost covers the cut set, $\tilde{S}_\alpha$. The optimal solution, $\boldsymbol{x}^*$, is very close to the

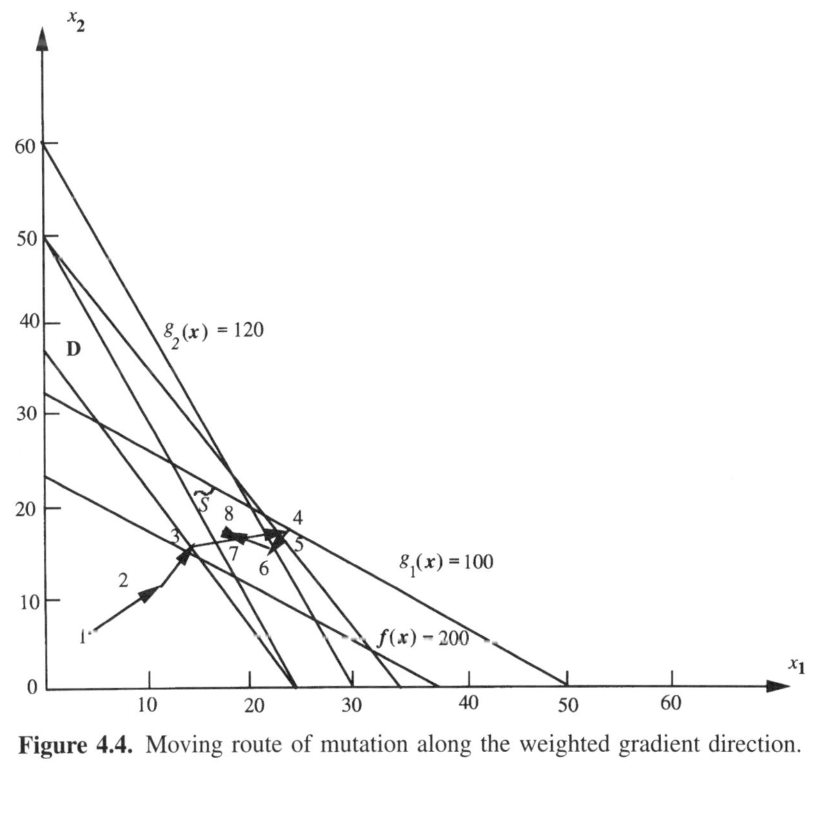

Figure 4.4. Moving route of mutation along the weighted gradient direction.

TABLE 4.6. Preferred Solutions for Selection

Solution	x_1	x_2	Objective Value	$\mu_{\tilde{S}}(x)$	Meaning
x^*	20.0001	14.9997	179.9991	0.49998	Exact optimal solution
$z_1(0)$	19.1533	14.2708	172.0026	0.30007	Minimum of objective
$z_2(0)$	20.0293	16.8886	187.7302	0.30528	Maximum of objective
$z_1(1)$	15.4291	19.9664	172.4402	0.30809	Minimum of product A
$z_2(1)$	25.6704	5.5495	176.2207	0.31096	Maximum of product A
$z_1(2)$	25.0125	5.4824	172.0044	0.30011	Minimum of product B
$z_2(2)$	15.4291	19.9664	172.4402	0.30809	Maximum of product B

exact optimal solution. The decision maker thinks that the solutions $z_2(0)$ and $z_2(1)$ are better. It means that he hopes to achieve a big profit and more product A. He selects the labor constraint as the comparison criterion for $z_2(0)$ and $z_2(1)$. The decision maker finds that there is no overtime if the combination coefficient $\lambda = 0.9116$. Then he selects the combined solution as

$$z = \lambda z_2(1) + (1 - \lambda)z_2(0) = [25.1715, 6.5523]^T$$

For the combined solution, z, the objective value, the membership function, and constraints are as follows:

$$f(z) = 177.2383, \qquad \mu_{\tilde{S}}(z) = 0.31047$$

$$g_1(z) = 70, \qquad g_2(z) = 113.7907$$

The decision maker is satisfied with this solution, so it is taken as the final decision for the production planning for the coming month.

4.3 FUZZY NONLINEAR PROGRAMMING

The research on fuzzy mathematical programming is largely limited to the range of linear programming [408, 519, 613, 655] and multiobjective programming [92,553], while fuzzy nonlinear programming [616], including fuzzy quadratic programming [616], is rarely involved. In many actual problems, however, especially in complex industrial systems, there exist many kinds of fuzzy nonlinear production planning and scheduling problems. They cannot be described as scheduling models. Thus research on modeling and optimization methods for nonlinear programming under fuzzy environments is not important only in fuzzy optimization theory but it also valuable in production planning and scheduling problems.

In this section we introduce an inexact method to solve a type of fuzzy nonlinear programming problem with fuzzy objective and resource constraints. Instead of finding an exact optimal solution, this approach aims at finding a neighboring domain of optimal solutions such that every solution in the neighboring domain can be acceptable: It is an optimal solution under the fuzzy environment. A special genetic algorithm with mutations along the weighted gradient direction is suggested to find a family of solutions with acceptable membership degree, which is desired by the decision maker, and then by means of human–computer interaction, preferred solutions under different types of criteria can be achieved.

4.3.1 Nonlinear Programming Model

Generally speaking, nonlinear programming problems with resource constraints have the following general form:

$$\begin{aligned} \max \quad & f(\boldsymbol{x}) \\ \text{s.t.} \quad & g_i(\boldsymbol{x}) \le b_i, \qquad i = 1,2,\ldots,m \\ & \boldsymbol{x} \ge 0 \end{aligned} \tag{4.22}$$

where b_i are crisp available resources.

In actual production planning problems, however, the available quantity of a resource in a period is uncertain and possesses different types of fuzziness. Also, the objective defined by the decision maker is an ill-defined goal, or there are some fuzzy parameters in the objective function due to a lack of full understanding of the problem or of the actual maximization and minimization limits. We discuss nonlinear programming problems with the following types of fuzzy objective and fuzzy resource constraints.

1. The objective value that the decision maker desires is not an actual maximum but a fuzzy value. The decision maker aspires to reach a level such as z_0, and not less than a lowest level z_0-p_0. Moreover, the decision maker's satisfaction degree increases as the objective value increases.
2. The available quantity of a type of resource $i(i = 1,2,\ldots,m_1)$ has some increments that are accepted by the decision maker by way of taking overtime work, putting to use the inventory quantity, and so on. Assuming that the planned available quantity of this type of resource i is b_i $(i = 1,2,\ldots,m_1)$, the largest increment acceptable to the decision maker is p_i, and the fuzzy available quantity is denoted by $\tilde{b}_i$. Moreover, it is attached to a monotonously nonincreasing membership function. For the decision maker this type of resource is utilized no more than the availability.
3. The available quantity of another type of resource $i(i = m_1 + 1, m_1 + 2,\ldots,m)$ is imprecise. Assume that the available $\tilde{b}_i$ $(i = m_1 + 1, m_1 + 2,\ldots,m)$ of this type of resource i is an estimate with means b_i and errors

p_i^- and p_i^+, respectively, and has an L-R (left-right) type of membership function. For the decision maker this type of resource is utilized as fully as possible.

Further assume that the objective function $f(\boldsymbol{x})$ and the resource constraints functions $g_i(\boldsymbol{x})$ are nonlinear, continuous, and derivable in $(R^n)^+$.

The problem is how to make a reasonable plan such that the objective can be optimal or the decision maker is "most" satisfied with his or her preferred criteria under the environment of the foregoing fuzzy objective and resource constraints. This type of problem, which belongs to the set of fuzzy optimization problems, has the following general form:

$$\begin{aligned} \widetilde{\max}\quad & f(\boldsymbol{x}) \\ \text{s.t.}\quad & g_i(\boldsymbol{x}) \le \tilde{b}_i, \qquad i = 1,2,\ldots,m_1 \\ & g_i(\boldsymbol{x}) = \tilde{b}_i, \qquad i = m_1 + 1, m_1 + 2,\ldots,m \\ & \boldsymbol{x} \ge 0 \end{aligned} \tag{4.23}$$

where $\boldsymbol{x}$ is an n-dimensional decision variable vector, $\boldsymbol{x} = [x_1 \;\; x_2 \;\; \cdots \;\; x_n]^T$ and $\tilde{\boldsymbol{b}}$ is a fuzzy available resource vector, $\tilde{\boldsymbol{b}} = [\tilde{b},\tilde{b}_2,\ldots,\tilde{b}_m]^T$.

Wang and Tang introduce the following type of membership function to describe the fuzzy numbers $\tilde{b}_i$, fuzzy objective, and fuzzy constraints [616, 649, 650]. For resource i ($i = 1,2,\ldots,m_1$), let $\mu_{\tilde{b}_i}(\boldsymbol{x})$ indicate the attainability of a fuzzy available resource $\tilde{b}_i$ and define $\mu_{\tilde{\mathbf{b}}_i}(\boldsymbol{x})$ as follows:

$$\mu_{\tilde{\mathbf{b}}_i}(\boldsymbol{x}) = \begin{cases} 1, & g_i(\boldsymbol{x}) \le b_i \\ 1 - \left(\dfrac{g_i(\boldsymbol{x}) - b_i}{p_i}\right)^r, & b_i \le g_i(\boldsymbol{x}) \le b_i + p_i \\ 0, & g_i(\boldsymbol{x}) > b_i + p_i \end{cases} \tag{4.24}$$

where $r > 0$. $\mu_{\tilde{\mathbf{b}}_i}(\boldsymbol{x})$ is a monotonously nonincreasing function that denotes the attainable degree of fuzzy available resource $\tilde{b}_i$.

For resource i ($i = m_1 + 1, m_1 + 2,\ldots,m$), let $\mu_{\tilde{\mathbf{b}}_i}(\boldsymbol{x})$ be defined as

$$\mu_{\tilde{\mathbf{b}}_i}(\boldsymbol{x}) = \begin{cases} 0, & g_i(\boldsymbol{x}) \le b_i - p_i \\ 1 - \left(\dfrac{b_i - g_i(\boldsymbol{x})}{p_i}\right)^r, & b_i - p_i^- \le g_i(\boldsymbol{x}) \le b_i \\ 1 - \left(\dfrac{g_i(\boldsymbol{x}) - b_i}{p_i^+}\right)^r, & b_i \le g_i(\boldsymbol{x}) \le b_i + p_i^+ \\ 0, & g_i(\boldsymbol{x}) > b_i + p_i^+ \end{cases} \tag{4.25}$$

$\mu_{\tilde{b}_i}(\boldsymbol{x})$ is an L-R type of function that denotes that accurate level of estimation for fuzzy available resource $\tilde{b}_i$. Similarly, let $\mu_i(\boldsymbol{x})$ be the membership function of the ith fuzzy constraint defined as follows:

$$\begin{aligned}\mu_i(\boldsymbol{x}) &= \max_{y \geq g_i(\boldsymbol{x})} \{\mu_{\tilde{b}_i}(y)\} = \mu_{\tilde{b}_i}(g_i(\boldsymbol{x})), \qquad i = 1,2,\ldots,m_1 \\ &= \mu_{\tilde{b}_i}(g_i(\boldsymbol{x})), \qquad i = m_1 + 1,\ldots,m\end{aligned}$$

$\mu_i(\boldsymbol{x})$ reflects the degree of the decision maker's satisfaction with the ith fuzzy constraint by point $\boldsymbol{x}$.

Let $\mu_0(\boldsymbol{x})$ describe the fuzzy objective $\widetilde{\max}\ f(\boldsymbol{x})$, defined as follows:

$$\mu_0(\boldsymbol{x}) = \begin{cases} 0, & f(\boldsymbol{x}) \leq z_0 - p_0 \\ 1 - \left(\dfrac{z_0 - f(\boldsymbol{x})}{p_0}\right)^r, & z_0 - p_0 < f(\boldsymbol{x}) \leq z_0 \\ 1, & f(\boldsymbol{x}) \geq z_0 \end{cases} \tag{4.26}$$

$\mu_0(\boldsymbol{x})$ is a monotonously nondecreasing continuous function and denotes the degree of satisfaction with the fuzzy objective function at point x.

Certainly, the types of membership functions that describe the fuzzy objective and fuzzy resource constraints may be determined by the decision maker to be other types, such as exponential, logarithmic, and so on.

A reasonable solution to this problem will make the decision maker generally satisfied with the fuzzy objective and fuzzy constraints. There is a balance of satisfaction between the fuzzy objective and fuzzy constraints. A usual approach to handle this kind of trade-off is that the highest degree of satisfaction is given to what the decision maker is most concerned with and the remains are assigned to different degrees of satisfaction. The most concerned objective is to be maximized and the remains are kept above a definite degree of satisfaction. Then the fuzzy objective and resource nonlinear program (FO/RNP) can be formulated as the following three kinds of models.

1. Maximize the minimum satisfaction degree (model FO/RNP-1):

$$\begin{aligned} \max \quad & \alpha \\ \text{s.t.} \quad & \mu_0(\boldsymbol{x}) \geq \alpha \\ & \mu_i(\boldsymbol{x}) \geq \alpha, \qquad i = 1,2,3,\ldots,m \\ & \boldsymbol{x} \geq 0 \end{aligned} \tag{4.27}$$

2. Maximize the sum of the weighted satisfaction degrees (model FO/RNP-2):

$$\begin{aligned} \max \quad & \Sigma_{i=0}^{m} \beta_i \mu_i(\boldsymbol{x}) \\ \text{s.t.} \quad & \mu_0(\boldsymbol{x}) \geq \alpha_0 \\ & \mu_o(\boldsymbol{x}) \geq \alpha_0, \qquad i = 1,2,3,\ldots,m \\ & \boldsymbol{x} \geq 0 \end{aligned} \tag{4.28}$$

where α_0 is an acceptable satisfaction degree preset by the decision maker, and β_i $(i = 0,1,2,\ldots,m)$ are the weights for the objective and resources constraints, respectively. They reflect the relative importance of the objective and resources.

3. Maximize the decision target (model FO/RNP-3)

$$\begin{aligned} \max \quad & \text{target} \\ \text{s.t.} \quad & \mu_0(\boldsymbol{x}) \geq \alpha_0 \\ & \mu_i(\boldsymbol{x}) \geq \alpha_0, \qquad i = 1,2,3,\ldots,m \\ & \boldsymbol{x} \geq 0 \end{aligned} \tag{4.29}$$

where the target may be the objective function, the decision variable, the resource constraints, and so on, and α_0 is an acceptable satisfaction degree preferred by the decision maker.

Then the solution to FO/RNP may be transformed into solutions of FO/RNP-1, FO/RN-2, or FO/RNP-3 according to various types of criteria.

Definition 4.1. The fuzzy optimal solution of FO/RP is fuzzy set $\tilde{S}$ defined by

$$\tilde{S} = \{(\boldsymbol{x},\mu_{\tilde{S}}(\boldsymbol{x}))|\boldsymbol{x} \in (R^n)^+,\ \mu_{\tilde{S}}(\boldsymbol{x}) = \min\{\mu_0(\boldsymbol{x}),\mu_i(\boldsymbol{x})|i = 1,2,\ldots,m\}\} \tag{4.30}$$

Let $S_\alpha = \{\boldsymbol{x} \in (R^n)^+ | \mu_{\tilde{S}}(\boldsymbol{x}) \geq \alpha\}$, $\alpha \in [0,1]$. Then S_α is a general set that is an *α-level cut set* of $\tilde{S}$.

Definition 4.2. α^* is the best balance degree. If α^* is such that $0 \leq \alpha \leq \alpha^*$, S_α is nonempty; if $\forall \alpha > \alpha^*$, S_α is empty.

In general, a unique optimal solution α^* can be found from FO/RNP-1. The corresponding solution $\boldsymbol{x}^*$, which is not usually unique, is the solution with the highest membership degree in the FO/RNP. Its meaning is that the best balance of objective and constraints might be achieved at point $\boldsymbol{x}^*$. However, the solution $\boldsymbol{x}^*$ is probably not the one most desired by the decision maker under some types of criteria. In addition, the exact optimal solution is meaningless to the decision maker under a fuzzy environment. The solution desired by the decision maker is actually a multiplicity of solutions that satisfy both the objective and resource constraints under various criteria preferred by the decision maker. The inexact approach, based on the concept of fuzzy optimal solutions, aimed at finding a neighboring domain of the optimal solution such that every solution $\boldsymbol{x}$ in the domain can be the "optimal" solution desired by the decision maker under the fuzzy environment.

In the following sections, we introduce the inexact approach based on a genetic algorithm to solve the FO/RNP-1 model, and an overall procedure for a solution to FO/RNP by means of human–computer interaction, which may supply a preliminary frameworrk for practical application of the model FO/RNP.

4.3.2 Inexact Approach to FO/RNP-1

Obviously, FO/RNP-1 is equivalent to the following optimization problem P:

$$\max_{x \in (R_n)^+} \quad \mu_{\tilde{S}}(\boldsymbol{x}) = \max \min \{\mu_0(\boldsymbol{x}), \mu_1(\boldsymbol{x}), \mu_2(\boldsymbol{x}), \ldots, \mu_m(\boldsymbol{x})\} \tag{4.31}$$

P is an unconstrained optimization problem, but its objective function is not continuous and derivable. It cannot be solved by traditional optimization methods, but it may be solved by genetic algorithms [303, 455]. In this section we discuss a special genetic algorithm with mutation along the weighted gradient direction developed by Wang and Tang [616, 649, 650]. It uses the mutation as the main operator; the arithmetic combinatorial crossover operator is used only in the later generations. The basic idea is as follows. First, randomly produce an initial population with a size of *pop_size* individuals, with each individual selected to reproduce children with incremental membership degree of both fuzzy objectives and constraints. Individuals with membership degree less than α_0 (the acceptable membership degree) may have less chance than others of being selected to reproduce children in a later generation. As the generations increase, individuals with membership degree less than α_0 gradually die to and others exist. After a number of generations, all individuals will have membership degree greater than α_0, namely, $S_{\alpha k} \subseteq S_{\alpha 0}$, and most individuals will be close to the optimal solution.

For individual $\boldsymbol{x}$, let $\mu_{\min}(\boldsymbol{x}) = \min\{\mu_0(\boldsymbol{x}), \mu_1(\boldsymbol{x}), \ldots, \mu_m(\boldsymbol{x})\}$. If $\mu_{\min}(\boldsymbol{x}) \leq \mu_0(\boldsymbol{x}) < 1$, move along the gradient direction of $\mu_0(\boldsymbol{x})$. The value of $\mu_0(\boldsymbol{x})$ may be improved. The smaller $\mu_0(\boldsymbol{x})$ is, the greater an improvement of $\mu_0(\boldsymbol{x})$ may be achieved. Similarly, if $\mu_{\min}(\boldsymbol{x}) \leq \mu_i(\boldsymbol{x}) < 1$, move along the gradient direction of $\mu_i(\boldsymbol{x})$. The value of $\mu_i(\boldsymbol{x})$ may be improved. The smaller $\mu_i(\boldsymbol{x})$ is, the greater the improvement in $\mu_i(\boldsymbol{x})$ that may be achieved. Based on the idea above, construct the direction as follows:

$$D(\boldsymbol{x}) = w_0 \nabla \mu_0(x) + \sum_{i=1}^{m} w_i \nabla \mu_i(\boldsymbol{x}) \tag{4.32}$$

where for $0 < \mu_0(\boldsymbol{x}) \leq 1$, $0 < \mu_i(\boldsymbol{x}) \leq 1$,

$$\begin{aligned}
\nabla \mu_0(\boldsymbol{x}) &= r(z_0 - f(\boldsymbol{x}))^{r-1} \frac{\nabla f(\boldsymbol{x})}{p_0^r} \\
\nabla \mu_i(\boldsymbol{x}) &= -r(g_i(\boldsymbol{x}) - b_i)^{r-1} \frac{\nabla g_i(\boldsymbol{x})}{p_i^r}, \qquad i = 1,2,\ldots,m_1 \\
\nabla \mu_i(\boldsymbol{x}) &= -r(g_i(\boldsymbol{x}) - b_i)^{r-1} \, sgn\{g_i(\boldsymbol{x}) - b_i\} \frac{\nabla g_i(\boldsymbol{x})}{q_i^r}, \\
& \quad i = m_1 + 1,\ldots,m
\end{aligned} \tag{4.33}$$

for $\mu_0(\boldsymbol{x}) = 0$, $\nabla \mu_0(\boldsymbol{x}) = \nabla f(\boldsymbol{x})$; for $\mu_i(\boldsymbol{x}) = 0$, $\nabla \mu_i(\boldsymbol{x}) = \nabla g_i(\boldsymbol{x})$.

$$q_i = \begin{cases} p_i^-, & g_i(x) \leq b_i \\ p_i^+, & g_i(x) \geq b_i \end{cases} \tag{4.34}$$

$D(\boldsymbol{x})$ is called the weighted gradient direction of $\mu_{\tilde{S}}(\boldsymbol{x})$. $sgn(\boldsymbol{x})$ is the sign function, and w_i is the gradient direction weight defined as follows:

$$w_i = \begin{cases} 0, & \mu_i = 1 \\ \dfrac{1}{\mu_i - \mu_{\min} + e}, & \mu_{\min} \leq \mu_i < 1 \end{cases} \tag{4.35}$$

where e is a sufficiently small positive number and $1/e$ is the largest weight.

The child $\boldsymbol{x}_j^{k+1}$ generated from $\boldsymbol{x}_i^k$ by mutation along the weighted gradient direction $D(\boldsymbol{x})$ can be described as follows:

$$\boldsymbol{x}_j^{k+1} = \boldsymbol{x}_i^k + \beta^k D(\boldsymbol{x}_i^k) \tag{4.36}$$

where β^k is a random step length of the Erlang distribution generated by a random number generator with declining means.

The membership degree $\mu_{\tilde{S}}(\boldsymbol{x}_j)$ is calculated as follows:

$$\mu_{\tilde{S}}(\boldsymbol{x}_j) = \begin{cases} \mu_{\min}(\boldsymbol{x}_j), & \text{if } \mu_{\min} \geq \alpha_0 \\ \epsilon \, \mu_{\min}(\boldsymbol{x}_j), & \text{else} \end{cases} \tag{4.37}$$

where α_0 is the acceptable satisfaction degree preferred by the decision maker, and $\epsilon \in \text{U}(0,1)$.

From the formulas of weight and mutation, equations (4.32) to (4.36), we can know that the constraints that cannot be met have a minimum membership of degree 0 and will get the largest weight $1/e$, so the mutation will lead the individual to the feasible area. When $\mu_{\tilde{S}}(\boldsymbol{x})$ is greater than 0, an objective or constraint with minimum membership degree than gets the largest weight. The weighted gradient direction will improve the minimum, which results in an improvement of $\mu_{\tilde{S}}(\boldsymbol{x})$. From equation (4.37) we may find that the $\boldsymbol{x}_j \notin S_{\alpha 0}$, we give them a lower membership degree (but nonzero), so that they have a small chance of being selected as parents to reproduce children. Therefore, the weighted gradient direction will lead all individuals to be close to the exact optimal solution, and the individuals form a neighboring domain, including the exact optimal solution.

In Wang and Tang's algorithm, mutation along the weighted gradient direction is the main operator, while the arithmetic combinatorial crossover operator is used only in later generations. The proportional selection strategy is adopted to select new population at each generation.

4.3.3 Interactive Approach

Wang and Tang designed an interactive procedure to help a decision maker to select a solution from the fuzzy optimal solution (i.e., the neighboring domain of the optimal solution) under different criteria. It can be described as follows [616, 649, 650]. First, the approach asks the decision maker the acceptable membership degree of the fuzzy optimal solution, α_0. Second, by means of interaction with a computer, the prepared membership degree that describes the fuzzy objective and resource constraints is displayed for the decision maker to select the type of membership degree. It then elicits the decision maker to find his or her preferred structure and point out which criteria are of most concern. The criteria may include the best balanced degree of the objective and resource constraints (FO/RNP-1), the sum of weighted satisfaction degree (FO/RNP-2), and the objective, decision variables, or resources (FO/RNP-3). Solutions in the α-level cut set of the fuzzy optimal solution by a genetic algorithm with mutation along the weighted gradient direction and with the highest and lowest values of these criteria are updated in each generation. When the iteration terminates, solutions with their criteria values will be shown for the decision maker to select. The step-by-step procedure can be described as follows:

Procedure: Overall Procedure for FO/RNP

Step 1. Initialize.

(*1.1*) Input an acceptable satisfaction degree α_0 and the largest *max_gen* and *pop_size*.

(*1.2*) Input the decision maker's set of criteria of concern, $C_I = \{0,1,2,...,n,n+1,...,n+m,n+m+1\}$, where 0 stands for the objective function, $j = 1,2,...,n$ for the decision variables, $j = n+1,n+2,...,n+m$ for the constraints, and $j = n+m+1$ for the sum of weighted satisfaction degree, respectively. Give the initial values and the upper and lower values of criteria r, $r \in C_i$.

(*1.3*) Input the type of membership function that describes the fuzzy objective and fuzzy constraints.

(*1.4*) Input the weights of objective and resource constraints β_i.

Step 2. Randomly produce the initial population and calculate their membership degrees by $x_i(j) = \xi \boldsymbol{x}_i^{\mathrm{UP}}$, where $\xi \in \mathrm{U}(0,1)$, $i = 1,2,...,n$, $j = 1,2,...,pop\_size$ and $\boldsymbol{x}_i^{\mathrm{UP}}$ is the upper bound of the ith element of variable x. The membership function $\mu_{\tilde{S}}(\boldsymbol{x}_j)$ is calculated with equation (4.37).

Step 3. Set iteration index $k = 1$.

Step 4. For individual $j(j = 1,2,...,pop\_size)$, calculate the fitness functions $F(j)$ and selection probabilities $P(j)$ as follows:

$$F(j) = \mu_{\tilde{S}}(\boldsymbol{x}_j) + e, \qquad P(j) = \frac{F(j)}{\sum_{i=1}^{pop\_size} F(i)}$$

Step 5. Produce new individuals x_j with the parent x_i as $x_j^k = x_i^{k-1} + \beta^k D(x_i^{k-1})$, where β is a random step length of the Erlang distribution with declining means generated by a random number generator, and $D(x)$ is defined by equations (4.32) to (4.35).

Step 6. For individual j, calculate the membership function $\mu_{\tilde{S}}(x_j)$ with equation (4.37), and update optimal membership degree $\mu_{\max}$ and the upper and lower values of criteria r.

Step 7. Set $k + 1 \rightarrow k$; if $k \leq max\_gen$, go to step 4; otherwise, go to step 8.

Step 8. Output the optimal membership degree $\mu_{\max}$ and the upper and lower values of criteria preferred by the decision maker, then stop.

4.3.4 Numerical Example

Wang and Tang used the following instance to test their method [616, 649, 650].

Example 4.4 A manufacturing factory is going to produce two types of product A and B, in a period such as 1 month. The production of A and B requires three types of resources: R_1, R_2, and R_3. The requirements of each type of resource to produce each product A are 2, 4, and 3 units, respectively. To produce each product B, 3, 2, and 2 units are required, respectively. The available resources of R_1 and R_2 are 50 and 44 units, respectively. But there are an additional 30 and 20 units of safety stores of material that are administered by the general manager. The estimated value of the quantity of resource R_3 available in 36 units, and the estimated error is 5 units. Assume that the planned production quantities of A and B are x_1 and x_2, respectively. Further assume that the unit costs and sale prices of products A and B are $\text{UC}_1 = c_1$, $\text{UC}_2 = c_2$ and $\text{US}_1 = k_1/x_1^{1/\alpha_1}$, $\text{US}_2 = k_2/x_2^{1/\alpha_2}$, respectively [650]. Then the decision maker hopes that the total profits reach an aspiration level z_0 but not less than a lower level $z_0 - p_0$. This is a typical FO/RNP problem. It can be described as follows:

$$\max \quad \tilde{f}(\boldsymbol{x}) = k_1 x_1^{1-1/\alpha_1} - c_1 x_1 + k_2 x_2^{1-1/\alpha_2} - c_2 x_2$$

$$\text{s.t.} \quad 2x_1 + 3x_2 \leq \widetilde{50}, \qquad 4x_1 + 2x_2 - \widetilde{44}$$

$$3x_1 + 2x_2 = \widetilde{36}, \qquad x_2, x_2 \geq 0$$

$$p_1 = 30, \quad p_2 = 20, \quad p_3^- = p_3^+ = 5.0, \quad r = 1, \quad k_1 = 50,$$

$$c_1 = 8.0, \quad k_2 = 45, \quad c_2 = 10, \quad a_1 = a_2 = 2$$

For the FO/RNP above, the decision maker cares about net profit (objective), production quantities of products A and B, and consumption quantities of resources R_1, R_2, and R_3. Set *pop_size* = 80. The results under various criteria after 52 generations are shown in Table 4.7. The maximum satisfaction degree of the generation is shown in Table 4.8. The results are generated by Wang and Tang's inexact approach, combining a genetic algorithm with an interactive procedure.

From Table 4.7 the decision maker may choose an appropriate solution. For example, he or she may choose a solution with maximum objective and the best balance between objective and all kinds of resources. From Table 4.8 we may find that individuals of each iteration can fast reach $\tilde{S}$, the fuzzy optimal solution. After only 18 generations, they are very close to the exact optimal solution (9.76222, 5.06271) with the best balance degree $\alpha^* = 0.29167$.

4.4 FUZZY NONLINEAR MIXED-INTEGER GOAL PROGRAMMING

The ideal of goal programming was developed by Charnes and Cooper in the 1960s. Since then, many researchers in various fields have devoted their efforts to this technique, such as Ijiri [316], Lee [403], Ignizio [315], and Gen and Ida, [222, 236]. In this section we attempt to apply genetic algorithms to the series–parallel system reliability optimization problem represented by

TABLE 4.7. Results Under Various Criteria as $z_0 = 150$, $z_0 - p_0 = 120$, and $\alpha = 0.25$

Criteria	x_1	x_2	$f(x)$	$\mu_{\tilde{S}}(x)$	Resource	Meaning
α^*	9.76222	5.06271	128.75000	0.29167		Optimal membership degree
0(1)	7.88529	6.00671	127.54300	0.25143		Minimum objective
0(2)	9.76222	5.06271	128.75000	0.29167		Maximum objective
1(1)	7.69458	4.80628	127.73060	0.25769		Minimum product A
1(2)	10.64834	3.80453	127.70040	0.25668		Maximum product A
2(1)	8.89719	3.76743	127.63340	0.25445		Minimum product B
2(2)	8.63742	6.40122	127.68870	0.25629		Maximum product B
3(1)	8.89719	3.76743	127.63340	0.25445	29.09666	Minimum resource R_1
3(2)	8.89301	6.31004	127.90030	0.26334	36.71612	Maximum resource R_1
4(1)	7.69458	4.80628	127.73060	0.25769	40.39088	Minimum resource R_2
4(2)	10.64834	3.80453	127.70040	0.25668	50.20244	Maximum resource R_2
5(1)	7.69458	4.80628	127.73060	0.25769	32.69630	Minimum resource R_3
5(2)	9.78006	5.20287	128.74040	0.25082	39.74592	Maximum resource R_3

TABLE 4.8. Maximum Satisfaction Degree of Each Iteration as $z_0 = 150$, $z_0 - p_0 = 120$, and $\alpha_0 = 0.25$

Iteration Number	Individual Number	$\mu_{\tilde{S}}(x)$	Iteration Number	Individual Number	$\mu_{\tilde{S}}(x)$	Iteration Number	Individual Number	$\mu_{\tilde{S}}(x)$
1	66	0.26351	19	51	0.29167	37	67	0.29167
2	66	0.26351	20	51	0.29167	38	80	0.29167
3	66	0.26351	21	51	0.29167	39	80	0.29167
4	28	0.26393	22	51	0.29167	40	80	0.29167
5	28	0.26393	23	68	0.29167	41	78	0.29167
6	76	0.28529	24	68	0.29167	42	78	0.29167
7	76	0.28529	25	68	0.29167	43	80	0.29167
8	17	0.28894	26	75	0.29167	44	80	0.29167
9	17	0.28894	27	76	0.29167	45	80	0.29167
10	17	0.28894	28	71	0.29167	46	80	0.29167
11	47	0.28955	29	71	0.29167	47	80	0.29167
12	30	0.29081	30	76	0.29167	48	80	0.29167
13	30	0.29081	31	72	0.29167	49	80	0.29167
14	48	0.29150	32	67	0.29167	50	80	0.29167
15	48	0.29150	33	68	0.29167	51	80	0.29167
16	7	0.29164	34	74	0.29167	52	80	0.29167
17	76	0.29164	35	78	0.29167			
18	62	0.29167	36	79	0.29167			

fuzzy nonlinear mixed-integer goal programming problems (f-nMIGP), which involve imprecise nonlinear mixed-interger information. The main idea for solving the system reliability optimization problem formulated by fuzzy nonlinear mixed-integer goal programming is to transform the original problem into the nonlinear mixed-integer programming problem and solve the problem using the genetic algorithm. The reliability design problem with fuzzy goals in Section 5.2 is a typical actual fuzzy nonlinear mixed-integer goal programming problem. The problem is also a multiobjective optimization problem.

4.4.1 Fuzzy Nonlinear Mixed-Integer Goal Programming Model

A typical formulation of fuzzy nonlinear mixed-integer goal programming (f-nMIGP) can be defined as follows. Find $\boldsymbol{x}^i$ and $\boldsymbol{x}^r$ to optimize the next fuzzy goals subject to m nonlinear system constraints:

$$\begin{aligned}
&f_k(\boldsymbol{x}^i,\boldsymbol{x}^r) && \lesssim b_k, && k = 1,2,\ldots,q_0 \\
&f_k(\boldsymbol{x}^i,\boldsymbol{x}^r) && \gtrsim b_k, && k = q_0 = 1,\ldots,q_1 \\
&f_k(\boldsymbol{x}^i,\boldsymbol{x}^r) && \approxeq b_k, && k = q_1 + 1,\ldots,q_2 \\
&\text{s.t.} \quad g_c(\boldsymbol{x}^i,\boldsymbol{x}^r) && \le G_c, && c = 1,2,\ldots,m
\end{aligned} \tag{4.38}$$

where $\boldsymbol{x}^i$ is an n-dimensional integer vector; $\boldsymbol{x}^r$ is an n-dimensional real vector; the symbol $\lesssim$ (fuzzy-min), referring to approximately less than or equal to the aspiration level b_k, signifies that the decision maker is satisfied even if greater than b_k up to a certain tolerance limit; the symbol $\gtrsim$ (fuzzy-max), referring to approximately greater than or equal to the aspiration level b_k, signifies that the decision maker is satisfied even if less than b_k up to a certain limit; the symbol $\approxeq$ (fuzzy-equal), meaning that $f_k(\boldsymbol{x}^i,\boldsymbol{x}^r)$ should be in the vicinity of the aspiration b_k, signifies that the decision maker is satisfied even if greater than (or less than) b_k up to a certain limit; $f_k(\boldsymbol{x}^i,\boldsymbol{x}^r)$ is the kth nonlinear goal constraint, $g_c(\boldsymbol{x}^i,\boldsymbol{x}^r)$ is the cth nonlinear system constraint; b_k is the target value according to goal k; G_c is the available resource of system constraint c; q_0 is the number of fuzzy-min goal constraints; $q_1 - q_0$ is the number of fuzzy-max constraints; $q_2 - q_1$ is the number of fuzzy-equal constraints; and m is the number of system constraints.

The membership functions of the fuzzy goals can be defined as follows: For the fuzzy-equal case,

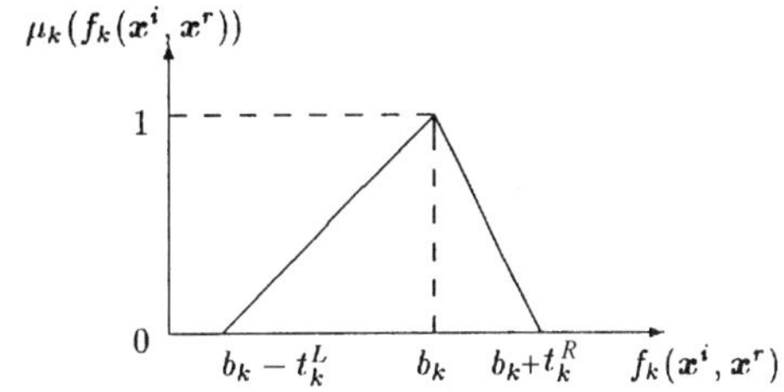

Figure 4.5. Membership function of fuzzy-equal.

$$\mu_k(f_k(x^i,x^r)) = \begin{cases} 0, & \text{if } f_k(x^i,x^r) < b_k - t_k^L \\ \dfrac{f_k(x^i,x^r) - (b_k - t_k^L)}{t_k^L}, & \text{if } b_k - t_k^L \le f_k(x^i,x^r) \le b_k \\ 1, & \text{if } f_k(x,^i x^r) = b_k \\ 1 - \dfrac{f_k(x^i,x^r) - b_k}{t_k^R}, & \text{if } b_k \le f_k(x^i,x^r) \le b_k + t_k^R \\ 0, & \text{if } f_k(x^i,x^r) > b_k + t_k^R \end{cases} \tag{4.39}$$

For the fuzzy-min case,

$$\mu_k(f_k(x^i,x^r)) = \begin{cases} 1, & \text{if } f_k(x^i,x^r) < b_k \\ 1 - \dfrac{f_k(x^i,x^r) - b_k}{t_k^R}, & \text{if } b_k \le f_k(x^i,x^r) \le b_k + t_k^R \\ 0, & \text{if } f_k(x^i,x^r) > b_k + t_k^R \end{cases} \tag{4.40}$$

For the fuzzy max case,

$$\mu_k(f_k(x^i,x^r)) = \begin{cases} 0, & \text{if } f_k(x^i,x^r) < b_k - t_k^L \\ \dfrac{f_k(x^i,x^r) - (b_k - t_k^L}{t_k^L}, & \text{if } b_k - t_k^L\, f_k(x^i,x^r) \le b_k \\ 1, & \text{if } f_k(x^i,x^r) > b_k \end{cases} \tag{4.41}$$

Membership functions are strictly monotonously decreasing (or increasing) and continuous functions with respect to $f_k(x^i,x^r)$, while t_k^L is a maximal left tolerance limit to b_k and t_k^R is a maximal right tolerance limit to b_k (Figures 4.5 to 4.7). Then formulation (4.38) is transformed into the following nonlinear mixed-integer programming (nMIP) problem:

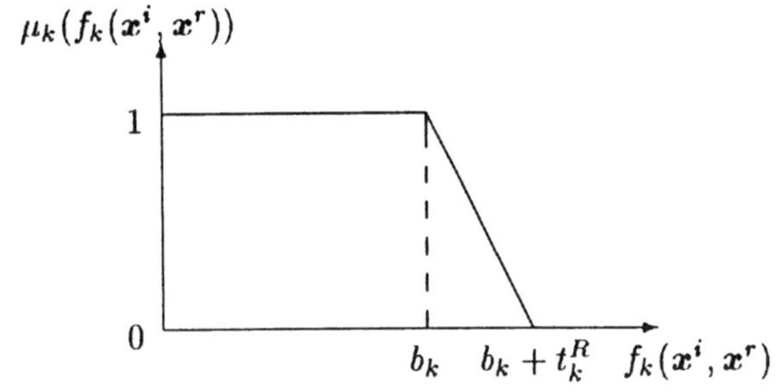

Figure 4.6. Membership function of fuzzy-min.

$$
\begin{aligned}
\max \quad & \sum_{k=1}^{q2} w_k \lambda_k \\
\text{s.t.} \quad & \lambda_k = \mu_k(f_k(\mathbf{x}^i, \mathbf{x}^r)), \qquad k = 1,2,\ldots,q_2 \\
& g_c(\mathbf{x}^i, \mathbf{x}^r) \leq G_c, \qquad c = 1,2,\ldots,m \\
& 0 \leq \lambda_k \leq 1, \qquad k = 1,2,\ldots,q_2
\end{aligned}
\tag{4.42}
$$

where w_k is the suitable weight factor that is assigned by the decision maker. One of the major difficulties is for the decision maker to set the relative importance of goals correctly.

4.4.2 Genetic Algorithm Approach

Gen, Ida, and Kim proposed a genetic algorithm for solving series–parallel system reliability optimization problems represented by the f-nMIGP model [224, 230].

Representation and Initialization. Let V denote the chromosomes in a population. The chromosomes can be denoted as follows:

$$
V = [(x_1^i, x_1^r) \quad (x_2^i, x_2^r) \quad \cdots \quad (x_n^i, x_n^r)]
$$

This representation of chromosomes is convenient to manipulate f-n MIGP

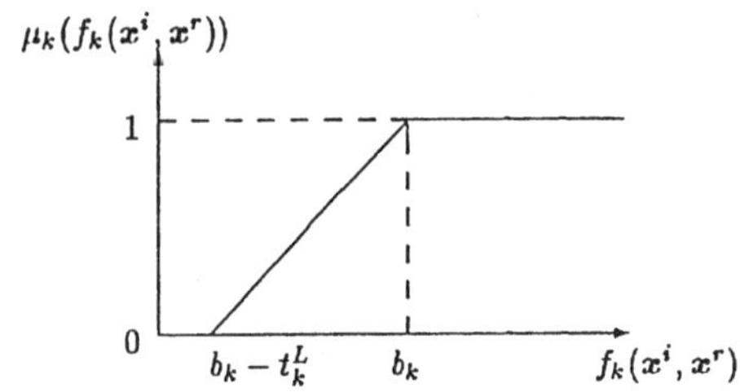

Figure 4.7. Membership function of fuzzy-max.

with n-dimensional mixed-integer decision vectors. Define *pop_size* as the number of chromosomes. Initial chromosomes are randomly produced within their domain.

Evaluation. The evaluation function is defined with consideration of weight factors as follows:

$$\text{eval}(\boldsymbol{V}_p) = \sum_{k=1}^{q_2} w_k \lambda_k, \qquad p = 1,2,\ldots,pop\_size$$

where λ_k are given by problem (4.42) and w_k are weight factors based on the relative importance of goals (or objectives) and assigned by the decision maker. With the following equations, we can keep the best chromosome $\boldsymbol{V}^*$ with the largest fitness value at each generation:

$$\boldsymbol{V}^* = \text{argmax}\{\text{eval}(\boldsymbol{V}_p) | p = 1,2,\ldots,pop\_size\}$$

Selection. The selection method combines the roulette wheel select and elitist approaches. *Roulette wheel selection* is a method to reproduce a new generation proportional to the fitness of each individual, and the *elitist* method is employed to preserve the best chromosome for the next generation and overcome the stochastic errors of sampling. The selection procedure is described below.

Step 1. Calculate a cumulative probability a_p for each chromosome $\boldsymbol{V}_p$ ($p = 1,2,\ldots,pop\_size$).

Step 2. Generate a random real number r in $[0,1]$.

Step 3. If $r \leq a_1$, select the first chromosome, $\boldsymbol{V}_1$; otherwise, select the pth chromosome, $\boldsymbol{V}_p (2 \leq p \leq pop\_size)$, such that $a_{p-1} < r \leq a_p$.

Step 4. Repeat Steps 2 and 3 *pop_size* times and obtain *pop_size* copies of chromosomes.

Step 5. If the best chromosome is not selected in the next generation, replace one chromosome randomly from the new population by the best one.

Crossover. An arithmetical crossover is used, which is defined as a convex combination of two chromosomes [224]. When the constraint set is convex, the arithmetical crossover operation ensures that if both parents are feasible, both children are feasible. The parameter of probability of crossover p_C defines the expected number $p_C \cdot pop\_size$ of chromosomes that undergo the crossover operation.

There are two types of variables in a chromosome for this problem due to its mixture nature: the real variable and the integer variables. Obviously, the arithmetical crossover will result in real numbers for the integer variables. We simply truncate the decimal part of the real numbers resulting from the crossover operation and just take its integer part.

Mutation. Uniform mutation is used. The parameter of probability of mutation p_M defines the expected number $p_M \cdot pop\_size$ of chromosomes that undergo the mutation operation. Generating a random number r in [0,1], we select the chromosome for mutation if $r < p_M$. For a chosen parent $\boldsymbol{V}$, if its element x_j is randomly selected for mutation, the resulting offspring is $\boldsymbol{V}' = [x_1 \cdots x_j' \cdots x_n]$, where x_j' is a random (uniform probability distribution) value from $[x_j^L, x_j^U]$. If x_j' is an integer variables, the mutation returns an integer random number. This operator ensures that the genetic algorithm can search freely throughout the search space from start to finish, but has the shortcoming of divergence at later steps.

Cauchy's Method. Gen, Ida, and Kim employed Cauchy's method, also called the steepest descent method, as a local optimization method, to improve the performance of the genetic algorithm [224]:

$$\boldsymbol{V}' = \boldsymbol{V} + \alpha \cdot \nabla f(\boldsymbol{V})$$

where α, the step-length parameter, is a positive number, $\nabla f(\boldsymbol{V})$ is a gradient direction vector; and $\boldsymbol{V}'$ is the chromosome after the selection operation.

The parameter α is calculated by the following method. For different cases there are different directions $\nabla f(\boldsymbol{V})$; if $\boldsymbol{V} + \alpha \cdot \nabla f(\boldsymbol{V})$ is not feasible for the inequality constraints, set α as a random number between 0 and α until it is feasible. If a feasible solution cannot be found using the process above in a predetermined number of interactions, take $\alpha = 0$. For the maximizing objective function it may be better to choose the gradient direction; for the minimizing objective function, it may be better to choose the negative of the gradient direction.

Because the problem has several objective functions, the gradient direction vector is calculated using the following equation:

$$\nabla f(\boldsymbol{V}) = \sum_{i=1}^{n} w_i \cdot \nabla f_i(\boldsymbol{V})$$

where n is the number of objective functions.

Overall Procedure. The algorithm for f-nMIGP problems is shown as follows:

Step 1. Set the parameters. Set the population size (*pop_size*), mutation rate (p_M), crossover rate (p_C), maximum generation (*max_gen*), and initialize generation number $t = 0$.

Step 2. Initialize. Randomly produce an initial population V_p ($p = 1,2,\ldots,pop\_size$).

Step 3. Perform the crossover. Generate a random number r in [0,1]. Select the given chromesome for crossover if $r < p_C$. Repeat this operation *pop_size* times. For each pair of parents $(\boldsymbol{V}_1,\boldsymbol{V}_2)$, the crossover operation operator will produce the following two offspring $(\boldsymbol{V}'_1,\boldsymbol{V}'_2)$:

$$\boldsymbol{V}'_1 = c_1 \cdot \boldsymbol{V}_1 + c_2 \cdot \boldsymbol{V}_2$$
$$\boldsymbol{V}'_2 = c_2 \cdot \boldsymbol{V}_1 + c_1 \cdot \boldsymbol{V}_2$$

where $c_1 + c_2 = 1$ and c_1 is a random number in [0,1].

Step 4. Perform the mutation. Generate a random number r in [0,1]. Select the chromosome $\boldsymbol{V}$ if $r < p_M$; repeat this operation *pop_size* times. Randomly select the element x_j of a given parent $\boldsymbol{V}$. Mutate x_j using the uniform mutation method.

Step 5. Calculate the fitness value. Set $t = t + 1$, and calculate eval$(\boldsymbol{V}_p)(p = 1,2,\ldots,pop_size)$ for each chromosome.

Step 6. Select *pop_size* chromosomes for the next generation according to the *roulette wheel* and *elitist* selection methods.

Step 7. Apply Cauchy's method. Use the following equation to move offspring to their local optima before injecting them into the next generation:

$$\boldsymbol{V}'_p = \boldsymbol{V}_p + \alpha \cdot \nabla f(\boldsymbol{V}_p), \qquad 1,2,\ldots,pop\_size$$

Step 8. Perform the terminating test. If $t < max\_gen$, go to step 3 for the next generation. If $t = max\_gen$, terminate.

4.4.3 Numerical Examples

Gen, Ida, and Kim used the following three instances to demonstrate the effectiveness of their method [224].

Example 4.5. This problem includes not only reliability maximization but also cost and weight minimization. It is formulated as follows: Find $\boldsymbol{x}^i$ and $\boldsymbol{x}^r$ to optimize the next fuzzy goals

$$f_1(\boldsymbol{x}^i,\boldsymbol{x}^r) = \prod_{j=1}^{4} \{1 - (1 - x_j^r)x_j^i\} \gtrsim b_1$$

$$f_2(\boldsymbol{x}^i,\boldsymbol{x}^r) = \sum_{j=1}^{4} C(x_j^r)\left[x_j^i + \exp\left(\frac{x_j^i}{4}\right)\right] \lesssim b_2$$

$$f_3(\boldsymbol{x}^i) = \sum_{j=1}^{4} d_j x_j^i \exp\left(\frac{x_j^i}{4}\right) \lesssim b_3$$

$$\text{s.t.} \quad g_1(\boldsymbol{x}^i) = \sum_{j=1}^{4} v_j (x_j^i)^2 \le V$$

$$1 \le x_j^i \le 10, \text{ positive integer}, \qquad j = 1,\ldots,4$$

$$0.5 \le x_j^r \le 1 - 10^{-6}, \text{ real number}, \qquad j = 1,\ldots,4$$

where x_j^i is the number of redundant components of subsystem j, x_j^r the reliability of the component of subsystem j, v_j the product of weight and volume per element of subsystem j, d_j the weight of each component of the subsystem j, F_1 is equal to $b_1 - t_1^L$, F_2 is equal to $b_2 + t_2^R$, F_3 is equal to $b_3 + t_2^R$, and $C(x_j^r)$ is the cost of each component with reliability x_j^r of subsystem j as follows:

$$C(x_j^r) = \alpha_j \left(\frac{-t_O}{\ln(x_j^r)}\right)^{\beta_j}, \qquad j = 1,\ldots,4$$

where α_j and β_j are constants representing the physical characteristic of subsystem j, and t_O is the operating time during which the component must not fail.

Dhingra's design data for this problem are given in Table 4.9 [162]. After parameter tuning, as shown in Figure 4.8 and Table 4.9, by running the genetic algorithm several times, the reasonable parameters setting are *pop_size* = 20, $p_c = 0.4$, $p_m = 0.1$, and $T = 2000$. Also, we set the parameters $t_1^L = 0.25$, $t_2^R = 300$, $t_3^R = 360$ and selected the relative weights of objectives (0.5, 0.25, 0.25) in order to give the first objective twice as much importance as other objectives. The evolutionary process with the different pop_sizes 10, 20, and 30 is shown in Figure 4.9.

The comparison results between Gen et al.'s and Yokota et al.'s methods [684] are shown in Figure 4.10 and Table 4.10. From the comparison we can see that Gen et al.'s method outperforms Yokota et al.'s method. Figure 4.10 illustrates that Gen et al.'s method is better than Yokotas' method, in convergence to the best compromise solution. But Gen et al.'s method has a longer average computing time than Yokota et al.'s method.

The best solution is found in the 1923th generation with objectives $f_1(\boldsymbol{x}^i, \boldsymbol{x}^r) = 0.982335$, $f_2(\boldsymbol{x}^i,\boldsymbol{x}^r) = 135.001259$, $f_3(\boldsymbol{x}^i) = 147.048541$, and solution

TABLE 4.9. Design Data of Example 4.5

Number of Subsystems				4
Limit on F_1				0.75
Limit on F_2				400.0
Limit on F_3				500.0
Upper limit on V				250.0
Operating time t_O				1000 h
Subsystem	$10^5 \cdot \alpha_j$	β_j	v_j	d_j
1	1.0	1.5	1	6
2	2.3	1.5	2	6
3	0.3	1.5	3	8
4	2.3	1.5	2	7

[(3,0.853070), (3,0.821421), (2,0.939903), (3,0825620)]. The optimal solution in [162] has $f_1(\boldsymbol{x}^i,\boldsymbol{x}^r) = 0.94478$, $f_2(\boldsymbol{x}^i,\boldsymbol{x}^r) = 104.472$, and $f_3(\boldsymbol{x}^i) = 128.727$ with [(2,0.86504), (3,0.81814), (2,0.84253), (3,0.80645)].

Example 4.6. This is a variation of the system reliability optimization problem with a five-subsystem series–parallel system and is another example of a nonlinear mixed-integer programming problem. It can be formulated as follows: Find $\boldsymbol{x}^i$ and $\boldsymbol{x}^r$ to optimize the fuzzy goals

$$f_1(\boldsymbol{x}^i,\boldsymbol{x}^r) = \prod_{j=1}^{5} \{1 - (1 - x_j^r)x_j^i\} \gtrsim b_1$$

$$f_2(\boldsymbol{x}^i) = \sum_{j=1}^{5} d_j x_j^i \exp\left(\frac{x_j^i}{4}\right) \lesssim b_2$$

$$\text{s.t.} \quad g_1(\boldsymbol{x}^i) = \sum_{j=1}^{5} v_j (x_j^i)^2 \le G_1$$

$$g_2(\boldsymbol{x}^i,x^r) = \sum_{j=1}^{5} C(x_j^r)\left[x_j^i + \exp\left(\frac{x_j^i}{4}\right)\right] \le G_2$$

$$x_j^i \ge 1, \text{ positive integer}, j = 1,\ldots,5$$

$$0 \le x_j^r \le 1, \text{ real number}, j = 1,\ldots,5$$

where $C(x_j^r)$ can be represented by the following equation:

$$C(x_j^r) = \alpha_j\left(\frac{-t_O}{\ln(x_j^r)}\right)^{\beta_j}, \qquad j = 1,\ldots,5$$

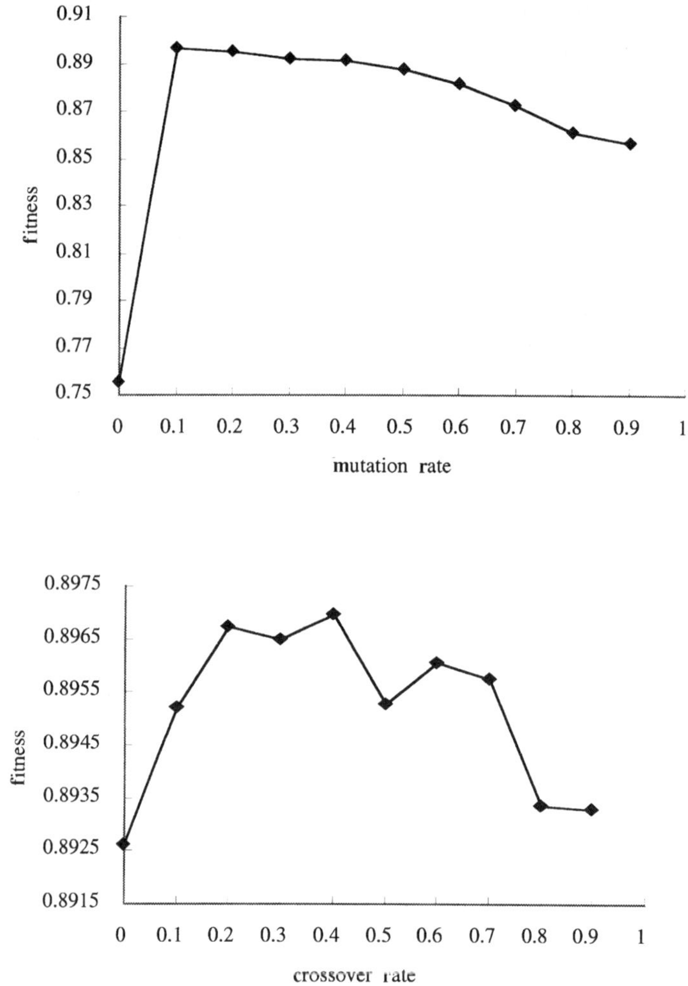

Figure 4.8. Parameters for a genetic operator.

The design data for this f-nMIGP instance are shown in Table 4.11. The parameters are *pop_size* = 20, $T = 2000$, $p_c = 0.4$, $p_m = 0.1$, $t_1^L = 0.25$, and $t_2^R = 120$. The relative weights of fuzzy goals are (0.75, 0.25). The results are shown in Table 4.12 and Figure 4.11.

Example 4.7. Consider a similar five-subsystem problem as in Example 4.6. The design data of Table 4.13 are used. The mathematical model is given as follows: Find $\boldsymbol{x}^i$ and $\boldsymbol{x}^r$ to optimize the next fuzzy goals:

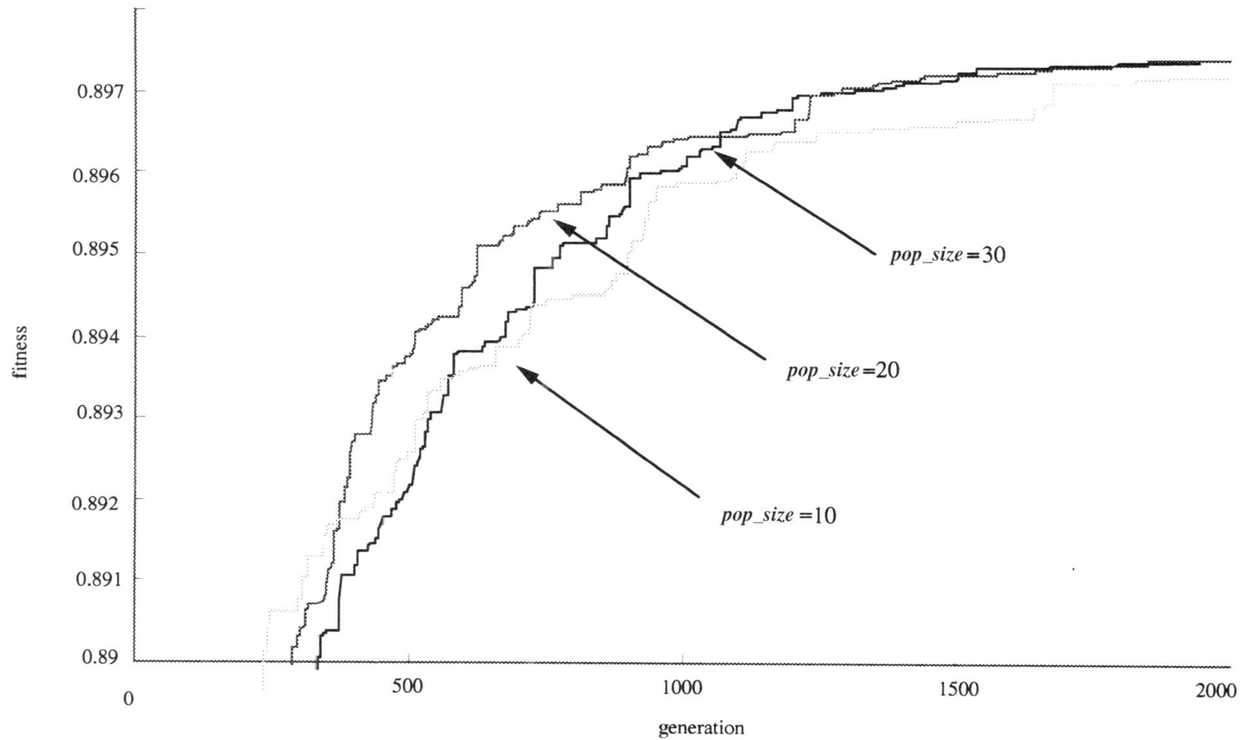

Figure 4.9. The *pop_size* for a genetic operator.

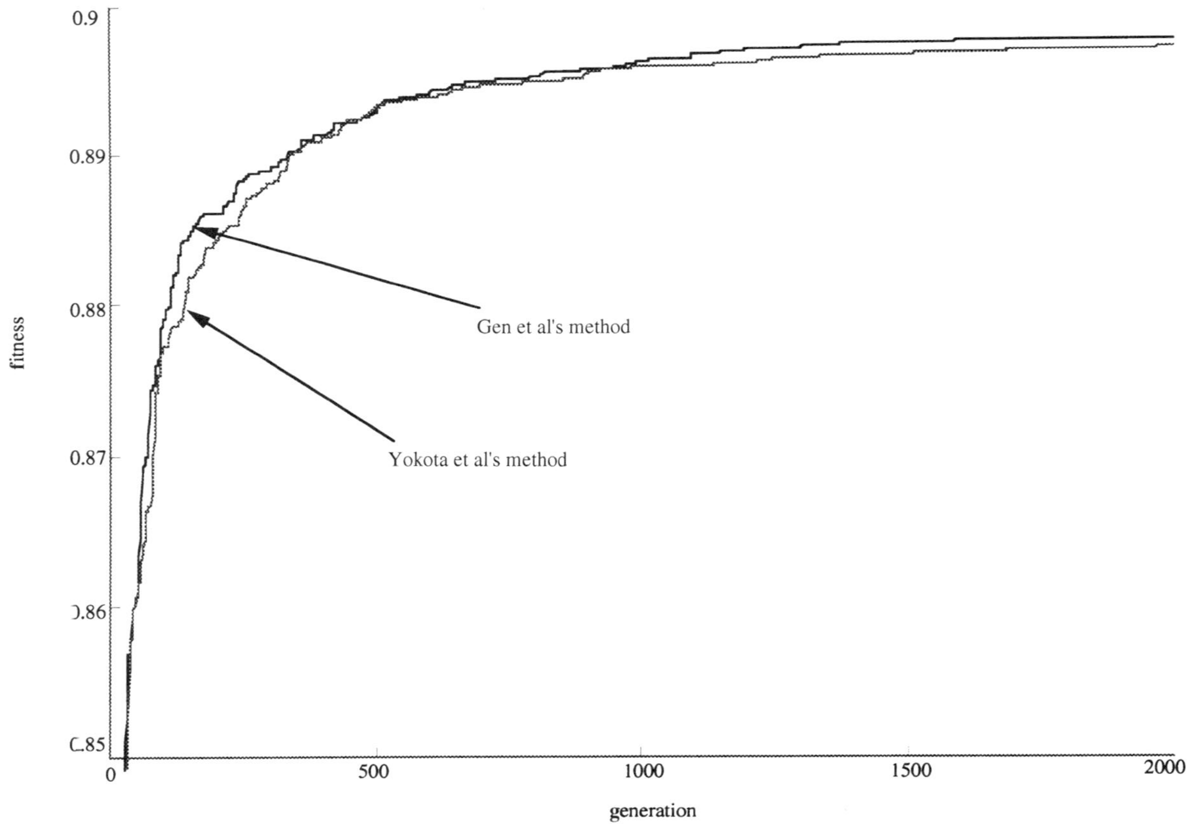

Figure 4.10. Genetic processes for Example 4.5.

TABLE 4.10. Performance Comparison for Example 4.5

	Yokota et al.'s Method	Gen et al.'s Method
Average		
CPU time (s)	4.427	18.7125
Fitness		
Average	0.896845	0.897456
Worst	0.894179	0.896436
Best	0.897654	0.897773
STD[a]	0.000894	0.000339

[a]STD, standard deviation.

$$f_1(\boldsymbol{x}^i,\boldsymbol{x}^r) = \prod_{j=1}^{5} \{1 - (1 - x_j^r)x_j^i\} \gtrsim b_1$$

$$f_2(\boldsymbol{x}^i,\boldsymbol{x}^r) = \sum_{j=1}^{5} C(x_j^r)\left[x_j^i + \exp\left(\frac{x_j^i}{4}\right)\right] \lesssim b_2$$

$$f_3(\boldsymbol{x}^i) = \sum_{j=1}^{5} d_j x_j^i \exp\left(\frac{x_j^i}{4}\right) \lesssim b_3$$

$$\text{s.t.}\quad g_1(\boldsymbol{x}^i) = \sum_{j=1}^{5} v_j(x_j^i)^2 \le G_1$$

$$x_j^i \ge 1, \text{ positive integer}, j = 1,\dots,5$$

$$0 \le x_j^r \le 1, \text{ real number}, \quad j = 1,\dots,5$$

where $C(x_j^r)$ can be represented by the equation

TABLE 4.11. Design Data of Example 4.6

Number of Subsystems				5
Limit on F_1				0.75
Limit on F_2				400.0
Limit on G_1				220.0
Limit on G_2				350.0
Operating time t_O				1000 h
Subsystem	$10^5 \cdot \alpha_j$	β_j	v_j	d_j
1	2.33	1.5	1	7
2	1.45	1.5	2	8
3	0.541	1.5	3	8
4	8.05	1.5	4	6
5	1.95	1.5	2	9

TABLE 4.12. Performance Comparison for Example 4.6

	Yokota et al.'s Method	Gen et al.'s Method
Average		
CPU time (s)	5.907	19.661
Fitness		
Average	0.957442	0.963852
Worst	0.946778	0.956742
Best	0.965478	0.968382
STD[a]	0.005282	0.003348

[a]STD, standard deviation.

$$C(x_j^r) = \alpha_j \left(\frac{-t_O}{\ln(x_j^r)} \right)^{\beta_j}, \qquad j = 1,\ldots,5$$

The parameters are set as follows: *pop_size* = 20, T = 2000, p_c = 0.4, p_m = 0.1, t_1^L = 0.25, t_2^R = 40, and t_3^R = 50. The relative weights of fuzzy goals are (0.5, 0.25, 0.25). The results are shown in Table 4.14 and Figue 4.12. From Table 4.14 and Figure 4.12, we can also see that Gen et al.'s method outperforms Yokota et al.'s method [684]. The best result is found in the 1914th generation with the objectives $f_1(\boldsymbol{x}^i,\boldsymbol{x}^r)$ = 0.924627, $f_2(\boldsymbol{x}^i,\boldsymbol{x}^r)$ = 165.229873, and $f_3(\boldsymbol{x}^i)$ = 192.481082, and the solution [(3,0.780943), (2,0.852900, (2,0.901031), (3,0.700890), (3,0.792657)].

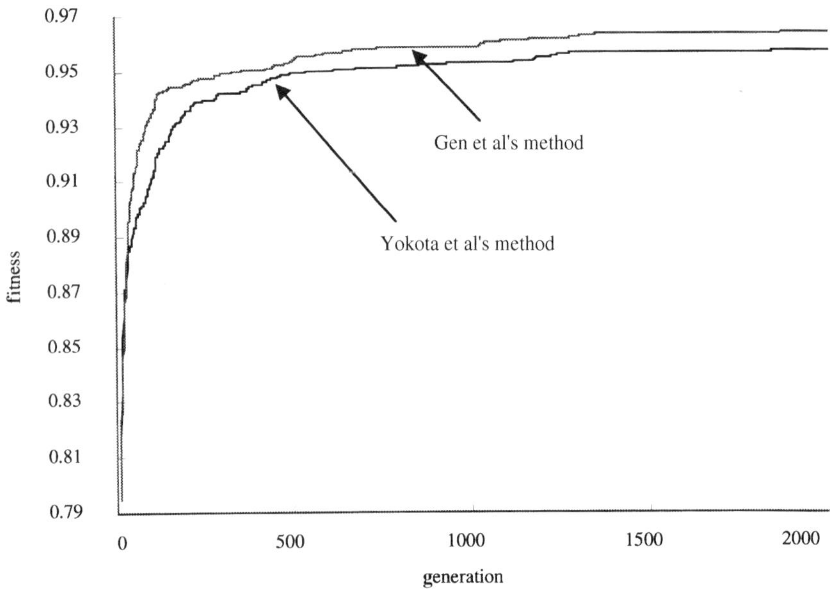

Figure 4.11. Genetic processes for Example 4.6.

TABLE 4.13. Design Data of Example 4.7

Number of Subsystems				5
Limit on F_1				0.75
Limit on F_2				205.0
Limit on F_3				240.0
Limit on G_1				110.0
Operating time t_O				1000 h
Subsystem	$10^5 \cdot \alpha_j$	β_j	v_j	d_j
1	2.33	1.5	1	7
2	1.45	1.5	2	8
3	0.541	1.5	3	8
4	8.05	1.5	4	6
5	1.95	1.5	2	9

4.5 FUZZY MULTIOBJECTIVE INTEGER PROGRAMMING

Sakawa, Shibano, and Kato developed a genetic algorithm for solving fuzzy multiobjective programming problems [549]. They introduced fuzzy goals into the multiobjective integer programming problem to handle the ambiguity of decision maker's judgment. By introducing decision maker's subjective preference in a form of reference membership, the problem can be transformed into an augmented minimax problem. A genetic algorithm was developed to solve the augmented minimax problem.

4.5.1 Problem Formulation

Generally, the multiobjective integer programming problem can be formulated as follows:

TABLE 4.14. Performance Comparison for Example 4.7

	Yokota et al.'s Method	Gen et al.'s Method
Average		
CPU time (s)	5.3005	23.6125
Fitness		
Average	0.699208	0.710613
Worst	0.654877	0.676798
Best	0.727734	0.730113
STD[a]	0.022590	0.013156

[a]STD, standard deviation.

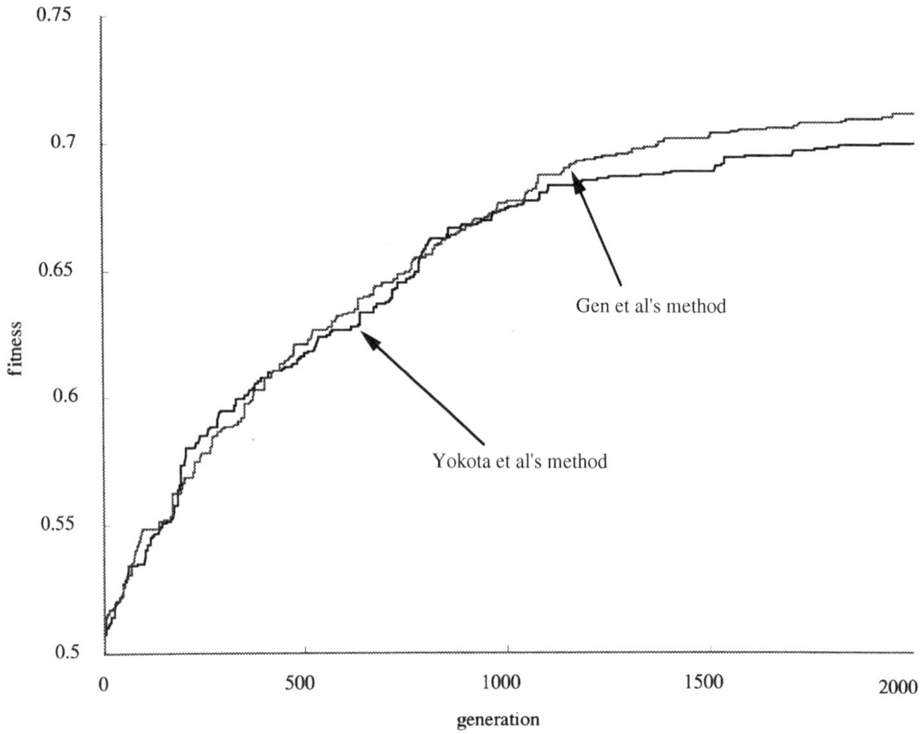

Figure 4.12. Genetic processes for Example 4.7.

$$
\begin{array}{ll}
\min & z_1(\boldsymbol{x}) = \boldsymbol{c}_1\boldsymbol{x} \\
 & \vdots \\
\min & z_k(\boldsymbol{x}) = \boldsymbol{c}_k\boldsymbol{x} \\
\text{s.t.} & \boldsymbol{A}\boldsymbol{x} \leq \boldsymbol{b} \\
 & x_j \in \{0,\ldots,v_j\}, \qquad j = 1,\ldots,n
\end{array}
\tag{4.43}
$$

where $\boldsymbol{c}_i = (c_{i1},\ldots,c_{in})$, $\boldsymbol{x} = (x_1,\ldots,x_n)^T$, $\boldsymbol{b} = (b_1,\ldots b_m)^T$, $\boldsymbol{A} = (a_{ij})$ is an $m \times n$ matrix, and v_i are positive integers. Assume that each element in matrix $\boldsymbol{a}$ and vector b is positive. A special case of the assumption is the multiobjective multidimensional integer knapsack problems. In general, however, for multiobjective programming problems, an optimal solution that minimizes all objective functions simultaneously does not always exist when the objective functions conflict with each other. Thus, instead of an optimal solution, Pareto optimality is in widespread use in multiobjective programming problems [547].

To consider the ambiguity of decision maker's judgments, fuzzy goals for objective functions such as "$z_i(\boldsymbol{x})$ should be substantially less than or equal

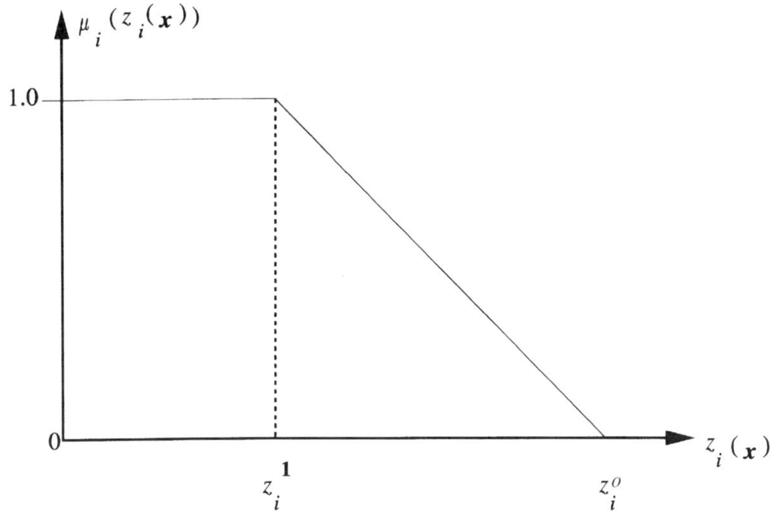

Figure 4.13. Linear membership function.

to a certain value" are incorporated. The membership functions for each fuzzy objective are defined by equation (4.44) and demonstrated in Figure 4.13.

$$\mu_i(z_i(\boldsymbol{x})) = \begin{cases} 0, & z_i(\boldsymbol{x}) > z_j^0 \\ \dfrac{z_i(\boldsymbol{x}) - z_i^0}{z_i^1 - z_i^0}, & z_i^1 < z_i(\boldsymbol{x}) \leq z_i^0 \\ 1, & z_i(\boldsymbol{x}) \leq z_i^1 \end{cases} \tag{4.44}$$

As one possible way to help decision makers to dettermine z_i^0 and z_i^1, the minimal value $z_i^{\min}$ and maximal value $z_i^{\max}$ of each objective function are calculated under given constraints. By taking into account the calculated minimum and maximum of each objective function, the decision maker is asked to assess z_i^0 and z_i^1 within the interval $[z_i^{\min}, z_i^{\max}]$, $i = 1,...,k$. Zimmermann suggested a way to determine the linear membership function $\mu_i(z_i(\boldsymbol{x}))$ [705]. Calculate the maximum and minimum for each objective as follows:

$$z_i^{\min} = \min_{x \in X} \{z_i(\boldsymbol{x}) | i = 1,...,k\} \tag{4.45}$$

$$z_i^{\mathrm{m}} = \max_{x \in X} \{z_i(\boldsymbol{x}^{1o}),...,z_i(\boldsymbol{x}^{io}),...,z_i(\boldsymbol{x}^{ko})\} \tag{4.46}$$

Then the linear membership function can be simply determined by choosing $z_i^1 = z_i^{\min}$ and $z_i^0 = z_i^{\mathrm{m}}$.

Having elicited the linear membership function $\mu_i(z_i(\boldsymbol{x}))$ from decision makers for each objective $z_i(\boldsymbol{x})$ and introduced a general conjunctive function,

$$\mu_D(\mu(z(\boldsymbol{x}))) = \mu_D(\mu_1(z_1(\boldsymbol{x})),\ldots,\mu_k(z_k(\boldsymbol{x}))) \tag{4.47}$$

the problem is transformed into the following fuzzy decision-making problem:

$$\begin{aligned} \max_{x \in X} \quad & \mu_0(\mu(z(\boldsymbol{x}))) \\ \text{s.t.} \quad & \boldsymbol{Ax} \le b \\ & x_j \in \{0,\ldots,v_j\}, \qquad j = 1,\ldots,n \end{aligned} \tag{4.48}$$

and the value of the conjunctive function $\mu_D(\mu(z(\boldsymbol{x})))$ is interpreted as the degree of satisfaction of decision makers for k fuzzy goals. For the conjunctive function, if we adopt the well-know fuzzy or minimum operator given by Bellman and Zadeh [52],

$$\min \{\mu(z_1(\boldsymbol{x})),\ldots,\mu(z_k(\boldsymbol{x}))\}$$

the fuzzy multiobjective integer programming problem can be transformed as

$$\begin{aligned} \min \quad & \max_{i=1,\ldots,k} \{\mu_i - \mu_i(z_i(\boldsymbol{x}))\} \\ \text{s.t.} \quad & \boldsymbol{Ax} \le b \\ & x_j \in \{0,\ldots,v_j\}, \qquad j = 1,\ldots,n \end{aligned} \tag{4.49}$$

4.5.2 Augmented Minimax Problems

Sakawa, Shibano, and Kao introduced an augmented minimax approach to solve the problem (4.49). Let $\bar{\mu}_i$, $i = 1,\ldots,k$, denote reference membership levels reflecting the aspiration levels of the decision maker for each membership function, which are specified by decision makers subjectively. The optimal solution can be obtained by solving the following minimax problem:

$$\begin{aligned} \min \quad & \max_{i=1,\ldots,k} \{\bar{\mu}_i - \mu_i(z_i(\boldsymbol{x}))\} \\ \text{s.t.} \quad & \boldsymbol{Ax} \le b \\ & x_j \in \{0,\ldots,v_j\}, \qquad j = 1,\ldots,n \end{aligned} \tag{4.50}$$

If the reference membership levels are attainable, the optimal solution above the reference membership levels can be obtained. If it is not attainable, it is desirable to obtain an optimal solution that is nearest to the reference membership levels in the minimax sense.

index	$s(1)$	$s(2)$	$\cdots$	$s(i)$	$\cdots$	$s(n)$
value	$x_{s(1)}$	$x_{s(2)}$	$\cdots$	$x_{s(i)}$	$\cdots$	$x_{s(n)}$

Figure 4.14. Double string.

Note that the optimal solution of the minimax problem may not be unique. Therefore, it is necessary to use other information or criteria to select one solution from the set of optimal solutions. Sakawa, Shibano, and Kato suggested an augmented minimax approach to handle this matter.

$$\begin{aligned} \min \quad & \max_{i=1,\dots,k} \{\bar{\mu}_i - \mu_i(z_i(\boldsymbol{x})) + \rho \sum_{i=1}^{k} (\bar{\mu}_i - \mu_i(z_i(\boldsymbol{x})))\} \\ \text{s.t.} \quad & \boldsymbol{Ax} \le b \qquad\qquad (4.51) \\ & x_j \in \{0,\dots,v_j\}, \qquad j = 1,\dots,n \end{aligned}$$

where ρ is a sufficiently small positive number. The augmented part reflects the total deviation of each objective from its aspiration level. By solving the augmented minimax problem (4.51), an optimal solution can be obtained which is nearest to the reference membership levels in the minimax sense with total best satisfaction [551, 552].

4.5.3 Genetic Algorithm Approach

Since the augmented minimax problem (4.51) is an integer programming problem whose objective and constraints are linear and whose coefficients are all positive, Sakawa, Shibano, and Kato developed a genetic algorithm to solve it. A double-string encoding method was proposed to guarantee generating only feasible solutions from an arbitrary chromosome. An interactive fuzzy satisfaction method was incorporated into the genetic algorithm to give decision makers a change in adjusting membership functions.

Representation. The double-string encoding method contains two parts: an upper part for the index of a variable and a lower part for the value of a variable. As shown in Figure 4.14, $s(i)$ corresponds to the index j of a variable x_j, and $x_{s(j)}$ corresponds to the value of the variable x_j.

A direct way to obtain a solution from a given chromosome will be

$$x_i = x_{s(i)}, \qquad \text{for } i = s(i), \quad i = 1,2,\dots,n$$

For the instance given in Figure 4.15, a solution may be

$$x_1 = 0, \qquad x_2 = 5, \qquad x_3 = 2, \qquad x_4 = 3, \qquad x_5 = 1, \quad x_5 = 2$$

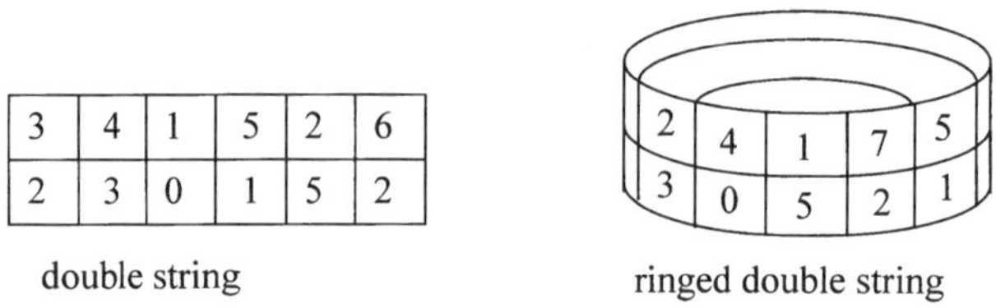

Figure 4.15. Double string and ringed double string.

Generally, a solution obtained in the direct way may not be feasible, because if we take all values specified by a give chromosome, the constraints of the problem may be broken. Therefore, a one-pass procedure was given to decode solutions, which scans the chromosome from left to right and fixes one decision variable value x_i at a time until the constraints are broken. Because the decoding procedure is deterministic, always beginning its search from the leftmost point of a given chromosome, the index string encoded in the double-string method has meaning. Compared with a single-string encoding method, the indexes of variables are fixed in a chromosome; for example, the leftmost corresponds to x_1 and the rightmost corresponds to x_n. If we decode a solution from such a single-string method using the same pone-pass scan procedure, x_1 is alway fixed first, then x_2, and so on. Then the genetic search spaces will be narrowed by the decoding method. In the double-string encoding method, the index of variables has an equal chance to be fixed at first. the overall decoding procedure is given below.

Procedure: Decoding Procedure for Double Strings

Step 1. Let $i = 1$, $sum_j = 0$ $(j = 1,\dots,m)$.
Step 2. The value of the variable with the index $s(j)$ is fixed to $p_{s(i)}$ by

$$p_{s(i)} = \min\left\{ \min_{\{j | a_{js(i)} \neq 0\}} \left[\frac{b_j - \text{sum}_j}{a_{js(i)}} \right], x_{s(i)} \right\}$$

where the $a_{js(i)}$'s are coefficients in the constraints.
Step 3. Let $sum_j = sum_j + a_{js(i)} p_{s(i)}$, $j = 1,\dots,m$ and $i = i + 1$.
Step 4. If $i > n$, stop; otherwise, return to step 2.

Note that according to the decoding procedure above, a gene has a higher priority to be fixed first than any one in its right position. To consider the imbalance of the influence between genes, ringed double strings were also suggested. Both are illustrated in Figure 4.15.

Fitness. The fitness $f(\mathbf{s})$ for each individual $\mathbf{s}$ is calculated by the following equation:

$$f(s) = 1.0 + k\rho - \max\left\{(\bar{\mu}_i - \mu_i(z_i(\mathbf{x}))) + \rho \sum_{i=1}^{k} (\bar{\mu}_i - \mu_i(z_i(\mathbf{x})))\right\}$$

A linear scaling method was used to adjust the fitness value of chromosomes such that the best chromosome gets a fixed number of expected offspring, thus preventing it from reproducing too much.

Reproduction. Sakawa and Shibano [551] have investigated the performances of six reproduction operators: ranking selection, elitist ranking selection, expected value selection, elitist expected value selection, roulette wheel selection, and elitist roulette wheel selection. It was confirmed that the elitist expected value selection is relatively efficient for multiobjective integer programming problems incorporating the fuzzy goals of decision makers [549, 551].

Crossover. If a single-point crossover or multipoint crossover is applied directly to individuals of the double-string type, the index $s(i)$ in the ith gene of an offspring may take the same number that an index $s(i')$ $(i \neq i')$ takes. The same violation occurs in solving the traveling salesman problem or the scheduling problem. To avoid this violation, the PMX method was modified to be suitable for the double strings.

Procedure: Revised PMX for Double Strings

Step 1. Choose two parent strings X and Y and let the ith column of X and Y be $(s_X(i), x_{sX(i)})^T$ and $(s_Y(i), x_{sY}(i))^T$, respectively. Then determine two crossover points h and k randomly, and let $i = h$.

Step 2. According to steps (2.1) and (2.2) below, make a new string X' by rearranging X so that the upper row of the substring of X between h and k will be identical with that of Y between h and k.

(*2.1*) After finding j satisfying $s_Y(i) = s_X(j)$, exchange the ith column $(s_X(i), x_{sX(i)})^T$ of X for the jth column $(s_X(j), x_{sX(j)})^T$ of X, and let $i = i + 1$.

(*2.2*) If $i > k$, stop; otherwise, return to step (2.1).

Then make Y' by rearranging Y according to the same procedures as above.

Step 3. Make an offspring X^* by exchanging the substring of X' between h and k for the corresponding substring of Y. Then make another offspring, Y^*, by the same operation for Y' as that for X'.

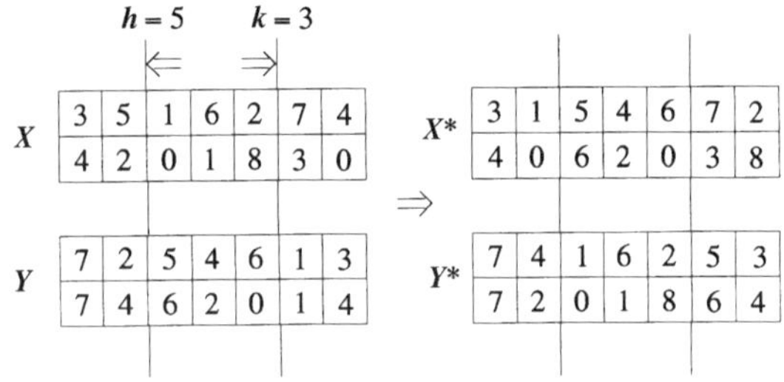

Figure 4.16. Example of PMX for double strings.

An example is illustrated in Figure 4.16. Note that the probability of the middle part of a string being exchanged is higher than for other parts of the string. Ringed double strings, where every part of a string will be exchanged equally, are proposed.

Procedure: Revised PMX for Ringed Double Strings

Step 1. Choose two parent strings X and Y and let the ith columns of X and Y be $(s_X(i),x_{sX(i)})^T$ and $(s_Y(i),x_{sY(i)})^T$, respectively.

Step 2. Determine two crossover points h and k randomly, and let $i = h$.

Step 3. If $i >$ length (*length* denotes the length of the string), let $n = i - length$. Otherwise, let $n = i$.

Step 4. After finding j satisfying $s_Y(n) = s_{sX(i)}$, exchange the ith column $(s_X$ (i), $x^T_{sX(i)}$ of X. Then exchange the substring of X between h and k for the corresponding substring of Y. Let $i = i + 1$.

Step 5. If $i > k$, stop; otherwise, return to step 3.

Step 6. Perform the same operation for Y. The strings X^* and Y^* obtained are regarded as offspring.

An example is illustrated in Figure 4.17.

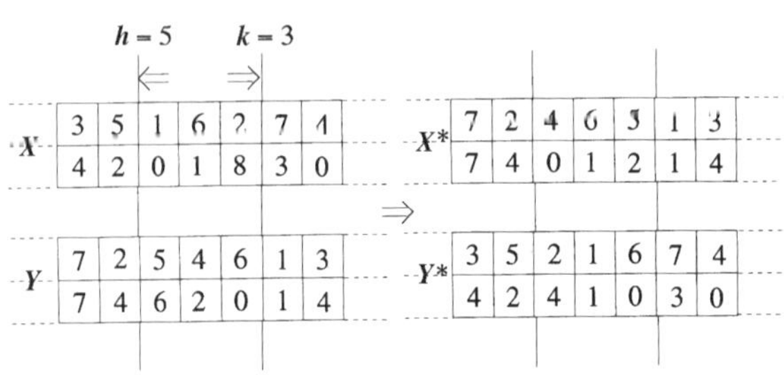

Figure 4.17. Example of PMX for ringed double strings.

Mutation and Inversion. In the mutation operation, after choosing the mutation point randomly, determine the value of the variable $x_{s(i)}$ randomly. Here $0 \leq x_{s(i)} \leq v_{s(i)}$ ($v_{s(i)}$ denotes the maximal value of $x_{s(i)}$ that satisfies all constraints). Furthermore, the inversion operation defined by the following algorithms was also adopted.

Procedure: Mutation

Step 1. For each column off a double string, generate a real random number rand(·) $\in [0,1]$.

Step 2. If $p_m \geq$ rand(·), determine the value of the lower element of the column. Here $0 \leq x_{s(i)} \leq v_{s(i)}$ ($v_{s(i)}$ denotes the maximal value of $x_{s(i)}$ satisfying all constraints).

Step 3. Perform steps 1 and 2 on all strings in the population.

Procedure: Inversion

Step 1. After determining two inversion points h and k (for ordinary double strings, the condition $h < k$ is added), pick out the part of the string from h to k.

Step 2. Arrange the substring in reverse order.

Step 3. Put the arranged substring back in the string.

4.5.4 Interactive Fuzzy Satisfaction Method

An interactive fuzzy satisfaction method was incorporated into the genetic algorithm to give decision makers a chance to readjust membership functions according to information provided by current genetic search.

Step 1. Set initial reference membership levels (if it is difficult to determine these values, set them to 1.0).

Step 2. Generate the initial population involving N individuals of (ringed) double-string type at random.

Step 3. Calculate the fitness for each individual and apply the reproduction operator based on the fitness.

Step 4. Insert into the current population the corresponding individual with the optimal solution for each objective function.

Step 5. Apply a crossover operator according to the probability of crossover p_c.

Step 6. Apply a mutation operator according to the probability of mutation p_m.

Step 7. Repeat steps 3 to 6 until the termination condition is satisfied. If it is satisfied, regard the individual with maximal fitness as the optimal individual and go to step 8.

Step 8. If the decision maker is satisfied with the current values of membership functions and objective functions given by the current optimal individual, stop. Otherwise, ask the decision maker to update reference membership levels by taking account of the current values of membership functions and objective functions and return to step 2.

In the generation of the initial population at each interaction, the (approximate) optimal individual in the preceding interaction is incorporated into the initial population. By using both the elitist expected value selection and the way of generating the initial population mentioned above, it s expected that the (approximate) optimal solution obtained in the current interaction will not be dominated by that in the preceding interaction.

4.5.5 Numerical Example

A multidimensional integer knapsack problem with 20 variables and 25 dimensional was generated through the following procedures.

1. Determine each of the a_{ij}'s by a random number according to the Gaussian distribution with mean $\mu = 300$ and standard deviation $\sigma = 50$.
2. Let each of the b_i's be a real number by multiplying a random number in [1.25, 1.75] by $\Sigma_{j=1}^{n} a_{ij}$.
3. Determine the c_{ij}'s in the same manner as the a_{ij}'s, where the c_{1j}'s and c_{2j}'s are supposed to be all positive and negative, respectively, while the c_{ij}'s are supposed to be positive and negative half and half.

An example of problems generated by the foregoing procedure is shown in Table 4.15.

Results of Simulations. For the numerical example shown in Table 4.15, simulations were performed 20 times for both ordinary and ringed double-string representation. The number of individuals is 100, the probability of crossover $p_c = 0.8$, the probability of mutation $p_m = 0.02$, the maximal generation is 300, and the maximal value of each integer decision variable is 10. The pairs of objective function values (z_1^0, z_1^1), (z_2^0, z_2^1), and (z_3^0, z_3^1) that make the linear membership function value become 0 or 1 were determined to be (9200,0), (0,−9600), and (3000,−9200), respectively, on the basis of the Zimmermann's method [705].

Table 4.16 shows the best value, the worst value, and the mean value of $\min(\mu_i(z_i(\boldsymbol{x})))$ obtained by two methods, each with 20 trials. As determined from Table 4.16, the performance of the method using ringed double-string representation was better with respect to the worst and mean values.

TABLE 4.15. Example of Integer Programming Problems with 3 Objectives and 15 Constraints

a_1	271	331	300	314	256	369	306	226	285	284
	333	379	307	182	312	323	264	288	325	192
a_2	316	202	324	346	328	348	286	263	362	311
	359	340	223	276	376	291	325	296	342	263
a_3	252	265	319	276	158	314	276	277	364	246
	240	316	311	395	371	339	411	300	266	360
a_4	188	288	285	313	306	227	284	273	317	254
	279	340	312	285	286	285	240	399	293	353
a_5	339	252	314	328	269	392	268	367	220	249
	281	295	323	283	286	242	295	270	313	312
a_6	302	267	340	239	377	287	279	282	215	264
	302	421	374	293	227	313	339	353	315	315
a_7	188	260	299	344	352	321	234	317	280	287
	374	353	330	272	297	227	332	291	311	273
a_8	246	349	350	230	297	370	252	265	290	276
	321	237	242	300	260	243	247	411	310	326
a_9	245	299	318	299	307	266	314	301	309	333
	341	315	324	297	344	327	250	367	309	296
a_{10}	211	361	248	217	197	249	300	231	373	320
	256	261	301	302	324	286	391	276	273	298
a_{11}	303	330	328	328	356	284	298	321	269	274
	301	346	289	239	403	252	230	315	192	391
α_{12}	232	277	315	331	262	286	314	369	434	290
	247	318	353	331	315	292	323	250	279	318
a_{13}	198	326	376	289	191	253	239	354	251	298
	315	329	272	255	293	368	320	371	396	293
a_{14}	248	264	275	225	353	310	329	423	306	248
	264	307	251	242	305	274	398	230	204	207

Interactive Processes. Based on the simulation results shown in Table 4.17, an interactive method through genetic algorithms using the ringed double-string representation was applied to the problem above. The same values of parameters in genetic algorithms and the same membership functions as in the preceding experiment were used. In this experiment, the hypothetical decision maker updated the reference membership levels $(\bar{\mu}_1, \bar{\mu}_2, \bar{\mu}_3)$ as $(1.0,1.0,1.0) \rightarrow (0.8,1.0,1.0) \rightarrow (0.8,1.0,0.9)$. Table 4.17 shows the (approximate) optimal solution for each interaction. In the first interaction, the membership values (0.5773,0.5755,0.5917) of the (approximate) optimal solution

TABLE 4.15. (*Continued*)

a_{15}	274	335	376	299	334	198	318	305	251	288
	286	273	310	293	226	344	352	236	305	306
c_1	281	327	305	344	371	247	335	336	364	392
	263	280	370	292	231	352	297	254	293	320
c_2	−271	−338	−282	−267	−319	−379	−245	−215	−187	−419
	−360	−302	−266	−324	−315	−307	−322	−366	−332	−273
c_3	−315	−293	−245	−291	282	316	226	297	360	182
	−330	394	−345	276	−321	−219	341	−372	250	324
b	8,012	7,871	8,660	8,297	9,144	10,289	8,200	7,670	10,091	8,522
	9,934	10,432	7,959	9,623	10,108					

TABLE 4.16. Results of 20 Trials

Double String	Best	Worst	Mean
Ordinary	0.575521	0.563804	0.571542
Ringed	0.575521	0.567717	0.572829

for the reference membership levels (1.0,1.0,1.0) were obtained, but the decision maker was not satisfied with the solution. Then the decision maker considered that the membership function values μ_2 and μ_3 should be improved upon sacrifice of the membership function value μ_1 and updated the reference membership levels from (1.0,1.0,1.0) to (0.8,1.0,1.0).

In the second interaction, the membership values (0.4917,0.6872,0.7373) of the (approximate) optimal solution for the reference membership levels (0.8,1.0,1.0) were obtained. Since the decision maker hoped for more improvement of μ_2, she updated the reference membership levels from (0.8,1.0,1.0) to (0.8,1.0,0.9). The decision maker was satisfied with the membership values (0.4927,0.6966,0.6310) obtained in the third interaction and the interactive procedure was finished.

In this example, after updating the reference membership levels twice, at the third interaction, a satisfactory solution that is also Pareto optimal was derived. Here, since it took about 69 s to solve the augmented minimax problem for each reference membership level, the proposed method is considered to be sufficiently fast as an interactive method. Unfortunately, however, at each interaction, (approximate) Pareto optimal solutions were obtained in only about 2 out of 10 trials. Concerning this point, further refinements of individual representation, decoding algorithm, and so on, will be required.

TABLE 4.17. Results of Interactions

	μ_1	μ_2	μ_3	$z_1(x)$	$z_2(x)$	$z_3(x)$	Number of Solutions
First interaction	0.5773	0.5749	0.5882	3898	−5525	−4219	2
$\bar{\mu}_1 = 1.0$	0.5763	0.5749	0.5882	3898	−5519	−4177	2
$\bar{\mu}_2 = 1.0$	0.5757	0.5742	0.5848	3907	−5513	−4135	1
$\bar{\mu}_3 = 1.0$	0.5740	0.5795	0.6522	3919	−5563	−4958	2
	0.5722	0.6378	0.5736	3923	−5494	−4781	1
	0.5827	0.5707	0.6331	3839	−5479	−4724	1
	0.5677	0.5762	0.5877	3977	−5532	−4170	1
Second interaction	0.4917	0.6872	0.7373	4676	−6598	−5939	2
$\bar{\mu}_1 = 0.8$	0.4908	0.6867	0.7292	4685	−6598	−5939	1
$\bar{\mu}_2 = 1.0$	0.4945	0.6899	0.6832	4651	−6623	−5335	2
$\bar{\mu}_3 = 1.0$	0.4829	0.6821	0.7422	4757	−6548	−6055	1
	0.4849	0.6768	0.7169	4727	−6497	−5746	1
	0.4753	0.6734	0.7484	4827	−6465	−6135	1
	0.4701	0.7227	0.6884	4875	−6938	−5398	1
	0.4824	0.6673	0.7579	4762	−6406	−6246	1
Third interaction	0.4927	0.6966	0.6310	4667	−6687	−4698	2
$\bar{\mu}_1 = 0.8$	0.4933	0.6900	0.6199	4622	−6624	−4563	2
$\bar{\mu}_1 = 1.0$	0.4945	0.6899	0.6832	4651	−6623	−5335	1
$\bar{\mu}_1 = 0.9$	0.4892	0.6926	0.7369	4669	−6649	−5990	1
	0.4908	0.6867	0.7293	4685	−6592	−5897	1
	0.4836	0.6845	0.6108	4751	−6571	−4452	1
	0.4827	0.6828	0.6824	4751	−6571	−4452	1
	0.4799	0.6935	0.7170	4785	−6658	−5747	1

As another application of fuzzy optimization based on genetic algorithms, bicriteria transportation problems with fuzzy coefficient [412, 414, 712] in Section 7.6. Multiobjective reliability design with fuzzy goals [224] in Section 5.4, fuzzy assembly line balancing problem by Gen, Tsujimura, and Li [716], and multiobjective knapsack problem with fuzzy goals and multiple choices by Sasaki, and Gen [724] developed respectively.

5
RELIABILITY DESIGN PROBLEMS

5.1 INTRODUCTION

Reliability optimizations appeared in the late 1940s and was first applied to communication and transportation systems. Much of the early work was confined to an analysis of certain performance aspects of systems. One goal of the reliability engineer is to find the best way to increase system reliability. The reliability of a system can be defined as the probability that the system has operated successfully over a specified interval of time under stated conditions. As systems have grown more complex, the consequences of their unreliable behavior have become severe in terms of cost, effort, lives, and so on. The interest in accessing system reliability and the need to improve the reliability of products and systems have become more and more important.

Generally reliability problems include reliability analysis, reliability optimization, reliability growth, reliability testing, reliability data analysis, accelerated testing, and life-cycle costing [525]. In the past two decades, numerous reliability design techniques have been proposed. These techniques can be classified as linear programming, dynamic programming, integer programming, geometric programming, heuristic method, Langrangian multiplier method, and so on [521]. A review of optimization techniques for system reliability can be found in [382] and [620].

Most of these techniques transform the original problems formulated as nonlinear integer programming problem into a 0–1 linear programming problem. But this transformation leads the original problems to increase the number of variables and constraints to be treated. This requires much computation time and more memory space. Therefore, it is better to solve the original reliability design problems without any transformation.

After proposing a simple genetic algorithm to solve reliability optimization problem by Gen, and Ida in 1993 and Smith, and Coit in 1994 [219, 229], many researchers developed various reliability design problems based on genetic algorithms [717, 719, 722, 723, 725, 727]. In this chapter we describe several genetic algorithm approaches for solving reliability design problems, such as networks reliability design problems, tree-based network reliability design problems, bicriteria reliability design problems of redundant system which are formulated as nonlinear integer programming (NIP) problems, and reliability design problems with fuzzy goals [229]. These network design problems that consider a system reliability constraint or objective have been the subject of many research efforts and have many applications in telecommunications and computer networking and related domains in electric, gas, and sewer networks [157].

5.2 NETWORK RELIABILITY DESIGN

Network reliability design problems have attracted many researchers' attentions, such as network designers, network analysts, and network administrators, in order to share expensive hardware and software resources and provide access to main systems from distant locations. The problems have many applications in areas of telecommunications and computer networking and related domains in electric, gas, and sewer networks. When designing network systems, one of important steps is to find the best layout of components to optimize some performance criteria, such as cost, transmission delay, throughput or reliability. The performance criteria of the systems are especially important and are largely determined by network topology. Among network design problems, the design of networks considering reliability is one of typical network design problems. The optimal design problem can be formulated as a combinatorial problem. Usually, it belongs to NP-hard problems. The reliabilities related to network design problems can be classified into the following two classes:

1. *All-terminal network reliability* (also called *overall network reliability*): the probability that every node in the network is connected to each other (i.e., every node in the network can communicate with every other node over a specified mission time).
2. *Source–sink network reliability:* the probability that the source is connected with the sink, so the source node in the network can communicate with the sink node over a specified mission time.

Solution approaches for the optimal network design problems considering these reliabilities can be classified as follows: (1) an enumeration-based approach, (2) a heuristic-based approach, or (3) a genetic algorithm-based approach. Enumeration-based approaches are applicable only for small network sizes. The enumeration-based approach is founded on the enumeration of states, minpaths, or mincuts. Roughly speaking, these methods use some combination of enumeration and reduction techniques. Enumeration-based methods enumerate a set of probabilistic events that are mutually exclusive and collectively exhaustive with respect to the measure. Jan, Hwang, and Chen proposed an algorithm using decomposition based on branch-and-bound to minimize edge costs with a minimum network reliability constraint [325]. Aggarwal and Rai proposed a method based on spanning trees to evaluate the reliability of computer networks [8]. Wilkov proposed a method to design computer networks with a new reliability measure [664].

Heuristic-based approaches, such as tabu search, simulated annealing, and so on, can be applied to larger networks but do not guarantee optimality. Aggarwal, Chopra, and Bajwa applied greedy heuristic approach to the computer network design problems considering the overall reliability [7]. Chopra et al. proposed a greedy heuristic approach for the design of network topology to a maximize the terminal reliability [120]. Venetsanopoulos and Singh used a two-step heuristic for minimizing cost subject to a reliability constraint [638]. Glover, Lee, and Ryan proposed a method using tabu search to design the network topology with least cost [246]. Koh and Lee applied a tabu search method to the reliable fiber optic communication network design [372]. Atiqullah and Rao proposed a method using simulated annealing to optimize the reliability of communication networks [22]. Fetterlof and Anandalingam applied simulated annealing to the optimization of local area network/wide area network (LAN–WAN) internetworking design [187]. Pierre et al. proposed an algorithm for the topology design of computer communication networks using simulated annealing [510].

Recently, genetic algorithm–based approaches have been receiving increasing attention as a new solution method for the optimal design of networks considering reliability. Kumar et al. proposed a genetic algorithm to design the distributed system topology [380]. Kumar, Pathak, and Gupta also developed a genetic algorithm considering diameter, distance, and reliability to design and expand computer networks [379]. Walters and Smith proposed an evolutionary algorithm for the optimal layout design of networks with tree structure [627]. Deeter and Smith presented a genetic algorithm-based approach to design of networks considering all-terminal reliability with alternative edge options, allowing edges to be chosen from different components with different costs and reliabilities [153, 154]. Dengiz, Altiparmak, and Smith attempted to apply genetic algorithms to the optimal design of networks considering all-terminal design [157, 159]. Dengiz, Altiparmak, and Smith focused on large backbone communication network design considering all-terminal network reliability and used a genetic algorithm, but customized it appreciably to the all-terminal design problem to give an effective, efficient optimization methodology [157, 158].

5.2.1 Problem Formulation

A communication network can be represented by an undirected graph $G = (N,E)$, in which nodes N and edges E represent computer sites and commu-

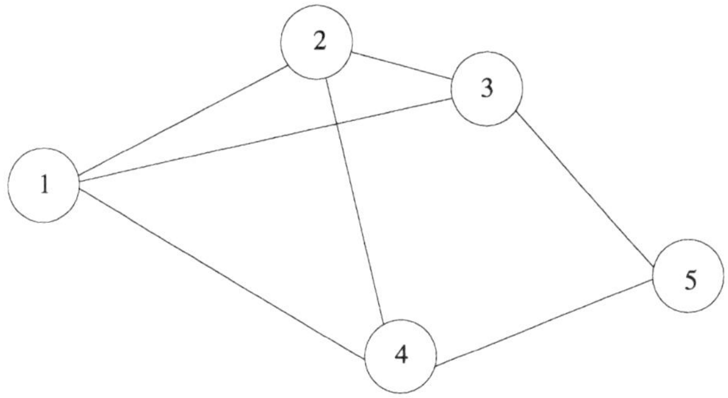

Figure 5.1. Sample network architecture.

nication cables, respectively. A graph G is connected if there is at least one path between every pair of nodes i and j, which minimally requires a spanning tree with $(n - 1)$ edges. The number of possible edges is $n(n-1)/2$. The following notation is defined to describe the optimal design problem of all-terminal reliable networks: n is the number of nodes; $x_{ij} \in \{0,1\}$ is a decision variable representing edges between nodes i and j; $\boldsymbol{x}$ $(= \{x_{12}, x_{13}, \ldots, x_{n-1,n}\})$ is a topology architecture of network design; $\boldsymbol{x}^*$ is the best solution found so far; p is the edge reliability for all edges; q is edge unreliability for all edges (i.e., $p + q = 1$); $R(\boldsymbol{x})$ is the all-terminal reliability of network design $\boldsymbol{x}$; $R_{\min}$ is a network reliability requirement; $R_U(\boldsymbol{x})$ is the upper bound of reliability of the candidate network; c_{ij} is the cost of the edge between nodes i and j; $c_{\max}$ is the maximum value of c_{ij}; δ has a value of 1 if $R(\boldsymbol{x}) < R_{\min}$, otherwise, has a value of 0; E' is a set of operational edges ($E' \subseteq E$); and Ω is all operational states (E').

Assume that the location of each node is given, nodes are perfectly reliable, each c_{ij} and p are fixed and known, each edge is bidirectional, there are no redundant edges in the network, edges are either operational or failed (non-operational), the failure of edges are mutually statistically independent, and there is no repair.

The optimal design of network can be represented as follows:

$$\min \quad Z(\boldsymbol{x}) = \sum_{i=1}^{n-1} \sum_{j=i+1}^{n} c_{ij} x_{ij}$$

$$\text{s.t.} \quad R(\boldsymbol{x}) \geq R_{\min}$$

At any mission time, only some edges of G might be operational. An operational state of G is a subgraph $G' = (V, E')$. The network reliability of state $E' \subseteq E$ is as follows:

$$\sum_{\Omega} \left(\prod_{e \in E'} p_e \right) \left(\prod_{e \in (E \setminus E')} q_e \right)$$

TABLE 5.1. Connectivity Matrix of Edges

$$\begin{bmatrix} - & 1 & 1 & 1 & 0 \\ 1 & - & 1 & 1 & 0 \\ 1 & 1 & - & 0 & 1 \\ 1 & 1 & 0 & - & 1 \\ 0 & 0 & 1 & 1 & - \end{bmatrix}$$

5.2.2 Dengiz, Altiparmak, and Smith's Approach

Genetic algorithms have been applied successfully to solving network reliability optimization problems. In this section we introduce the Dengiz, Altiparmak, and Smith approach for network reliability optimization in detail [153, 157].

Representation. A genetic algorithm lends itself to this problem because each network design $\boldsymbol{x}$ is easily formed into a binary string that can be used as a chromosome for genetic algorithms. Each element of the chromosome represents a possible edge in the network design problem, so there are $n \times (n - 1)/2$ string components in each candidate architecture $\boldsymbol{x}$. The value of each of these elements tells whether or not the specific edge connects with the pair of nodes. Consider the example with five nodes shown in Figure 5.1. Note that there are $(5 \times 4)/2 = 10$ possible edges for this example but only seven are included; the other three are not connected. The chromosome representation of the instance given in Figure 5.1 and Table 5.1 is shown in Figure 5.2. We can see that the index of $\boldsymbol{x}$ is represented by the following equations:

$$\text{index of } k = \frac{n\,(n-1)}{2} - \frac{(n-i)(n-i+1)}{2} + (j-i)$$

Note that the solution space of possible network architectures is $2^{[n \times (n-1)]/2}$.

Initial Population. The initial population is considered as a set of connected networks that are also 2-connected but is otherwise generated randomly with preference for combinations yielding high reliability. The 2-connectivity check is used as a preliminary screening, since it is usually a property of highly reliable networks. This 2-connectivity check is made by counting the

k	1	2	3	4	5	6	7	8	9	10
	1	1	1	0	1	1	0	0	1	1

Figure 5.2. Chromosome of Figure 5.1.

TABLE 5.2. Probability Values to Generate the Initial Population

n	Probability Values
10	0.15–0.60
20	0.15–0.50
30	0.10–0.30

node degree, (i.e., all nodes have at least degree 2, where degree is defined as the number of edges incident with node).

Selection of the probability values that are used in deciding whether or not an edge exists is an important stage for efficient generation of such an initial population. Table 5.2 shows the probability intervals used, established by exploratory research.

Therefore, the initial population can be obtained by the following procedure:

Step 1. Generate the initial population randomly according to the edge probabilities in Table 5.2.

Step 2. If the chromosome generated in step 1 fails to meet the 2-connectivity requirement, discard it and repeat step 1.

Step 3. If *pop_size* chromosomes are generated (where *pop_size* means the size of population), stop.

Fitness. The objective of the penalty function is to lead the optimization algorithm to nearly optimal feasible solutions. The objective function here is the sum of the total cost for all edges in the network, plus a quadratic penalty function for networks that fail to meet the minimum reliability requirement. The objective function considering infeasibility is defined as follows:

$$Z(\mathbf{x}) = \sum_{i=1}^{n-1} \sum_{j=i+1}^{n} c_{ij}x_{ij} + \delta[c_{\max}(R(\mathbf{x}) - R_{\min})]^2 \tag{5.1}$$

The fitness function eval($\mathbf{x}$) is defined by the equation

$$\text{eval}(\mathbf{x}) \equiv Z_{\max} - Z(\mathbf{x}) \tag{5.2}$$

where $Z_{\max}$ is the largest (worst) value of equation (5.1) in the current population.

Selection. The selection process is based on spinning the roulette wheel *pop_size* times, and each time a single chromosome is selected as a new offspring.

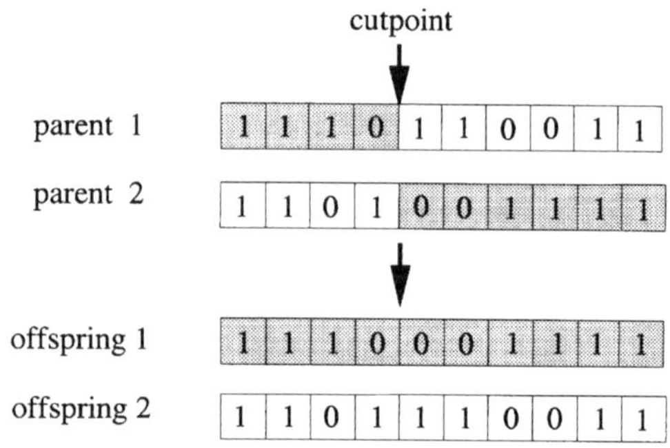

Figure 5.3. Single-point crossover operator.

Crossover. One-cut point crossover operation is used. An illustration of this method is shown in Figure 5.3.

Mutation. The bit-flip mutation operation is employed, which is performed on a bit-by-bit basis. Suppose that the point to be mutated is an element of index 2. Then the offspring generated by bit-flip mutation is shown in Figure 5.4.

Reliability Estimation. Another active area of research related to the network design problem is the calculation or estimation of network reliability. There are two main approaches: (1) exact calculation through analytic methods and (2) estimation through variations of Monte Carlo simulation.

All known analytic methods for all-terminal network reliability calculation have worst-case computation times that grow exponentially with the size of the network. Hence these analytic methods are not recommended for large networks. Ball and Van Slyke proposed a method of enumerating modified cutsets by backtracking to compute the overall reliability of an undirected graph [41]. Hanzhong and Dongkui proposed a new method for computing complex network reliability [283]. Mandaltsis and Kontolen gave a method for calculating overall reliability of computer networks that have hierarchical routing strategies [435]. Liu et al. proposed a method for calculation of overall network reliability [420].

For the all-terminal network reliability problem, simulation is especially difficult because these methods generally lose efficiency as a network ap-

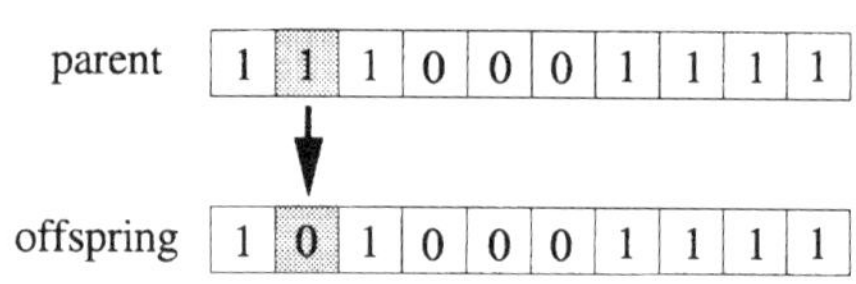

Figure 5.4. Bit-flip mutation operator.

proaches a fully connected state. However, simulation methods are suitable for large networks because the computation time required grows only slightly faster than linearly with network size. Cancela and Khadiri used a recursive variance-reduction algorithm with the Monte Carlo method to estimate communication network reliability [83]. Fishman studied a Monte Carlo sampling plan to estimate network reliability and analyzed about four Monte Carlo methods to estimate the probability of source–sink connectedness [190, 191]. Kamat and Riley examined a determination method of reliability with event-based Monte Carlo simulation [343]. Kumamoto, Tanaka, and Inoue proposed an efficient evaluation method of system reliability using a Monte Carlo method [378]. Yeh, Lin, and Yeh proposed a new Monte Carlo method of estimating network reliability [682].

Since the genetic algorithm is an iterative algorithm and many candidate networks must be evaluated during each generation, conventional computation intensive methods are not suitable for use together with genetic algorithms. Therefore, the following screening procedure for candidate network designs is employed.

Step 1. Connectivity check. A connectivity check for a spanning tree is made on all new network designs using depth-first or breath-first search, where a graph is connected if every pair of its nodes is connected.

Step 2. 2-connectivity check. For network designs that pass step 1, a 2-connectivity check is performed. The 2-connectivity measure is made by making sure that all nodes have at least degree 2.

Step 3. Computation of the upper bound. For networks that pass both of these preliminary checks, Jan's upper bound is used to compute $R_U(\boldsymbol{x})$ [324, 325].

This upper bound is used in the calculation of objective functions for all networks except $\boldsymbol{x}^*$. Networks that have $R_U(\boldsymbol{x}) \geq R_{\min}$ and the lowest cost so far are sent to a Monte Carlo simulation procedure for more precise estimation of network reliability using an efficient Monte Carlo technique. The simulation is done for 3000 iterations for each candidate network.

Jan's Upper Bound. Let G be a network with n nodes and degree sequence $\boldsymbol{d}$. So Jan's upper bound is given as follows:

$$R(G) \leq H(\boldsymbol{d})$$

$$H(\boldsymbol{d}) \equiv 1 - \left[\sum_{i=1}^{n} q^{d_i} \prod_{k=1}^{m_i} (1 - q^{d_k - 1}) \prod_{k=m_i+1}^{i-1} (1 - q^{d_k}) \right]$$

$$m_i \equiv \min\{d_i, i - 1\} \qquad \text{for all } i$$

For example, let $\boldsymbol{d} = \{2,2,2,2\}$ and $q = 0.1$.

$$
\begin{aligned}
R(G) &\leq 1 - (0.1^2 + 0.1^2 \cdot 0.9 + 0.1^2 \cdot 0.9 \cdot 0.9 + 0.1^2 \cdot 0.9 \cdot 0.9 \cdot 0.99) \\
&= 0.9649
\end{aligned}
$$

Overall Algorithm

Step 1. Set the parameters. Set the population *size* (*pop_size*), mutation rate (p_m), crossover rate (p_c), and maximum generation (*max_gen*), and initialize the generation number $gen = 0$.

Step 2. Initialize.

(2.1) Randomly generate the initial population according to the edge probability in Table 5.2. If the chromosome generated fails to meet the 2-connectivity requirement, discard it and regenerate.

(2.2) Calculate the fitness of each candidate network in the population using equations (5.1) and (5.2) and Jan's upper bound as $R(\boldsymbol{x})$, except for the lowest-cost network with $R_U(\boldsymbol{x}) \geq R_{\min}$.

(2.3) For the lowest-cost network ($\boldsymbol{x}^*$), simulate $R(\boldsymbol{x})$ using Monte Carlo estimation.

Step 3. Select two candidate networks from the current population by the roulette wheel selection mechanism.

Step 4. Perform the crossover and mutation. To obtain two children candidate networks, apply the crossover operator and mutation operator to the selected chromosomes using p_C and p_M.

Step 5. Determine the 2-connectivity of each new child. Discard any that do not satisfy 2-connectivity.

Step 6. Calculate Jan's upper bound. Calculate the $R_U(\boldsymbol{x})$ for each child and compute its objective function.

Step 7. Check the number of children. If $n < pop\_size - 1$, go to step 3; otherwise, go to step 8, where n means the number of new children.

Step 8. Establish the new population. Replace the parents with children, retaining $\boldsymbol{x}^*$ from the previous generation.

Step 9. Evaluate.

(9.1) Sort the new generation in increasing order of $Z(\boldsymbol{x})$.

(9.2) If $Z(\boldsymbol{x}_p) < Z(\boldsymbol{x}^*)$, $p = 1,2,\ldots,pop\_size$, estimate the system reliability of this network using Monte Carlo simulation and $\boldsymbol{x}^* = \boldsymbol{x}_p$; else go to step (9.3).

(9.3) Calculate eval($\boldsymbol{x}$) for each network in the new population.

Step 10. Perform the terminating test. If $gen < max\_gen$, $gen = gen + 1$ and return to step 3 for the next generation. If $gen = max\_gen$, terminate.

Numerical Example. The test problems used by Dengiz, Altiparmak, and Smith are summarized in Table 5.3 [156, 157]. These problems are both fully connected and non-fully connected networks (only a subset of E is possible

TABLE 5.3. Comparison of Results for B&B and Genetic Algorithms[a]

						Genetic Algorithms		
Problem	N	E	p	R_{min}	B&B Optimal Cost	Best Cost	Mean Cost	Coefficient of Variation
1	5	10	0.90	0.95	201	201	201.0	0
2	8	28	0.95	0.95	179	179	180.3	0.0228
3	10	45	0.90	0.95	197	205	206.6	0.0095
4	20	190	0.95	0.95	—	926	956.0	0.0304
5	14	21	0.90	0.90	1063	1063	1076.1	0.0129
6	16	24	0.90	0.95	1022	1022	1032.0	0.0204
7	20	30	0.95	0.90	596	596	598.6	0.0052

[a]Problems 1 to 4 are fully connected networks; problems 5 to 7 are non-fully connected networks.

for selection). Nodes for the networks range from 5 to 20. The available edges of the non-fully connected networks were randomly generated and were $1.5N$. The edge costs for all networks were randomly generated over [1, 100]. Each problem for the genetic algorithm was run 10 times, each time with a different random number seed. The optimal solutions obtained by the branch-and-bound (B&B) method of Jan, Hwang, and Chen [325] are given in Table 5.3.

Table 5.4 lists the search space for each problem along with the proportion actually searched by the genetic algorithm during a single run of $pop\_size \cdot max\_gen$. Depending on problem size, max_gen ranged from 30 to 20,000. This proportion is an upper bound because genetic algorithms can (and often do) revisit solutions considered earlier in the evolutionary search. The genetic algorithm examines only a very tiny fraction of possible solutions for the larger problems, yet still yields optimal or nearly optimal solutions. Table 5.4 also compares the efficiency of Monte Carlo estimation of network reliability. The exact network reliability is calculated using a backtracking algorithm given by Jan, Hwang, and Chen [325] and compared to the estimated counterpart for the final network for those problems where the genetic algorithm was optimal. The reliability estimation of Monte Carlo method is unbiased and always within 1% of exact network reliability. Since the computation time for the Monte Carlo method is constant with network size and estimation accuracy does not degrade with an increase in E, this simulation estimation is a very effective surrogate for an exact network reliability calculation.

5.2.3 Deeter and Smith's Approach

Deeter and Smith [153, 154] proposed a genetic algorithm for the design of networks when considering all-terminal reliability. A common assumption in this problem is that reliability of all edges in the network are identical, while cost depends on which two nodes are connected. Only one reliability and cost alternative is available for each pair of nodes. The approach proposed by Deeter and Smith significantly expands previous research by allowing edges to be chosen from different components with different costs and reliabilities. Define the following notations to describe optimal design of the network, allowing edges to be chosen from different edge options: k is the number of options for the edge connection, t is the option between nodes, x_{ij} ($x_{ij} \in \{0,1,2,...,k-1\}$) is an edge option for the edge between nodes i and j, $p(x_{ij})$ is the reliability of edge option, and $c(x_{ij})$ is a unit cost of the edge option.

Representation. Because each network design $\boldsymbol{x}$ is easily formed into an integer vector, it can be used as a chromosome for the genetic algorithm. Each element of the chromosome represents a possible edge in the network design problem, so there are $n \times (n-1)/2$ vector components in each candidate architecture $\boldsymbol{x}$. The value of each element tells what type of connection the specific edge has with the pair of nodes it connects. consider an example with

TABLE 5.4. Comparison of Search Effort Versus Reliability Estimation

Problem	Search Space	Solutions Searched	Fraction Searched	$R_{\min}$	$R(\boldsymbol{x})$	$\hat{R}(\boldsymbol{x})$	Percent Difference
1	1.02×10^{6}	6.00×10^{2}	5.86×10^{-1}	0.95	0.9579	0.9604	0.261
2	2.68×10^{8}	2.00×10^{4}	7.46×10^{-5}	0.95	0.9637	0.9645	0.083
3	3.52×10^{13}	3.00×10^{4}	2.27×10^{-9}	0.95	0.9516	[a]	
4	1.57×10^{27}	2.00×10^{5}	1.27×10^{-52}	0.95	[b]	0.9925	
5	2.10×10^{6}	1.50×10^{4}	7.14×10^{-3}	0.90	0.9035	0.9035	0.000
6	1.68×10^{7}	2.00×10^{4}	1.19×10^{-3}	0.95	0.9538	0.9550	0.126
7	1.07×10^{9}	3.00×10^{4}	2.80×10^{-5}	0.90	0.9032	0.9027	−0.055

[a]Optimal not found by genetic algorithms.
[b]Network is too large to calculate reliability exactly.

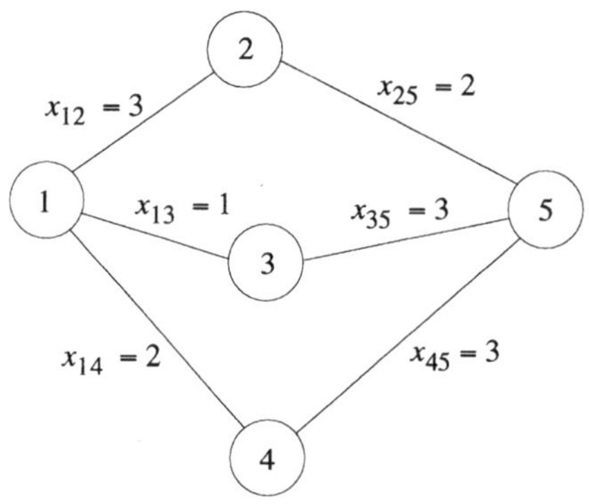

Figure 5.5. Sample network architecture.

five nodes and $k = 4$ options of connection, shown in Figure 5.5 and Table 5.5. Note that there are $(5 \times 4)/2 = 10$ possible edges for this example but only six are included; the other four are at option of connection $k = 0$. On the edge matrix in the Table 5.5, it is seen that the matrix is symmetric. Thus all the information needed to record this particular architecture lies in the upper triangular half of the matrix. This information is placed in a chromosome (vector) of length $(5 \times 4)/2 = 10$ by copying and concatenating each row into the chromosome, as shown in Figure 5.6. Note that the only possible values allowed in each position of the chromosome are $0,1,...,k - 1$. The solution space of possible network architectures is $k^{\{n\times(n-1)\}/2}$.

Fitness. The genetic algorithm attempts to find the minimum-cost network architecture that meets or exceeds a prespecified network reliability, $R_{\min}$. The fitness function is constructed so that it may consider infeasible network architectures. Infeasible solutions may contain beneficial parts. Consequently, breeding two infeasible solutions or an infeasible with a feasible solution can yield a good feasible solution. Also, since only one constraint will be active or nearly active for a minimum-cost network, the optimal design will lie on the boundary between feasible and infeasible designs. The fitness function is given as follows:

TABLE 5.5. Connectivity Matrix of Edge

$$\begin{bmatrix} — & 3 & 1 & 2 & 0 \\ 3 & — & 0 & 0 & 2 \\ 1 & 0 & — & 0 & 3 \\ 2 & 0 & 0 & — & 3 \\ 0 & 2 & 3 & 3 & — \end{bmatrix}$$

3	1	2	0	0	0	2	0	3	3

Each element represents a possible edge
Its value represents what type of connection

Figure 5.6. Chromosome for a sample network.

$$Z_p(\boldsymbol{x}) = Z(\boldsymbol{x}) + Z(\boldsymbol{x}^*)[1 + R_{\min} - R(\boldsymbol{x})]^{r_p + (pop\_size*gen)/50}$$

where $Z_p(\boldsymbol{x})$ is the penalized cost, $Z(\boldsymbol{x})$ the unpenalized cost, $Z(\boldsymbol{x}^*)$ the cost of the best feasible solution in the population, r_p the penalty rate, *pop_size* the population size, and *gen* the generation number.

Initial Population. The initial population is randomly generated by the following procedure:

Step 1. Generate the initial population randomly according to the edge options of reliability and cost.

Step 2. Send the initial population to the reliability calculation procedure.

Step 3. Send the initial population to the cost calculation procedure. If infeasible chromosomes exist, the chromosomes are penalized.

Step 4. Check for the best initial chromosome, and if no chromosome is feasible, the best infeasible chromosome is recorded; otherwise, record the best chromosome.

Selection. Two parents are selected to breed using a rank-based quadratic procedure as follows:

Step 1. Sort chromosomes according to their level of fitness, with 1 being the highest level of fitness (descending order).

Step 2. Generate a random number between 0 and $\sqrt{pop\_size}$.

Step 3. Square a generated random number.

Step 4. Truncate a squared number.

Step 5. Add one to a truncated number.

Step 6. Select the resulting chromosome. Then to choose the second parent, go to step 2. If the same parent is chosen, repeat until a distinct parent is chosen.

For example, suppose that a population size is 20. Then selection is processed as follows:

Step 1. Sort 20 chromosomes in descending order of fitness.

Step 2. Generate a random number between 0 and $\sqrt{20}$.

Step 3. Suppose that the number generated is 2.5; this number is then squared, giving 6.25.

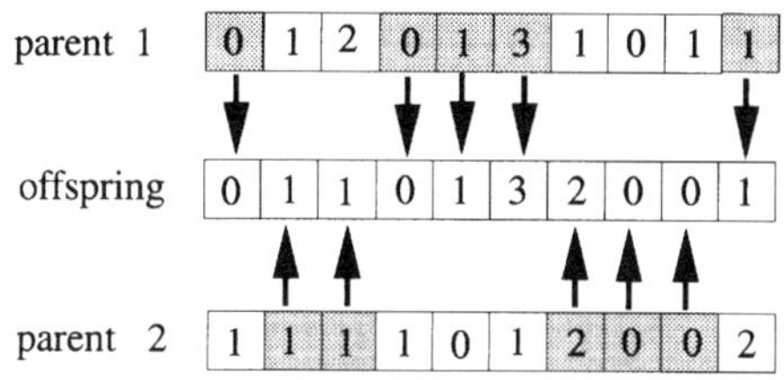

Figure 5.7. Uniform crossover operator.

Step 4. If 6.25 is truncated, 6 is yielded.
Step 5. Add 1 to 6; the number then becomes 7.
Step 6. Select the seventh chromosome.

Crossover. Uniform crossover is employed. This type of crossover is accomplished by selecting two parent solutions and randomly taking a component from one parent to form the corresponding component of the child, as shown in Figure 5.7.

Step 1. Take two chromosomes chosen by the selection operation.
Step 2. Randomly take a component from one chromosome to form the corresponding component of the child.
Step 3. Repeat step 2 until the components of the child fill up perfectly.

Mutation. Mutation is a way to add new genetic material to a population. This is done to help the genetic algorithm avoid getting stuck at a local optimum. A child undergoes mutation according to the percentage of population mutated, p_{m1}. Once a child is chosen to be mutated, the probability of mutation for each vector component is equal to the mutation rate, p_{m2}. If the $p_{m2} = 0.3$, each component of a selected child will be mutated with 0.3 probability.

Step 1. Generate a random real number $r_1, r_1 \in [0,1]$.
Step 2. If $r_1 < p_{m1}$, the chromosome is chosen to mutate and go to step 3; otherwise, go to step 6.
Step 3. Generate a random number $r_2, r_2 \in [0,1]$.
Step 4. If $r_2 < p_{m2}$, go to step 5; otherwise, go to step 6.
Step 5. If $x_{ij} = 0$, randomly choose one in $\{1,2,\ldots,k-1\}$. If $x_{ij} = t\ (\neq 0)$, randomly choose one in $\{1,2,\ldots,k-1\} - \{t\}$ with 0.5 probability or $x_{ij} = 0$ with 0.5 probability.
Step 6. Finish the mutation operation.

A mutation might be as shown in Figure 5.8.

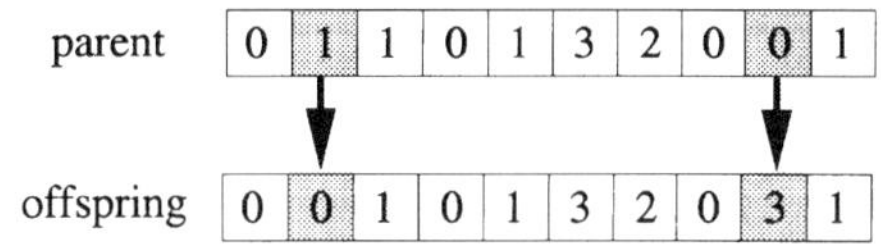

Figure 5.8. Mutation operator.

Reliability Calculation. A backtracking algorithm is used to calculate exactly the system unreliability, $1 - R(\boldsymbol{x})$, for problems due to their computationally tractable size. The following backtracking algorithm is used where the probability of all edges within a stack is the product of the failure probabilities of all inoperative edges times the product of 1 minus the failure probabilities of all operative edges.

Step 1. Initialize. Mark all edges as free and create a stack S that is initially empty.

Step 2. Generate a modified cutset.

(2.1) Find a set of free edges that together with all inoperative edges will form a network cut.

(2.2) Mark all the edges found in step (2.1) inoperative and add them to the stack.

(2.3) Now the stack represents a modified cutset; add its probability to a cumulative sum.

Step 3. Backtrack.

(3.1) If the stack is empty, go to step 4; otherwise, go to step (3.2).

(3.2) Take the edge off the top of the stack.

(3.3) If the edge is inoperative and if when making it operative, a spanning tree of operative edges exists, mark it free and go to step (3.1).

(3.4) If the edge is inoperative and the condition tested in step (3.3) does not hold, mark it operative, put it back on the stack, and go to step 2.

(3.5) If the edge is operative, mark it free and go to step (3.1).

Step 4. Return the network unreliability and end the procedure.

Overall Algorithm

Step 1. Set the parameters. Set the population size (*pop_size*), the percentage of population mutated (p_{m1}), the mutation rate (p_{m2}), the penalty rate (r_p), and the maximum generation (*max_gen*), and initialize the generation number $gen = 0$.

Step 2. Initialize.

(2.1) Randomly generate the initial population.

(2.2) Send the initial population to the reliability calculation function.

TABLE 5.6. Reliability and Edge Unit Cost

Connection Type	Reliability	Cost/Unit Distance
0	0.00	0
1	0.85	4
2	0.90	16
3	0.95	48

(2.3) Send the initial population to the cost calculation function. If infeasible chromosomes exist, they are penalized.

(2.4) Check for the best initial solution. If no chromosome is feasible, the best infeasible chromosome is recorded.

Step 3. Make the selection.

(3.1) Insert the best chromosome into the new population.

(3.2) Select two distinct candidate chromosomes from the current population by the rank-based quadratic selection mechanism.

Step 4. Perform the crossover and mutation. To obtain two children, apply a uniform crossover operator, and after a child is created, mutate it.

Step 5. Check the number of children. If $n < pop\_size - 1$, go to step 3; otherwise, go to step 5, where n represents the number of new children.

Step 6. Establish the new population. Replace parents with children that have been created.

Step 7. Evaluate.

(7.1) Send the new population to the reliability calculation function.

(7.2) Calculate eval($\boldsymbol{x}$) for each chromosome in the new population. If infeasible chromosomes exist, they are penalized.

Step 8. Check for the best new chromosome. Save the best new chromosome; if no chromosome is feasible, the best infeasible chromosome is recorded.

Step 9. Perform the terminating test. If $gen < max\_gen$, $gen = gen + 1$, and return to step 3 for the next generation. If $gen = max\_gen$, terminate.

Numerical Example. A five-node test problem with multiple levels was used by Deeter and Smith [153, 154]. The problem is based on Jan, Hwang, and Chen's test problem given in Table 5.6 [325]. Note that unit cost increases greater than linearly with reliability. This problem has a search space of 1,048,576 and is considered at seven different system reliability levels. Since this is a relatively small problem, it is possible to enumerate all solutions to obtain optimal solutions for the seven different system reliability levels. These appear in Table 5.7. Note that each optimal solution makes use of two or more edge connection levels. This indicates that designs improve by allowing differing edge reliabilities.

TABLE 5.7. Optimal Solution to Test Problems

R_{min}	Cost	$R(\boldsymbol{x})$	Chromosome
0.99900	5522	0.99908	3323333323
0.99500	4352	0.99518	3113311313
0.99000	3754	0.99052	3203322303
0.95000	2634	0.95353	3003302303
0.93125	2416	0.93361	2003301303
0.90000	2184	0.91854	3003300303
0.85000	1904	0.85536	3003200203

After exploratory trial runs, the genetic algorithm's parameters were established as follows: $pop\_size = 40$, $p_{m1} = 0.25$, $p_{m2} = 0.25$, $max\_gen = 6000$, and $r_p = 6$. Ten runs were then made for each system reliability using 10 randomly selected random number seeds. This examines the variability of the genetic algorithm since it is a stochastic search method.

5.3 TREE-BASED NETWORK RELIABILITY AND LAN DESIGN

Topology design problems for communication networks have been investigated by many researchers. The problems of how to design a network effectively with certain constraints is very important in many real-world applications. Especially in computer network systems, LANs are commonly used as a communication infrastructure that meets the demands of users in a local environment. These computer networks typically consist of several LAN segments connected via bridges. Use of these transparent bridges requires *loop-free* paths between LAN segments [174]. Therefore, spanning tree topologies can be used as active LAN configurations. Also, in designing the topology of these computer network systems, one important step is to find the best layout of components to optimize some performance criteria, such as cost, message delay, traffic, or reliability. The network topology design problem includes two primary issues: clustering and routing. The clustering problem consists of issues of how many segments (clusters) the LAN should be divided into and how to allocate users (stations) to the LAN segment (cluster). The routing problem is defined as the determination of segment interconnection spanning tree topology.

Genetic algorithms have received a great deal of attention regarding their potential as optimization techniques for network optimization problems and are often used to solve many real-world problems [219, 455]. The selection of optimal LAN topology is a very complicated combinatorial optimization problem and belongs to NP-hard problems. Therefore, a heuristic algorithm that is based on genetic algorithms gets attention. Solving the network to-

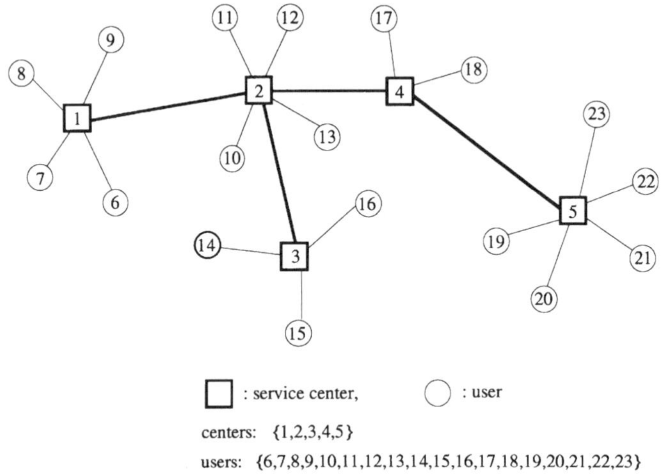

Figure 5.9. Sample LAN structure.

pology design problem using genetic algorithms was proposed in several papers. Elbaum and Sidi used genetic algorithms based on Huffman tree to solve the topological design of LANs [174]. Gen, Ida, and Kim proposed a genetic algorithm for solving bicriteria network design problems considering connection cost and average message delay [223, 225].

5.3.1 Bicriteria Network Topology Design

Consider a LAN that connects m users (stations). For example, Figure 5.9 shows LANs with 5 service centers and 18 users. We also assume an $n \times n$ *service center topology matrix* $\boldsymbol{X}_1$, which represents the connection between service centers. An element x_{1ij} is represented as

$$x_{1ij} = \begin{cases} 1, & \text{if the centers } i \text{ and } j \text{ are connected} \\ 0, & \text{otherwise} \end{cases}$$

Assume that the LAN is partitioned into n segments (service centers or clusters). The users are distributed over those n service centers. The $n \times m$ *clustering matrix* $\boldsymbol{X}_2$ specifies which user belongs to which center. Thus

$$x_{2ij} = \begin{cases} 1, & \text{if user } j \text{ belongs to center } i \\ 0, & \text{otherwise} \end{cases}$$

A user can belong only to one center: thus $\forall j = 1,2,\ldots,m$, $\Sigma_{i=1}^{n} x_{2ij} = 1$. We define an $n \times (n + m)$ matrix $\boldsymbol{X}$ called the *spanning tree matrix* ($[\boldsymbol{X}_1 \quad \boldsymbol{X}_2]$).

Service centers are interconnected via bridges. The traffic from source center i to destination center j is forwarded through the bridges. The path between centers i and j may include a single bridge that connects both centers, or multiple bridges in case there is no direct connection between centers. In the latter case, the traffic travels through bridges and centers on the connecting path. As stated before, only the use of transparent bridges is considered. The selection of transparent bridges forces the network to be configured in a spanning tree topology.

A center is a LAN segment with known capacity. A LAN segment can be token ring, Ethernet, or other kinds of architectures. It is clear that the behavior and performance of each type of LAN segment differ from each other. Nevertheless, the average delay in all those architectures responds qualitatively to load growth in the same manner. The average delay increases as the load over the segment builds up. When the load approaches the segment capacity, the delay approaches infinity.

The bicriteria LAN topology design problem can be formulated as the following nonlinear 0–1 programming model:

$$\max \quad R(\boldsymbol{X}) \tag{5.3}$$

$$\min \quad \sum_{i=1}^{n-1} \sum_{j=i+1}^{n} w_{1ij} x_{1ij} + \sum_{i=1}^{n} \sum_{j=1}^{m} w_{2ij} x_{2ij} \tag{5.4}$$

$$\text{s.t.} \quad \sum_{j=1}^{m} x_{1ij} \le g_i, \qquad i = 1,2,\ldots,n \tag{5.5}$$

$$\sum_{i=1}^{n} x_{2ij} = 1, \qquad j = 1,2,\ldots,m \tag{5.6}$$

where $R(\boldsymbol{X})$ is the network reliability, w_{1ij} the weight of link between centers i and j, w_{2ij} the weight of link between center i and user j, and g_i the maximum number capable of connecting to center i.

Representation and Initialization. The genetic representation is a type of data structure that represents the candidate solutions of problems. Usually, different problems have different data structures or genetic representations. Two types of data structure are suitable to represent active LAN configurations: (1) both service centers and users are represented by Prüfer number [223, 225], and (2) service centers are represented by Prüfer number and users are depicted by a clustering string that describes the distribution of users into service centers.

One of the classical theorems in graphical enumeration is Cayley's theorem that there are $k^{(k-2)}$ distinct labeled trees on a complete graph with k nodes. Prüfer provided a constructive proof of Cayley's theorem by establishing a

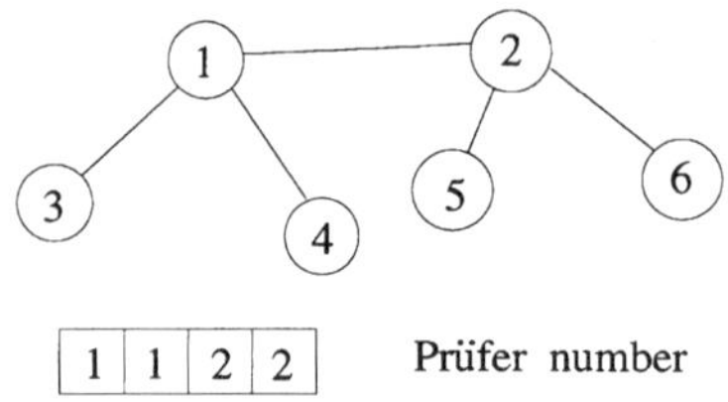

Figure 5.10. Tree and its Prüfer number.

one-to-one correspondence between such trees and the set of all strings of $k - 2$ digits [699]. This means that we can use only $(k - 2)$-digit permutation to uniquely represent a tree where each digit is an integer between 1 and k inclusive. This permutation is usually known as a Prüfer number. For any tree there are always at least two leaf nodes [576]. Based on this observation, we can easily construct an encoding as follows:

Procedure: Encoding of Prüfer Number

Step 1. Let Node i be the smallest labeled leaf node in a labeled tree T.

Step 2. Let j be the first digit in the encoding, as the code j incident to node i is uniquely determined. The encoding is built by appending digits from left to right.

Step 3. Remove node i and the link from i to j; thus we have a tree with $k - 1$ nodes.

Step 4. Repeat the steps above until one link is left. We produce a Prüfer number or an encoding with $k - 2$ digits between 1 and k inclusive.

It is also possible to generate a unique tree from a Prüfer number (Figure 5.10) via the following procedure.

Procedure: Decoding of Prüfer Number

Step 1. Let P be the original Prüfer number and let $\overline{P}$ be the set of all nodes not included in P which are designated as eligible nodes for consideration in building a tree.

Step 2. Let i be the eligible node with the smallest label. Let j be the left most digit of P. Add the edge from i to j into the tree. Remove i from $\overline{P}$ and j from P. If j does not occur anywhere in P, put it into $\overline{P}$. Repeat the process until no digits are left in P.

Step 3. If no digits remain in P, there are exactly two nodes, r and s, still eligible for consideration. Add a link from r to s into the tree and form a tree with $k - 1$ links.

This representation has two kinds of genotype: The service centers are represented by Prüfer number, and users are described by clustering string.

center

2	2	4

users

6	7	8							. . .								23
1	1	1	1	2	2	2	2	3	3	3	4	4	5	5	5	5	5

Figure 5.11. Representation for Figure 5.9.

The chromosomes are generated randomly in the initialization process (Figure 5.11). The service centers are composed of $n - 2$ digits (Prüfer number) randomly generated in the range $[1,n]$, and users are made up of m digits (clustering string) randomly generated in the range $[1,n]$, which indicate how to allocate users to service centers so that each user belongs to a specific center.

Evaluation. A single-objective LAN topology problem can easily be manipulated by calculating the fitness value of each chromosome according to the objective function. But in the bicriteria LAN topology design problem, we must handle Pareto solutions. The weighted-sum method is used to construct the fitness function, which combines the bicriteria objective functions into one overall fitness function.

Step 1. Convert chromosomes represented by the Prüfer number and clustering string to spanning tree matrix $\boldsymbol{X}$.

Step 2. Calculate the objective values $[f_i(\boldsymbol{X}),\ i = 1,2]$.

Step 3. Choose solution points that contain the minimum $f_i^{\min}$ (or maximum $f_i^{\max}$) corresponding to each objective function value, and compare with the stored solution points in the preceding generation and select the best points to save again.

$$f_i^{\min(t)} = \min_k \{(f_i^{\min(t-1)}, f_i^{(t)}(\boldsymbol{X}_k)) \mid k = 1,2,\ldots,chr\_size\}, \qquad i = 1,2$$

$$f_i^{\max(t)} = \max_k \{(f_i^{\max(t-1)}, f_i^{(t)}(\boldsymbol{X}_k)) \mid k = 1,2,\ldots,chr\_size\}, \qquad i = 1,2$$

where $f_i^{\min(t)}$ and $f_i^{\max(t)}$ are the minimum and maximum value of the ith objective function in generation t, respectively; $f_i^{(t)}$ is the ith objective function value of the kth chromosome in generation t; and *chr_size* is equal to the *pop_size* plus the offspring generated after genetic operations.

Step 4. Calculate the weight coefficient as follows:

$$\beta_i = \frac{\alpha_i}{\sum_{i=1}^{2} \alpha_i}, \qquad i = 1,2$$

where

$$\alpha_i = \frac{f_i^{\max(t)} - f_i^{\min(t)}}{f_i^{\max(t)}}, \qquad i = 1,2$$

Step 5. Calculate the fitness function value for each chromosome as follows:

$$\text{eval}(\boldsymbol{X}_k) = \sum_{i=1}^{2} \beta_i d_i(\boldsymbol{X}_k), \qquad k = 1,2,\ldots,chr\_size$$

where $d_i(\boldsymbol{X}_k)$ is the normalized value of the *i*th objective function and is represented as follows:

$$d_i(\boldsymbol{X}_k) = \begin{cases} \dfrac{f_i^{(t)}(\boldsymbol{X}_k) - f_i^{\min(t)} + \gamma}{f_i^{\max(t)} - f_i^{\min(t)} + \gamma}, & \text{for maximization} \\[2ex] \dfrac{f_i^{\max(t)} - f_i^{(t)}(\boldsymbol{X}_k) + \gamma}{f_i^{\max(t)} - f_i^{\min(t)} + \gamma}, & \text{for minimization} \end{cases}$$

γ is a small positive real number that is usually restricted within the open interval (0,1), used to prevent the equation from zero division, and makes it possible to adjust the selection behavior from fitness-proportional selection to pure random selection.

For more detailed information of this normalizing method and weighted-sum evaluation, readers can refer to [219].

Tree-Based Reliability Calculation. Consider as a reliability measure the probability of all operative nodes (centers and users) being connected. Now calculate the reliability of a spanning tree network, assuming that the reliability of its elements, nodes, and links are known. Assume the tree to be a rooted tree. A state vector is associated with the root of each subtree. The state vector contains all information about that node relevant to calculation. Define a set of recursion relations that yield the state vector of a rooted tree given the state of its subtrees. For subtrees involving single nodes, the state is obvious. Then the rooted subtrees become larger and larger rooted subtrees using recursion relations until the state of the entire network is obtained [359].

Deriving the recurrence relations is somewhat mechanical. It comes simply from considering the situation depicted in Figure 5.12. We have two subtrees, one with root *i* and the other with root *j*. Assume that the states of nodes *i* and *j* are known and we wish to compute the state of *j* relative to the tree obtained by joining node *i* and node *j* by the link (*i*,*j*), where node *j* is the father of node *i*.

Assume that the probability of node failure p_i^f and the probability of the node being operative p_i^o $(=1 - p_i^f)$ are associated with node *i*. Similarly, link (*i*,*j*) has probabilities t_i^f and l_i^o of the link (*i*,*j*) failing and being operative,

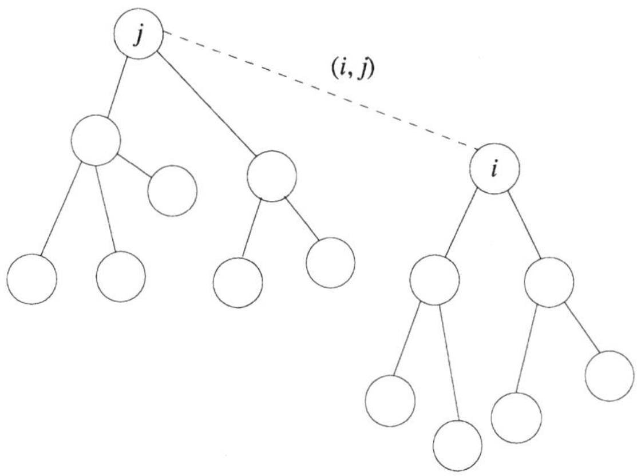

Figure 5.12. Recurrence relation.

respectively. Also define the following state vectors for each subtree: e_i is the probability that all nodes in the subtree are failed; o_i is the probability that the set of operative nodes, including the root of the subtree, are connected; and r_i is the probability that the root of the subtree is failed and the set of operative nodes in the subtree is connected.

For a tree with root node 1 and n nodes, we can calculate the reliability of the tree [i.e., the probability $R(\boldsymbol{X})$ of all operating nodes communicating] as follows:

Procedure: Reliability Calculation

Step 1. Set $r_i = 0$, $o_i = p_i^o$, and $e_i = p_i^f$, $i = 1,2,\ldots,n$. Set $i = n$.

Step 2. If node j is the father of node i, using the following recurrence relations, recalculate r_i, o_i, and e_i:

$$r_j' = r_j e_i + r_i e_j + o_i e_j$$

$$o_j' = o_i o_j l_i^o + o_j e_i$$

$$e_j' = e_i e_j$$

Step 3. Set $i = i - 1$. If $i = 1$, go to step 4; otherwise, go to step 2.

Step 4. Return $r_1 + o_1 + e_1$.

Selection. The selection used here combines the roulette wheel and elitist approaches as follows:

Step 1. Calculate a cumulative probability a_p for each chromosome $\boldsymbol{X}_p$ ($p = 1,2,\ldots,chr\_size$).

Step 2. Generate a random real number r in [0,1].

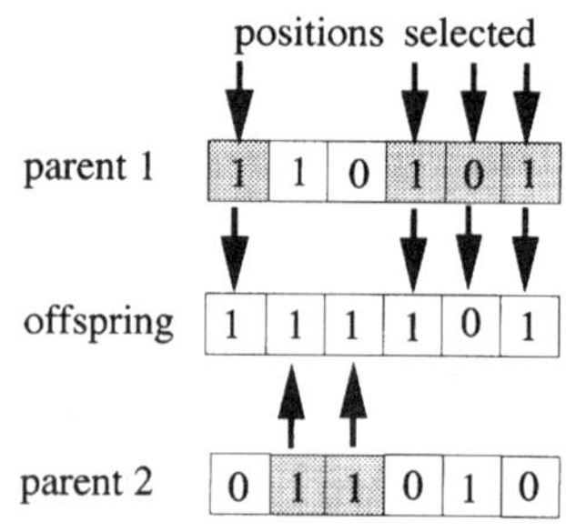

Figure 5.13. Uniform crossover operator.

Step 3. If $r \leq a_1$, select the first chromosome $\boldsymbol{X}_1$; otherwise, select the pth chromosome $\boldsymbol{X}_p$ $(2 \leq p \leq chr\_size)$ such that $a_{p-1} < r < a_p$.

Step 4. Repeat steps 2 and 3 *pop_size* times and obtain *pop_size* copies of chromosomes.

Step 5. If the best chromosome is not selected in the next generation, replace one randomly with the best one from the new population.

Crossover. Uniform crossover is used. This type of crossover is accomplished by selecting two parent solutions and randomly taking a component from one parent to form the corresponding component of the offspring (Figure 5.13).

Mutation. Swap mutation is used, which simply selects two positions at random and swaps their contents as shown in Figure 5.14.

Repairing Chromosomes. Because of the existence of a constraint on the maximum number that is capable of connecting each center, the chromosomes generated randomly in the initial population and the offspring produced by crossover may be illegal in the sense of violating the maximum number of connections for each center. There are two strategies to deal with this violation: the penalizing strategy and the repairing strategy. For some combinatorial optimization problems, it is really difficult to provide a reasonable penalizing factor for the illegal chromosomes when the illegality is not easily

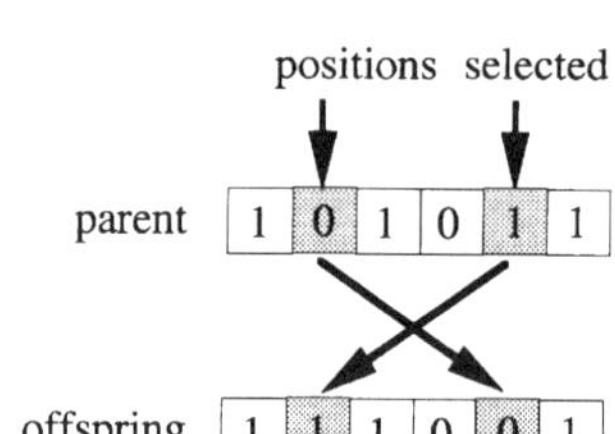

Figure 5.14. Swap mutation operator.

TABLE 5.8. Weight Matrix $[w_{1ij}]$ of Link Between Centers i and j

$$\begin{bmatrix} — & 125 & 110 & 210 \\ & — & 221 & 134 \\ & & — & 123 \\ & & & — \end{bmatrix}$$

measured quantitatively. Because it is relatively easy to repair an illegal chromosome of the spanning tree representation, the repairing strategy is used to modify the connection number for the center in an illegal chromosome.

Let $\overline{G}$ be the set of centers whose maximum number of connections has not been checked and modified in a chromosome. If center i violates the constraint with the maximum number g_i of connection for center i, this means that the number of this center in the chromosome is more than $g_i - 1$. Then decrease the number of the center by checking the extra center and replace it randomly with another center from $\overline{G}$.

5.3.2 Numerical Examples

Gen and Kim [228, 363] used the following two instances to demonstrate the performance of their genetic algorithm

Example 5.1. The first problem has 4 service centers ($n = 4$), 8 users ($m = 8$), and $g_i = 3$. The weight matrix of the link between centers i and j (w_{1ij}) is given in Table 5.8. the weight matrix of link between center i and user j (w_{2ij}) is given in Table 5.9.

Assume that the operative probability of centers is 0.95, the operative probability of users is 0.9, the operative probability between centers is 0.9, and the operative probability between center and user is 0.85.

The parameters for genetic algorithm are set as follows: $pop\_size = 100$, $max\_gen = 500$, $p_c = 0.3$, $p_m = 0.7$, and the experiment was run 20 times. Pareto optimal solutions can generally be found illustrated in Figure 5.15.

From Figure 5.15 and Table 5.10 we can see that Gen and Kim's method obtained better result than the method proposed by Dengiz, Altiparmak, and Smith [156, 157].

TABLE 5.9. Weight Matrix $[w_{2ij}]$ of Link Between Center i and User j

$$\begin{bmatrix} 34 & 62 & 7 & 46 & 26 & 31 & 19 & 29 \\ 26 & 15 & 16 & 68 & 37 & 43 & 15 & 58 \\ 10 & 49 & 86 & 72 & 36 & 49 & 31 & 78 \\ 28 & 22 & 98 & 80 & 8 & 15 & 98 & 7 \end{bmatrix}$$

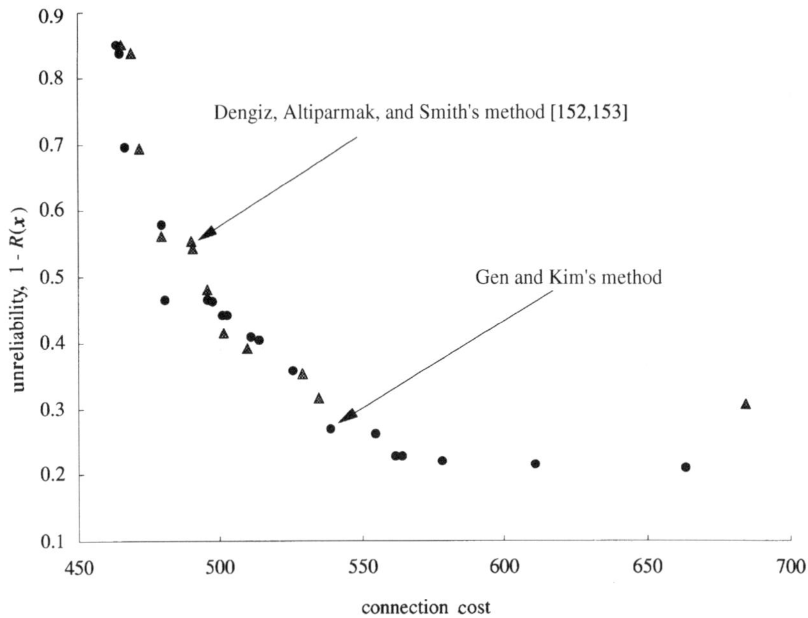

Figure 5.15. Pareto solutions for Example 5.1.

Example 5.2. The second problem has 6 service centers ($n = 6$), 30 users ($m = 30$), and $g_i = 10$. The weight matrix of link between centers i and j (w_{1ij}) is given in Table 5.11. The weight matrix of link between center i and user j (w_{2ij}) is given in Table 5.12.

Assume that the operative probability of centers is 0.95, the operative probability of users is 0.9, the operative probability between centers is 0.9, and the operative probability between center and user is 0.85.

The parameters for genetic algorithm are set as follows: $max\_gen = 500$, $p_c = 0.3$, $p_m = 0.7$. pop_size is set by 50, 100, and 200, and the experiment is run 20 times. The Pareto optimal solutions obtained are illustrated in Figure 5.16.

TABLE 5.10. Comparison of Proposed Encoding and Edge-Based Encoding[a]

Example	Gen and Kim Method		Dengiz, Altiparmak, and Smith Method	
	ACT (s)	Memory Size, $n + m - 2$	ACT (s)	Memory Size, $n \times (n - 1)/2$
5.1	5.16	10	9.37	66
5.2	33.38	34	[b]	630

[a]ACT, average computation time.
[b]Not executable.

TABLE 5.11. Weight Matrix $[w_{1ij}]$ of Link Between Centers i and j

$$
\begin{bmatrix}
— & 125 & 110 & 210 & 221 & 105 \\
 & — & 107 & 202 & 154 & 167 \\
 & & — & 175 & 163 & 231 \\
 & & & — & 217 & 206 \\
 & & & & — & 157 \\
 & & & & & —
\end{bmatrix}
$$

Gen and Kim [228, 363] employed the TOPSIS method of Yoon and Hwang to determine the best compromise solution among Pareto solutions. TOPSIS (technique for order preference by similarity to ideal solution) is based on the concept that the chosen alternative should have the shortest distance from the positive ideal solution and the longest from the negative ideal solution. More detailed information about TOPSIS can be obtained in [219] and [313]. Figure 5.17 depicts the best compromise solution obtained by the TOPSIS method. The chromosome of the best compromise solution is

Centers: 6 5 6 2
Users: 5 5 5 6 6 4 6 4 5 2 4 2 1 6 6 2 1 2 5 4 6 5 3 6 3 1 2 5 1 4

The connection cost is 1595 and the network reliability is 0.79654.

5.4 MULTIOBJECTIVE RELIABILITY DESIGN

5.4.1 Bicriteria Reliability Design

The bicriteria reliability design problem considered in this section tries to maximize the reliability of a series system and simultaneously minimize the total cost of the system. The problem is a variation of the optimal reliability allocation problem given by Dhingra [162], which can be formulated as a nonlinear mixed-integer programming problem as follows [227, 410]:

$$\max \quad f_1(\boldsymbol{m},\boldsymbol{x}) = \prod_{i=1}^{n} [1 - (1 - x_i)^{m_i}] \tag{5.7}$$

$$\min \quad f_2(\boldsymbol{m},\boldsymbol{x}) = \sum_{i=1}^{n} C(x_i)\left[m_i + \exp\left(\frac{m_i}{4}\right)\right] \tag{5.8}$$

$$\text{s.t.} \quad G_1(\boldsymbol{m}) = \sum_{i=1}^{n} w_i m_i \exp\left(\frac{m_i}{4}\right) \le W_s \tag{5.9}$$

$$G_2(\boldsymbol{m}) = \sum_{i=1}^{n} v_i (m_i)^2 \le V_s \tag{5.10}$$

$$1 \le m_i \le 10, \qquad i = 1,\ldots,4 \tag{5.11}$$

$$0.5 \le x_i \le 1 - 10^{-6}, \qquad i = 1,\ldots,4 \tag{5.12}$$

TABLE 5.12. Weight Matrix $[w_{2ij}]$ of Link Between Center i and User j

$$
\left[\begin{array}{rrrrrrrrrrrrrrrrrrrrrrrrrrrrrr}
34 & 62 & 7 & 46 & 26 & 19 & 31 & 29 & 26 & 15 & 16 & 68 & 37 & 15 & 100 & 58 & 10 & 49 & 86 & 100 & 36 & 31 & 49 & 78 & 100 & 22 & 98 & 80 & 8 & 98 \\
15 & 7 & 80 & 35 & 54 & 85 & 22 & 71 & 81 & 5 & 29 & 46 & 37 & 29 & 79 & 17 & 20 & 39 & 25 & 84 & 5 & 36 & 22 & 12 & 86 & 96 & 9 & 79 & 15 & 54 \\
42 & 27 & 25 & 39 & 52 & 90 & 80 & 35 & 58 & 19 & 2 & 34 & 43 & 43 & 51 & 64 & 19 & 36 & 26 & 36 & 16 & 71 & 41 & 51 & 52 & 50 & 13 & 34 & 58 & 73 \\
66 & 76 & 84 & 81 & 20 & 45 & 10 & 61 & 34 & 86 & 50 & 18 & 21 & 94 & 25 & 27 & 50 & 61 & 81 & 33 & 54 & 18 & 93 & 7 & 62 & 18 & 75 & 28 & 12 & 37 \\
73 & 62 & 34 & 89 & 44 & 85 & 96 & 78 & 7 & 57 & 50 & 43 & 48 & 78 & 25 & 53 & 16 & 45 & 55 & 71 & 11 & 69 & 50 & 93 & 86 & 62 & 18 & 23 & 9 & 73 \\
22 & 44 & 24 & 32 & 31 & 3 & 50 & 47 & 76 & 11 & 92 & 63 & 44 & 24 & 30 & 63 & 46 & 66 & 44 & 70 & 23 & 10 & 72 & 7 & 63 & 9 & 17 & 87 & 41 & 64
\end{array}\right]
$$

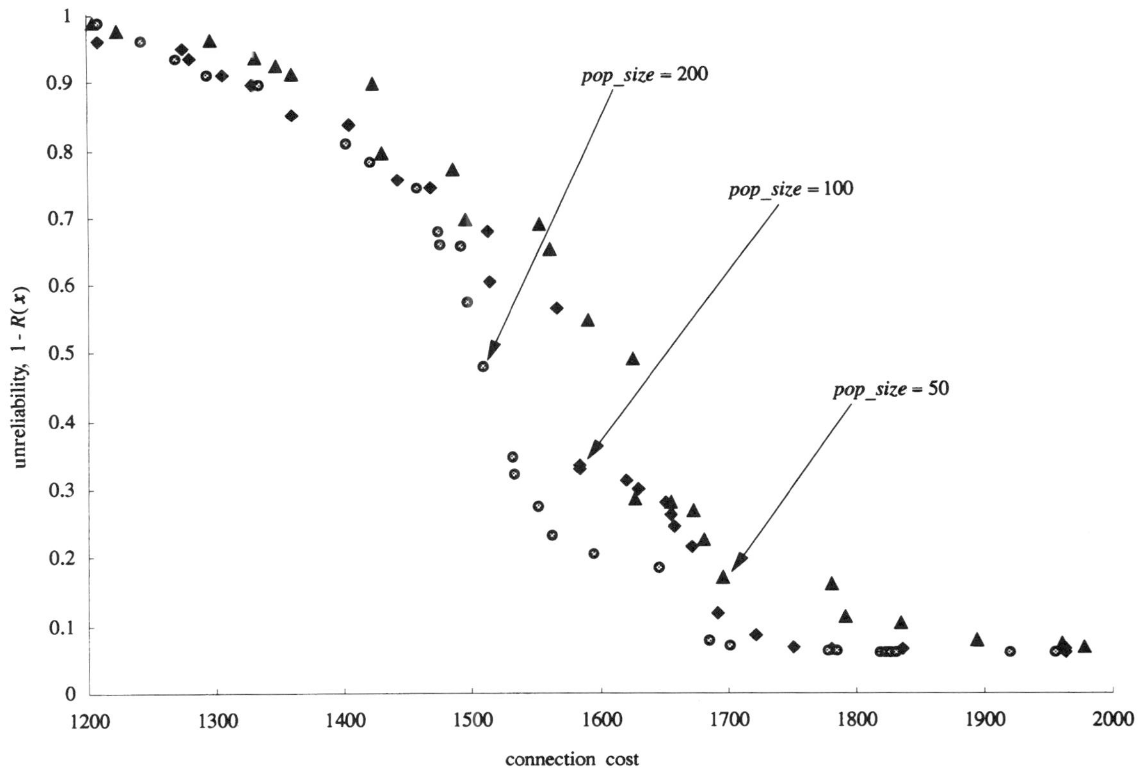

Figure 5.16. Pareto solutions for Example 5.2.

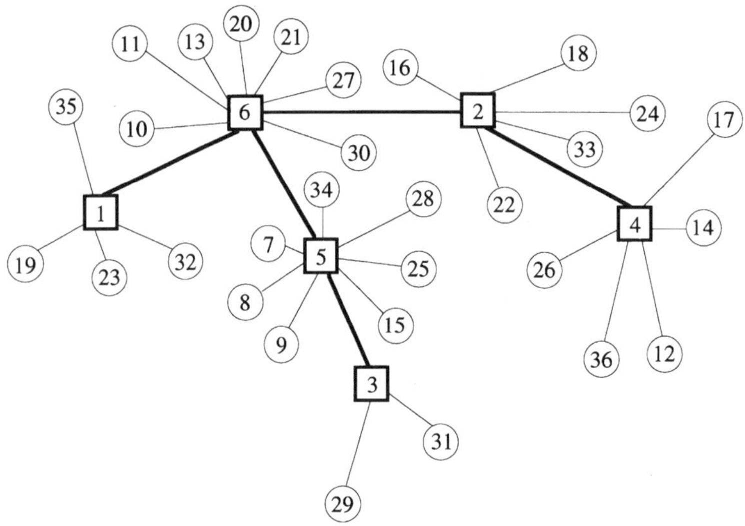

Figure 5.17. Best compromise solution for Pareto solutions.

where m_i is the number of redundant components in subsystem i and $\boldsymbol{m} = [m_1 \quad m_2 \cdots m_n]$, x_i is the level of component reliability for the ith subsystem and $\boldsymbol{x} = [x_1 \quad x_2 \cdots x_n]$, $f_1(\boldsymbol{m},\boldsymbol{x})$ is the reliability of the system with redundant components $\boldsymbol{m}$ and component reliabilities $\boldsymbol{x}$, $f_2(\boldsymbol{m},\boldsymbol{x})$ is the total cost of the system with component allocation $\boldsymbol{m}$ and component reliability $\boldsymbol{x}$, v_i is the product of weight and volume per element in subsystem i, w_i is the weight of each components in subsystem i, and $C(x_i)$ is the cost of each component with reliability x_i at subsystem i as follows:

$$C(x_i) = \alpha_i \left(\frac{-O_T}{\ln(x_i)} \right)^{\beta}, \qquad i = 1,\dots,4$$

where α_i and β_i are constants representing the physical characteristics of each component in subsystem i, and O_T is the operating time during which the component must not fail.

5.4.2 Genetic Algorithm Approach

Gen, Kim and Li developed a genetic algorithm to solve this problem [230, 410].

Representation and Initialization. Let $\boldsymbol{v}_k$ denote the kth chromosome in a population as follows:

$$\boldsymbol{v}_k = [(m_{k1},x_{k1})(m_{k2},x_2)(m_{k3},x_{k3})(m_{k4},x_{k4})] \qquad k = 1,2,\dots,pop\_size$$

The initial population is produced such that each gene in a chromosome is generated randomly within its domain.

Evaluation and Selection. The fitness of chromosomes is calculated by a ranking method described as follows:

Procedure: Rank-Based Evaluation

Step 1. Calculate each objective value for each chromosome.

Step 2. Ranking chromosomes based on their objective function values and obtaining the order $r_i(\boldsymbol{v}_k)$. $r_i(\boldsymbol{v}_k)$ is the rank value of the ith objective value of chromosome $\boldsymbol{v}_k$ and is obtained by setting a value of 1 on the best objective function value of the present populations and *pop_size* on the worst objective function. If the objective function of maximization type, $r_i(\boldsymbol{v}_k)$ is set as 1 on the largest objective function value and *pop_size* on the smallest objective function. On the contrary, for the minimization type, $r_i(\boldsymbol{v}_k)$ is set as 1 on the smallest objective function value and *pop_size* on the largest objective function.

Step 3. Calculate the fitness value using the following equation:

$$\text{eval}(\boldsymbol{v}_k) = \sum_{i=1}^{Q} r_i(\boldsymbol{v}_k)$$

where Q is the number of objective functions.

Elitist selection [219, 455] was adopted. The procedure works as follows: Calculate the evaluation function eval($\boldsymbol{v}_k$), and select chromosomes among the parents and offspring that are superior to the others. The number to be selected is *pop_size*. Duplication of selection is prohibited.

Crossover. An arithmetic crossover operator is used, which is defined as a linear combination of two chromosomes [219, 455]. Let $\boldsymbol{v}_{k1}$ and $\boldsymbol{v}_{k2}$ denote two chromosomes selected randomly for crossover in generation k. The offspring will be

$$\boldsymbol{o}_{k1} = \lfloor c\boldsymbol{v}_{k1} + (1 - c)\boldsymbol{v}_{k2} \rfloor$$

$$\boldsymbol{o}_{k2} = \lfloor c\boldsymbol{v}_{k2} + (1 - c)\boldsymbol{v}_{k1} \rfloor$$

where $\lfloor x \rfloor$ denotes the maximum integer smaller than the real number x and c is a random number in the range [0,1].

Mutation. Uniform mutation is used here. For a chosen parent $\boldsymbol{v}_k$, if its element x_{k3} is randomly selected for mutation, the resulting offspring is

$$\boldsymbol{v}'_k = [(m_{k1}, x_{k1})(m_{k2}, x_{k2})(m_{k3}, x'_{k3})(m_{k4}, x_{k4})]$$

where x'_{k3} is a random (uniform probability distribution) value in the range $[0.5, 1 - 10^{-6}]$. If m_{k3} is selected, the integer random number is then returned

TABLE 5.13. Constant Coefficients for Reliability Apportionment Problem

Number of subsystems			4	
Limit on W_s			500.0	
Limit on V_s			250.0	
Operating time t_O			1000 h	
Subsystem	$10^5 \cdot \alpha_j$	β_j	v_j	w_j
1	1.0	1.5	1	6
2	2.3	1.5	2	6
3	0.3	1.5	3	8
4	2.3	1.5	2	7

within its domain. This operator ensures that the genetic algorithm can search the solution space freely but has the shortcomings of divergence at later steps.

Numerical Example and Results. Gen, Kim and Li used Dhingra's example to test their algorithm. The data are given in Table 5.13. The parameters are set as $pop\_size = 200$, $p_c = 0.4$, $p_m = 0.6$, and $max\_gen = 1000$. The Pareto solutions obtained are shown in Table 5.14.

5.4.3 Hybrid Genetic Algorithm Approach

Gen and Kim proposed a hybrid genetic algorithm to solve the bicriteria reliability design problem [229]. A Hooke–Jeeves method, a type of direct search [526], was embedded in the main loop of genetic algorithms and applied to each newly generated offspring to move it to its local optima. The hybrid genetic algorithm was applied to the problem given in equations (5.7) to (5.12).

Representation and Initialization. Because there are two types of decision variables in the problem, the following representation of chromosome is used:

$$\mathbf{v}_k = [(m_{k1},x_{k1})\ (m_{k2},x_{k2})\ (m_{k3},x_{k3})\ \cdots\ (m_{kn},x_{kn})], \qquad k = 1,2,\ldots,pop\_size$$

where m_{ki} is an integer corresponding to the number of redundant components and x_{ki} is a real number corresponding to the level of component reliability. The initial population is produced such that each gene in the chromosome is randomly generated within its domain.

Evaluation and Selection. The fitness function takes a weighted sum form while combining multiple objectives into a single objective. The weight coefficients are determined by the variance of objective function values in the current generation. The details are described below.

TABLE 5.14. Pareto Solutions

$R(\boldsymbol{m},\boldsymbol{x})$	$C(\boldsymbol{m},\boldsymbol{x})$	$G_1(\boldsymbol{m})$	$G_2(\boldsymbol{m})$	Chromosome
0.998395	398.559	396.25	217.00	[(0.777820, 5) (0.697710, 6) (0.916213, 4) (0.748629, 6)]
0.995562	313.737	391.43	244.00	[(0.791324, 6) (0.705309, 5) (0.682558, 6) (0.743228, 5)]
0.992422	253.985	369.69	235.00	[(0.764636, 4) (0.607286, 6) (0.801730, 5) (0.713641, 6)]
0.989157	206.800	364.25	206.00	[(0.670458, 6) (0.689409, 5) (0.832996, 4) (0.574555, 6)]
0.987313	186.650	386.00	244.00	[(0.694460, 5) (0.649614, 6) (0.769595, 5) (0.561921, 6)]
0.972312	136.093	386.00	244.00	[(0.602327, 5) (0.576630, 6) (0.729846, 5) (0.529785, 6)]

Procedure: Evaluation

Step 1. Calculate the objective function values $[f_i(\boldsymbol{v}_k),\ i = 1,2,\ k = 1,2,...,chr\_size]$, where *chr_size* is the total number of parents and offspring.

Step 2. Standardize the objective function values as follows:

$$\hat{f}_i(\boldsymbol{v}_k) = \frac{f_i(\boldsymbol{v}_k) - \bar{f}_i}{\sigma_i}, \qquad i = 1,2, \quad k = 1,2,...,chr\_size$$

where $\bar{f}_i$ is the mean value of the ith objective function among the entire population and σ_i is the standard deviation of the ith objective function.

Step 3. Calculate the scaled value $c_i(\boldsymbol{v}_k)$ with mean value h and standard deviation d of the entire population as follows:

$$c_i(\boldsymbol{v}_k) = d\hat{f}_i(\boldsymbol{v}_k) + h, \qquad i = 1,2, \quad k = 1,2,...,chr\_size$$

Step 4. Determine the maximum and minimum standardized values for each objective function:

$$f_i^{\min(t)} = \min\{(f_i^{\min(t-1)}, f_i^{(t)}(\boldsymbol{v}_k))|k \in [1,chr\_size]\}$$

$$f_i^{\max(t)} = \max\{(f_i^{\max(t-1)}, f_i^{(t)}(\boldsymbol{v}_k))|k \in [1,chr\_size]\}, \qquad i = 1,2$$

where $f_i^{\min(t)}$ and $f_i^{\max(t)}$ are the minimum and the maximum of the ith objective function in generation t, respectively. $f_i^{(t)}(\boldsymbol{v}_k)$ is an ith objective function value of the kth chromosome in generation t.

Step 5. Calculate the weight coefficients as follows:

$$\lambda_i = \frac{\delta_i}{\sum_{i=1}^{2} \delta_i}, \qquad i = 1,2$$

where

$$\delta_i = \frac{f_i^{\max(t)} - f_i^{\min(t)}}{f_i^{\max(t)}}$$

Step 6. Calculate the fitness value for each chromosome as follows:

$$\text{eval}(\boldsymbol{v}_k) = \sum_{i=1}^{2} \lambda_i c_i(\boldsymbol{v}_k), \qquad k = 1,2,...,chr\_size$$

Crossover and Mutation. An arithmetic crossover operator is used. Let $\boldsymbol{v}_{k1}$ and $\boldsymbol{v}_{k2}$ be two chromosomes randomly selected for crossover in generation k. The offspring will be

$$\boldsymbol{o}_{k1} = \lfloor c\boldsymbol{v}_{k1} + (1 - c)\boldsymbol{v}_{k2} \rfloor$$
$$\boldsymbol{o}_{k2} = \lfloor c\boldsymbol{v}_{k2} + (1 - c)\boldsymbol{v}_{k1} \rfloor$$

where $\lfloor x \rfloor$ is defined as the maximum integer smaller than the real number x and c is a random number within the range [0,1].

For a chosen parent $\boldsymbol{v}_k$, suppose that element x_{k3} is randomly selected for mutation. The resulting offspring then is

$$\boldsymbol{v}'_k = [(m_{k1},x_{k1})(m_{k2},x_{k2})(m_{k3},x'_{k3})(m_{k4},x_{k4})]$$

where x'_{k3} is a random (uniform probability distribution) value in the range $[0{,}5,\ 1 - 10^{-6}]$. If m_{k3} is selected, a random integer number is returned within its domain.

Hooke–Jeeeves (HJ) Method. This method [526] is adopted as a local search method to improve the performance of genetic algorithms. The Hooke–Jeeves method is a sequential technique of which each step consists of two types of moves: exploratory and pattern. The exploratory move examines the local behavior of the function to be optimized and seeks to locate the direction of any sloping valleys that might be present. The pattern move utilizes the information generated in the exploration move to progress rapidly along the valleys. The following version of the Hooke–Jeeves method is used by Gen and Kim to solve the bicriteria reliability design problem [227].

Procedure: Hooke–Jeeves Method

Step 1. Start with a chromosome $\boldsymbol{v}$, called the starting base point, and prescribed step lengths δ_i in each coordinate direction $\boldsymbol{d}_i$, $i = 1,2,\ldots,n$.

Step 2. Compute $f_s(\boldsymbol{v})$, where $f_s(\boldsymbol{v})$ is $-f_1(\boldsymbol{v})$ if $s = 1$; otherwise, $f_s(\boldsymbol{v})$ is $f_2(\boldsymbol{v})$ and s is a random integer number in the range [1,2]. Set $i = 1$, $\boldsymbol{u}_0 = \boldsymbol{v}$, and start the exploratory move as stated in step 3.

Step 3. Each variable x_i is perturbed around the current temporary base point $\boldsymbol{u}_{i-1}$ to obtain a new temporary base point as follows:

$$\boldsymbol{u}_i = \begin{cases} \boldsymbol{u}_{i-1} + \delta_i \boldsymbol{d}_i, & \text{if } F^+ < F \\ \boldsymbol{u}_{i-1} - \delta_i \boldsymbol{d}_i, & \text{if } F^- < F < F^+ \\ \boldsymbol{u}_{i-1}, & \text{if } \max\{F^+,F^-\} < F \end{cases} \tag{5.13}$$

where the point $\boldsymbol{u}_i$ indicates the temporary base point obtained from the base point $\boldsymbol{u}_{i-1}$ by perturbing the ith component of $\boldsymbol{v}$, $F^+ = f_s(\boldsymbol{u}_{i-1} + \delta_i \boldsymbol{d}_i)$, $F^- = f_s(\boldsymbol{u}_{i-1} - \delta_i \boldsymbol{d}_i)$, $F = f_s(\boldsymbol{u}_{i-1})$. The process of finding a new temporary base point is continued until x_n is perturbed to find $\boldsymbol{u}_n$.

Step 4. If the point $\boldsymbol{u}_n$ remains the same as $\boldsymbol{v}$, reduce the step lengths δ_i divided by 2, set $i = 1$, and go to step 3. if $\boldsymbol{u}_n$ is different from $\boldsymbol{v}$, obtain the new base point as $\boldsymbol{v}' = \boldsymbol{u}_n$ and go to step 5.

Step 5. With the help of base points $\boldsymbol{v}$ and $\boldsymbol{v}'$, establish a pattern direction $\boldsymbol{s} = \boldsymbol{v}' - \boldsymbol{v}$ and calculate a new chromosome:

$$\boldsymbol{v}_{\text{new}} = \boldsymbol{v}' + \boldsymbol{s} \tag{5.14}$$

The procedure above is employed after selection operation to over the newly selected offspring to its local optima before the next reproduction.

Numerical Example and Result. This numerical example is a variation of the optimal reliability allocation problem given by Dhingra [162], which is a nonlinear mixed-integer programming problem. This example includes two objectives: reliability maximization and cost minimization. The data for this problem are shown in Table 5.13.

For this problem, the parameters were set as $h = 50$, $d = 10$, $pop\_size = 40$, $p_c = 0.5$, $p_m = 0.1$, and $max\_gen = 2000$. The Pareto solutions obtained are shown in Figure 5.18. Also, the results of several experiments show that the hybrid genetic algorithm outperforms a simple genetic algorithm without Hooke–Jeeves [227, 229].

5.4.4 Reliability Design with Fuzzy Goals

Gen, Ida, and Kim proposed a genetic algorithm approach for solving reliability design problems with fuzzy goals [224]. The problem is a variation of optimal reliability allocation problem given by Dhingra [162], which is formulated as the following nonlinear mixed-integer programming problems:

$$\max \quad R(\boldsymbol{m},\boldsymbol{x}) = \prod_{i=1}^{4} [1 - (1 - x_i)^{m_i}] \gtrsim b_1 \tag{5.15}$$

$$\min \quad C(\boldsymbol{m},\boldsymbol{x}) = \sum_{i=1}^{4} C(x_i)\left[m_i + \exp\left(\frac{m_i}{4}\right)\right] \lesssim b_2 \tag{5.16}$$

$$\min \quad W(\boldsymbol{m}) = \sum_{i=1}^{4} w_i m_i \exp\left(\frac{m_i}{4}\right) \lesssim b_3 \tag{5.17}$$

$$\text{s.t.} \quad G_1(\boldsymbol{m}) = \sum_{i=1}^{4} v_i (m_i)^2 \le 250 \tag{5.18}$$

$$1 \le m_i \le 10, \text{ integer}, \; i = 1,\ldots,4 \tag{5.19}$$

$$0.5 \le x_i \le 1 - 10^{-6}, \text{ real number}, \qquad i = 1,\ldots,4 \tag{5.20}$$

where v_i is the product of weight and volume per element at stage i, w_i the weight of each component at stage i, and $C(x_i)$ the cost of each component with reliability x_i at stage i as follows:

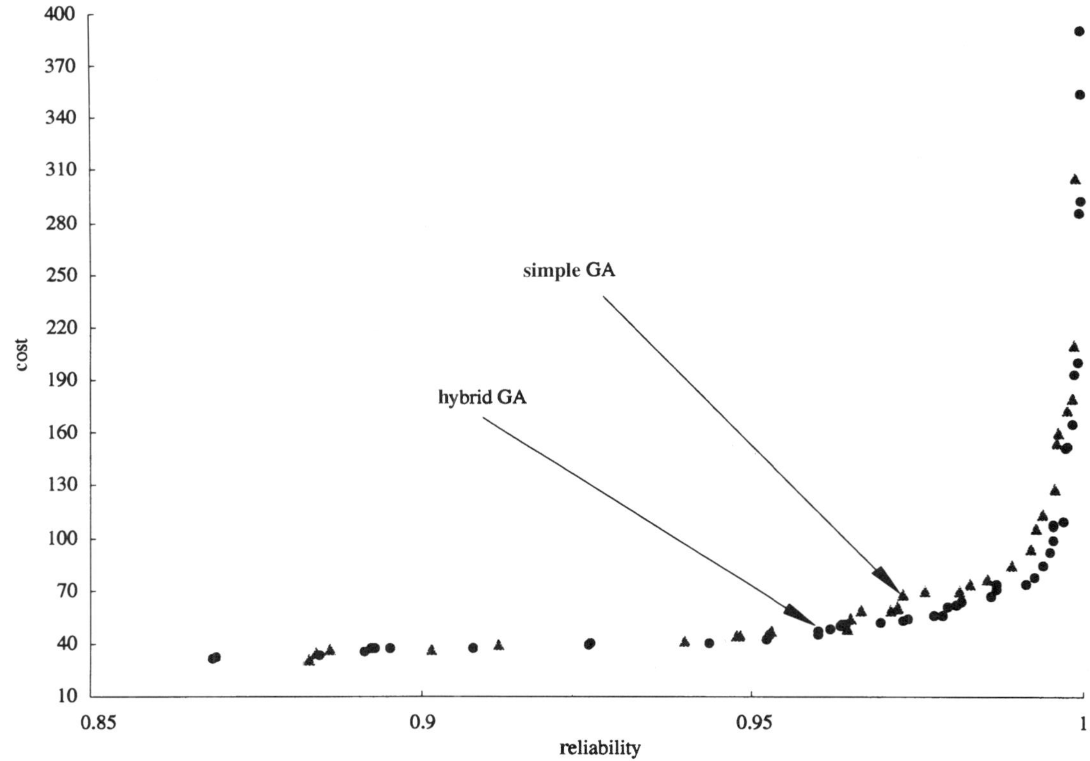

Figure 5.18. Pareto solutions.

$$C(x_i) = \alpha_i\left(\frac{-O_T}{\ln(x_i)}\right)^{\beta_i}, \qquad i = 1,\ldots,4$$

where α_i and β_i are constants representing the physical characteristic of each component in subsystem i, O_T is the operating time during which the component must not fail, F_1 is equal to $b_1 + t_1^L$, F_2 is equal to $b_2 + t_2^R$, and F_3 is equal to $b_2 + t_3^R$. The symbol $\gtrsim$ (fuzzy-max) signifies that the decision maker is satisfied even if the result is less than an aspiration level b_k up to a certain tolerance limit t_k^L, and the symbol $\lesssim$ (fuzzy-min) signifies that the decision maker is satisfied even if the result is greater than b_k up to a certain tolerance limit t_k^R.

According to Tiwari, Dharmar and Rao, the problem can be transformed into the following nonlinear mixed-integer programming problem [122, 622]:

$$\max \quad q_R\lambda_R + q_C\lambda_C + q_W\lambda_W \tag{5.21}$$

$$\text{s.t.} \quad \lambda_R = \mu_R(R(\boldsymbol{m},\boldsymbol{x})) \tag{5.22}$$

$$\lambda_C = \mu_C(C(\boldsymbol{m},\boldsymbol{x})) \tag{5.23}$$

$$\lambda_W = \mu_W(W(\boldsymbol{m})) \tag{5.24}$$

$$G_1(\boldsymbol{m}) = \sum_{i=1}^{4} v_i m_i^2 \leq 250 \tag{5.25}$$

$$0 \leq \lambda_R \leq 1, \quad 0 \leq \lambda_C \leq 1, \quad 0 \leq \lambda_W \leq 1 \tag{5.26}$$

where q_R, q_C, and q_W are weight factors assigned by decision makers [622]. The membership functions [705] of the fuzzy goals can be defined as follows:

$$\mu_R(r(\boldsymbol{m},\boldsymbol{x})) = \begin{cases} 0, & \text{if } R(\boldsymbol{m},\boldsymbol{x}) < b_1 - t_1^L \\ \dfrac{R(\boldsymbol{m},\boldsymbol{x}) - (b_1 - t_1^L)}{t_1^L}, & \text{if } b_1 - t_1^L \leq R(\boldsymbol{m},\boldsymbol{x}) \leq b_1 \\ 1, & \text{if } R(\boldsymbol{m},\boldsymbol{x}) > b_1 \end{cases}$$

$$\mu_C(C(\boldsymbol{m},\boldsymbol{x})) = \begin{cases} 1, & \text{if } C(\boldsymbol{m},\boldsymbol{x}) < b_2 \\ 1 - \dfrac{C(\boldsymbol{m},\boldsymbol{x}) - b_2}{t_2^R}, & \text{if } b_2 \leq C(\boldsymbol{m},\boldsymbol{x}) \leq b_2 + t_2^R \\ 0, & \text{if } C(\boldsymbol{m},\boldsymbol{x}) > b_2 + t_2^R \end{cases}$$

$$\mu_W(W(\boldsymbol{m},\boldsymbol{x})) = \begin{cases} 1, & \text{if } W(\boldsymbol{m},\boldsymbol{x}) < b_3 \\ 1 - \dfrac{W(\boldsymbol{m},\boldsymbol{x}) - b_3}{t_3^R}, & \text{if } b_3 \leq W(\boldsymbol{m},\boldsymbol{x}) \leq b_3 + t_3^R \\ 0, & \text{if } W(\boldsymbol{m},\boldsymbol{x}) > b_3 + t_3^R \end{cases}$$

TABLE 5.15. Constant Coefficients for Reliability Design with Fuzzy Goals

Number of subsystems	4
Limit on F_1	0.75
Limit on F_2	400.0
Limit on F_3	500.0
Upper limit on V	250.0
Operating time t_O	1000 h

Subsystem	$10^5 \cdot \alpha_j$	β_j	v_j	d_j
1	1.0	1.5	1	6
2	2.3	1.5	2	6
3	0.3	1.5	3	8
4	2.3	1.5	2	7

The constant coefficients for this problem are given in Table 5.15 [162].

Representation and Initiation. Let $\boldsymbol{v}_k$ denote the chromosomes in a population. The chromosomes can be represented as follows:

$$\boldsymbol{v}_k = [(m_{k1},x_{k1})(m_{k2},x_{k2})(m_{k3},x_{k3})(m_{k4},x_{k4})], \qquad k = 1,2,\ldots,pop\_size$$

where m_{ki} corresponds to the number of redundant components and x_{ki} corresponds to the level of component reliability. The initial population is produced such that each gene is generated randomly within its domain.

Evaluation of Chromosomes. The fitness function takes the following form:

$$\text{eval}(\boldsymbol{v}_k) = q_r\lambda_R + q_c\lambda_C + q_W\lambda_W, \qquad k = 1,2,\ldots,pop\_size$$

The best chromosome $\boldsymbol{v}^*$ is selected by the following equation at each generation:

$$\boldsymbol{v}^* = \{\text{eval}(\boldsymbol{v}_k)\{k = 1,2,\ldots,pop\_size\}$$

Crossover. An arithmetic crossover operator was used. Let $\boldsymbol{v}_{k1}$ and $\boldsymbol{v}_{k2}$ be two chromosomes randomly selected for crossover in generation k. The offspring will be

$$\boldsymbol{o}_{k1} = \lfloor c\boldsymbol{v}_{k1} + (1 - c)\boldsymbol{v}_{k2} \rfloor$$
$$\boldsymbol{o}_{k2} = \lfloor c\boldsymbol{v}_{k2} + (1 - c)\boldsymbol{v}_{k1} \rfloor$$

where c is a random number within the range [0,1]. For the integer variables in a chromosome, the outcomes after the arithmetic crossover operation will become real numbers. We simply take their integer part as the resulting gene.

Mutation. Uniform mutation was used. For a chosen parent v_k, if its element x_{k3} is randomly selected for the mutation, the resulted offspring is

$$\boldsymbol{v}'_k = [(m_{k1},x_{k1})(m_{k2},x_{k2})(m_{k3},x'_{k3})(m_{k4},x_{k4})]$$

where x'_{k3} is a random (uniform probability distribution) value within the range $[0.5,1 - 10^{-6}]$. If m_{k3} is selected, an integer random number within its domain will be returned. This operator ensures that the genetic algorithm can search the search space freely from start to end but has the shortcomings of divergence at later steps.

Selection. The selection operation combines roulette wheel selection with an elitist approach. Roulette wheel selection is used, essentially, to reproduce a new generation randomly; the elitist method is employed to preserve the best chromosome for the next generation and overcome stochastic errors of sampling.

Numerical Example and Results. The parameters were set as follows: *pop_size* = 20, $p_c = 0.4$, $p_m = 0.1$, and $T = 2000$, as well as $t_1^L = 0.25$, $t_2^R = 300$, and $t_3^R = 360$. The relative weights of objectives (0.5,0.25,0.25) were selected.

The best solution was found in the 1923th generation with objectives $R(\boldsymbol{m},\boldsymbol{x}) = 0.982335$, $C(\boldsymbol{m},\boldsymbol{x}) = 135.001259$, $W(\boldsymbol{m}) = 147.048541$, and $(\boldsymbol{m},\boldsymbol{x})$ = [(3,0.853070) (3,0.821421) (2,0.939903) (3,0.825620)]. The optimal solutions in [158] are $R(\boldsymbol{m},\boldsymbol{x}) = 0.94478$, $C(\boldsymbol{m},\boldsymbol{x}) = 104.472$, and $W(\boldsymbol{m}) = 128.727$ with $(\boldsymbol{m},\boldsymbol{x})$ = [(2,0.86504) (3,0.81814) (2,0.84253) (3,0.80645)].

6
SCHEDULING PROBLEMS

6.1 INTRODUCTION

Scheduling problems, broadly speaking, concern the allocation of limited resources over time to perform tasks to satisfy certain criteria. Resources can be very different in nature: for example, labor, money, machines, tools, materials, energy, and so on. Also, tasks can have a variety of interpretations, from machining parts in manufacturing systems to processing information in computer systems. A task is usually characterized by such factors as ready time, due date, relative urgency weight, processing time, and resource consumption. Moreover, a structure of a set of tasks, reflecting precedence constraints among them, can be defined in different ways. In addition, different criteria that measure the quality of the performance of a schedule can be taken into account.

Scheduling problems exist almost everywhere in real-world situations, especially in the industrial engineering world. Many scheduling problems from manufacturing industries are quite complex in nature and very difficult to solve by conventional optimization techniques. Usually, these difficult-to-solve problems are characterized as a combinatorial optimization problem (the problem with a finite number of feasible solutions) subject to highly complex constraints.

They belong to the class of NP-hard problem. This has led to the recent interest in using genetic algorithms to address the problem. In the following sections we explain how to solve them with genetic algorithms, including job-shop scheduling, grouped job scheduling, parallel machine scheduling, resource-constrained project scheduling, and multiprocessor scheduling.

6.2 JOB-SHOP SCHEDULING

In the job-shop scheduling problem, we are given a set of jobs and a set of machines. Each machine can handle at most one job at a time. Each job consists of a chain of operations, each of which needs to be processed during an uninterrupted time period of given length on a given machine. The purpose is to find a schedule, that is, an allocation of the operations to time intervals on the machines, that has the minimum duration required to complete all jobs [37].

This problem is one of the best known of the difficult combinatorial optimization problems. During the last three decades, the problem has captured the interest of a significant number of researchers, and many solution methods have been proposed, ranging from simple and fast dispatching rules to sophisticated branch-and-bound algorithms. However, with the rapid increase in the speed of computing and the growing need for efficiency in scheduling, it becomes increasingly important to explore ways of obtaining better schedules at some extra computational cost.

6.2.1 Basic Approaches

The essence of the job-shop scheduling problem is to establish a permutation of operations on each machine subject to precedence constraints to minimize production time. If such a permutation can be determined, a solution can then easily be derived with a problem-specific procedure. A general approach to applying genetic algorithms to the job-shop scheduling problem is to:

1. Use genetic algorithms to evolve an appropriate permutation
2. Use a heuristic method to construct a subsequent solution according to the permutation

Because genetic algorithms are not well suited for fine tuning of structures close to optima, various methods of hybridization have been suggested to compensate for this shortcoming. The hybridization methods for job-shop scheduling problem can be classified into the following three categories [147, 533]: (1) adapted genetic operators, (2) heuristic-featured genetic operators, and (3) hybrid genetic algorithms. The first approach is to revise or invent genetic operators so as to meet the features of a given encoding representation. The second approach is to create new genetic operators inspired from conventional heuristics. The third approach involves hybridizing conventional heuristics into the main loop of genetic algorithms where possible.

6.2.2 Encodings

In general, a crossover operator is regarded as a main genetic operator, and the performance of genetic algorithms depends to a great extent on the performance of the crossover operator used. Conceptually, crossover operates on two chromosomes at a time and generates offspring by combining both chromosomes' features. A simple way to achieve crossover is to use the well-known one- or two-cutpoint crossover methods, which work well with bit string encodings. For many complex problems, it is usually very difficult to represent a solution using bit string encodings. Therefore, many non-bit string encodings, especially various literal string encodings, have been suggested and many new genetic operators have been adapted to cope with these non-bit string encodings. Essentially, the crossover methods adapted can be viewed as revised versions of one- or two-cutpoint crossovers from some non-bit string encodings by incorporating repair procedures to avoid yielding illegal or infeasible solutions.

For ease of explanation, we first define two types of literal strings: pure and general. A *pure literal string encoding* consists of distinctive symbols; a *general literal string encoding* allows a symbol to repeat a prescribed number. In other words, duplication is prohibited in a pure literal string, whereas the same symbol can coexist in a general literal string.

During the past 10 years, the following nine representations for the job-shop scheduling problem have been proposed [114]:

1. Operation-based representation
2. Job-based representation
3. Preference list–based representation
4. Job pair relation–based representation
5. Priority rule–based representation
6. Disjunctive graph–based representation
7. Completion time–based representation
8. Machine–based representation
9. Random keys representation

Among these nine methods, job-based and machine-based encodings are of pure literal string types; operation-based, preference list–based, and priority rule–based encodings are of general literal string type. Note that preference list–based encoding is used successfully in many studies. It consists of several substrings, and one substring is a pure literal string corresponding to an operation sequence for a machine. The usual way in practice is to apply genetic operators to each substring for this encoding. With multiple use of genetic operators, this encoding is treated essentially as a pure literal string. Therefore, many types of genetic operators for a pure literal string can be used directly for this encoding.

As we know, there are two types of order relations in the job-shop scheduling problem: (1) the operation sequence on each machine and (2) the precedence constraints among operations for a job. The first type must be determined by a solution method, while the second must be maintained in a schedule. Accordingly, several procedures have been used to handle order relations. One is that the information with respect to both operation sequence and precedence constraints is preserved in the encoding simultaneously. In such a case, all crossover methods for literal permutation encodings are not directly applicable. A chromosome may be either *infeasible* in the sense that some precedence constraints are violated, or *illegal* in the sense that the repetitions of some symbols are not equal to the prescribed numbers. For pure literal string encoding, illegality means that some symbols are repeated more than once while other symbols get lost. For general literal string encoding, illegality means that some symbols may appear more than necessary, others less than necessary. Special attention must be given to how to handle infeasibility and illegality when designing a new crossover operator for such types of encodings so as not to disorder precedence relations. Among literal string encodings, operation-based encoding is the only type that retains information on both operation sequence and precedence constraints in a chromosome.

In another situation only information with respect to operation sequence is encoded; a special decoder or schedule builder procedure is used to resolve the precedence constraints. In such cases the chromosome is one type of literal permutation, with attention directed to preventing a crossover operator from producing illegal offspring. Among literal string encodings, preference list–based, job-based, and machine-based encodings are of this type.

In the final case, neither order relation, but some guild information, is encoded in the chromosome. Such encoding is a pure permutation of literal strings. Although the duplication of some genes results in illegality, as in the case above, the order of genes has no direct correspondence to the operational sequence of jobs. Priority rule–based encoding is of this type.

6.2.3 Adapted Genetic Operators

During the past two decades, various crossover operators have been proposed for literal permutation encodings, such as partially mapped crossover (PMX), order crossover (OX), position-based crossover, order-based crossover, and cycle crossover (CX). Note that there are two basic considerations when designing crossover operators for literal permutation encodings:

1. To make less change when crossing over so as to inherit parents' features as much as possible. All variations of two-cutpoint crossover operator belong to this class.
2. To make more change when crossing over so as to explore new patterns of permutation and thereby enhance the search ability [115]. All variations of uniform crossover belong to this class.

Partially Mapped Crossover (PMX). Partially mapped crossover was proposed by Goldberg and Lingle [252]. It can be viewed as a variation of two-cutpoint crossover by incorporating a special repair procedure to resolve possible illegitimacy. PMX has the following major steps:

Step 1. Select two cutpoints along the string at random. The substrings defined by the two cutpoints are called *mapping sections.*

Step 2. Exchange two substrings between parents to produce protochildren.

Step 3. Determine the mapping relationship between two mapping sections.

Step 4. Legalize offspring using the mapping relationship.

Order Crossover (OX). Order crossover was proposed by Davis [603]. It can be viewed as a kind of variation of PMX that uses a different repair procedure. OX has the following major steps:

Step 1. Select a substring from one parent at random.

Step 2. Produce a protochild by copying the substring into positions corresponding to those in the parent.

Step 3. Delete all the symbols from the second parent which are already in the substring. The resulting sequence contains the symbols the protochild needs.

Step 4. Place the symbols into unfixed positions in the protochild from left to right according to the order of the sequence to produce an offspring.

Position-Based Crossover. Position-based crossover was proposed by Syswerda [603]. It is essentially a uniform crossover for literal permutation encodings incorporated with a repair procedure. A uniform crossover operator was proposed for bit string encoding by Syswerda [603]. It first generates a random mask and then exchanges relative genes between parents according to the mask. A crossover makes is simply a binary string with the same size of chromosome. For each corresponding bit in an offspring, the parity of each bit in the mask determines from which parent it will receive that bit. Because uniform crossover will produce illegal offspring for literal permutation encodings, position-based crossover uses a repair procedure to resolve the illegitimacy. Position-based crossover has the following major steps:

Step 1. Select a set of positions from one parent at random.

Step 2. Produce a protochild by copying the symbols on these positions into the corresponding positions in the protochild.

Step 3. Delete the symbols already selected from the second parent. The resulting sequence contains only the symbols the protochild needs.

Step 4. Place the symbols into unfixed positions in the protochild from left to right according to the order of the sequence used to produce one offspring.

Order-Based Crossover. Order-based crossover was also proposed by Syswerda [603]. It is a slight variation of position-based crossover in which the order of symbols in the position selected in one parent is imposed on the corresponding position in the other parent.

Cycle Crossover (CX). Cycle crossover was proposed by Oliver, Smith, and Holland [490]. As in position-based crossover, it takes some symbols from one parent and the remaining symbols from the other parent. The difference is that symbols from the first parent are not selected randomly and only those symbols are selected that define a cycle according to the corresponding positions between parents. CX works as follows:

Step 1. Find the cycle that is defined by the corresponding positions of symbols between parents.

Step 2. Copy the symbols in the cycle to a child with positions corresponding to those of one parent.

Step 3. Determine the remaining symbols for the child by deleting those symbols that are already in the cycle from the other parent.

Step 4. Fulfill the child with the remaining symbols.

Linear Order Crossover (LOX). Falkenauer and Bouffoix proposed a modified version of order crossover, linear order crossover [183]. Order crossover tends to transmit the relative positions of genes rather than the absolute positions. In order crossover, the chromosome is considered to be circular since an operator is devised for the traveling salesman problem (TSP). In the job-shop problem, the chromosome cannot be considered to be circular. For this reason Falkenauer and Bouffoix developed a variant of the OX called linear order crossover (LOX), in which the chromosome is considered linear instead of circular. The LOX works as follows:

Step 1. Select sublists from parents randomly.

Step 2. Remove $sublist_2$ from parent p_1, leaving some "holes" and then slide the holes from the extremities toward the center until they reach the cross section. Similarly, remove $sublist_1$ from parent p_2 and slide the holes to the cross section.

Step 3. Insert $sublist_1$ into the holes of parent p_2 to form offspring o_1 and insert $sublist_2$ into the holes of parent p_1 to form the offspring o_2.

The crossover operator can preserve as much as possible both the relative positions between genes and the absolute positions relative to the extremities of parents. The extremities correspond to the high- and low-priority operations.

Subsequence Exchange Crossover. Kobayashi, Ono, and Yamamura [371] proposed a subsequence exchange crossover method, inspired by similar ideas of Brady [74] and Mühlenbein, Schlouter, and Kraemer [470] for the TSP. A job-sequence matrix is used as an encoding. For an n-job m-machine problem, the encoding is an $m \times n$ matrix where each row specifies an operation sequence for each machine. A subsequence is defined as a set of jobs that are processed consecutively on a machine for both parents but not necessarily in the same order.

Step 1. Identify subsequences machine by machine for the parents.

Step 2. Exchange these subsequences machine by machine among parents to create offspring.

Because it is difficult to maintain the precedence relation among operations in either initial population or offspring by use of job-sequence matrix encoding, a Giffler and Thompson algorithm is used to adjust job orders carefully on each machine to resolve the infeasibility and to convert offspring into active schedules.

Job-Based Order Crossover. Ono, Yamamura, and Kobayashi [493] proposed a variation of their subsequence exchange crossover method, called job-based order crossover, by relaxing the requirement that all jobs in the subsequence must be processed consecutively. Job-based order crossover is also designed for the encoding of a job-sequence matrix.

Step 1. Identify sets of jobs from parents, one set for one machine.

Step 2. Copy the selected jobs of the first parent onto the corresponding positions of the first child machine by machine. Do the same thing for the second child.

Step 3. Fulfill the unfixed position of the first child by the unselected jobs from left to right according to their order of appearance in the second parent. Do the same for the second child.

Partial Schedule Exchange Crossover. Gen, Tsujimura, and Kubota proposed a partial schedule exchange crossover for operation-based encoding [240]. They consider partial schedules to be the natural building blocks and intend to use such crossover to maintain building blocks in offspring in much the same manner as Holland described [303].

Step 1. Identify a partial schedule in one parent randomly and in the other parent accordingly.

Step 2. Exchange the partial schedules to generate protooffspring.

Step 3. Determine gene missed and exceeded for the protooffspring.

Step 4. Legalize offspring by deleting genes exceeded and adding genes missed.

The partial schedule is identified by the same job in the head and tail of the partial schedule.

Substring Exchange Crossover. Cheng, Gen, and Tsujimura presented another version of partial schedule exchange crossover, called substring exchange crossover [115]. It can be viewed as an adaptation of two-cutpoint crossover for general literal string encodings.

Step 1. Select two cutpoints along the string at random. Exchange two substrings defined by the two cuts between two parents to produce protochildren.

Step 2. Determine the missed and exceeded genes for each protochild by making a comparison between two substrings.

Step 3. Legalize the protochildren by random replacement of the genes exceeded by the genes missed.

Mutation. It is relatively easy to make some mutation operators for permutation representation. During the last decade, several mutation operators have been proposed for permutation representation, including inversion, insertion, displacement, reciprocal exchange mutation, and shift mutation [219]. *Inversion mutation* selects two positions within a chromosome at random and inverts the substring between these two positions. *Insertion mutation* selects a gene at random and inserts it in a random position. *Displacement mutation* selects a substring at random and inserts it in a random position. Insertion can be viewed as a special case of displacement in which the substring contains only one city. *Reciprocal exchange mutation* selects two positions at random and swaps the genes on these positions. In *shift mutation* a gene is chosen randomly and then shifted to a random position right or left of the gene's position.

6.2.4 Heuristic-Featured Genetic Operators

Inspired by some successful heuristic methods, several heuristic-featured genetic operators have been created for job-shop scheduling problems. The kernel procedures within this type of genetic operator are heuristic methods. A method can be identified as a crossover simply because it will merge two parents to produce two offspring, and as a mutation because it will alter a certain number of genes from one parent to produce offspring.

Giffler and Thompson Algorithm–Based Crossover. Yamada and Nakano proposed a crossover operator based on the Giffler–Thompson algorithm [676]. The generation procedure of the Giffler–Thompson algorithm is a tree search approach. At each step it essentially identifies all processing conflicts (operations competing for the same machine), and an enumeration procedure is used to resolve these conflicts. Yamada and Nakano's crossover operator is essentially a one-pass procedure, not a tree search approach. When generating offspring, at each step it identifies all processing conflicts, as in the

Giffler–Thompson method, and then chooses one operation from a conflicting set of operations according to one of their parents' schedules. Let

o_{ji} = ith operation of job j

S = set of schedulable operations for a given partial schedule

ϕ_{ji} = earliest completion time of the ith operation of job j in S

G_r = set of conflicting operations in S on machine r

The procedure to generate offspring from two parents works as follows:

Procedure: Giffler and Thompson Algorithm–Based Crossover

Step 1. Let S include all operations with no predecessors initially.

Step 2. Determine $\phi^* = \min\{\phi_{ji} \mid o_{ji} \in S\}$ and machine r^* on which ϕ^* could be realized.

Step 3. Let G_{r^*} include all operations $o_{ji} \in S$ that require machine r^*.

Step 4. Choose one operation from G_{r^*} as follows:

(4.1) Generate a random $\epsilon \in [0,1)$ and compare it with mutation rate p_m if ($\epsilon < p_m$), then choose an arbitrary operation from G_{r^*} as o_{ji}^*.

(4.2) Otherwise, select one of two parents with equal probability, say p_s, and find the operation o_{ji}^* that was scheduled earliest in p_s among all the operations in G_{r^*}.

(4.3) Schedule o_{ji}^* in the offspring according to ϕ_{ji}.

Step 5. Update S as follows:

(5.1) Remove operation o_{ji}^* from S.

(5.2) Add the direct successor of operation o_{ji}^* to S.

Step 6. Return to step 2 until a complete schedule is generated.

In step (4.1), conflict is resolved by choosing an operation randomly, whereas in step (4.2), conflict is resolved by giving priority to the operation scheduled earliest in one of its parents p_s among all conflicting operations in G_{r^*}. The parent p_s is selected randomly with equal probability. So the Yamada–Nakano approach is based essentially on priority dispatching heuristics, not a pure Giffler–Thompson approach. At each step, one operation is selected to add into the partial schedule of offspring, and conflicts among operations are resolved by specifying priority to operations according to their parents' schedules.

Neighborhood Search–Based Mutation. In conventional genetic algorithms, mutation is a background operator used to produce small perturbation on chromosomes to maintain population diversity. Cheng, Gen, and Tsujimura proposed a mutation inspired by a neighborhood search technique [115]. It is then not a background operator and is used to perform intensive search to find improved offspring.

Many definitions can be considered for the neighborhood of a schedule. For operation-based representation, the neighborhood for a given chromosome can be considered as the set of chromosomes (schedules) transformable from a given chromosome by exchanging the positions of λ genes (randomly selected, nonidentical genes). A chromosome (schedule) is said to be *λ-optimum* if it is better than any others in the neighborhood according to some measure.

```
Procedure: Neighborhood Search-Based Mutation
begin
 i ← 0;
 while (i ≤ pop size × p_m) do
  choose an unmutated chromosome randomly;
  pick up from the chromosome λ nonidentical genes
  randomly;
  make its neighbors on all permutations of the genes;
  evaluate all neighbor schedules;
  select the best neighbor as offspring;
  i ← i + 1;
 end
end
```

Tsujimura and Gen employed the concept of information entropy as a measure of diversity of a chromosome population [628]. This entropy-based genetic algorithm can provide both fast convergability and a good-quality solution.

6.2.5 Hybrid Genetic Algorithms

The role of local search in the context of genetic algorithms has been receiving serious consideration, and many successful applications are strongly in favor of such a hybrid approach. One of the most common forms of hybrid genetic algorithm is to incorporate local search techniques as an add-on to the main genetic algorithm's loop of recombination and selection. With the hybrid approach, genetic algorithms are used to perform global exploration among populations, while heuristic methods are used to perform local exploitation around chromosomes. Because of the complementary properties of genetic algorithms and conventional heuristics, a hybrid approach often outperforms either method operating alone. Hybridization can be done in a variety of ways, such as the following:

1. Incorporate heuristics into initialization to generate a well-adapted initial population. In this way, a hybrid genetic algorithm with elitism can guarantee to do no worse than the conventional heuristic does.

2. Incorporate heuristics into an evaluation function to decode chromosomes to schedules.
3. Incorporate a local search heuristic as an add-on to the basic loop of the genetic algorithm, working together with mutation and crossover operators to perform quick, localized optimization to improve offspring before returning it to be evaluated.

Combining Genetic Algorithm with Local Search. A common form of hybrid genetic algorithm is the combination of local search and genetic algorithm. Genetic algorithms are good at global search but slow to converge; local search is good at fine tuning but often falls into local optima. The hybrid approach complements the properties of genetic algorithms and local search heuristic methods. Genetic algorithms are used to perform global search to escape from local optima; local search is used to conduct fine tuning.

Local search in this context can be thought of as being analogous to a type of learning that occurs during the lifetime of an individual string. With a standard genetic algorithm, chromosome selection is based on instantaneous fitness at birth; with hybrid methods, selection is based on fitness at the end of the individual life, the life being determined by local search. The improved offspring will pass on to future offspring, through common crossover, traits acquired during learning (local optimization) [353], a phenomenon termed *Lamarkian evolution.* So this hybrid approach can be viewed as a combination of Darwinian and Larmarckian evolution [194, 468].

```
Hybrid Procedure: Genetic Algorithms Combined with Local
Search
begin
 t ← 0;
 initialize P(t) (job sequences);
 perform local search to improve chromosomes;
 evaluate P(t);
 while (not termination condition) do
 begin
  recombine P(t);
  perform local search to improve chromosomes;
  evaluate P(t);
  select to next population P(t);
  t ← t + 1;
 end
end
```

Combining Genetic Algorithms with Giffler and Thompson's Method. Dorndorf and Pesch proposed a priority rule–based encoding method for the job-shop scheduling problem and developed a hybrid version of genetic algorithms by incorporating the well-known Giffler–Thompson algorithm [165]. In the hybrid approach, the genetic algorithm is used to evolve a sequence of priority dispatching rules, and the Giffler–Thompson algorithm is used to deduce a schedule from the encoding of priority dispatching rules. The overall procedure of the hybrid method is given as follows:

```
Hybrid Procedure: Genetic Algorithm Combined with Giffler
and Thompson's Algorithm
begin
 t ← 0;
 initialize P(t) (job sequences);
 deduce schedules with Giffler-Thompson algorithm;
 evaluate P(t);
 while (not termination condition) do
 begin
  recombine P(t);
  deduce schedules with Giffler-Thompson algorithm;
  evaluate P(t);
  select to next population P(t);
  t ← t + 1;
 end
end
```

Priority rules are probably the most frequently applied heuristics for solving scheduling problems in practice because of their ease of implementation and their low time complexity. The algorithms of Giffler and Thompson can be considered as the common basis of all priority rule–based heuristics [590]. The generation procedure of Giffler and Thompson is a tree-structured approach. The nodes in the tree correspond to partial schedules, the arcs represent the possible choices, and the leaves of the tree are the set of enumerated schedules. For a given partial schedule, the algorithm essentially identifies all processing conflicts (i.e., operations competing for the same machine), and an enumeration procedure is used to resolve these conflicts in all possible ways at each stage [37]. Instead of enumerative tree search, Dorndorf and Pesch used priority dispatching rules to resolve these conflicts; that is, a gene specifies a priority rule for selecting one operation at a time among conflicting operations [165]. Therefore, they in fact used a one-pass priority dispatching heuristic, not a pure version of the Giffler–Thompson algorithm.

Generation procedures operate with a set of schedulable operations at each stage. Schedulable operations are operations not yet scheduled that have immediately scheduled predecessors, and this set can be determined simply from precedence structure. The number of stages in a one-pass procedure is equal to the number of operations, $m \times n$. At each stage, one operation is selected to add to a *partial schedule*. Conflicts among operations are solved by priority dispatching rules specified by a corresponding gene. Following the notations of Baker [37], we have:

PS_t = partial schedule containing t scheduled operations

S_t = set of schedulable operations at stage t, corresponding to a given PS_t

σ_i = earliest time at which operations $i \in S_t$ could be started

ϕ_i = earliest time at which operation $i \in S_t$ could be completed

For a given active partial schedule, the potential start time σ_i is determined by the completion time of the direct predecessor of operation i and the latest completion time on the machine required by operation i. The larger of these two quantities is σ_i. The potential finishing time ϕ_i is simply $\sigma_i + t_i$, where t_i is the processing time of operation i. The procedure to generate an active schedule works as follows:

Procedure: Deduce Offspring for Priority Rule–Based Encoding

Step 0. Let $[p_1, p_2, \ldots, p_{mn}]$ be a given chromosome.

Step 1. Let $t = 1$ and begin with PS_t as the null partial schedule, and let S_t include all operations with no predecessors.

Step 2. Determine $\phi_t^* = \min_{i \in S_t}\{\phi_i\}$ and the machine m^* on which ϕ_t^* could be realized. If more than one such machine exists, the tie is broken by a random choice.

Step 3. Form a conflicting set C_t which includes all operations $i \in S_t$ with $\sigma_i < \phi_t^*$ that requires machine m^*. Select one operation from C_t by priority rule p_t and add this operation to PS_t as early as possible, thus creating new partial schedule PS_{t+1}. If more than one operation exists according to priority rule p_t, the tie is broken by a random choice.

Step 4. Update S_t by removing from it the operation selected and adding the direct successor of the operation. Increment t by one.

Step 5. Return to step 2 until a complete schedule is generated.

The remaining problem is to identify an effective priority rule. For an extensive summary and discussion, see Panwalkar and Iskander [501], Haupt [290], and Blackstone, Phillips, and Hogg [67]. Table 6.1 consists of some of the priority rules commonly used in practice.

Kobayashi, Ono, and Yamamura proposed another version of hybrid genetic algorithms with the Giffler–Thompson algorithm [371, 493]. The job-sequence matrix encoding method is used in their studies. The Giffler–Thompson algorithm is used to deduce an active schedule from the encoding. The procedure to generate an active schedule works as follows:

TABLE 1.1. Job-Shop Dispatch Rules

Rule	Description
SPT (shortest processing time)	Select the operation with the shortest processing time.
LPT (longest processing time)	Select the operation with the longest processing time.
MWR (most work remaining)	Select an operation for the job with the most total processing time remaining.
LWR (least work remaining)	Select an operation for the job with the least total processing time remaining.
MOR (most operations remaining)	Select an operation for the job with the greatest number of operations remaining.
LOR (least operations remaining)	Select an operation for the job with the smallest number of operations remaining.
EDD (earliest due date)	Select the job with the earliest due date.
FCFS (first come, first served)	Select the first operation in the queue of jobs for the same machines.
RANDOM (random)	Select an operation at random.

Procedure: Deduce an Offspring for Job-Sequence Matrix Encoding

Step 1. Let $t = 1$ and S_t include all operations with no predecessors.

Step 2. Determine $\phi_t^* = \min_{i \in S_t}\{\phi_i\}$ and the machine m^* on which ϕ_t^* could be realized. If more than one such machine exists, a tie is broken by random choice.

Step 3. Form conflict set C_t, which includes all operations $i \in S_t$ with $\sigma_i < \phi_t^*$ that require machine m^*.

Step 4. Check the first unscheduled operation on machine m^*. If it belongs to the conflict set C_t, keep it unchanged; otherwise, swap it with an operation selected randomly from C_t.

Step 5. Update S_t by removing the scheduled operation and adding the direct successor of the operation. Increment t by one.

Step 6. Return to step 2 until all operations are scheduled.

Combining a Genetic Algorithm with the Shifting Bottleneck Heuristic. Dorndorf and Pesch proposed a machine-based encoding for the job-shop problem and developed a hybrid genetic algorithm by incorporating the well-known bottleneck shifting heuristic [165]. In this hybrid approach, a genetic algorithm is used to evolve a sequence of machines, and the bottleneck shifting heuristic is used to deduce a schedule from the machine sequence encoding.

The *shifting bottleneck heuristic* proposed by Adams, Balas, and Zawack [4] is probably the most powerful procedure known for the job-shop scheduling problem. It sequences machines one by one, successively, each time selecting the machine identified as a bottleneck among the machines not yet sequenced. Each time a new machine is sequenced, all previously established sequences are locally reoptimized. Both the bottleneck identification and local reoptimization procedures are based on repeatedly solving a certain one-machine scheduling problem that is a relaxation of the original problem. The main contribution of their approach is in using this relaxation to decide on the order in which the machines should be sequenced.

The shifting bottleneck heuristic is based on the classic idea of giving priority to bottleneck machines. Different measures of the bottleneck quality of machines will yield different sequences of bottleneck machines. The quality of the schedules obtained by a shifting bottleneck heuristic depends strongly on the sequence of bottleneck machines. Adams, Balas, and Zawack also proposed that an enumerative version of the shifting bottleneck heuristic be used to consider different sequences of machines.

Instead of an enumerative tree search. Dorndorf and Pesch proposed a genetic strategy to determine the best machine sequence for use with the shifting bottleneck heuristic. A chromosome is a list of ordered machines. Genetic algorithms here are used to evolve those chromosomes to find out a better sequence of machines for the shifting bottleneck heuristic. The difference between the shifting bottleneck heuristic and genetic algorithms is that the bottleneck is no longer a decision criterion in the choice of the next machine, which is determined by a given chromosome, and there is no need to identify one bottleneck machine at each iteration.

Let M_0 be the set of machines already sequenced and a given chromosome be $[m_1,m_2,...,m_m]$. The procedure for deducing a schedule from the chromosome works as follows:

Procedure: Deduce a Schedule for Machine-Based Encoding

Step 1. Let $M \leftarrow \emptyset$, $i \leftarrow 1$, and the chromosome $[m_1,m_2,...,m_m]$.

Step 2. Sequence machine m_i optimally. Update the set $M_0 \leftarrow M_0 \cup \{m_i\}$.

Step 3. Reoptimize the sequence of each critical machine $m_i \in M_0$ in turn while keeping the other sequences fixed.

Step 4. Let $i \leftarrow i + 1$. Then if $i > m$, stop; otherwise, go to step 2.

The details of step 3 can be found in Adams, Balas, and Zawack [4] or in Applegate and Cook [17].

Combining Genetic Algorithms with Beam Search. The hybrid approach proposed by Holsapple et al. combines genetic algorithms with a beam search technique [304]. Roughly, beam search is used to convert a chromosome (job-based representation) to a schedule. Beam search is a heuristic refinement of

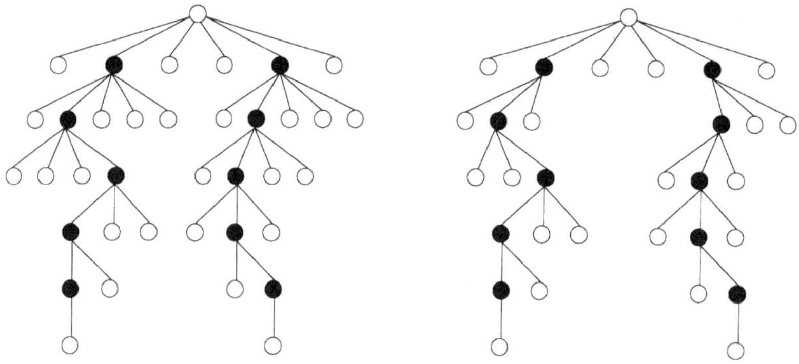

(a) beam search with $w = 2$ (b) filtered beam search with $w = 2$ and $f = 3$

Figure 6.1. Beam search and filtered beam search.

breadth-first search that relies on the notion of *beam width* to restrict the number of nodes that we branch from at each stage (or level) of the search tree. At each level of the search process, all nodes are evaluated using a predefined evaluation function. Then the beam width w is used to pick the w best nodes (in terms of their evaluations) to branch out from (i.e., expand) at the current level to generate the offspring nodes at the next level in the tree. The remaining nodes at the current level are pruned out permanently. If there are only fewer than w nodes available to begin with, all of these are chosen for expansion. From each of the w nodes chosen, all possible successor nodes are generated. The process continues until no further expansion is possible. Figure 6.1(a) illustrates beam search with a simple example.

Filtered beam search is a refinement of the beam search method. It uses one other construct, called *filter width,* to further prune the state space. For each w node selected for expansion at any level, the filter width f determines the maximum number of successor nodes that could be generated. That is, we could generate as many successor nodes as possible but not exceeding f. For a given value of f, the choice of which f successors (from the set of possible successor nodes) to generate is random. Both w and f are user-specified quantities. The search process is shown in Figure 6.1(b).

Holsapple et al. adopt a job-related representation. A genetic algorithm is used to manipulate the job sequence, and filtered beam search is used to generate the "best" schedule for a given job sequence (or a chromosome). We can consider beam search as part of the evaluation function, that is, converting a job sequence (the genotype of chromosome) to a schedule (the phenotype of chromosome), but it does more than converting. Holsapple et al. considered a type of static scheduling problem in flexible manufacturing contexts. The difference between this problem and the classic job-shop problem is that any job operation can be processed on any machine. In such a case, a chromosome of the job-based representation corresponds to many feasible schedules, so they use beam search to find the "best" among all feasible schedules for a given chromosome. In a classical job-shop scheduling problem, the processing machine for each operation is predetermined, so a job-based chromosome corresponds to only one feasible schedule, as illustrated earlier. In this case we do not need a tree search technique such as beam search to convert a chromosome into a feasible schedule. The overall procedure of Holsapple et al. is given below.

```
Hybrid Procedure: Genetic Algorithm Combined with Beam
Search
begin
 t ← 0;
 initialize P(t) (job sequences);
 perform beam search to generate best schedule for each
 chromosome;
 evaluate each schedule;
 while (not termination condition) do
 begin
  recombine P(t);
  perform beam search to generate best schedule for each
  chromosome;
  evaluate each schedule;
  select next population P(t);
  t ← t + 1;
 end
end
```

6.2.6 Discussion

During the past decade, job-shop scheduling has become a hot topic in the genetic algorithm field. This is because this problem exhibits all aspects of constrained combinatorial problems and serves as a paradigm for testing new algorithmic ideas. Note that the solution to a genetic algorithm is not necessarily the solution to a given problem, but it may contain the necessary information for constructing a solution to a given problem. As we know, there are two types of order relations in the job-shop scheduling problem: (1) the operation sequence on each machine and (2) precedence constraints among operations for a job. The first must be determined by a solution method; the second must be maintained in a schedule. If these two types of order relations are mingled together in the chromosomes, it will impose a heavy burden on genetic operators as to how to handle the infeasibility and illegality of offspring. How to represent a solution to a job-shop scheduling problem in genetic algorithms is a well-studied issue. Because of the existence of precedence constraints among operations, the permutation of operations usually corresponds to infeasible solutions. A way suggested by many researchers is to cope with the trouble of precedence constraints in a special schedule builder; then the job-shop scheduling problem is treated as a permutation problem. Genetic algorithms are used to evolve an appropriate permutation of operations on each machine, and a schedule builder is used to produce a feasible solution according to both the permutation and precedence requirements. Tsujimura et al. undertook comparative studies of three encoding methods—order-based representation, random key–based representation, and operation-based representation—for on open-shop scheduling problem, a special case of a job-shop scheduling problem [630].

Most encoding methods for job-shop scheduling problems are of general literal string type in the sense that a symbol can be repeated in a chromosome. Among them, preference list–based encoding, consisting of several substrings, is best used. Because each substring of the encoding is a pure literal string, a common method suggested by many studies is to apply genetic operators at each substring. In this way, the general literal string is treated essentially as several pure literal strings, and many genetic operators developed for pure literal strings can then be used directly. This multiple use of genetic operators will increase the amount of computation greatly as problem size increases. It is then hoped that genetic search ability can be increased accordingly. Note that because multiple genetic operators are used independently in each substring, one of the consequences of this kind of multiple use is that precedence constraints among operations may be violated. This is why we need a builder to readjust the order of operations to obtain a feasible solution. The trade-off between complexity of encoding and the builder should thus be taken into account.

A trend in genetic job-shop scheduling practice is to incorporate local search techniques into the main loop of a genetic algorithm to convert each offspring into an active schedule. In general, feasible permutation of operations corresponds to a set of semiactive schedules. As the size of the problem increases, the size of the set of semiactive schedules will become larger and larger. Because genetic algorithms are not very good at fine tuning, search within the huge space of semiactive schedules will make genetic algorithms less effective. We know that the optimal solution to a job-shop scheduling problem resides within the set of active schedules, which is much smaller than the set of semiactive schedules. By using a permutation encoding, we have no way to confine the genetic search within the space of the active schedules. One possible way would be to leave genetic search in the entire search space of the semiactive schedules while we convert each chromosome into an active schedule by an add-on procedure, the schedule builder. One way to do this is to incorporate the Giffler–Thompson algorithm into a genetic algorithm to perform the conversion. The Giffler–Thompson algorithm originally was an enumeration method. In most such hybrid approaches, it is used in a one-pass heuristic way to readjust operation order, to resolve precedence constraints, and to convert each offspring into an active schedule.

Recently, Bierwith and Mattfeld proposed a genetic algorithm for solving a general job-shop scheduling problem describing static, dynamic, and nondeterministic production environments [709].

6.3 GROUPED JOB SCHEDULING PROBLEM

Since group technology (GT) has become widely used in industrial manufacturing systems, production scheduling of grouped jobs has been an active area of research [657]. In GT, jobs are divided into several families according to their similarity. A job does not require a setup when it follows another job from the same family, but a known family setup is required when it follows a member of another family. The computational complexity of a grouped job scheduling problem remains open as to when a family can be split. This means that computation is very complicated without a simplifying GT assumption requiring precisely m setups in the schedule [461, 514].

Grouped job scheduling problems arise from industrial practice. For example, grouped parts are processed in a machine tool works; steel pins of different sizes are milled by different rollers in a steelworks, and multiple products are produced in batch production mode in a chemical plant. The production scheduling of both can be described by grouped job scheduling problems [15, 459]. Approaches to problems of grouped job scheduling include primarily dynamic programming and heuristic approaches [15, 518]. Wang, Gen, and Cheng have developed a genetic algorithm to solve this problem [651].

6.3.1 Problem Description and Necessary Condition

The problem of minimizing total flow time on a single machine for scheduling grouped jobs can be described as follows: Assume that there are N jobs waiting to be processed on a machine center. These jobs can be divided into m groups according to similarity. For group i there are n_i jobs. The jth job in group i is noted as the pair (i,j). For job (i,j) the process time is p_{ij}. The family setup time for group i is $s_i > 0$, $i = 1,2,\ldots,m$. A schedule consists of a series of *runs*. A run is the job sequence that is processed between two family setups [657]. For a given schedule, let η be the number of runs. Denote the ith run as R_i. Let β_i be the number of jobs in R_i, let $[i]$ denote the index of the family processed in R_i, let $[i,j]$ denote the job pair of the jth job processed in R_i, and define the position of R_i in the schedule as K_i:

$$K_i = \sum_{k=1}^{i-1} \beta_k + 1, \qquad i = 1,2,\ldots,\eta \tag{6.1}$$

The problem is to find an optimal schedule S to minimize the total flow time $F(S)$ as follows:

$$\min \quad F(S) = \sum_{i=1}^{\eta} \left[(N - K_i + 1)s_{[i]} + \sum_{j=1}^{\beta_i} (N - K_i - j + 2)p_{[i,j]} \right] \tag{6.2}$$

For run i, define its machine occupied time as q_i, and the mean of machine occupied time by a job in R_i as v_i. We have

$$q_i = s_{[i]} + \sum_{j=1}^{\beta_i} p_{[i,j]}, \qquad i = 1,2,\ldots,\eta, \tag{6.3}$$

$$v_i = q_i / \beta_i, \qquad i = 1,2,\ldots,\eta. \tag{6.4}$$

The problem is equivalent to minimizing the total weighted flow time with all weights equal to 1. Following Property 2 in Webster and Baker [657], Wang has proven the necessary conditions for an optimal solution to the grouped job scheduling problem [648]:

Property 6.1. To minimize the total flow time of grouped jobs on a single machine, the optimal solution satisfies the following conditions:

1. If $p_{ij} \leq p_{ik}$, job (i,j) is scheduled ahead of job (i,k).
2. The runs are in SMMOT (shortest mean of machine occupied time) order; that is,

$$v_1 \leq v_2 \leq \cdots \leq v_\eta \tag{6.5}$$

3. If $[i] = [k]$ and $i < k$, then

$$p_{i\beta_i} \leq v_{i+1} \leq \cdots \leq v_{k-1} \leq p_{k1} \tag{6.6}$$

4. If

$$\frac{s_i + \sum_{k=1}^{j} p_{ik}}{j} \geq p_{ij} \tag{6.7}$$

then jobs $(i,1)$ through $(i, j + 1)$ are processed consecutively.

It is clear that Property 6.1 gives the necessary conditions for an optimal solution. To discuss how to maintain the necessary conditions, the following definition is given:

Definition 6.1. A schedule that satisfies Property 6.1 is called a *reasonable schedule.* Reasonable schedules have the following properties.

Property 6.2. In a reasonable schedule, R_i and R_j are runs of same family, and R_i is ahead of R_j. If run R_k is generated by a combination of R_i and R_j, its mean of machine occupied time by a job, v_k, satisfies

$$v_k \leq v_j \tag{6.8}$$

Proof: Because $v_i \leq v_j$,

$$\begin{aligned} v_k &= \frac{1}{\beta_i + \beta_j}(\beta_i v_i + \beta_j v_j - s_{[j]}) \\ &\leq \frac{1}{\beta_i + \beta_j}(\beta_i v_j + \beta_j v_j - s_{[j]}) \\ &= v_j - \frac{s_{[j]}}{\beta_i + \beta_j} \end{aligned}$$

Due to the fact that $s_{[j]} \geq 0$, we have $v_k \leq v_j$.

Property 6.2 tells us that in a reasonable schedule the combination of two runs of the same family will be processed ahead of the last run.

Property 6.3. If a combination of two runs of the same family moves to kth place in the schedule the condition that keeps it reasonable is that there are no other runs of the same family between places i,k and k,j.

It is clear by Property 6.3 that the combining and moving of runs in a reasonable schedule cannot jump past any runs of the same family because the necessary condition 1 of Property 6.1 would not be kept if that occurred.

6.3.2 Fundamental Runs

Fundamental runs are defined based on the necessary conditions and properties of reasonable schedules. A step-by-step procedure for generalizing fundamental runs is described below.

Procedure: Fundamental Runs

Step 1. Let $\eta = N$, and take all identical jobs as independent runs. Sequence these runs in increasing order of $s_i + p_{ij}$. Clearly, it is an initial reasonable schedule. It has $v_1 \leq v_2 \leq \cdots \leq v_\eta$, where, $v_i = s_{[i]} + p_{[i,1]}$.

Step 2. Let $\eta' = \eta$. For $k = 1$ to η, check: If R_{k+1} is a run of the same family of R_k, combine R_{k+1} with R_k, and let $\eta \leftarrow \eta - 1$.

Step 3. If $\eta = \eta'$, stop; otherwise, update v_i, $i = 1,2,\ldots,\eta$.

Step 4. For $k = 1$ to η, check: if $v_{k+1} < v_k$, move R_{k+1} ahead until $v_{k+1} \geq v_i$ or $[i] = [k + 1]$ (i.e., R_i and R_{k+1} are runs of the same family). Locate R_{k+1} just after R_i, and relabel all runs as the sequence. Go to step 2.

Definition 6.2. Define the schedule generated by the algorithm of fundamental runs as the *fundamental schedule,* the runs in the fundamental schedule as *fundamental runs.*

It is easy to see that the fundamental schedule can meet the necessary conditions for optimal solutions in Property 6.1. For most practical problems, the number of groups is usually far less than the number of jobs. Experience shows that the number of fundamental runs is less than the mumber of jobs in the case of $m << N$. It can be proven that the optimal solution can be achieved by combining the fundamental runs. Thus it is only necessary to consider η fundamental runs rather than N jobs in the solution procedure.

Lemma 6.1. The incremental contribution to total flow time by moving a run R_i from K_i to K is $(K_i - K)q_i$.

Proof: Denote the incremental contribution as $\Delta F_i(K_i,K)$. From the objective function (6.2), we have

$$\begin{aligned}\Delta F_i(K_i,K) &= (N - K + 1)s_{[i]} + \sum_{j=1}^{\beta_i} (N - K - j + 2)p_{[i,j]} \\ &\quad - (N - K_i + 1)s_{[i]} - \sum_{j=1}^{\beta_i} (N - K_i - j + 2)p_{[i,j]} \\ &= (K_i - K)s_{[i]} + (K_i - K) \sum_{j=1}^{\beta_i} p_{[i,j]} \\ &= (K_i - K)q_i\end{aligned}$$

Theorem 6.1. The optimal solution of a grouped job scheduling problem to minimize total flow time on a single machine is a combination of fundamental runs.

Proof: Assume that the optimal schedule is not a combination of fundamental runs. Then there is a fundamental run that is split into at least two parts in the optimal schedule. Assume that the fundamental schedule S_0 is constructed as

$$\text{S0:} \quad * + R_1 + R_0 + R_2 + *$$

where R_0 is a fundamental run, R_1 and R_2 are parts of the neighboring fundamental runs, * represents other parts of the schedule, and + signifies the neighboring relation of runs.

If the optimal schedule inserts R_1 and R_0 and splits R_0 up into two subruns, R_{01} and R_{02}, it is

$$\text{S1:} \quad * + R_1' + R_{01} + R_1 + R_{02} + R_2 + *$$

where R_1' is the complementary part of R_1 to make a fundamental run.

From Property 6.1 we have

$$v(R_1') \leq v(R_{01}) \leq v(R_1) \leq v(R_{02}) \leq v(R_2) \tag{6.9}$$

where $v(R)$ stands for the mean machine occupied time of a job in run R. According to procedure Fundamental Runs (FR), there are three possible cases in the composition process of the fundamental schedule:

C1: $* + R_1' + R_1 + R_{01} + R_{02} + R_2 + *$
C2: $* + R_{01} + R_1' + R_1 + R_{02} + R_2 + *$
C3: $* + R_{01} + R_{02} + R_1' + R_1 + R_2 + *$

For case 1, from (C1) we have $v(R_1) \leq v(R_{01})$. Comparing this with (6.9), we get

$$v(R_{01}) = v(R_1)$$

Exchanging the positions of R_{01} and R_1 in optimal schedule S1, it is easy to see that the exchange makes R_{01} move backward β_1 jobs and R_1 ahead β_{01}. Denote the new schedule as SE. By Lemma 6.1, we have

$$\begin{aligned} F(\mathrm{SE}) - F(\mathrm{S1}) &= \beta_{01}q_1 - \beta_1 q_{01} - s_0K_{02} \\ &= (q_1/\beta_1 - q_{01}/\beta_{01})\beta_1\beta_{01} - s_0K_{02} \\ &= (v(R_1) - v(R_{01}))\beta_1\beta_{01} - s_0K_{02} \\ &= -s_0K_{02} < 0 \end{aligned}$$

which is contradictory to the optimality of schedule S1.

For case 2, from (C2) and (6.9), we know that $v(R_1') = v(R_{01})$. By exchanging R_1' and R_{01}, it can be proven similarly that S1 is not optimal. For case 3, from (C3) and (6.9), we know that $v(R_1) = v(R_{02})$. By exchanging R_1 and R_{02}, it can also be proven similarly that S1 is not optimal. Where R_0 is split by R_2 in the optimal schedule, we can also prove that the schedule would not be optimal. Based on Theorem 6.1, we produce the fundamental runs using procedure FR. Then the optimal schedule can be achieved by combining the fundamental runs. We can compose the runs by condition 4 in Property 6.1, but it is more serious than is necessary. Thus the number of runs generated by condition 4 would be much higher than the number of fundamental runs.

6.3.3 Representation

Wang, Gen, and Cheng developed a genetic algorithm for the grouped job scheduling problem. A primary binary string is adopted to represent a combination of fundamental runs. Assume that a fundamental schedule is given as follows:

$$R_{\{1\}} + R_{\{2\}} + \cdots + R_{\{\eta\}}$$

There are η_i fundamental runs in family i, $i = 1,2,\ldots,m$. Certainly,

$$\eta = \sum_{i=1}^{m} \eta_i \tag{6.10}$$

Define $\bar{\eta}_0 = 0$, and $\bar{\eta}_i$ $(i = 1,2,\ldots,m)$ as

$$\bar{\eta}_i = \sum_{k=1}^{i} \eta_k \tag{6.11}$$

Then the runs can be grouped into families and labeled as follows:

Family 1: $R_1, R_2,\ldots,R_{\bar{\eta}_1}$
Family 2: $R_{\bar{\eta}_1} + 1, R_{\bar{\eta}_1+2},\ldots,R_{\bar{\eta}_2}$
$\vdots$
Family m: $R_{\bar{\eta}_{m-1}+1}, R_{\bar{\eta}_{m-1}+2},\ldots,R_{\bar{\eta}}$

Since it is impossible to combine the runs in different families, there are no bits for a combination of the last run in family k and the first run in family $k + 1$, $k = 1,2,\ldots m - 1$. Thus a binary number with $\eta - m$ bits can be used for the gene representation of fundamental run combinations. Define the length of the binary string for gene representation L as

$$L = \eta - m \tag{6.12}$$

Then if runs R_{i-1} and R_i belong to the same family k, we define the binary bits for the combination between runs as follows:

$$b_{i-k} = \begin{cases} 1, & R_{i-1} \text{ and } R_i \text{ are combined} \quad k = 1,2,3,\ldots,m,\ i \in [\bar{\eta}_{k-1} + 2,\ \bar{\eta}_k]. \\ 0, & \text{otherwise} \end{cases} \tag{6.13}$$

The relationship between b_{i-k} amd R_{i-1}, R_i is shown as follows:

$$\underbrace{R_1 - R_2}_{b_1}\underbrace{- R_3}_{b_2} - \cdots - \underbrace{R_{\eta_1-1} - R_{\eta_1}}_{b_{\eta_1-1}}|\underbrace{R_{\eta_1+1} - R_{\eta_1+2}}_{b_{\eta_1}} - \cdots - \underbrace{R_{\eta-1} - R_{\eta}}_{b_{\eta-m}}$$

Then

$$B = [b_1, b_2, \ldots, b_L] \tag{6.14}$$

is a combinational representation of the fundamental runs. For a simple example, the fundamental schedule is

$$R_1 + R_4 + R_2 + R_5 + R_3$$

grouped follows:

Family 1: R_1, R_2, R_3
Family 2: R_4, R_5

Thus a binary string of $5 - 2 = 3$ bits can describe the combination of these runs. The relationship between bits and runs is

$$\underbrace{R_1 - R_2}_{b_1}\underbrace{- R_3}_{b_2}|\underbrace{R_4 - R_5}_{b_3}$$

For example, (1 0 1) says that R_1 and R_2 are combined; R_4 and R_5 are also combined; but R_2 and R_3 are not combined.

6.3.4 Evaluation

Let $B(j)$, $j = 1,2,\ldots,pop\_size$, denote individuals of the current population. Their fitness value can be calculated by the following procedure:

Procedure: Fitness Function

Step 1. For individual j, $j = 1,2,\ldots,L$, combine its runs with bits equal to 1, and calculate the mean of the machine-occupied time of the job, $v(C_i)$, for all combinations of runs, where C_i stands for the ith combination.

Step 2. Sequence and relabel the combinations as

$$v(C_1) \le v(C_2) \le \cdots \le v(C_{NC}) \tag{6.15}$$

where NC is the number of combinations in individual j. It is easy to see that $NC = L -$ number of bits equal to 1 in $B(j)$.

Step 3. Calculate the objective function, $F(j)$, of $B(j)$ by equation (6.2).

Step 4. Calculate the fitness function of individual j by

$$f(j) - F_{max} - F(j) \tag{6.16}$$

where

$$F_{\max} = \max\{F(j) | j = 1,2,...,pop\_size\} \tag{6.17}$$

It is clear from equation (6.16) that the better the schedule, the greater the fitness.

6.3.5 Genetic Operators

One-cutpoint crossover is used. For crossover operation, use a pair of parents, one selected by the roulette wheel proportional method and another randomly. Let $P(j)$ be the selection probability for individual j; then we define

$$P(j) = \frac{f(j)}{\sum_{k=1}^{pop\_size} f(k)}, \qquad j = 1,2,...,pop\_size \tag{6.18}$$

where $f(k)$ is defined by equation (6.16). It is evident that individuals with a greater fitness function are selected with higher probability. To select a child s, $s = 1,2,...,N$, of the next generation, define $S(j) = P(1) + P(2) + \cdots + P(j)$, for $j = 1,2,...,N$, and generate a random number $\xi_s \in U(0,1)$, where $U(0,1)$ is the uniform distribution on (0,1) [68]. If $S(j - 1) < \xi_s < S(j)$, select individual j as the parent of child s [455].

6.3.6 Overall Procedure

The overall procedure is described as follows:

Procedure: Genetic Algorithm

Step 1. Specify the maximum number of generations, *max_gen*; the population size, *pop_size*; and the probabilities of crossover and mutation, p_c and p_m. Input the job data, s_i and p_{ij}, $\forall i,j$.

Step 2. Produce the fundamental schedule, FS, and label all fundamental runs as a family sequence. Determine the length of binary string, L, by equation (6.12).

Step 3. Produce an initial population. For $B(j)$, $j = 1,2,...,pop\_size$, produce its bits by

$$b_i = \begin{cases} 1, & \xi_i > 0.5 \\ 1, & \text{otherwise} \end{cases} \qquad i = 1,2,...,L$$

where $\xi_i \in U(0,1)$ and $U(0,1)$ is the uniform distribution on (0,1).

Step 4. Find the initial best individual by

$$j^* = \arg\min\{F(j)|j = 1,2,\ldots,pop\_size\}$$

Let $F^* = F(j^*)$ and $B^* = B(j^*)$. Set the initial iteration index $k = 0$.

Step 5. Set the iteration index $k \leftarrow k + 1$. If $k = max\_gen$, output results F^* and B^*, then stop; otherwise, continue the iteration.

Step 6. Calculate the objective and fitness functions of all individuals FF. Select the best individual by finding

$$j^* = \arg\min\ \{F(j)|j = 1,2,\ldots,\mathrm{NP}\} \tag{6.19}$$

If $F(j^*) < F^*$, let $F^* = F(j^*)$ and $B^* = B(j^*)$.

Step 7. Calculate the selection probability of all individuals by equation (6.18) and set

$$S(j) = P(1) + P(2) + \cdots + P(j), \qquad j = 1,2,\ldots,N \tag{6.20}$$

Step 8. For $s = 1,3,5,\ldots,pop\_size$, reproduce the children (new individuals). Generate a random number ξ_s; $\xi_{s+1} \in U(0,1)$. If $S(j - 1) < \xi_s < S(j)$, select individual j as a parent and take $k = \mathrm{int}[\mathrm{NP} * \xi_{s+1}] + 1$ as another parent of children s and $s + 1$. Carry out the crossover or mutation with probability P_c or P_m.

Step 9. Update $B^k(j) \leftarrow B^{k+1}(j)$, $j = 1,2,\ldots,pop\_size$, and go to step 5.

6.3.7 Numerical Example

Consider a small test problem in which a mill receives 20 orders. Each order (job) must be rolled by one of five shapes of roller. The rolling time of the orders and the setup time of the rollers are shown in Table 6.2. According to the algorithm of fundamental runs, take all identical jobs as independent runs and sequence them by increasing order of $S_i + p_{ij}$. This is

$$11 + 6 + \underbrace{1 + 2 + 3} + 10 + 12 + 5 + \underbrace{8 + 15} + 4 + 9 + 20 + 7$$
$$+ \underbrace{17 + 14 + 16 + 19} + 13 + 18$$

The objective value is 4876. Combine the neighboring jobs in the same groups—(1 + 2 + 3), (8 + 15), and (17 + 14 + 16 + 19)—and sequence the runs by increasing order of v_i. Repeat this procedure. Finally, we get the fundamental schedule, which consists of 10 fundamental runs:

TABLE 6.2. Rolling and Setup Time of Orders

Order	Rolling Time	Roller	Setup Time	Order	Rolling Time	Roller	Setup Time
1	8			11	8		
2	10	1	3.0	12	19	4	1.0
3	15			13	95		
4	25			14	19		
5	18	2	7.0	16	21		
6	4			17	8	5	31.0
7	19			18	75		
8	12			19	23		
9	17	3	31.0	20	1		
10	5						
15	12						

$$R_6 + R_2 + R_1 + R_4 + R_7 + R_9 + R_3 + R_5 + R_8 + R_{10}$$

broken down as follows:

1. Family 1:
   R_1: job 1, job 2, job 3
2. Family 2:
   R_2: job 6; R_3: job 5, job 4
3. Family 3:
   R_4: job 10, job 8, job 15; R_5: job 9, job 7
4. Family 4:
   R_6: job 11; R_7: job 12; R_8: job 13
5. Family 5:
   R_9: job 20, job 17, job 14, job 16, job 19; R_{10}: job 18

The objective value of the fundamental schedule is 3551.
From equation (6.12), the length of binary string is $10 - 5 = 5$. Thus the gene representation is

$$R_1 | \underbrace{R_2 - R_3}_{b_1} | \underbrace{R_4 - R_5}_{b_2} | \underbrace{R_6 - R_7}_{b_3}\underbrace{- R_8}_{b_4} | \underbrace{R_9 - R_{10}}_{b_5}$$

Let *pop_size* = 10 and *max_gen* = 50. By running the algorithm, the best scheduling is achieved:

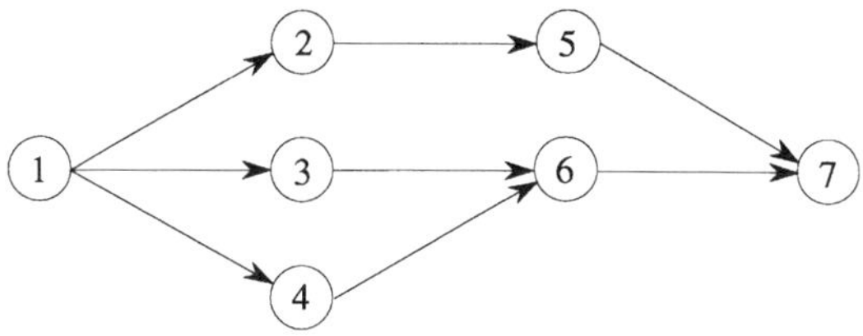

Figure 6.2. Network representation of a project.

$$B^* = 11001, \qquad F^* = 3393$$

The best schedule is

$$R_6 + R_1 + \underbrace{R_4 + R_5} + \underbrace{R_2 + R_3} + R_7 + \underbrace{R_9 + R_{10}} + R_8$$

or

$$11| + 1 + 2 + 3| + 10 + 8 + 15 + 9 + 7| + 6 + 5$$
$$+ 4| + 12| + 20 + 17 + 14 + 16 + 19 + 18| + 13$$

This is the optimal solution as verified by a branch-and-bound algorithm. It is formed by combining the fundamental runs $R_2 + R_3$, $R_4 + R_5$, and $R_9 + R_{10}$.

6.4 RESOURCE-CONSTRAINED PROJECT SCHEDULING

The problem of scheduling activities under resource and precedence restrictions with the objective of minimizing the project duration is referred to the literature as a resource-constrained project scheduling problem [37]. The basic type of problem can be stated as follows: A project consists of a number of interrelated activities, each characterized by a known duration and given resource requirements. Resources are available in limited quantities but are renewable from period to period. There is no substitution between resources, and activities cannot be interrupted. A solution is to determine the start times of activities with respect to the precedence and resource constraints so as to optimize the objective.

As an example, consider the simple project containing five activities shown in Figure 6.2. The nodes denote the activities and the direct arcs denote the precedence constraints. Suppose that the total amount of resources available is 4 units. The duration and resource consumption for each activity are given in Table 6.3.

TABLE 6.3. Duration and Resource Consumption of Each Activity

Activity	Duration	Resource Consumption	Parent Activities
1	(dummy activity)		
2	4	2	1
3	3	3	1
4	2	2	1
5	4	1	2
6	2	2	3,4
7	(dummy activity)		

If the project is scheduled as shown in Figure 6.3, the amount of resources required over time, the resource profile, is as shown in Figure 6.4.

The problem can be stated mathematically as follows:

$$\min \quad t_n \tag{6.21}$$

$$\text{s.t.} \quad t_j - t_i \geq d_i, \qquad \forall_j \in S_i \tag{6.22}$$

$$\sum_{t_i \in A_{t_i}} r_{ik} \leq b_k, \qquad k = 1,2,\ldots,m \tag{6.23}$$

$$t_i \geq 0, \qquad i = 1,2,\ldots,n \tag{6.24}$$

where t_i is the starting time of activity i, d_i the duration (processing time) of activity i, S_i the set of successors of activity i, r_{ik} the amount of resource k required by activity i, b_k the total availability of resource k, A_{t_i} the set of activities in process at time t_i, and m the number of different resource types. Activities 1 and n are dummy activities that mark the beginning and end of the project. The objective is to minimize total project duration. Constraint (6.22) ensures that none of the precedence constraints is violated. Constraint (6.23) ensures that the amount of resource k used by all activities does not exceed its limited quantity in any period. The major attention in this study is placed on how to handle the prcedence constraints existing in the problem.

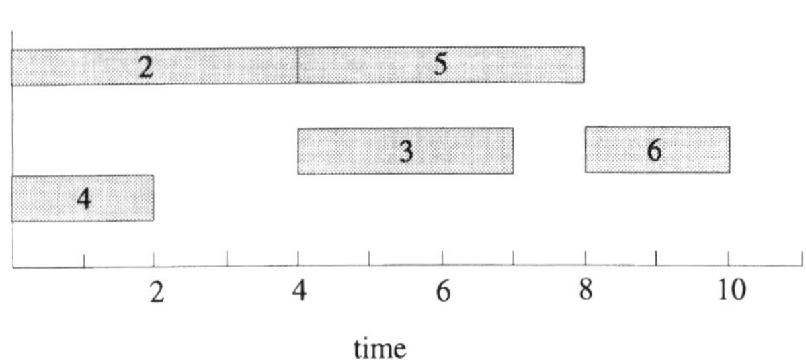

Figure 6.3. Gantt chart schedule.

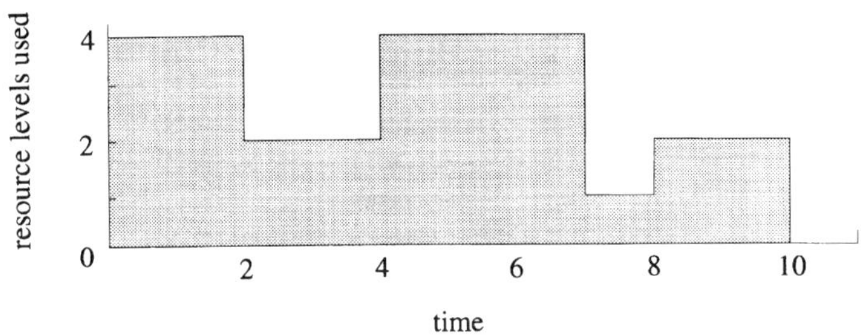

Figure 6.4. Resource profile.

A new encoding method is proposed, which is essentially capable of representing all feasible permutations of activities for a given instance [105, 106, 110].

The earliest attempts were made to find an exact optimal solution to the problem by using standard solution techniques of mathematical programming [121, 143, 506, 516, 589, 611]. Because the resource-constrained project scheduling problem in NP-hard, for large projects the size of the problem may render optimal methods computationally impracticable. In such cases, the problem is most amenable to heuristic problem solving using fairly simple scheduling rules capable of producing reasonably good suboptimal schedules. Over the past 30 years, a larger number of heuristic algorithms have been developed and tested; we refer the reader to Alvarez-Valdés and Tamarit for a survey and computational comparison [12].

Most of the heuristic methods known so far can be viewed as priority dispatching rules that assign activity priorities in making sequencing decisions for resolution of resource conflicts according to either temporally or resource-related heuristic rules. Recently, Cheng and Gen have proposed a hybrid genetic algorithm for solving the resource-constrained project scheduling problem. Essentially, the problem consists of the following two basic issues: (1) to determine the processing order of activities without violating the precedence constraints, and (2) subsequently to determine the start time for each activity without violating the resource constraint. How to determine the order of activities is critical to the problem because if the order of activities is determined, a schedule can then be constructed easily according to the order. The basic idea of the approach proposed by Cheng and Gen is to (1) use genetic algorithms to evolve an appropriate activity processing order, and (2) use a fit-in-best procedure to calculate the start times of activities.

6.4.1 Priority-Based Encoding

How to encode a problem solution into a chromosome is a key issue which conditions all subsequent steps in genetic algorithms. In Cheng and Gen's approach, a genetic algorithm is used to evolve an appropriate activity proc-

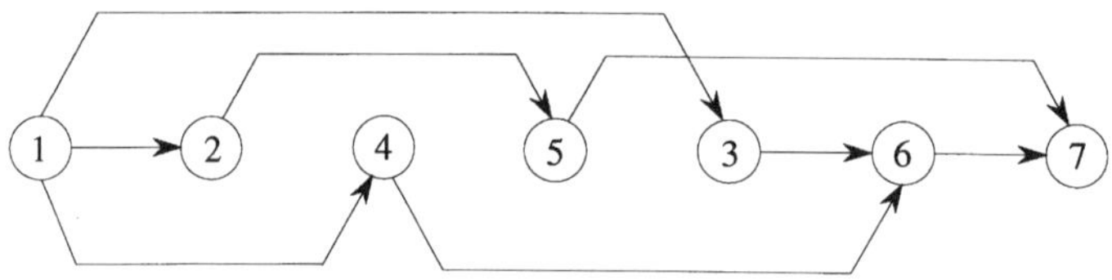

Figure 6.5. Topological sort of the DAG in Figure 6.2.

essing order and thus only the activity order needs to be encoded into a chromosome. This is a permutation problem in nature. Because of the existence of precedence constraints among activities, an arbitrary permutation may yield an infeasible processing order. How to produce encoding that can treat the precedence constraint efficiently is a critical step and conditions all subsequent steps. A priority-based encoding method is proposed to handle this difficulty, based on the concepts of a directed acyclic graph model.

Directed Acyclic Graph. A instance of a project can be represented using a directed acyclic graph (DAG) [9]. A *directed acyclic graph* $G = (V,A)$ consists of a set of nodes V representing activities and a set of directed edges A representing the precedence constraints among activities. The terms *node* and *activity* will be used interchangeably in the following sections. Figure 6.2 gives an example of a DAG to represent a project. The problem of ranking all activities in an appropriate order to meet the precedence constraint then is equivalent to the problem of generating a *topological sort* of the DAG as defined below [134].

Definition 6.3 (Topological Sort). For a given directed graph $G = (V,A)$, a *topological sort* is a linear ordering of all its nodes such that for any directed edge $(u,v) \in A$, node u appears before node v in the ordering.

Figure 6.5 shows a topological sort of the example given in Figure 6.2. Nodes are arranged such that all directed edges go from left to right.

It is easy to see that a random permutation of activities usually does not correspond to the topological sort of a given DAG, and therefore the permutation encoding, one of the most common encoding methods in the practice of genetic algorithms, is not suitable for the problem. A new encoding method, *priority-based encoding,* is suggested by Cheng and Gen. It will be shown later that (1) the encoding is capable of representing all possible topological sorts for a given instance, and (2) any permutation of the encoding always corresponds to a topological sort, that is, a feasible solution.

Priority-Based Encoding. Recall that a gene contains two kinds of information: the *locus,* the position of a gene located within the structure of a chromosome, and the *allele,* the value taken by the gene. Thus there are two

1	2	3	4	5	6	7	position: activity ID
3	7	1	6	4	5	2	value: priority of activity

Figure 6.6. Priority-based encoding.

possible ways to represent an activity: using the *locus* and using the *allele*. Here the position is used to denote an activity ID, and the value is used to denote the priority associated with the activity, as shown in Figure 6.6. The value of a gene is an integer exclusively within $[1,n]$. The larger the integer, the higher the priority.

A one-pass procedure is used to generate a topological sort from a chromosome: to determine an activity from left to right in a single pass. When making a decision regarding a position, several activities may compete for the position, and the one with the highest priority wins the position. The encoding does not explicitly represent a topological sort for a given DAG. It just contains some message for the resolution of conflicts. A topological sort can be determined uniquely according to the encoding in most cases. Any changes in priorities usually result in a different topological sort. Therefore, this encoding is essentially capable of representing all possible topological sorts for a given DAG.

Let us see how to generate a topological sort from the encoding. Consider the example shown in Figure 6.2. An array $A[\cdot]$ is used to store the topological sort generated. At the beginning $A[1] = 1$. Then three activities 2, 3, and 4 compete for $A[2]$. Their priorities are 7, 1, and 6, respectively. Activity 2 wins the position because it has the highest priority. After fixing $A[2] = 2$, the candidates for the next position, $A[3]$, are activities 3, 4, and 5. Activity 4 wins the position and fixes $A[3] = 4$. Repeat these two steps:

Step 1. Construct the set of candidates for current position.

Step 2. Select the highest-priority activity until obtaining a topological sort as shown in Figure 6.5.

Topological Sort. The procedure for making a topological sort works in a one-pass fashion: generating the topological sort from left to right and fixing one node's order at a time. At each iteration of the procedure, all nodes are in one of the following three states:

1. *Sorted node:* the nodes in the topological sort constructed so far
2. *Eligible node:* the nodes all of whose parent nodes are sorted nodes
3. *Free node:* all others

The tricky part is, of course, how to find a set of eligible nodes. The following definitions and theorems provide a better explanation of how to make such a set and how the procedure works.

For an edge (u,v) of a directed graph $G = (V,A)$, we say that the edge is *incident from u* and *incident to v*. Moreover, u is said to be *adjacent to v* and v is *adjacent to u*. The *indegree* of node u, denoted $d^{\mathrm{IN}}(u)$, is the number of nodes adjacent to u. A *cut* in G is a set of edges of the form $(X, V - X)$ such that $u \in X$ and $v \in V - X$. We often write $\overline{X}$ in lieu of $V - X$. A *partial topological sort* is a sort being developed, which contains only the first $t(t < |V|)$ nodes with fixed orders. Let $\mathrm{PS}_t \subset V$ be the set of nodes corresponding to a given partial topological sort, where the subscript t denotes the cardinal number of the set, that is, $|\mathrm{PS}_t| = t$. Let $C(\mathrm{PS}_t, V - \mathrm{PS}_t) = \{(i,j) | i \in \mathrm{PS}_t, j \in V - \mathrm{PS}_t\}$ be the cut of the directed graph with respect to the given partial topological sort. Then we have the following lemma:

Lemma 6.2 (Eligible Node). For a given partial topological sort with nodes PS_t, a node $j \in V - \mathrm{PS}_t$ is eligible if and only if we can have a set $S(j) = \{(i,j) | (i,j) \in A\}$ of edges incident to j such that $S(j) \subseteq C(\mathrm{PS}_t, V - \mathrm{PS}_t)$.

Proof: For a given node $j \in V - \mathrm{PS}_t$, if there is an edge (x,j) incident to j, the node x is a parent node of j. If all such edges belong to the cut $C(\mathrm{PS}_t, V - \mathrm{PS}_t)$, it means that all the parent nodes of j belong to the set PS_t; that is, they are sorted nodes, so that node j is eligible.

Assume that there is an eligible node j and not all the edges incident to j belong to the cut. That is, $|C(\mathrm{PS}_t, V - \mathrm{PS}_t) \cup S(j)| > |C(\mathrm{PS}_t, V - \mathrm{PS}_t)|$. Then at least one of its parent nodes belongs to set $V - \mathrm{PS}_t$; that is, at least one of its parent nodes is not a sorted node. This is in contradiction to the definition of eligible node.

Theoretically, we can check whether a node is eligible with the lemma, but it is usually not easy for programming to check if a set is a subset of others. The following theorem provides a criterion to determine an eligible node.

Theorem 6.2 (Criterion of Eligible Node). For an eligible node $j \in V - \mathrm{PS}_t$, let $S_t(j)$ be a proper subset of cut $C(\mathrm{PS}_t, V - \mathrm{PS}_t)$ containing all edges incident to j. We then have $|S_t(j)| = d^{\mathrm{IN}}(j)$.

Proof: For an eligible node, we have, according to Lemma 6.2, $S_t(j) \cap S(j) = S(j)$ and $S_t(j) \cup S(j) = S(j)$. Because $|S(j)| = d^{\mathrm{IN}}(j)$, we have proved the theorem.

Now we can identify an eligible node simply by checking whether the number of edges incident to it in the cut equals its indegree. This criterion is easy for programming. Let us consider the example given in Figure 6.7. The partial topological sort is $\mathrm{PS}_3 = \{1,2,3\}$ and the cut contains the directed edges $C(\mathrm{PS}_3, V - \mathrm{PS}_3) = \{(1,4), (2,5), (2,6), (3,6), (3,7)\}$. Node 6 is eligible because its indegree $d^{\mathrm{IN}}(6) = 2$ and the two edges incident to node 6 belong to the cut. Node 5 is a free node because its indegree is 2 and only one edge

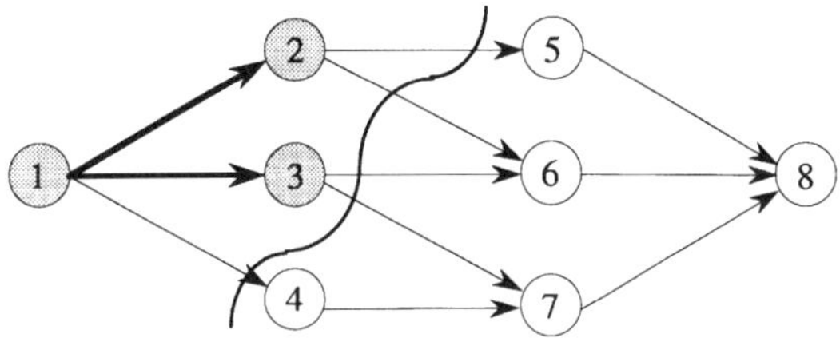

Figure 6.7. Partial topological sort, cut, and eligible nodes.

incident to it belongs to the cut. Its other parent node is 4, which is eligible but not sorted.

The basic idea of the topological sort procedure is, at each step as the procedure progresses, to (1) identify the set of eligible nodes with Theorem 6.2, (2) remove the one with the highest priority from the set, and (3) fix the removed node in the partial topological sort. The largest portion of operations in the procedure are those used to maintain the set of eligible nodes, which can be implemented efficiently with a *priority queue* [134]. A priority queue is an abstract data type for efficiently storing and manipulating a set of items when each item has an associated value called a *key* or a *priority*. The data structure supports the insertion of items into the queue and removal from the queue of the item with the highest priority: operations that we need to maintain a set of eligible nodes. A special implementation for a priority queue is given by means of heaps. A *heap* is a binary tree with the special property that the priority of every node is greater than or equal to the priority of its children [134]. A heap can be stored in any array such that for a node with index i, its parent has index $[i/2]$, its left child has index $2i$, and its right child has index $2i + 1$. Suppose that the eligible node set contains nodes 3, 4, 5, 6, and 7. Each node is associated with priority 2, 6, 9, 3, and 8, respectively. The corresponding heap represented by a binary tree and its implementation of arrays are given in Figure 6.8. The number within the circle

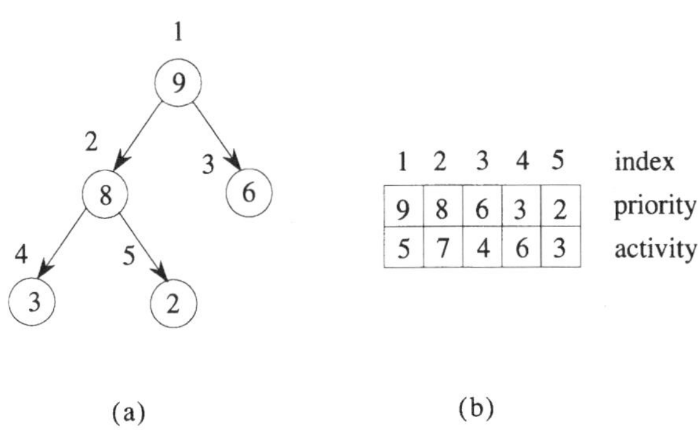

Figure 6.8. Heap viewed as (a) a binary tree and (b) an array.

at each node in the tree is the priority value. The number next to a node is the corresponding index in the array.

Let t be the iteration index of the procedure, V be the set of all nodes, Q_i be the set of all direct successors of activities i, PS[·] be the array for storing topological sorts, CUT[i] be the number of edges incident to node i in the cut, and S_t be the set of eligible nodes at step t. The procedure for generating a topological sort from a chromosome is given below.

Procedure: Topological Sort

Step 1. Initialize

$$t \leftarrow 1 \text{ (iteration index)}$$

$$\text{PS}[t] \leftarrow 1 \text{ (initial topological sort)}$$

$$S_t \leftarrow Q_1 \text{ (initial priority queue)}$$

$$\text{CUT}[i] \leftarrow 1, \forall_i \in Q_1 \text{ (initial number of edges in the cut)}$$

$$\text{CUT}[i] \leftarrow 0, \forall_i \in V - Q_1$$

Step 2. Perform the termination test. If PS[t] = n, go to step 6; otherwise, $t \leftarrow t + 1$, and continue.

Step 3. Fix the tth node, Remove the highest-priority node i^* from priority queue S_t and put it in array PS[t].

Step 4. Perform the cut set update.

$$\text{CUT}[i] \leftarrow \text{CUT}[i] + 1, \forall i \in Q_{i^*}$$

Step 5. Update the eligible node set. For all $i \in Q_{i^*}$, if CUT[i] = $d^{\text{IN}}(i)$, put i in priority queue S_t. Go back to step 2.

Step 6. Return a complete topological sort PS[·].

6.4.2 Genetic Operators

Genetic search is implemented through genetic operators and directed by selection pressure. Usually, a crossover operator is used as the main genetic operator and the performance of a genetic system is heavily dependent on it; a mutation operator used as a background operator produces spontaneous random changes in various chromosomes.

In Cheng and Gen's implementation, an alternative approach was taken to designing genetic operators: One operator is designed to perform a widespread search to explore the area beyond local optima; the other is designed to perform an intensive search to hunt for an improved solution. Two types of search approaches, intensive and widespread, form mutually complementary components of genetic search. With this approach, crossover and mutation operators play the same important role in the genetic search.

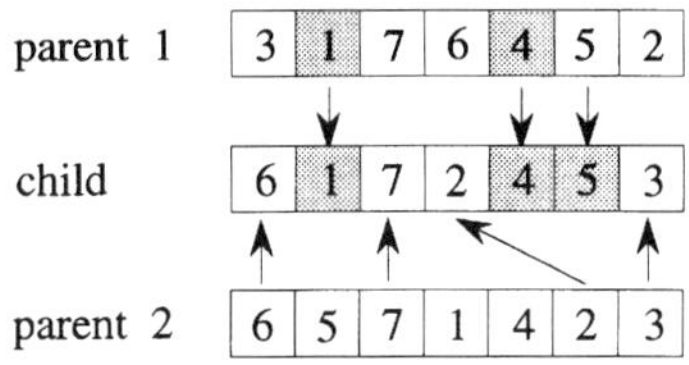

Figure 6.9. Position-based crossover operator.

Position-Based Crossover. The nature of the proposed encoding can be viewed as a type of permutation encoding. A number of recombination operators have been investigated for permutation representation. The position-based crossover operator proposed by Syswerda [604] was adopted as shown in Figure 6.9. Essentially, it takes some genes from one parent at random and fills vacuum positions with genes from the other parent by a left-to-right scan.

Swap Mutation. A swap mutation operator was used here, which simply selects two positions at random and swaps their contents as shown in Figure 6.10.

Local Search–Based Mutation. Local search methods seek improved solutions to a problem by searching in the neighborhood of an incumbent solution. The implementation of local search requires an initial incumbent solution, the definition of a neighborhood for an incumbent solution, and a method for choosing the next incumbent solution. The idea is that hunting for an improved solution by making a small change can be used in a mutation operator. Observing the swap mutation, the chromosome generated by pairwise interchange can be viewed as a neighbor of the original chromosome. A *neighborhood* of a chromosome is then defined as a set of chromosomes generated by such pairwise interchanges. For a pair of genes, one, called a *pivot*, which is fixed for a given neighborhood, and the other is selected at random as shown in Figure 6.11. For a given neighborhood, a chromosome is called a *local optimum* if it is better than any other chromosome according to the fitness value. The size of a neighborhood affects the quality of the local optimum. There is a clear trade-off between small and large neighborhoods: If the number of neighbors is larger, the probability of finding a good neighbor may be higher, but looking for it takes more time. A scheme for the mutation operator is described below.

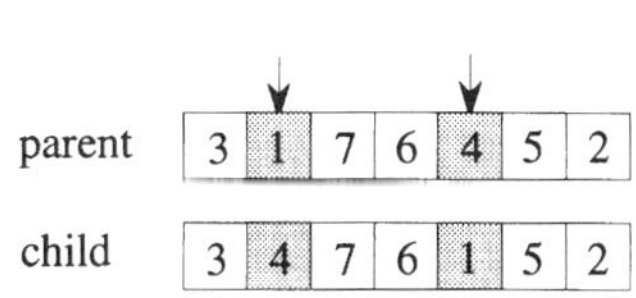

Figure 6.10. Swap mutation operator.

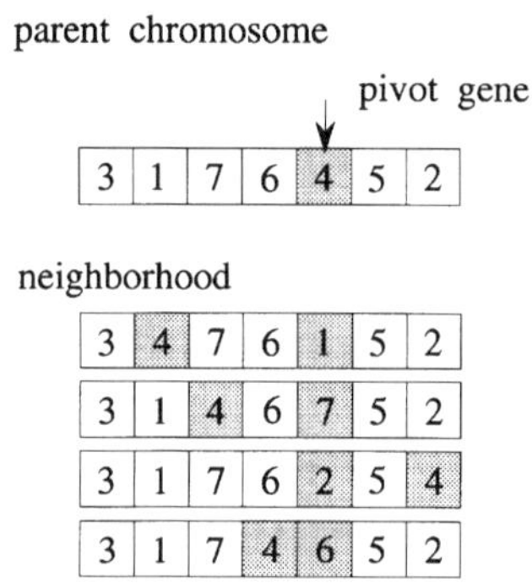

Figure 6.11. Incumbent chromosome and its neighborhood.

```
Procedure: Local Search-Based Mutation
inputs: P (parent)
output: C (offspring)
begin
 i ← 1;
 let P be a current incumbent chromosome;
 select a pivot gene;
 while (i < specified_number) do
  pick up a gene randomly;
  generate a neighbor by swapping it with pivot;
  evaluate the neighbor;
  let the neighbor be current incumbent if it is better;
  i ← i + 1;
 end
 save the incumbent in C;
end
```

6.4.3 Evaluation and Selection

During each generation, chromosomes are *evaluated* using some measure of fitness. The following four major steps are included in the evaluation phase:

Step 1. Convert chromosomes to topological sorts.

Step 2. Generate schedules from the topological sorts.

Step 3. Calculate objective values for each schedule.

Step 4. Convert objective values into fitness values.

The procedure for step 1 was discussed in Section 6.4.2. Because a topological sort gives a feasible order of activities, we construct a schedule by selecting activities in order of their appearance in the topological sort and scheduling them one at a time as early as resource availabilities permit. Let

i be the iteration index of the procedure, let V be the set of all node, P_i be the set of all direct predecessors of activities i, PS[·] be the array storing topological sort, σ_j and ϕ_j be start and finish times associated with activity j, $b_k[l]$ be the array for storing the amount of resource k available in time l, d_j be the duration associated with activity j, and r_{jk} be the consumption of resource k associated with activity j. The procedure for determining start and finish times for each activity from a given topological sort is given below.

Procedure: Start and Finish Times of Activities

Step 1. Initialize.

$$i \leftarrow 1 \text{ (iteration index)}$$

$$j \leftarrow \text{PS}[i] \text{ (initial activity)}$$

$$\sigma_j \leftarrow 0,\ \phi_j \leftarrow 0 \text{ (start and finish times for initial activity)}$$

$$b_k[l] \leftarrow b_k, \qquad l = 1,2,\ldots,\Sigma_{j=1}^{n} d_j, \quad k = 1,2,\ldots,m \text{ (initial resources)}$$

Step 2. Perform the termination test. If $i = n$, go to step 5; otherwise, $i \leftarrow i + 1$, and continue.

Step 3. Determine the start and finish times.

$$j \leftarrow \text{PS}[i]$$

$$\sigma_j^{\min} \leftarrow \max\{\phi_l | l \in P_j\}$$

$$\sigma_j \leftarrow \min\{t | t \geq \sigma_j^{\min},\ b_k[l] \leq r_{jk},\ l = t, t+1,\ldots,t+d_j,\ k = 1,2,\ldots,m\}$$

$$\phi_j \leftarrow \sigma_j + d_j$$

Step 4. Update the available resources.

$$b_k[l] \leftarrow b_k[l] - r_{jk}, \qquad l = t, t+1,\ldots,t+d_j, \quad k = 1,2,\ldots,m$$

Go back to Step 2.

Step 5. Stop. Return σ_j and ϕ_j.

Step 3 is trivial. Because we use the objective of project duration, the finish time of the last activity is the objective value. Since we are dealing with a minimization problem, we have to convert the original objective value to a fitness value to ensure that the fitter individual has a larger fitness value.

Let $\boldsymbol{v}_k$ be the kth chromosome in the current generation, $g(\boldsymbol{v}_k)$ be the fitness function, $f(\boldsymbol{v}_k)$ be the objective value (i.e., the project duration), $f_{\max}$ and $f_{\min}$ be the maximum and minimum values of the objective values in current generation. The transformation is given as follows:

$$g(\boldsymbol{v}_k) = \frac{f_{\max} - f(\boldsymbol{v}_k) + \gamma}{f_{\max} - f_{\min} + \gamma} \tag{6.25}$$

where γ is a positive real number that is usually restricted within the open interval (0,1). The purpose of using it is twofold: (1) to prevent Equation (6.25) from zero division and (2) to make it possible to adjust the selection behavior from fitness-proportional selection to pure random selection. When the fitness difference among chromosomes is relatively large, the selection is fitness proportional; when the difference becomes too small, the selection tends to pure randomness among relatively competitive chromosomes.

In Cheng and Gen's experiment, the *roulette wheel* approach, one of fitness-proportional selection, was adopted as the selection procedure. The *elitist selection* method was combined with this approach to preserve the best chromosome in the next generation and overcome the stochastic errors of sampling. With elitist selection, if the best individual in the current generation is not reproduced in the new generation, one individual is removed randomly from the new population and the best one is added to the population.

6.4.4 Experimental Results

Primary computational experiments were conducted in two parts: (1) tuning the parameters of genetic algorithms to investigate how they affect performance, and (2) testifying to the behavior of local search–based mutation in genetic search.

Parameter Turning. To investigate how population size affects the performance of the method proposed, maximum generation was fixed as 100, and crossover and mutation ratios were both fixed as 0.1. With a lower ratio of crossover and mutation, population size becomes a leading factor in the performance of genetic algorithms. Population size was varied from 10 to 100. Figure 6.12 shows the best, worst, and average values of the objective over 100 random runs for each parameter setting. From the results we can see that when *pop_size* is larger than 50, any increase has no significant influence on the performance of the genetic algorithm.

The aim of the following tests is to make a comparison between crossover and mutation operators to confirm which plays a more important role in the genetic search. Genetic algorithms are tested in the following two cases:

1. Fix the mutation ratio at 0 and vary the crossover ratio from 0.1 to 0.9.
2. Fix the crossover ratio at 0 and vary the mutation ratio from 0.1 to 0.9.

In both cases, fix *max_gen* = 100 and *pop_size* = 20. The best values of the objective function over 200 random runs for each parameter setting are given

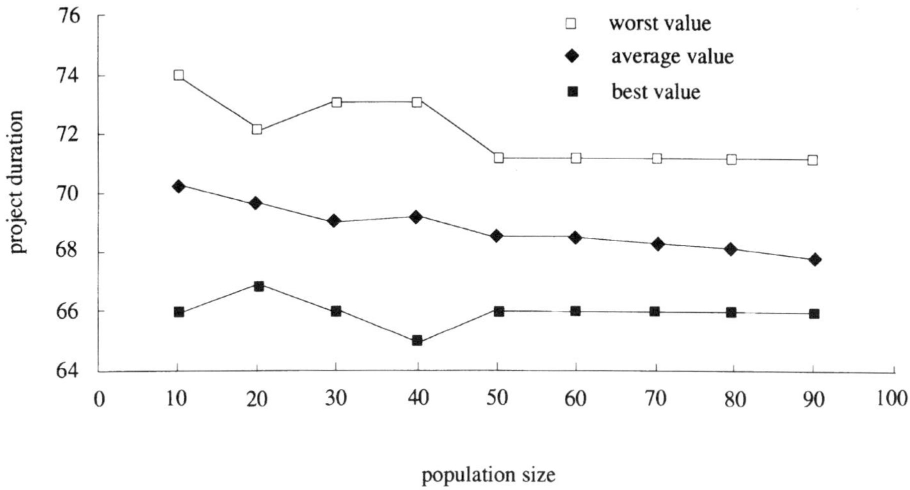

Figure 6.12. Comparison of the best, worst, and average values of an objective under different values of *pop_size*.

in Figure 6.13, the average values in Figure 6.14, and the worst values in Figure 6.15.

From the results we can see that for average performance there is no obvious difference between crossover and mutation, but the chance of obtaining an optimal solution is much higher when running genetic algorithms with mutation only than when running with crossover only. The results reveal that mutation plays a critical role in genetic systems, which contradicts conven-

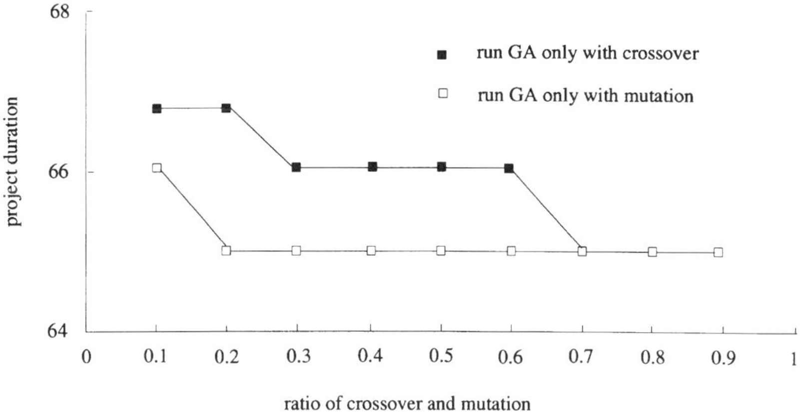

Figure 6.13. Best values over 200 random runs under different ratios of crossover and mutation.

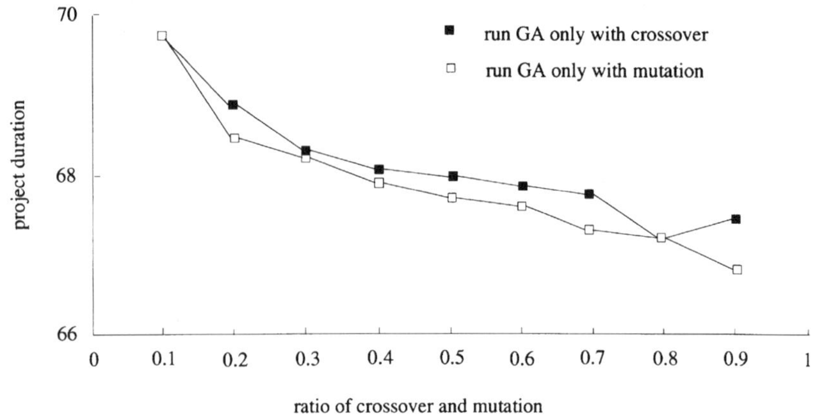

Figure 6.14. Average values over 200 random runs under different ratios of crossover and mutation.

tional beliefs. In conventional genetic algorithms, crossover is used as the main operator, with mutation used simply as a subsidiary means. Although the mechanism of swap mutation is very simple, it allows exploitation in the neighborhood of a given chromosome. This is why mutation can have a high probability of yielding the optimal solution.

Behavior of Local Search–Based Mutation. Experiments in the following three cases demonstrate the behavior of local search–based mutation in genetic search: (1) with position-based crossover only, (2) with swap mutation only, and (3) with local search–based mutation only.

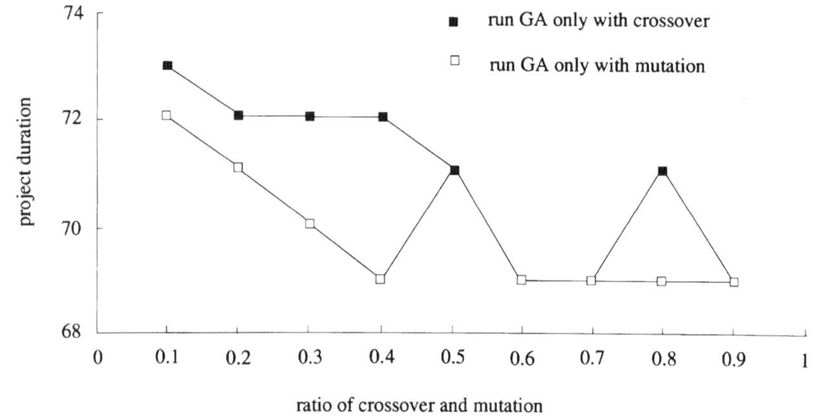

Figure 6.15. Worst values over 200 random runs under different ratios of crossover and mutation.

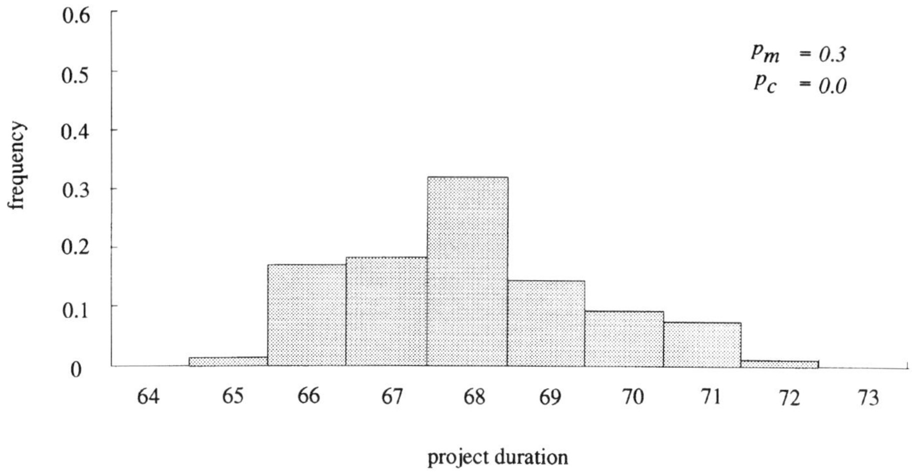

Figure 6.16. Solution distribution over 200 runs with swap mutation only.

To permit a fair comparison, for position-based crossover and swap mutation, population size was fixed at 50 and maximum generation at 200; while for local search–based mutation, population size was fixed at 20, maximum generation at 100. For each case, the ratio was fixed as 0.3. The size of the neighborhood for local search–based mutation was fixed at 6 so that the total number of chromosomes examined for each case would be nearly the same. The solution distribution over 200 runs is given in Figure 6.16 when only swap mutation was used, in Figure 6.17 when only position-based crossover was used, and in Figure 6.18 when only local search–based mutation was

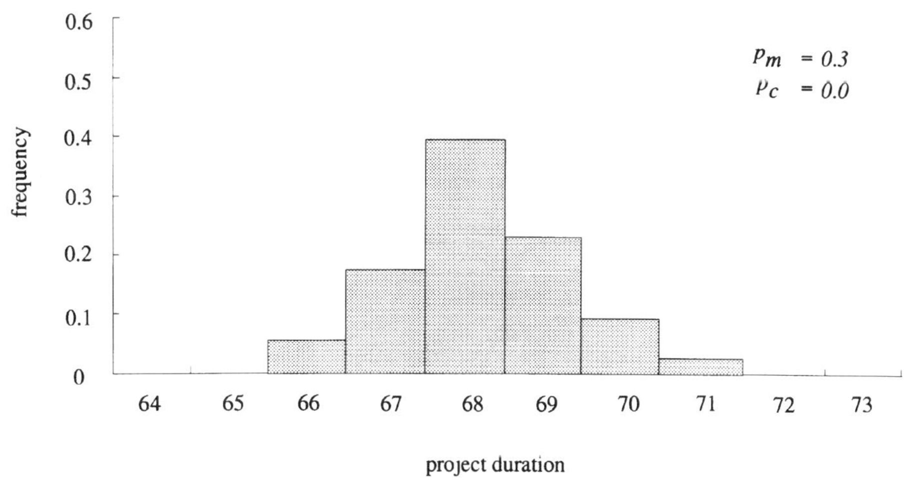

Figure 6.17. Solution distribution over 200 runs with position-based crossover only.

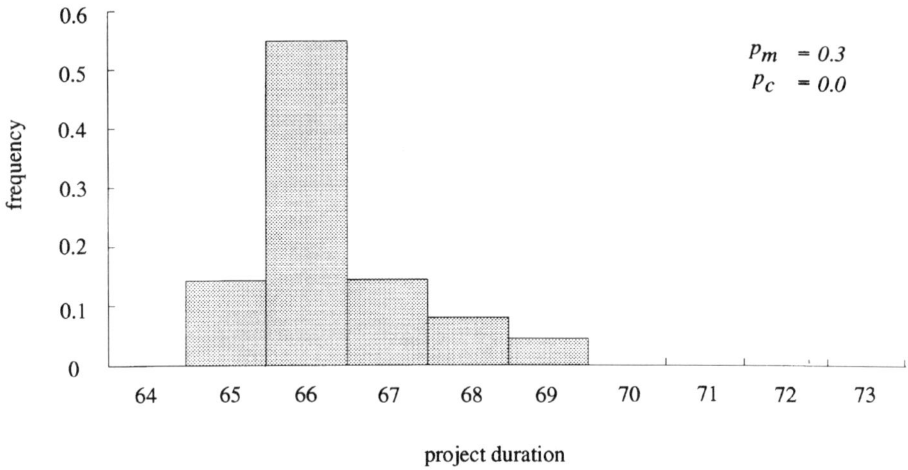

Figure 6.18. Solution distribution over 200 runs with local search–based mutation only.

used. It is easy to see that local search–based mutation has a significant impact on the performance of genetic algorithms.

Recently, Tsujimura and Gen developed an evolutionary algorithm for solving a project scheduling under resource constraint [736] and Özdamar proposed a genetic algorithm approach to a general category project scheduling problem [727].

6.5 PARALLEL MACHINE SCHEDULING

Next we consider a parallel machine scheduling problem:

- There are m ($m < n$) identical parallel machines and n independent jobs with known weights $w_1, w_2, \ldots, w_n$ as well as processing times $p_1, p_2, \ldots, p_n$.
- The jobs are immediately available for processing and can be processed by any one of m machines.
- No job can be preempted once its processing has begun.
- An unrestricted due date d, that is,

$$d \geq \sum_{j=1}^{n} p_j$$

  is common to all jobs.

It is easy to verify that an optimal schedule for the problem has no idle time between jobs. Let Π denote the set of feasible schedules without idle

TABLE 6.4. Three-Job/3-Machine Example

Job	1	2	3	4	5	6	7	8	9
Processing time	2	4	4	1	5	2	1	3	2
Weight	5	8	6	4	4	1	9	10	2

times between jobs. For a given schedule $\sigma \in \Pi$, let c_j be the completion time of job j under schedule σ for $j = 1,2,\ldots,n$, and let $f(\sigma)$ denote the corresponding objective function value. The problem is to minimize the maximum weighted absolute lateness as follows:

$$\min_{\sigma\in\Pi} \quad f(\sigma) = \max\{w_j|c_j - d|; j = 1,\ldots,n\} \tag{6.26}$$

The objective is known as a nonregular performance measure.

Usually, all machines are assumed to be identical such that the processing time of a job is independent of machine. A job is characterized by a processing time and a weight. Each job can be processed by at most one machine at a time, while each machine can process at most one job at a time. A common due date is associated with each job. Objectives can be separated into two classes: *regular measure* and *nonregular measure,* the latter being much harder to solve than the former, in general. As an example, consider the sample 9-job/3-machine problem given in Table 6.4. Figure 6.19 shows a feasible schedule.

We know that there are two essential issues to be dealt with in parallel machine scheduling problems: (1) job partition among machines (combination nature) and (2) job sequence within each machine (permutation nature). That is, the problem involves both permutation and combination components. Most parallel machine scheduling problems have been shown to be NP-complete [117].

Cheng and Gen have investigated how to apply the memetic algorithm, a kind of hybridization of genetic algorithms and the local search method, to solve the parallel machine scheduling problem [100, 104]. There are two essential issues to be dealt with for all kinds of parallel machine scheduling problems: job partition among machines and job sequence within each ma-

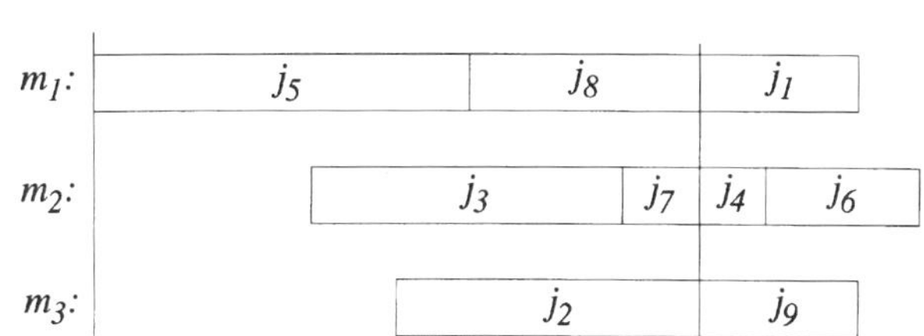

Figure 6.19. Feasible schedule for 9-job/3-machine example.

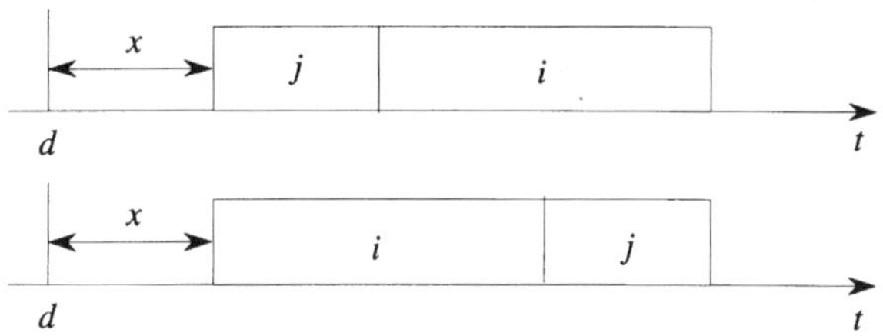

Figure 6.20. Two possible orders for jobs i and j in the tardy job set.

chine. In their proposed method, a genetic algorithm is used to evolve job partition and a heuristic procedure is used to adjust job permutation to push each chromosome to its local optimum. Several dominance conditions for an optimal solution of the problem were established. A fast heuristic algorithm for sequencing jobs on one machine was proposed based on the dominant condition.

6.5.1 Dominance Condition

A common way to determine job order on a machine is to establish some dominance properties among jobs. Dominance properties establish precedence relations among jobs in an optimal schedule. These are usually called *posteriori* precedence relations because they are not a part of the original problem statement. This type of precedence relation will be used to build a fast heuristic algorithm for job sequencing on a machine.

Let J be the set of jobs. A job belongs to either an early job set or a tardy job set. The early job set is defined as $E = \{j | c_j \leq d \text{ and } j \in J\}$ and the tardy job set is defined as $T = \{j | c_j > d \text{ and } j \in J\}$. For a given schedule, a job is called a *dominant job* if it has the maximal weighted absolute lateness; that is, if job i is the dominant job, we have $w_i|c_i - d| = \max_{j \in J}\{w_j|c_j - d|\}$. A machine is called a *dominant machine* if it processes the dominant job.

Property 6.4. For a pair of jobs i and j in the tardy job set T, if $w_i \geq w_j$, job i precedes job j in at least one optimal schedule.

Proof: Consider two possible orders of jobs i and j as shown in Figure 6.20, where x denotes the total length of jobs scheduled in set T until now. We have

$$\frac{w_i(p_i + p_j + x)}{w_j(p_i + p_j + x)} = \frac{w_i}{w_j} \geq 1 \tag{6.27}$$

Therefore, job i must precede job j.

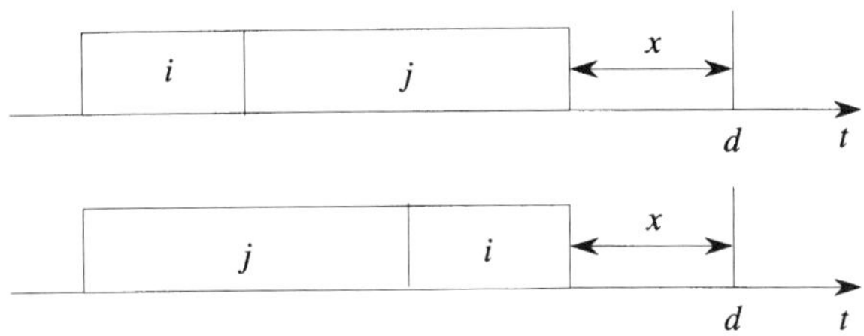

Figure 6.21. Two possible orders for jobs i and j in the early job set.

This property gives the fact that a weightier job precedes a lighter job in the tardy job set T in at least one optimal schedule. In other words, jobs in set T are in nonincreasing order of weights.

How to order jobs in the early job set E is not as simple as in the tardy job set T. A job is characterized by two factors: weight and processing time. Given two jobs i and j, there are four basic patterns of ordering their relations:

1. $w_i \geq w_j$ and $p_i \geq p_j$
2. $w_i \geq w_j$ and $p_i < p_j$
3. $w_i < w_j$ and $p_i \geq p_j$
4. $w_i < w_j$ and $p_i < p_j$

We only need to examine patterns 1 and 2, because for the precedence relation of a given pair of jobs, pattern 4 describes the same case as pattern 1, and pattern 3 as pattern 2.

Property 6.5. For a pair of jobs i and j in the early job set E, if $w_i \geq w_j$ and $p_i < p_j$, job j must precede job i in at least one optimal schedule.

Proof: Consider two possible orders of jobs i and j shown in Figure 6.21, where x denotes the total length of jobs scheduled until now in set E. We have

$$\frac{w_i(p_j + x)}{w_j(p_i + x)} \geq \frac{w_i(p_j + x)}{w_j(p_j + x)} = \frac{w_i}{w_j} \geq 1 \tag{6.28}$$

Therefore, job j precedes job i.

This property shows that a longer and lighter job precedes a shorter and weightier job in set E in at least one optimal schedule.

Property 6.6. For a pair of jobs i and j in the early job set E, if $w_i \geq w_j$, $p_i \geq p_j$, and $w_i/p_i \geq w_j/p_j$, job j precedes job i.

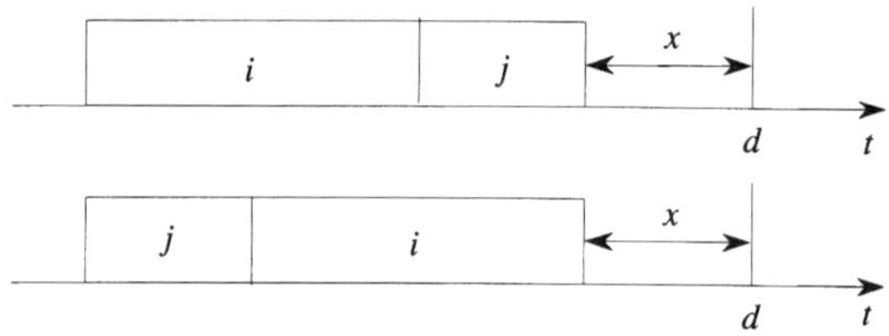

Figure 6.22. Two possible orders for jobs i and j in the early job set.

Proof: Consider two possible orders of job i and job j shown in Figure 6.22. We have

$$\frac{w_i(p_j + x)}{w_j(p_i + x)} \geq \frac{w_i p_j + w_j x}{w_j(p_i + x)} \geq \frac{w_j p_i + w_j x}{w_j(p_i + x)} = 1 \tag{6.29}$$

Therefore, job j must precede job i in an optimal schedule.

This property gives one of the necessary conditions that a shorter and lighter job precedes a longer and weightier job in set E in at least one optimal schedule.

Property 6.7. For a pair of jobs i and j of set E associated with $w_i \geq w_j$, $p_i \geq p_j$, and $w_i/p_i < w_j/p_j$, if the total length of scheduled jobs x in the set E is

$$x \geq \frac{w_j p_i - w_i p_j}{w_i - w_j} \tag{6.30}$$

then job j precedes job i in at least one optimal schedule.

Proof: Consider two possible orders of jobs i and j shown in Figure 6.22. If job j precedes job i in set E, we have

$$\frac{w_i(p_j + x)}{w_j(p_i + x)} \geq 1 \tag{6.31}$$

This implies that

$$x \geq \frac{w_j p_i - w_i p_j}{w_i - w_j} \tag{6.32}$$

This property gives another necessary condition that a shorter and lighter job precedes a longer and weightier job in an optimal schedule. From the foregoing properties we can know that there exists at least one optimal schedule where jobs in set T of the dominant machine are in nondecreasing order

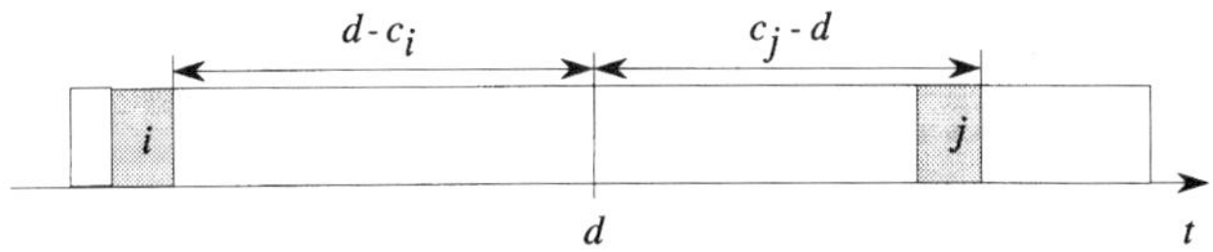

Figure 6.23. Two dominant jobs i and j in an optimal schedule.

of weights, while jobs in set E of the dominant machine are in nonincreasing order of weights for most cases. An exception may occur if a weightier and longer job does not satisfy the condition given in Property 6.7.

Property 6.8. For a given optimal schedule, there exist two dominant jobs on the dominant machine, one in set T and the other in set E.

Proof: Suppose that for a given optimal schedule, there is only one dominant job. Without loss of generality, assume that job $i \in E$ is the dominant job. Then there must exist a job $j \in T$ that dominates set T, and $w_j|c_j - d| < w_i|c_i - d|$, as shown in Figure 6.23. If we delay all jobs by the amount δ, which is determined by

$$\delta = \frac{w_i|c_i - d| - w_j|c_j - d|}{w_i + w_j} \tag{6.33}$$

These two jobs i and j have the equal value of absolute lateness, and the objective value of the given schedule is reduced by the amount $w_i\delta$. This is a contradiction to the precondition of optimal schedule. Therefore, we have proved the property.

Property 6.9. There exists at least an optimal schedule where the following condition holds true for the dominant machine:

$$\sum_{i \in E} p_i \geq \sum_{j \in T} p_j \tag{6.34}$$

Proof. There are four possible patterns in which dominant jobs may occur on the dominant machine.

1. One job is in the head of E while the other is in the tail of T.
2. One job is in the head of E while the other is in the middle of T.
3. One job is in the middle of E while the other is in the tail of T.
4. One job is in the middle of E while the other is in the middle of T.

For cases 1 and 3, as shown in Figure 6.24, suppose that the total length of T is greater than the length of E. Then we can make a much better schedule by moving the rightmost section of T to the leftmost section of E, which

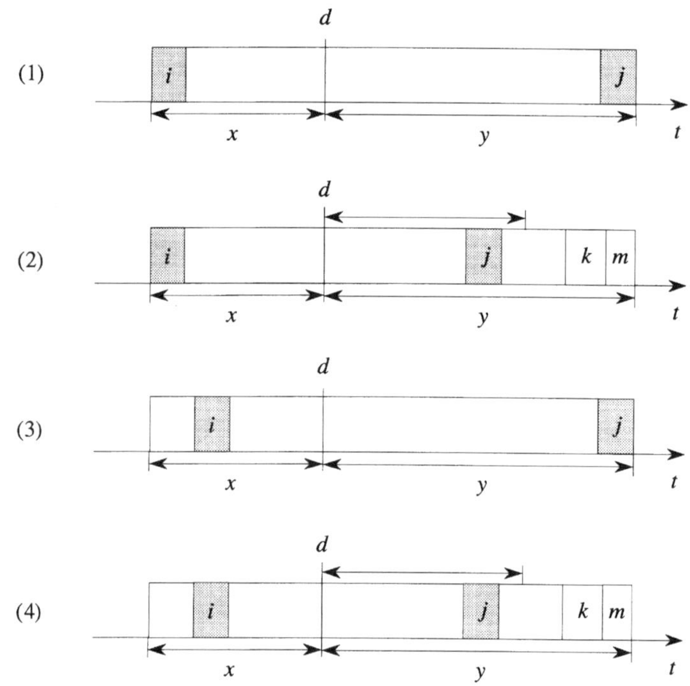

Figure 6.24. Four possible patterns of dominant jobs.

makes the given schedule nonoptimal. This is a contradiction to the precondition for an optimal schedule. For cases 2 and 4 shown in Figure 6.24, without loss of generality, assume that if the rightmost jobs k and m are removed from the tardy job set T, the two sets have nearly the same length. In such a case we can make a new schedule by putting job m at the leftmost of E and job k at the position following job k so that the total length of E is larger than T without increasing the maximal absolute lateness. Therefore, we have proved the property.

6.5.2 Memetic Algorithms

Memetic algorithms can be thought of as a special kind of genetic search over the subspace of local optima, within which a local optimizer is added to genetic algorithms and applied to every child before it is inserted into the population [468, 520]. Recombination and mutation will usually produce solutions that are outside this space of local optima, but a local optimizer can then repair such solutions to produce final children that lie within this subspace, yielding a memetic algorithm. With the hybrid approach, genetic algorithms are used to perform global exploration among populations, while heuristic methods are used to perform local exploitation around chromosomes.

1	2	3	4	5	6	7	8	9	position: job ID
3	1	1	3	3	2	2	3	1	value: machine ID

Figure 6.25. Proposed encoding method.

Because of the complementary properties of genetic algorithms and conventional heuristics, the hybrid approach often outperforms either method operating alone.

Representation. We just need to encode the component of *job partition* into a chromosome. Let $J = \{1,2,...,n\}$ be the set of jobs and $M = \{1,2,...,m\}$ be the set of machines. A chromosome consists of n genes and each gene takes an integer number from the set M. Thus the position of a gene represents the job ID and the value of the gene represents a machine ID. Figure 6.25 illustrates the encoding method using the simple example of 3 machines and 9 jobs. In this example, jobs 2, 3, and 9 are processed by machine 1, jobs 6 and 7 are processed by machine 2, and jobs 1, 4, 5, and 8 are processed by machine 3.

Genetic Operators. The uniform crossover operator proposed by Syswerda is used here [603]. This method first generates a random mask and then exchanges relative genes under the mask between parents. A mask is simply a binary string with the same-size chromosome. The parity of each bit in the mask determines, for each corresponding bit in an offspring, from which parent it will receive that bit. It is easy to see that the crossover operators can adjust job partitions among machines.

Observing the second child in Figure 6.26, we can know that machine 3 disappears from the chromosome. This means that no job is assigned to machine 3. Such a disappearance certainly produces a bad schedule. To enable

random mask

0	0	1	0	0	1	1	0	0

parents under mask

3	1	1	3	3	2	2	3	1
1	1	2	1	2	3	3	2	2

children after exchanging relative genes

3	1	2	3	3	3	3	3	1
1	1	1	1	2	2	2	2	2

Figure 6.26. Uniform crossover operation.

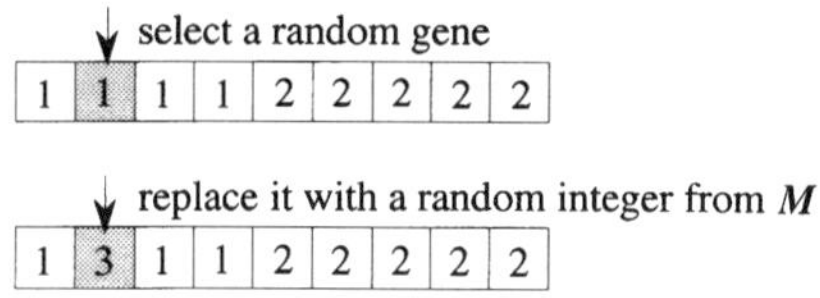

Figure 6.27. Random replacement mutation operation.

genetic search to recover such a machine for a chromosome, a *random replacement* mutation operator is used. The mutation operator first selects a gene randomly and then replaces it with a random integer from set M as shown in Figure 6.27. Essentially, the mutation operator can cause a gene either to recover or to disappear and can adjust the job partition among machines.

Heuristic Procedure. The heuristic procedure is used to sequence jobs on each machine, based on the dominance conditions of Properties 6.4 to 6.7.

Let $d - x$ denote the start time of the earliest job and $d + y$ the completion time of the latest job that have been scheduled on a machine, as shown in Figure 6.28. Let J be the set of jobs, E be the early job set, $E[i]$ be the ith element in E, and T be the tardy job set. The heuristic procedure works as follows:

```
Procedure: Job Sequencing
input: a set of jobs J on a machine
output: a sequence of jobs in the set E an the set T
begin
 sort all jobs in J on nonincreasing order of weight;
 save the ordered job into array S[·];
 x ← 0, y ← 0;
 i ← 0;
 while (i ≤| J |) do
  p ← p_S[i];
  if (y + p > x)
   then put job i in the head of set E_i
   x ← x + p;
   if p_E[2] ≥ p_E[1] and w_E[1]p_E[2] > w_E[2]p_E[1] and
    y < (w_E[1]p_E[2] − w_E[2]p_E[1]) / (w_E[2] − w_E[1]) − p_E[2]
```

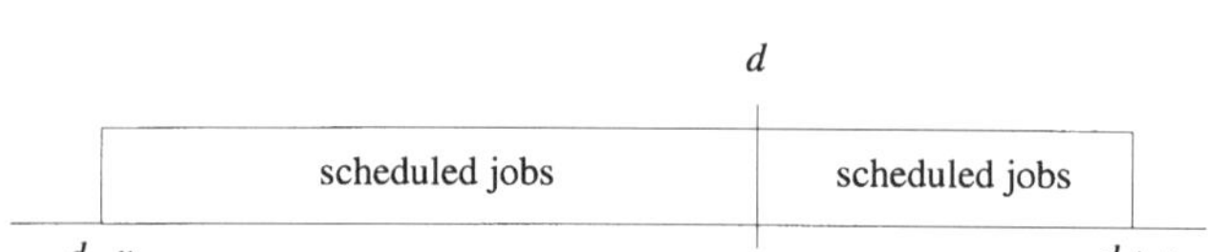

Figure 6.28. Start and completion times for scheduled jobs.

```
      then swap E[1] and E[2];
     end
    else
     put job i in the tail of set T;
     y ← y + p;
    end
    i ← i + 1;
  end
end
```

Evaluation and Selection. The two major steps are stated as follows:

1. Calculate the objective values for each chromosome.
2. Convert the objective values to fitness values.

From Equation (6.26) we can see that the objective is a function of completion times of jobs. Essentially, the completion time (start time) can be calculated according to the order given by the heuristic procedure described above. However, if the completion times are calculated in this way, only one dominant job may occur in the dominant machine. According to Property 6.8, we can slide each job properly to yield a much better schedule. Therefore, for a given chromosome, we first slide each job on each machine according to equation (6.33) and then calculate the objective of maximal absolute lateness for the chromosome.

Since the problem is a minimization problem, we have to convert the original objective value to a fitness value to ensure that a fitter individual has a larger fitness value. This is done by a transformation through the following fractional linear function:

$$\text{eval}(v_t) = \frac{1}{f(v_t)}, \qquad t = 1,2,\ldots,pop\_size \tag{6.35}$$

where $\text{eval}(v_t)$ is the fitness function for the tth chromosome and $f(v_t)$ is the objective function value.

The roulette wheel selection method was used as the basic selection mechanism to reproduce the next generation based on the current enlarged population. The elitist method was combined with it to preserve the best chromosome in the next generation and overcome the stochastic errors of sampling.

6.5.3 Experimental Results

Primary computational experiments were conducted on randomly generated test problems. First, the algorithm proposed was run under different parameter settings to investigate how they affect performance. It was further compared

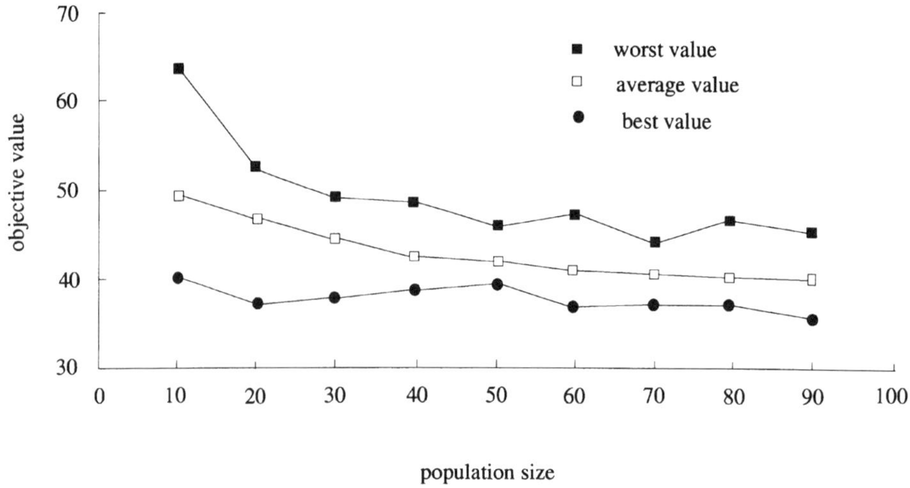

Figure 6.29. Comparison of the best, worst, and average values of the objective found by the memetic algorithm using variations in *pop_size*.

with Li and Cheng's heuristic algorithm to show the effectiveness of the algorithm proposed.

Parameter Tuning. How population size affects the performance of a memetic algorithm is investigated. A randomly generated problem with 30 jobs and 5 machines was used for the test, fixing *max_gen* at 500 and p_m and p_c at 0.1. With a lower ratio of crossover and mutation, the population size becomes a leading factor for the performance of memetic algorithms. Figure 6.29 shows the best, worst, and average values of objective over 100 random runs for *pop_size* from 10 to 90. From the results we can see that when *pop_size* is larger than 50, any increase has no significant influence on performance.

Comparison with Heuristic. A comparison was made with randomly generated problems of different job and machine size. The basic setting of parameters for the memetic algorithm is *pop_size* = 50, p_c = 0.4, p_m = 0.4, and *max_gen* = 500. The results over 20 runs for each problem are given in Table 6.5, where *heuristic* stands for Li and Cheng's heuristic. The results show that the memetic algorithm outperforms the heuristic algorithm.

6.6 MULTIPROCESSOR SCHEDULING

Multiprocessor systems have been employed in a wide variety of computer processing applications not only in the field of information processing but

TABLE 6.5. Results for Random Test Problems

			Memetic Algorithms		
Problem	Size	Heuristic	Best	Worst	Average
1	$j30 \times m5$	38.50	32.00	35.20	33.52
2	$j50 \times m5$	100.15	94.50	102.86	98.59
3	$j50 \times m10$	49.09	38.77	43.31	40.91
4	$j60 \times m5$	117.78	114.55	120.00	117.37
5	$j60 \times m10$	132.00	125.36	140.00	133.59
6	$j70 \times m8$	88.00	84.00	93.17	87.06
7	$j70 \times m10$	150.86	142.22	156.80	147.88
8	$j80 \times m5$	292.50	284.39	304.00	292.43
9	$j80 \times m10$	97.07	92.00	99.47	95.75
10	$j80 \times m12$	98.18	92.31	97.78	94.72
11	$j80 \times m14$	81.67	73.85	82.13	76.84
12	$j90 \times m7$	150.00	145.38	151.20	148.47
13	$j90 \times m11$	77.50	71.11	78.24	74.05
14	$j100 \times m20$	113.75	106.36	116.31	110.61

also for robot control, real-time high-speed simulation of dynamical systems, and so on, because they provide excellent responsiveness and cost-effectiveness.

Multiprocessor scheduling problems (MSPs) can be stated as assigning the tasks of a precedence-constrained task graph onto a set of processors and determining the sequence of execution of the tasks by each processor to minimize some cost function, such as overall task execution (completion) time. Such problems are, however, extremely difficult to solve and are generally intractable; that is, it is well known that relaxed or simplified subproblems constructed from the original scheduling problem by imposing a variety of restricting conditions still fall into the class of NP-hard problems.

Hou, Ren, and Ansari developed suitable genetic encoding of this problem [308]. Their genotype is divided into variable-length partitions or subchromes. Yue and Lilja proposed a parallel distributed genetic algorithm [688]. A simple genetic algorithm running on a local workstation is used to generate possible scheduling algorithms. Tsujimura and Gen proposed a height function–based genetic algorithm for solving this problem [627].

6.6.1 Problem Description and Assumptions

The multiprocessor scheduling is to assign n tasks onto m processors in such a way that precedence constraints are maintained and to determine the start and finish times of each task with the objective of minimizing completion time. The mathematical formulation of the problem is given as follows:

$$\min \quad \{\max_j \{x_j y_{ij}\}\}$$

$$\text{s.t.} \quad x_k - p_k \geq x_j, \qquad T_j < T_k \tag{6.36}$$

$$t_{\max} - \sum_{j=1}^{n} p_j y_{ij} \geq 0, \qquad i = 1,\ldots,n \tag{6.37}$$

$$\sum_{i=1}^{m} y_{ij} = 1, \qquad j = 1,\ldots,m \tag{6.38}$$

$$y_{ij} = 0 \text{ or } 1, \qquad i = 1,\ldots,m, \quad j = 1,\ldots,n \tag{6.39}$$

where

$$y_{ij} = \begin{cases} 1, & \text{if task } T_j \text{ is assigned to processor } P_i \\ 0, & \text{otherwise} \end{cases}$$

and where $t_{\max} = \max_i\{t_i\}$, x_j is the completion time of task T_j, p_j the processing time of task T_j, t_i the time required to process all tasks assigned to process P_i, and $<$ represents a precedence relation; a precedence relation between tasks, $T_j < T_k$, means that T_k preceeds T_j. We assume that the communication system is contention free and permits the overlap of communication started only after all dates have been received from predecessors. Duplication of the same task is not allowed.

6.6.2 Genetic Algorithm for MSP

For chromosome representation scheme and genetic operations, we adopt the concept of *height function* [308], which considers the precedence relations among tasks in implementation of a genetic algorithm. The construction of genetic operators is based on the height function for the tasks. Three genetic operators are proposed based on height function.

Height Function. To facilitate generation of the schedule and the construction of genetic operators, we define the *height* of each task in the task graph as

$$height(T_i) = \begin{cases} 0, & \text{if } pred(T_i) = \emptyset \\ 1 + \max\limits_{T_j \in pred(T_i)} \{height(T_j)\}, & \text{otherwise} \end{cases}$$

where $pred(T_j)$ is the set of predecessors of T_j and $succ(T_j)$ is the set of successors of T_j. One major drawback of this definition is that the optimal

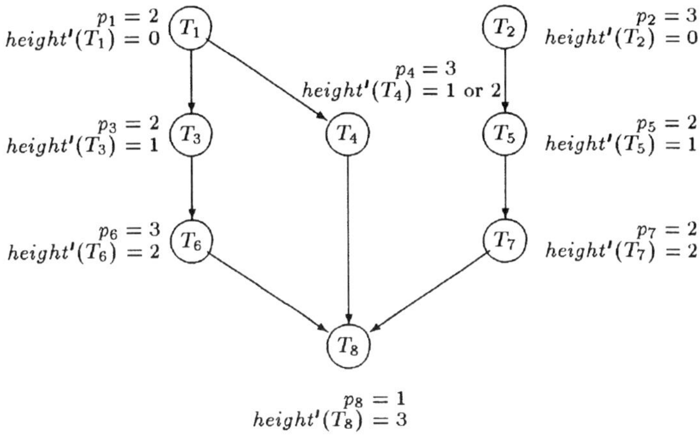

Figure 6.30. Task graph of Example 6.1.

schedule may not satisfy it. To reduce the likelihood of this happening, we can modify the definition of *height* as

$$height'(T_j) = rand \in [\max\{height(T_i)\} + 1, \min\{height'(T_k)\} - 1]$$
$$\text{over all} \quad T_i \in pred(T_j) \quad \text{and} \quad T_k \in succ(T_j) \tag{6.40}$$

where *rand* is a random integer number. This height function in a way conveys the precedence relations between tasks. In fact, if there is a path from T_i to T_j (i.e., there exists a series of directed edges leading from T_i to T_j), T_i must be executed before T_j and $height(T_i) < height(T_j)$. However, if there is no path between the two tasks, the order of execution between them can be arbitrary. However, the precedence relations within the processor must still be maintained.

Example 6.1. Let us consider an instance with 8 jobs and 2 processors; the *height'* of each task is shown in Figure 6.30.

Representation. The chromosome representation used here is based on the schedule of the tasks in each processor. This representation eliminates the need to consider the precedence relations between tasks scheduled to different processors. A representation of the schedule for genetic algorithms must accommodate the precedence relations between computational tasks. This is resolved by representing the schedule as several lists of computation tasks. Each list represents the computational tasks executed on a processor, and the order of the tasks in the list indicates the order of execution. This ordering allows us to maintain the precedence relations for the tasks executed in a

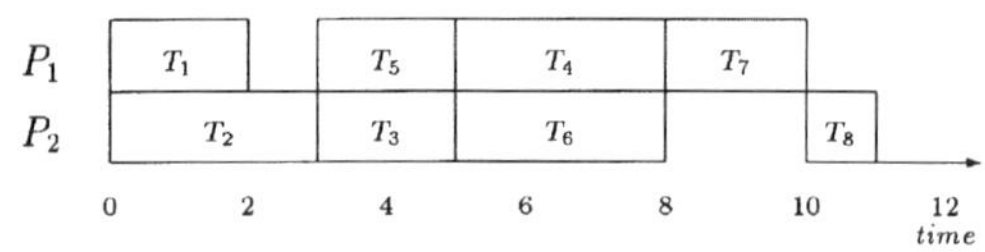

Figure 6.31. Schedule for two processors.

processor (*intraprocessor precedence relation*) and ignore the precedence relation between tasks executed in different processors (*interprocessor precedence relation*).

To simplify construction of the genetic operators, we imposed the following condition on the schedules: The list of tasks within each processor of the schedule is ordered in ascending order of their *height'*s.

Example 6.2. Let us consider the MSP shown in Figure 6.30. A chromosome includes two lists P_1 and P_2 for two processors, as follows:

$$P_1(T_1T_5T_4T_7): \quad height'(T_1) \leq height'(T_5) \leq height'(T_4) \leq height'(T_7)$$
$$P_2(T_2T_3T_6T_8): \quad height'(T_2) \leq height'(T_3) \leq height'(T_6) \leq height'(T_8)$$

The Gantt chart is shown in Figure 6.31. The representation of chromosome v for this schedule is as follows:

$$v = \begin{cases} P_1(T_1 \ T_5 \ T_4 \ T_7) \\ P_2(T_2 \ T_3 \ T_6 \ T_8) \end{cases}$$

Evaluation Function. The evaluation function is essentially the objective function for the problem. It provides a means of evaluating the search nodes and controls the selection process. For the MSP we can consider factors such as throughput, finishing time, and processor utilization for the evaluation function.

The evaluation functions used for the genetic algorithm proposed here is based on the makespan $C_{\max}$ of the schedule. Because we use roulette wheel selection, we define the evaluation function using makespan as

$$\text{eval}(v_k) = \frac{1}{C^k_{\max}}, \qquad i = 1,\ldots,pop\_size \tag{6.41}$$

where $C^k_{\max}$ is the makespan of the kth chromosome and *pop_size* is the number in the population.

Genetic Operators. The function of the genetic operators is to create new search nodes based on the current population of search nodes. New search nodes are typically constructed by combining or rearranging parts of the old search nodes. For the MSP, the genetic operators used here must enforce the intraprocessor precedence relations and the completeness and uniqueness of the tasks in the schedule. This would ensure that the new chromosome generated always represents legal search nodes. Tsujimura and Gen use roulette wheel selection and the task graph shown in Figure 6.30 as an example [627].

Operator 1. Operator 1 is performed in the following steps:

Step 1. Generate a random number c from range $[1, \max\{height'\}]$.
Step 2. Place the cutpoint at each processor such that the tasks' *height'* before the cutpoint is less than c and more than or equal to c after the cutpoint. Then we obtain two partial schedules in each processor.
Step 3. Exchange the second partial schedules.

Example 6.3. $c = 2$; $height'(T_6)$, $height'(T_7)$, $height'(T_8) \geq 2$.

$$v = \begin{cases} P_1T_1 \ T_5 \ T_4 | \mathbf{T_7}) \\ P_2(T_2 \ T_3 | \mathbf{T_6} \ \mathbf{T_8}) \end{cases}$$

$$v' = \begin{cases} P_1(T_1 \ T_5 \ T_4 | \mathbf{T_6} \ \mathbf{T_8}) \\ P_2(T_2 \ T_3 | \mathbf{T_7}) \end{cases}$$

Operator 2. Operator 2 is performed in the following steps:

Step 1. Generate a random number c from range $[1, \max\{height'\}]$.
Step 2. At each processor, pick all tasks whose *height'* is c.
Step 3. Replace the position of all tasks randomly.

Example 6.4. $c = 1$; $height'(T_3)$, $height'(T_4)$, $height'(T_5) = 1$.

$$v = \begin{cases} P_1(T_1 \ \mathbf{T_5} \ \mathbf{T_4} \ T_6 \ T_8) \\ P_2(T_2 \ \mathbf{T_3} \ T_7) \end{cases}$$

$$v' = \begin{cases} P_1(T_1 \ \mathbf{T_3} \ T_6 \ T_8 \\ P_2(T_2 \ \mathbf{T_4} \ \mathbf{T_5} \ T_7) \end{cases}$$

Operator 3. When we generate the initial population, we fix the *height'* for each task, and its value is not altered for all generations since there are a broad range. To eliminate this inconvenience, Tsujimura and Gen proposed a

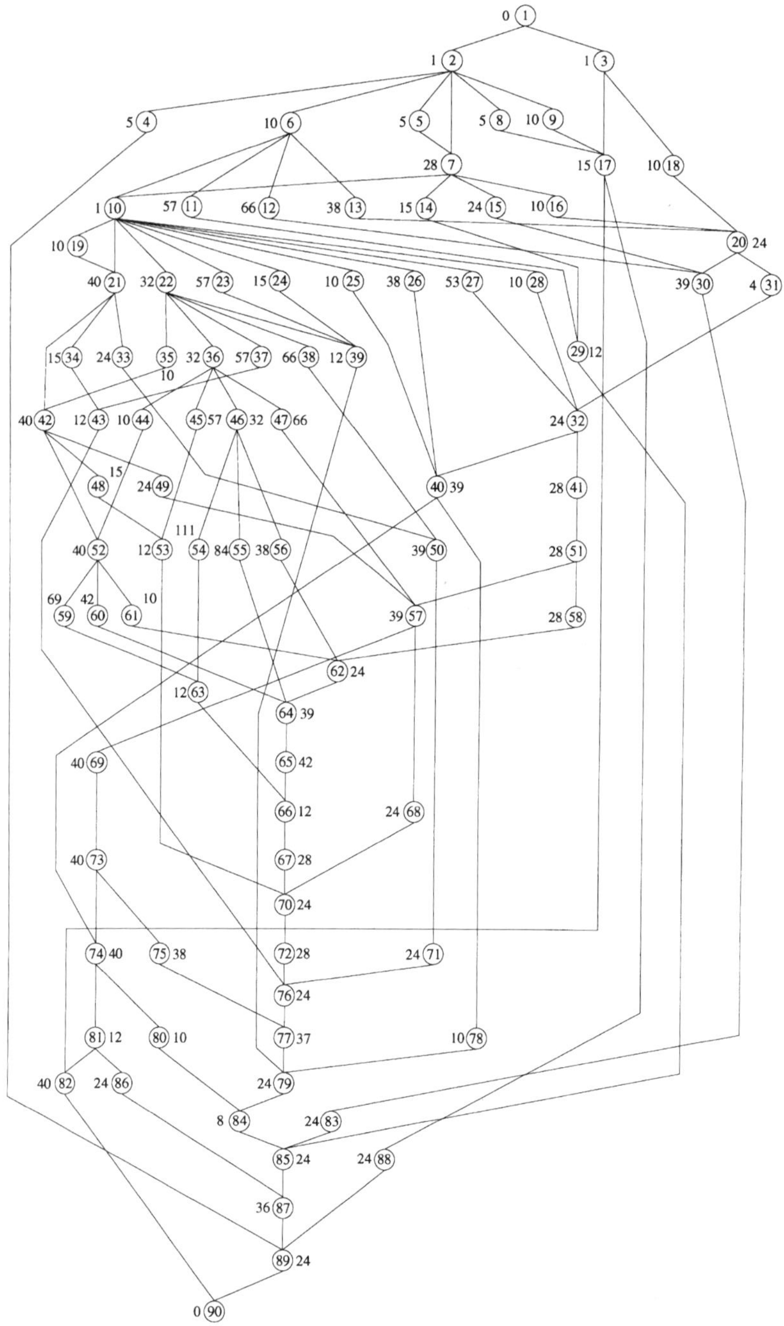

Figure 6.32. Feasible schedule for the 9-job/3-machine example.

TABLE 6.6. Setting Genetic Algorithm Parameters

Number of Processors	p_1	p_2	p_3	p_4
2	0.6	0.3	0.5	0.1
4	0.7	0.3	0.3	0.1
6	0.7	0.3	0.4	0.05
10	0.6	0.3	0.5	0.1

third operator, whose aim is to create a larger search space, to find a better chromosome [627]. Operator 3 is performed in the following steps:

Step 1. Pick tasks who *height'* can be altered.

Step 2. Select among these tasks randomly with a low probability.

Step 3. Alter the value of *height'* of the tasks selected to the value the tasks can take, and move them to appropriate positions according to the altered *height'*.

Example 6.5. It can select task T_4 as a candidate to be altered (i.e., the value of *height'* of task T_4 can be altered from 1 to 2). After alteration, exchange the position of task T_4 from before task T_5 to after it.

$$height'(T_4) = 1 \rightarrow 2$$

$$height'(T_5) = 1$$

$$v = \begin{cases} P_1(T_1 \ T_3 \ T_6 \ T_8) \\ P_2(T_2 \ \mathbf{T_4} \ \mathbf{T_5} \ T_7) \end{cases}$$

$$v' = \begin{cases} P_1(T_1 \ T_3 \ T_6 \ T_8) \\ P_2(T_2 \ \mathbf{T_5} \ \mathbf{T_4} \ T_7) \end{cases}$$

TABLE 6.7. Comparative Results

Number of Processors	T-G's GA	Hou et al.'s GA [295]	Optimal Solution [295]
2	1247	1246	1242
4	759	774	659
6	619	627	573
10	570	590	570

6.6.3 Numerical Example

In a numerical experiment, Tsujimura and Gen (T-G) use the data of the 88-task Stanford manipulator graph [347] shown in Figure 6.32 as a large-scale MSP. In Figure 6.32 it does not need to take nodes 1 and 90 into consideration because they are additional dummy nodes. It sets $pop\_size = 30$ and $gen_{\max} = 2000$. The probabilities for genetic operations, p_1, p_2, p_3, and p_4 were set at appropriate values determined experimentally and summarized in Table 6.6.

The comparative results for various numbers of processors are summarized in Table 6.7. Each result is obtained as the best among 20 runs. Table 6.7 presents the results ranging from 0.0 to 13.2% greater than the optimal solutions, with the genetic algorithm proposed by Tsujimura and Gen [627] dominating Hou's genetic algorithm [308]. From the comparative result it is confirmed that the genetic algorithm proposed can provide good solutions for large-scale MSPs.

Recently, Hadj-Alouane, Bean, and Marty [720] developed a hybrid genetic/optimization algorithm for a task allocation problem to different processors based on genetic algorithm for the multiple-choice integer program [719].

7
ADVANCED TRANSPORTATION PROBLEMS

7.1 INTRODUCTION

The transportation problem is a basic network problem. Many scholars have refined and extended the basic transportation model to include not only the determination of optimum transportation patterns but also an analysis of production scheduling problems, transshipment problems, and assignment problems. Recently, there have been many investigations of evolutionary approaches to solving a variety of transportation problems. Michalewicz et al. [455, 457] first discussed the use of genetic algorithms for solving linear and nonlinear transportation problems. The matrix representation was used to construct a chromosome, and they designed matrix-based crossover and mutation in their investigation. Gen and Li [234, 412] discussed genetic algorithms using a spanning tree representation for solving transportation problems. They used the Prüfer number, which is tree encoding, to design the criterion of feasibility of the chromosome. The merit of using spanning tree encoding is that the memory required for the chromosome to be designed is greatly reduced. In a problem with m plants and n warehouses, for a chromosome, the matrix-based representation requires $m \times n$ memories in the evolutionary process. Using Prüfer number representation, only $m + n - 2$ memories are required for a chromosome in the transportation problem. Matrix-based representation requires many more memories in the implementation and more computational time in problem solving.

7.1.1 Transportation Model

Production Scheduling and Inventory Control. The problem of scheduling production and controlling inventory over several periods can be formulated as the transportation model. The typical case is the manufacturer with a sea-

TABLE 7.1. Transportation Table for Example

Origin (Source of Production)	Destination (Demand in Period) 1	2	3	4	Dummy	Supply (Production Capacity)
1. Regular period 1	4.0	5.5	7.0	8.5	—	50
2. Overtime period 1	6.0	7.5	9.0	10.5	—	20
3. Regular period 2	—	4.0	5.5	7.0	—	50
4. Overtime period 2	—	6.0	7.5	9.0	—	20
5. Regular period 3	—	—	4.0	5.5	—	50
6. Overtime period 3	—	—	6.0	7.5	—	20
7. Regular period 4	—	—	—	4.0	—	50
8. Overtime period 4	—	—	—	6.0	—	20
Demand	40	60	75	60	45	280

sonal sales pattern, but production can remain completely stable, or production can vary to meet the sales pattern (no inventories), or production can follow some midground approach between these two extremes.

The most straightforward involves a trade-off between overtime costs and inventory costs. For each period, such as a month, we define regular and overtime capacity in units, such as production hours. We also estimate the unit costs of regular and overtime production and the cost per period of holding a unit in inventory. The transportation formulation identifies regular and overtime production capacities in each period as *origins* and demands for each period as *destinations*. The decision variables x_{ij} are defined in terms of production quantities in a given period manufactured by a given method (regular or overtime) for delivery in the present or a future period. The cost c_{ij} associated with each cell is determined as the sum of production cost per unit and the cost of holding a unit in inventory.

For example, consider a four-period model with projected (or contracted) demands of 40, 60, 75, and 60 units in periods 1, 2, 3, and 4, respectively. The costs are 4 per unit for regular time, 6 per unit for overtime, and 1.5 for holding a unit in inventory for one period. For simplicity we assume that these costs are identical in each period. Furthermore, in each period, regular production capacity is 50 units and overtime production capacity is 20 units. Table 7.1 illustrates the transportation table for this example.

Since the total supply is greater than the total demand, a dummy destination is created to pick up the slack. Each dummy in this case represents unused production capacity. It is tempting to assign zero cost values uniformly to the dummy volume. That would infer an ability to hold regular production costs completely variable. Alternatively, if product costing is based on direct labor plus raw materials plus overhead, dummy costs should be 4. This problem can be formulated as a transportation model as in Section 7.1.2.

Production Plan Problem. The production plan problem is to determine production on a machine so that the total machining cost is minimized. It can be stated as follows: Given n targets (e.g., production goals) and a set of m

means (fixed amounts of resources, instruments, etc.) to fulfill those targets, and assuming that each target can be fulfilled by several means, the goal is to assign fulfillment of the various targets to the various means so as to minimize the cost function. Such problems arise in a variety of fields on a factory scale as well as on larger scales. This production plan problem can be formulated as the generalized transportation model given in the next section.

7.1.2 Formulation of Transportation Problems

Production Scheduling and Inventory Control. The problem of production scheduling and inventory control over several periods described in Section 7.1.1 can be formulated as follows:

$$\begin{aligned}
\min\quad & 4.0x_{11} + 6.0x_{21} + 5.5x_{12} + 7.5x_{22} + 4.0x_{32} + 6.0x_{42} \\
& + 7.0x_{13} + 9.0x_{23} + 5.5x_{33} + 7.5x_{43} \\
& + 4.0x_{53} + 6.0x_{63} \\
& + 8.5x_{14} + 10.5x_{24} + 7.0x_{34} + 9.0x_{44} \\
& + 5.5x_{54} + 7.5x_{64} + 4.0x_{74} + 6.0x_{84} \\
\text{s.t.}\quad & x_{11} + x_{12} + x_{13} + x_{14} + x_{15} = 50 \\
& x_{21} + x_{22} + x_{23} + x_{24} + x_{25} = 20 \\
& x_{32} + x_{33} + x_{34} + x_{35} = 50 \\
& x_{42} + x_{43} + x_{44} + x_{45} = 20 \\
& x_{53} + x_{54} + x_{55} = 50 \\
& x_{63} + x_{64} + x_{65} = 20 \\
& x_{74} + x_{75} = 50 \\
& x_{84} + x_{85} = 20 \\
& x_{11} + x_{21} = 40 \\
& x_{12} + x_{22} + x_{32} + x_{42} = 60 \\
& x_{13} + x_{23} + x_{33} + x_{43} + x_{53} + x_{63} = 75 \\
& x_{14} + x_{24} + x_{34} + x_{44} + x_{54} + x_{64} + x_{74} + x_{84} = 60 \\
& x_{15} + x_{25} + x_{35} + x_{45} + x_{55} + x_{65} + x_{75} + x_{85} = 45 \\
& x_{ij} \geq 0, \qquad i = 1,2,\ldots,8, \quad j = 1,2,\ldots,5
\end{aligned}$$

where x_{i5}, $i = 1,2,\ldots,8$, is variable for dummy destination 5.

Generally, this problem with n given periods can be formulated as the following transportation model:

$$\min \sum_{j=1}^{n} \sum_{i=1}^{2j} c_{ij} x_{ij} \tag{7.1}$$

$$\text{s.t.} \quad \sum_{j=k}^{n+1} x_{ij} = a_i, \qquad i = 2k - 1, \quad k = 1,2,\ldots,n \tag{7.2}$$

$$\sum_{j=k}^{n+1} x_{ij} = a_i, \qquad i = 2k, \quad k = 1,2,\ldots,n \tag{7.3}$$

$$\sum_{i=1}^{2j} x_{ij} = b_j, \qquad j = 1,2,\ldots,n \tag{7.4}$$

$$x_{ij} \geq 0, \qquad i = 1,2,\ldots,2n, \quad j = 1,2,\ldots,n+1 \tag{7.5}$$

with dummy destination $n + 1$, where a_i is production capacity in the ith regular or overtime period, and b_j is the sum of production manufactured in regular and overtime periods in the jth period. This problem can be viewed as a balanced transportation problem with the following balanced condition:

$$\sum_{i=1}^{2n} a_i = \sum_{j=1}^{n+1} b_j$$

Linear Transportation Problem. Transportation problems contain two main sets of constraints: one set of m constraints associated with the source nodes, denoted as S, and one set of n constraints associated with the destination nodes, denoted as D. Also, there will be mn variables in the problem, each corresponding to an arc from a source node to a destination node. Therefore, the underlying graph is a direct graph and it is characterized as a *bipartite graph;* that is, the nodes of this network model can be partitioned into two sets, S and D. This is illustrated graphically in Figure 7.1. The objective of the transportation problem is to find the minimal cost pattern of shipment. The transportation problem with m plants and n warehouses can be formulated as follows:

$$\min \quad z(x) = \sum_{i=1}^{m} \sum_{j=1}^{n} c_{ij} x_{ij} \tag{7.6}$$

$$\text{s.t.} \quad \sum_{j=1}^{n} x_{ij} \leq a_i, \qquad i = 1,2,\ldots,m \tag{7.7}$$

$$\sum_{i=1}^{m} x_{ij} \geq b_j, \qquad j = 1,2,\ldots,n \tag{7.8}$$

$$x_{ij} \geq 0, \qquad \forall i,j \tag{7.9}$$

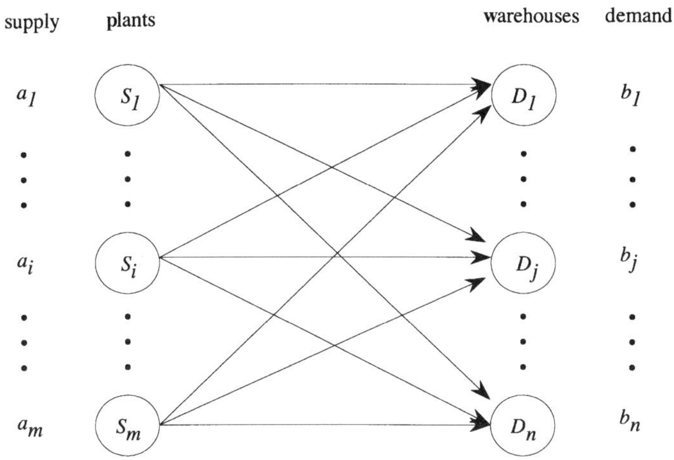

Figure 7.1. Network of the transportation problem.

where x_{ij} is the unknown quantity shipped from source i to destination j, c_{ij} the cost of shipping 1 unit from source i to destination j, a_i the number of units available at source i, and b_j the number of units demanded at destination j.

Transportation problems have some common characteristics in solutions:

1. There are $m + n - 1$ basic variables, and each variable corresponds to a cell in the transportation tableau.
2. The basis must be a transportation tree; that is, there must be at least one basic cell in each row and in each column of the transportation tableau.
3. The basis should not contain a cycle.

The characteristics of transportation problems are illustrated with one of the basis solutions in Figure 7.2.

Due to the complex nature of transportation planning problems, the following nonlinear objective functions can be considered.

- Nonlinear function 1:

$$f_{ij}(\boldsymbol{x}) = c_{ij}x_{ij}^2$$

- Nonlinear function 2:

$$f_{ij}(\boldsymbol{x}) = c_{ij}\sqrt{x_{ij}}$$

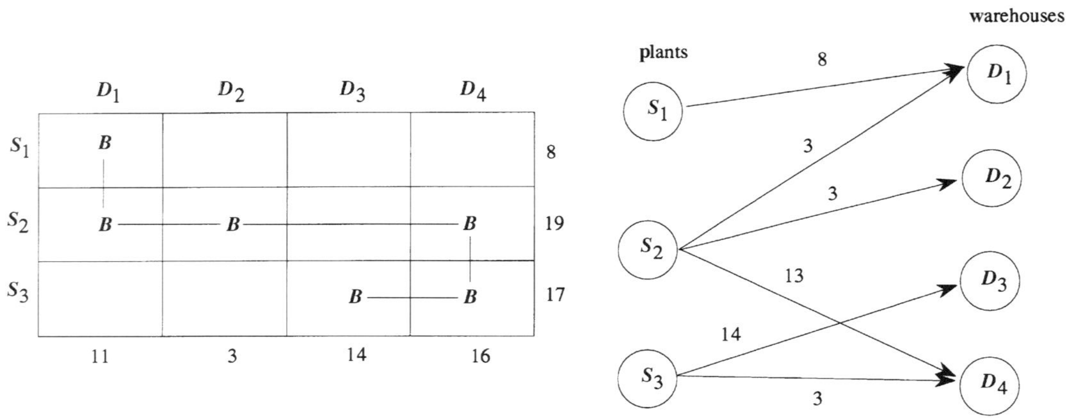

Figure 7.2. Basis solutions to the transportation tableau and transportation graph.

- Nonlinear function 3:

$$f_{ij}(\boldsymbol{x}) = \begin{cases} c_{ij}\left(\dfrac{x_{ij}}{S}\right), & \text{if } 0 \leq x_{ij} \leq S \\ c_{ij}, & \text{if } S < x_{ij} \leq 2S \\ c_{ij}\left(1 + \dfrac{x_{ij} - 2S}{S}\right), & \text{if } 2S < x_{ij} \end{cases}$$

where S is the order of a typical x value.

- Nonlinear function 4:

$$f_{ij}(\boldsymbol{x}) = c_{ij}x_{ij}\left[\sin\left(x_{ij}\frac{5\pi}{4S}\right) + 1\right]$$

where S is the order of a typical x value.

For the transportation problem with multiple commodities, we can also easily formulate some models and solve subproblems of the original problem.

Generalized Transportation Problem. The generalized transportation problem is an extended transportation problem. This problem usually is formulated in the following form:

$$\min \quad z(\boldsymbol{x}) = \sum_{i=1}^{m}\sum_{j=1}^{n} c_{ij}x_{ij} \tag{7.10}$$

$$\text{s.t.} \quad \sum_{j=1}^{n} x_{ij} = a_i, \qquad i = 1,2,\ldots,m \tag{7.11}$$

$$\sum_{i=1}^{m} p_{ij}x_{ij} = b_j, \qquad j = 1,2,\ldots,n \tag{7.12}$$

$$x_{ij} \geq 0, \qquad \forall i,j \tag{7.13}$$

where p_{ij} is a unit of the quantity to be transported.

Capacitated Transportation Problem. The capacitated transportation problem is an extended transportation problem. It has the transportation capacity on each route, which differs with transportation problem. The mathematical model is formulated as follows:

$$\min \quad z(\boldsymbol{x}) = \sum_{i=1}^{m} \sum_{j=1}^{n} c_{ij} x_{ij} \tag{7.14}$$

$$\text{s.t.} \quad \sum_{j=1}^{n} x_{ij} = a_i, \qquad i = 1,2,\ldots,m \tag{7.15}$$

$$\sum_{i=1}^{m} x_{ij} = b_j, \qquad j = 1,2,\ldots,n \tag{7.16}$$

$$0 \le x_{ij} \le u_{ij}, \qquad \forall i,j \tag{7.17}$$

where u_{ij} is the upper bound of transportation capacity.

7.2 SPANNING TREE–BASED APPROACH

As a special type of the network problems, transportation problems have a special data structure characterized as a transportation graph in their solution. The spanning tree-based genetic algorithm incorporating this transportation problem data structure was proposed by Gen and Li [234, 412]. This genetic algorithm utilized the Prüfer number [699] encoding based on a spanning tree, which is adopted because it is capable of representing all possible trees. Using the Prüfer number representation, the memory requires only $m + n - 2$ for a chromosome in the transportation problem. Because a transportation tree is a special type of spanning tree, Prüfer number encoding may correspond to an infeasible solution. From this point, Gen and Li designed a criterion for checking the feasibility of chromosomes. The spanning tree-based genetic algorithm can find the optimal or near-optimal solution for transportation problems in solution space.

7.2.1 Tree Representation

Denote the plants $1,2,\ldots,m$ as components of the set $S = \{1,2,\ldots,m\}$, and define warehouses $1,2,\ldots,n$ as components of the set $D = \{m + 1,\ldots,m + n\}$. Obviously, the transportation problem has $m + n$ nodes and $m \times n$ edges in its transportation tree.

The transportation graph in Figure 7.2 can be represented as a spanning tree, as in Figure 7.3. For a complete graph with p nodes, there are $p^{(p-2)}$ distinctly labeled trees. This means that we can use permutations of only $p - 2$ digits to uniquely represent a tree with p nodes, where each digit is an integer between 1 and p inclusive. For any tree there are always at least two leaf nodes.

The Prüfer number, which is one of the nodes encoding for the tree, can be used to encode a transportation tree. The constructive procedure for a Prüfer number according to a tree is as follows:

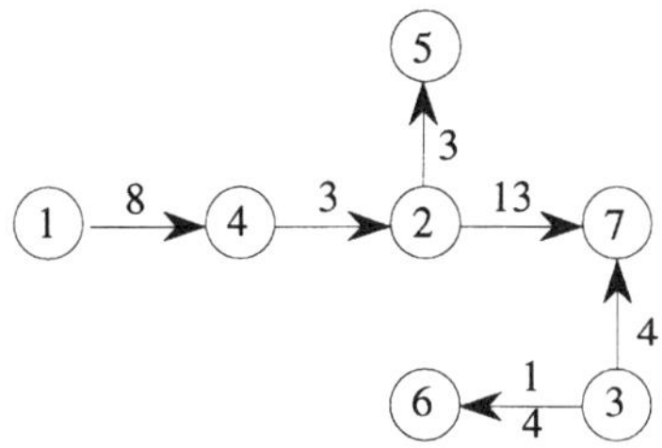

Prüfer number: $P(T) = [4\ 2\ 2\ 7\ 3]$

Figure 7.3. Spanning tree and its Prüfer number.

Procedure: Convert a Tree to a Prüfer Number

Step 1. Let i be the lowest-numbered leaf node in tree T. Let j be the node that is the predecessor of i. Then j becomes the rightmost digit of the Prüfer number $P(T)$. $P(T)$ is built up by appending digits to the right; thus $P(T)$ is built and read from left to right.

Step 2. Remove i and edge (i,j) from further consideration. Thus, i is no longer considered at all and if i is the only successor of j, then j becomes a leaf node.

Step 3. If only two nodes remain to be considered, $P(T)$ has been formed, so stop; otherwise, return to step 1.

The transportation tree illustrated in Figure 7.3, converted to the corresponding Prüfer number, is shown in Figure 7.4.

Each node in the transportation problem has its supply or demand quantity, which are characterized as constraints. Therefore, to construct a transportation tree, the constraint of nodes must be considered. From a Prüfer number, a unique transportation tree can also be generated by the following procedure.

Procedure: Convert a Prüfer Number to a Transportation Tree

Step 1. Let $P(T)$ be the original Prüfer number, and let $\overline{P}(T)$ be the set of all nodes that are not part of $P(T)$ and are designed as eligible for consideration.

Step 2. Repeat steps (2.1) to (2.5) until no digits are left in $P(T)$.

(2.1) Let i be the lowest-numbered eligible nodes in $\overline{P}(T)$ and j be the leftmost digit of $P(T)$.

(2.2) If i and j not in the same set S or D, add edge (i,j) to tree T. Otherwise, select the next digit k from $P(T)$ that is not included in the same set with i, exchange j with k, and add edge (i,k) to the tree T.

(2.3) Remove j (or k) from $P(T)$ and i from $\overline{P}(T)$. If j (or k) does not occur anywhere in the remaining part of $P(T)$, put it into $\overline{P}(T)$. Designate i as no longer eligible.

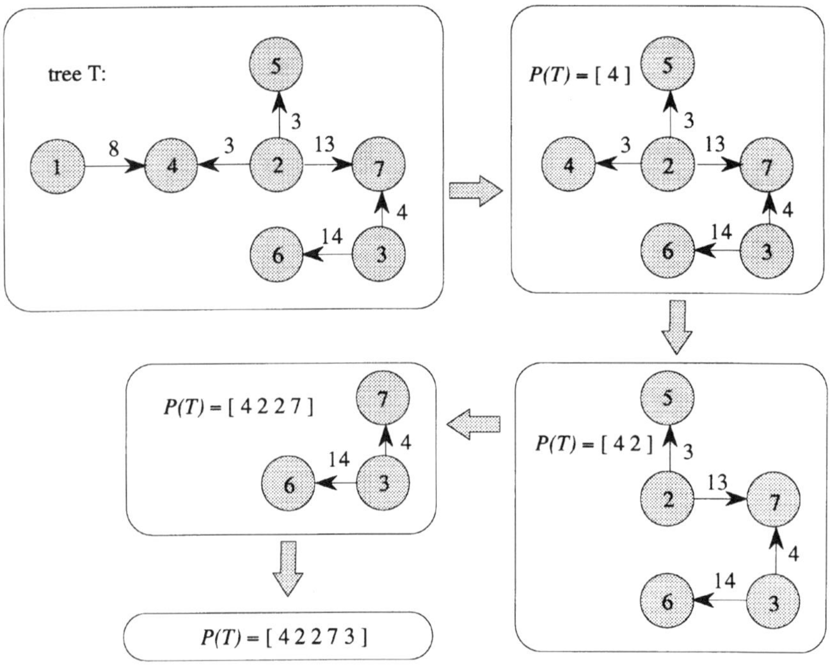

Figure 7.4. Encoding procedure.

(2.4) Assign the available amount of units to $x_{ij} = \min\{a_i, b_j\}$ (or $x_{ik} = \min\{a_i, b_k\}$) to edge (i,j) [or (i,k)], where $i \in S$ and $j, k \in D$.

(2.5) Update availability $a_i = a_i - x_{ij}$ and $b_j = b_j - x_{ij}$ (or $b_k = b_k - x_{ik}$).

Step 3. If no digits remain in $P(T)$, there are exactly two nodes, i and j, still eligible in $\overline{P}(T)$ for consideration. Add edge (i,j) to tree T and form a tree with $m + n - 1$ edges.

Step 4. If there no available units to assign, stop; otherwise, supply r and demand s remain. Add edge (r,s) to the tree and assign the available amount $x_{rs} = a_r = b_s$ to the edge. If there exists a cycle, remove the edge that is assigned zero flow. A new spanning tree is formed with $m + n - 1$ edges.

The decoding procedure from a Prüfer number to a transportation tree is illustrated in Figure 7.5.

7.2.2 Initialization

The initialization of a chromosome (a Prüfer number) is performed from randomly generated $m + n - 2$ digits in the range $[1, m + n]$. However, it is possible to generate an infeasible chromosome, which is not adapted to gen-

Prufer number: $P(T) = [4\ 2\ 2\ 7\ 3]$

Set of all nodes that are not part of $P(T)$: $\bar{P}(T) = \{1, 5, 6\}$

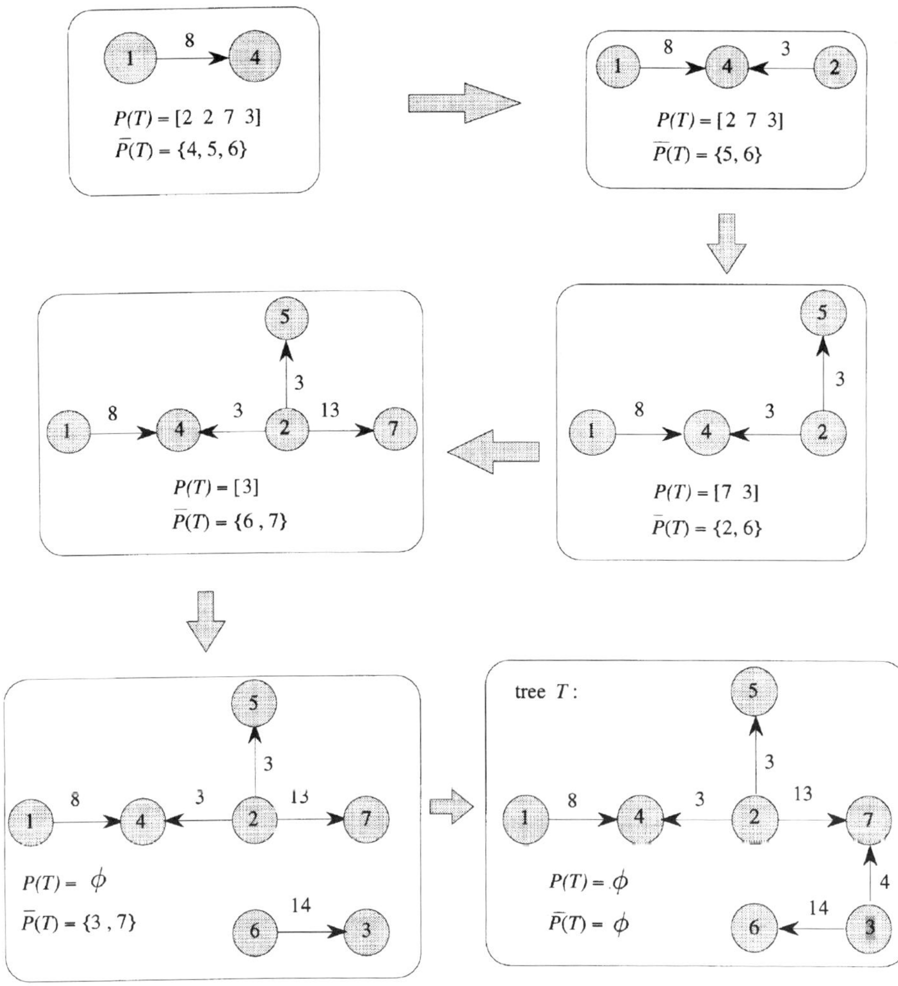

Figure 7.5. Decoding procedure.

erating a transportation tree. Prüfer number encoding not only is capable of representing all possible spanning trees equally and uniquely, but also explicitly contains the information as to node degree; that is, any node with degree d will appear exactly $d - 1$ times in the encoding. This means that when a node appears d times in a Prüfer number, the node has $d + 1$ connections with other nodes.

Gen and Li designed a criterion for checking the feasibility of chromosomes. Denote $\overline{P}(T)$ as the set of nodes that are not included in the Prüfer number $\overline{P}(T)$. Let L_S be total number of edges incident to plant nodes $|S \cap P(T)|$ in a spanning tree defined by the Prüfer number $P(T)$. L_D the total number of edges incident to warehouse nodes $|D \cap P(T)|$ in the same spanning tree. $\overline{L}_S$ the number of plant nodes included in $\overline{P}(T)$, that is, $\overline{L}_S = |S \cap \overline{P}(T)|$. $\overline{L}_D$ the number of warehouse nodes included in $\overline{P}(T)$, that is, $\overline{D}_S = |D \cap \overline{P}(T)|$. We have the following criterion:

Feasibility of a Chromosome. If $L_S + \overline{L}_S = L_D + \overline{L}_D$, then $P(T)$ is feasible; otherwise, $P(T)$ is infeasible.

For instance, the Prüfer number $P(T)$ is [4 2 2 7 3] and $\overline{P}(T)$ is {1,5,6} in Figure 7.5. Now, L_S is the sum of connections of nodes 2 and 3 included in set S from $P(T)$. There are 2 + 1 for node 2 and 1 + 1 for node 3, so $L_S = 5$. Similarly, L_D is the sum of connections for nodes 4 and 7 included in set D, and $L_D = 4$. $\overline{L}_S$ is the times of appearance of node 1 in $\overline{P}(T)$, so $\overline{L}_S = 1$. $\overline{L}_D$ is the times of appearance of nodes 5 and 6, so $\overline{L}_D = 2$. So $L_S + \overline{L}_S = 6$ and $L_D + \overline{L}_D = 6$; therefore, this Prüfer number satisfies the criterion of feasibility.

7.2.3 Genetic Operators

Crossover. The one-point crossover operation is used, as shown in Figure 7.6. To avoid unnecessary decoding from which an infeasible chromosome (a Prüfer number) may be generated after using the crossover operator, a checking procedure is embeded within the crossover operation to guarantee producing feasible offspring, that is, via this checking procedure, offspring always corresponds to a transportation tree.

Mutation. Inversion and displacement mutation are used. With inversion mutation, two positions within a chromosome are selected at random, and then the substring between these two positions is inverted as illustrated in Figure 7.7. With displacement mutation, a substring is selected at random and is inserted at a random position, as illustrated in Figure 7.8. These two mutation operators always generate feasible chromosomes if the parents are feasible, because the criterion $L_S + \overline{L}_S = L_D + \overline{L}_D$ is unchanged after these operations.

7.2.4 Evaluation and Selection

The evaluation procedure consists of two steps:

1. Convert a chromosome into a tree.

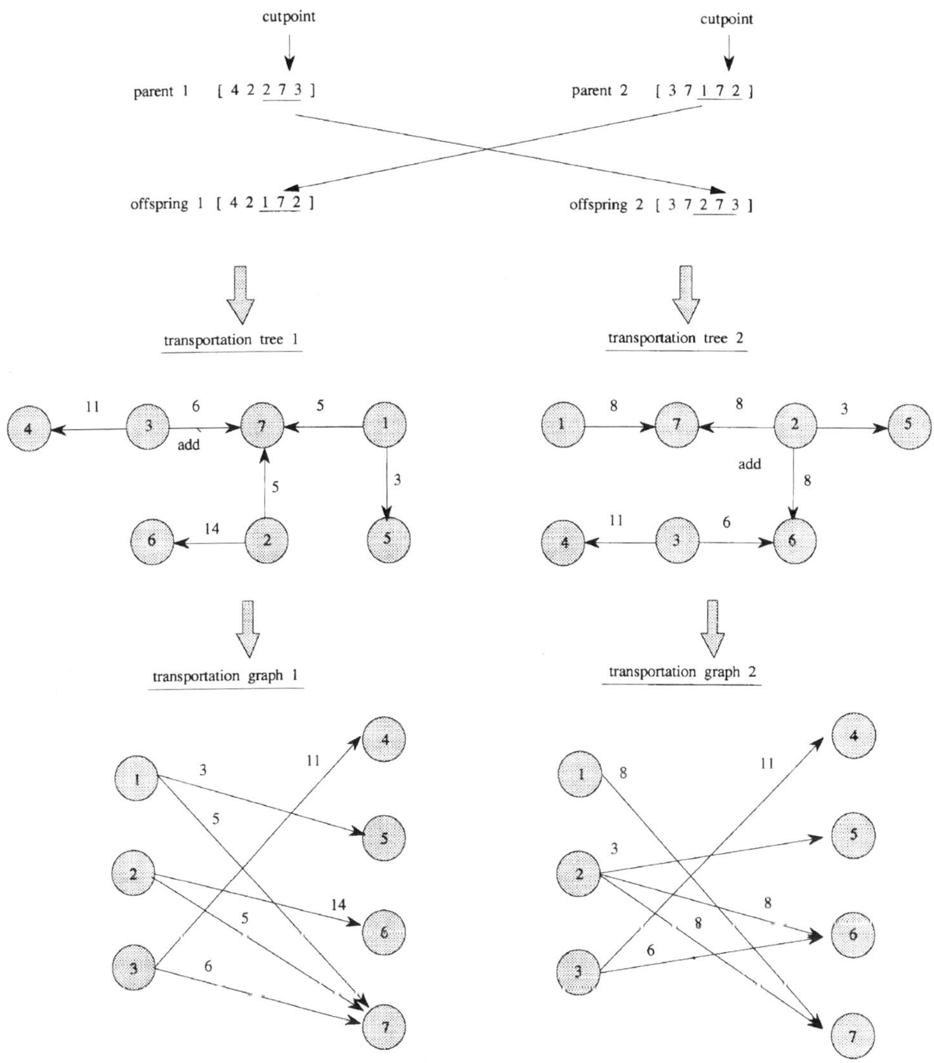

Figure 7.6. One-point crossover and offspring.

2. Calculate each objective function.

Procedure: Evaluation

Step 1. Let T be a tree and set it as an empty tree, $T = \{\emptyset\}$. Let p be the digit counter and set it as zero, $p = 0$. Set each objective function value as zero.

Step 2. Define $\overline{P}(T)$ according to the Prüfer number $P(T)$.

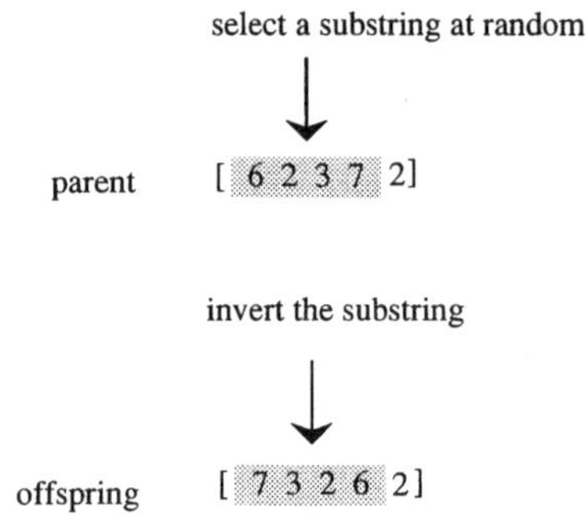

Figure 7.7. Inversion mutation.

Step 3. Repeat steps (3.1) to (3.6) until $p \leq m + n - 2$.

(3.1) Select the leftmost digit i from $P(T)$. Select the eligible node with the lowest number from $\overline{P}(T)$, and denote it as j.

(3.2) If $i,j \in S$ or $i,j \in D$, select the next digit from $P(T)$ not in the set with j. Denote it as k, exchange k with i, and take $i = k$. Add edge (i, j) to tree T.

(3.3) Remove i from $P(T)$ and j from $\overline{P}(T)$. If j does not occur anywhere in the remaining part of $P(T)$, put it into $\overline{P}(T)$. Designate j as no longer eligible.

(3.4) Assign the flow to edge (i,j) as $x_{ij} = \min\{a_i,b_j\}$, where $i \in S$, $j \in D$. Update the value of the objective function for edge (i,j).

(3.5) Update the available amounts as $a_i = a_i - x_{ij}$ and $b_j = b_j - x_{ij}$.

(3.6) Set $p = p + 1$.

Step 4. If no digits remain in $P(T)$, there are exactly two nodes, r and s, still eligible in $\overline{P}(T)$ for consideration. Add edge (r,s) to tree T and form a tree with $m + n - 1$ edges. Assign the flow to edge (r,s) as $x_{rs} = \min\{a_r,b_s\}$, where $r \in S$, $s \in D$. Update the value of the objective function for edge (r,s).

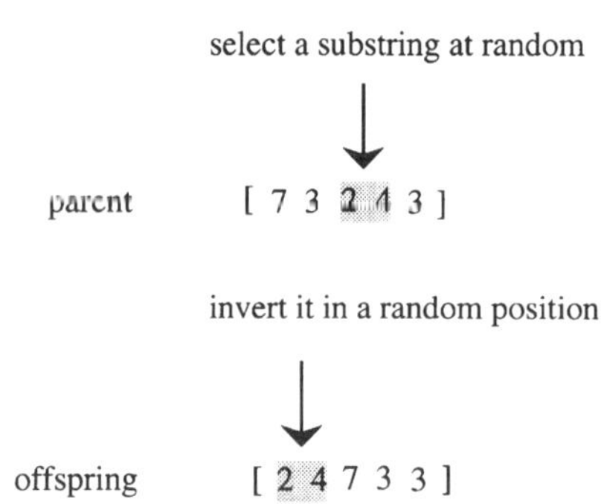

Figure 7.8. Displacement mutation.

Step 5. Repeat steps (5.1) and (5.2) until there is no available amount in any node.

(5.1) If $a_i > 0$, $\forall i$ and $b_j > 0$, $\forall j$, add edge (i,j) to tree T as $T = T \cup \{x_{ij}\}$, and assign the flow $x_{ij} = \min\{a_i, b_j\}$, where $i \in S$, $j \in D$.

(5.2) Update the available amounts as $a_i = a_i - x_{ij}$ and $b_j = b_j - x_{ij}$. Update the value of the objective function for edge (i,j).

Step 6. If there exists a cycle, find the edge with zero flow and remove it.

The selection procedure combines $(\mu + \lambda)$-selection with roulette wheel selection. The mixed selection strategy selets the μ best chromosomes from μ parents and λ offspring. If there are not μ different chromosomes available, the vacant pool of the population is filled up with roulette wheel selection [455]. The deterministic $(\mu + \lambda)$-selection can enforce the best chromosomes into the next generation and avoid the premature convergence of the evolutionary process [414].

7.2.5 Overall Procedure

Let $P(t)$ be a population of chromosomes and $C(t)$ be the chromosomes generated in the current generation t. The overall procedure is as follows:

```
Procedure: st-GA/TP
begin
  t ← 0;
  initialize P(t);
  evaluate P(t) based on spanning tree;
  while (not termination condition) do
  begin
    recombine P(t) to generate C(t);
    evaluate C(t) based on spanning tree;
    select P(t + 1) from P(t) and C(t);
    t ← t + 1;
  end
end
```

7.3 MULTIOBJECTIVE TRANSPORTATION PROBLEM

Often, representation of more than one goal or objective in models of economic decisions is desired. In a general sense we can think of the maximization of the utility framework as encompassing all objectives. However, the operational use of models often requires the specification of instrumental or intermediate goals. Or perhaps, the relative importance of the goals, or the specification of a minimum (or maximum) level of attainment required (or allowed) that is to be minimized with respect to one or more of the objectives, is designed.

In the general transportation problem, the objective is to minimize total transportation costs. The basic assumption underlying this method is that management is concerned primarily with cost minimization. However, this assumption is not always valid. In the transportation problem, there may be multiple objectives, such as the fulfillment of transportation schedule contracts, fulfillment of union contracts, provision for a stable employment level at various plants and transportation fleets, balancing of work among a number of plants, minimization of transportation hazards, and of course, minimization of cost by Gen, Li, and Ida [226, 233, 716].

Generally, a multiple objective linear program is written:

$$\max \quad z_1(\boldsymbol{x}) = \boldsymbol{c}^1\boldsymbol{x} \tag{7.18}$$

$$\max \quad z_2(\boldsymbol{x}) = \boldsymbol{c}^2\boldsymbol{x} \tag{7.19}$$

$$\vdots \tag{7.20}$$

$$\max \quad z_q(\boldsymbol{x}) = \boldsymbol{c}^q\boldsymbol{x} \tag{7.21}$$

$$\text{s.t.} \quad \boldsymbol{x} \in S \tag{7.22}$$

where q is the number of objectives, c^i the vector of the ith objective function coefficients (gradient), S the feasible region, and $z_i(\boldsymbol{x})$ the ith objective function value (criterion value). max indicates that the purpose is to maximize all objectives simultaneously.

Since multiple objective problems rarely have points that maximize all of the objectives simultaneously, we are typically in a situation of trying to maximize each objective to the "greatest extent possible." Many researchers have an interest in the multiple-objective transportation problem, and a number of methods have been proposed for solving it.

7.3.1 Problem Description

In the transportation problem, multiple objectives are required in practical situations such as minimizing transportation cost, minimizing average shipping time to priority customers, maximizing production, minimizing fuel consumption, and so on. The traditional multiobjective transportation problem (mTP) with m plants and n warehouses can be formulated as follows:

$$\min \quad z_q = \sum_{i=1}^{m} \sum_{j=1}^{n} c_{ij}^q x_{ij} \qquad q = 1,2,\ldots,Q \tag{7.23}$$

$$\text{s.t.} \sum_{j=1}^{n} x_{ij} \leq a_i, \qquad i = 1,2,\ldots,m \tag{7.24}$$

$$\sum_{i=1}^{m} x_{ij} \geq b_j, \qquad j = 1,2,\ldots,n \tag{7.25}$$

$$x_{ij} \geq 0, \qquad \forall i,j \tag{7.26}$$

where q indicates the qth objective function.

7.3.2 Spanning Tree–Based Genetic Algorithm for Multiobjective Transportation Problem

The Pareto optimal solutions are usually characterized as solutions of the multiobjective programming problem. Therefore, in implementation of genetic algorithms, a module for handling Pareto optimal solutions is added [233]. It consists of two steps:

1. Evaluate chromosomes by the objective functions.
2. Select Pareto solutions based on evaluation values.

Let E be the Pareto solution set generated up to current iteration t; then the procedure for Pareto solutions is given as follows:

Procedure: Pareto Solutions

Step 1. Set iteration $k = 1$, and $E(t) = \{\phi\}$.

Step 2. If $k > i\_size$, then stop. Otherwise, go to step 3.

Step 3. Evaluate chromosome T_k and obtain solution vector $z_k = [z_1(T_k)\; z_2(T_k) \cdots z_Q(T_k)]$.

Step 4. Compare it with all Pareto solutions in E.

(4.1) If it is dominated by one Pareto solution, go to step 5.

(4.2) If it dominates some Pareto solutions, add it into E and delete solutions dominated by it.

(4.3) If it is a new Pareto solution and dominates none of exiting ones, simply add it into E.

Step 5. Set $k = k + 1$ and go back to step 2.

The weighted sum method is used to construct the fitness function for survival. The multiple objective functions z_q, $q = 1,2,\ldots,Q$, are combined into one overall objective function.

Handling for the Fitness Function

1. Choose the solution points that contain the minimum $z_q^{\min}$ (or maximum $z_q^{\max}$) corresponding to each objective function value, compare them with the stored solution points from the preceding generation, and again select the best points to save.

$$z_q^{\min(t)} = \min_k \{z_q^{\min(t-1)}, z_q^{(t)}(T_k) \mid k = 1,2,\ldots,i\_size\}, \qquad q = 1,2,\ldots,Q$$

$$z_q^{\max(t)} = \max_k \{z_q^{\max(t-1)}, z_q^{(t)}(T_k) \mid k = 1,2,\ldots,i\_size\}, \qquad q = 1,2,\ldots,Q$$

   where $z_q^{\max(t)}$ [$z_q^{\min(t)}$] is the maximum (minimum) value of the objective function q at generation t, and *i_size* is the offspring generated after genetic operations.

2. Solve the following equations to get weights for the fitness function:

$$\delta_q = z_q^{\max(t)} - z_q^{\min(t)}, \qquad q = 1,2,\ldots,Q$$

$$\beta_q = \frac{\delta_q}{\sum_{q=1}^{Q} \delta_q}, \qquad q = 1,2,\ldots,Q$$

3. Calculate the fitness function value for each chromosome as follows:

$$\text{eval}(T_k) = \sum_{q=1}^{Q} \beta_q z_q(T_k), \qquad k = 1,2,\ldots,i\_size$$

Overall Procedure. Let $P(t)$ be a population of chromosomes, $C(t)$ the chromosomes generated in the current generation t, and $E(t)$ the Pareto solution set generated up to the current generation t. The overall procedure is summarized as follows:

```
Procedure: st-GA/mTP
begin
  t ← 0;
  initialize P(t);
  evaluate P(t) using fitness function;
  determine E(t) by module of Pareto solutions;
  while (not termination condition) do
  begin
    recombine P(t) to generate C(t);
    evaluate C(t) using fitness function;
    update E(t) by module of Pareto solutions;
    select P(t + 1) from P(t) and C(t);
    t ← t + 1;
  end
end
```

Hybridized Genetic Algorithm. To improve the effectiveness of the spanning tree-based genetic algorithm, Gen and Li proposed a hybrid approach by combining a local improvement procedure with reduced costs into the spanning tree-based genetic algorithm [232].

The reduced cost is originally used to check optimality of solutions in the simplex method for the transportation problem. In the single objective case, if the reduced costs $\bar{c}_{ij}$ for all nonbasis cells (i, j), the current solution is known as the optimal solution. Inspired by this idea, a local improvement procedure is designed. In the multiple objective case, there are Q reduced costs $\bar{c}_{ij}^q$, $q = 1,2,...,Q$ for each nonbasis cell (variables x_{ij}). Let S be the set of plant nodes, and D be the set of warehouse nodes. The local improvement procedure is given below:

Step 1. Calculate the reduced costs $\bar{c}_{ij}^q$, $q = 1,2,...,Q$, for each nonbasis cell (i,j), where $i \in S$, $j \in D$.

Step 2. Select Q cells (k,l), $q = 1,2,...,Q$, where $\bar{c}_{kl}^q = \max\{\bar{c}_{ij}^q | \bar{c}_{ij}^q > 0$, for all nonbasis arcs $(i,j)\}$, $q = 1,2,...,Q$.

Step 3. Add those Q arcs, respectively, to the current spanning tree to generate Q incomplete trees U_q, $q = 1,2,...,Q$.

Step 4. Find the minimum flow Δ_q, $q = 1,2,...,Q$, in the cycle created on the incomplete tree U_q, $q = 1,2,...,Q$.

Step 5. Induce a flow of $+\Delta_q$, $q = 1,...,Q$, around the cycle created in the same direction and a flow of $-\Delta_q$, $q = 1,...,Q$, in the different direction on the incomplete tree U_q, $q = 1,...,Q$, and remove an arc where flow is decreased to zero so that new trees T_q, $q = 1,...,Q$, are formed.

Step 6. Calculate objective function values by T_q, $q = 1,2,...,Q$, and compare the improvement values of each objective value.

Step 7. Select one tree from T_q, $q = 1,2,...,Q$, which has exceptionally good improvement of an objective value.

After this operation, we can always get a new solution that must be better than the previous solution of the transportation tree.

The overall procedure of the hybridized genetic algorithm approach is as follows:

```
Procedure: st-HGA/mTP
begin
  t ← 0;
  initialize P(t);
  evaluate P(t) by fitness function
  determine E(t) by module of Pareto solutions;
```

TABLE 7.2. Average Results by the m-GA and st-GA Approaches for the Transportation Problem[a]

Problem Size,	Parameter		m-GA		st-GA	
$m \times n$	*pop_size*	*max_gen*	Percent	ACT (min)	Percent	ACT (min)
3×4	10	200	100	0.017	100	0.017
4×5	20	50	100	0.212	100	0.146
5×6	20	500	100	0.313	100	0.166
6×7	100	1000	100	17.304	100	6.893
8×9	200	2000	100	74.536	100	14.135

[a]Percent, the percentage of running times to obtain the optimal soluition; ACT, the average computing time.

```
 while (not termination condition) do
 begin
  recombine P(t) to generate C(t);
  evaluate C(t) by fitness function
  update E(t) by module of Pareto solutions;
  select P(t + 1) from P(t) and C(t);
  renew P(t + 1) by local improvement;
  t ← t + 1;
 end
end
```

7.3.3 Numerical Examples

The proposed spanning tree-based genetic algorithm (st-GA) and hybridized genetic algorithm (st-HGA) approaches were implemented in *C* language and run on an NEC EWS 4800 under EWS-UX/V release 4.0 UNIX OS to carry out simulations. First, to confirm the effectiveness of the spanning tree-based genetic algorithms on the transportation problem, five randomly generated numerical examples were used in the computational studies; the first example is from [14]. The solutions were obtained considering only a single objective function. In the randomly generated examples, the supplies, demands, and transportation costs were generated randomly and distributed uniformly over [10,100] and [10,50].

The simulations were carried out on each example with their best parameter setting. Table 7.2 shows average computational results from 10 runs by the spanning tree-based genetic algorithm and matrix-based genetic algorithm (m-GA) on each problem. From Table 7.2 we know that the optimal solution was found by the st-GA in less time than by the m-GA. This means that the st-GA is more appropriate than the m-GA for solving transportation problems.

The performance of the st-GA and the st-HGA approaches for solving bicriteria transportation problems were tested on two numerical examples. In

TABLE 7.3. Supplies, Demands, and Costs for the First Bicriteria Transportation Problem

Destination, b_j	c^1_{ij}				c^2_{ij}				
Origin, a_i	4	5	6	7	4	5	6	7	Supply
1	1	2	7	7	4	4	3	4	8
2	1	9	3	4	5	8	9	10	19
3	8	9	4	6	6	2	5	1	17
Demand	11	3	14	16	11	3	14	16	44

the first example (from [14]) there are 3 plants and 4 warehouses, with supplies, demands and costs as indicated in Table 7.3.

The parameters for the st-GA and st-HGA were set as follows: *pop_size* = 30; $p_c = 0.2$; $p_m = 0.4$; *max gen* = 1000; and run = 20 times. Pareto optimal solutions were usually found. The solutions obtained were (143,265), (156,200), (176,175), (186,171), (208,167). Known as Pareto optimal solutions, they form the Pareto frontier. The two extreme points are (143,265) and (208,167), shown in Figure 7.9. The average computing times of the st-GA, m-GA, and st-HGA were 13.50, 14.21 and 18.16 s, respectively. This means that for small-scale problems there is no great difference in computation times for the st-GA, m-GA, and st-HGA cases.

Example 2 has 8 plants and 9 warehouses. Both coefficients of the two objectives are given in Table 7.4. The parameters for the algorithm were set as *pop_size* = 100, *max_gen* = 500, and a run of 20 with the better mech-

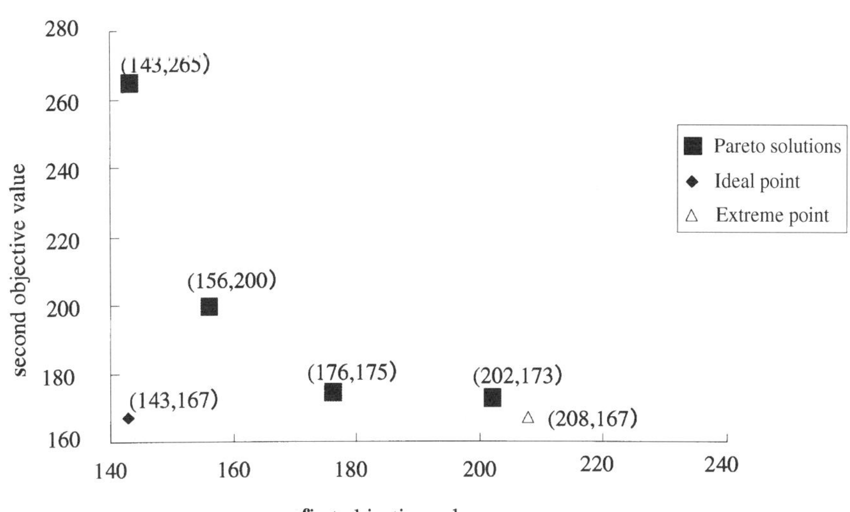

Figure 7.9. Pareto optimal solutions obtained by the st-GA.

TABLE 7.4. Supplies, Demands, and Costs for the Second Bicriteria Transportation Problem

D_j	c^1_{ij}									c^2_{ij}									Supply,
S_i	9	10	11	12	13	14	15	16	17	9	10	11	12	13	14	15	16	17	a_i
1	1	2	7	7	8	10	9	2	5	4	4	3	4	5	8	9	10	2	10
2	1	9	3	4	3	5	7	1	1	6	2	5	1	7	4	12	4	4	8
3	8	9	4	6	4	1	6	9	2	2	9	1	8	9	1	4	0	1	12
4	2	4	5	5	3	2	3	2	9	3	5	5	3	2	8	3	3	2	16
5	5	4	5	1	9	9	1	6	1	1	4	12	2	1	5	4	9	1	21
6	8	3	3	2	2	3	6	7	6	2	23	4	4	6	2	4	6	7	15
7	1	2	6	4	5	9	3	5	2	1	2	1	9	0	13	2	3	2	7
8	13	3	3	5	1	5	6	3	2	14	3	4	2	1	8	5	3	1	9
Demand, b_j	9	7	15	10	13	16	7	10	11	9	7	15	10	13	16	7	10	11	98

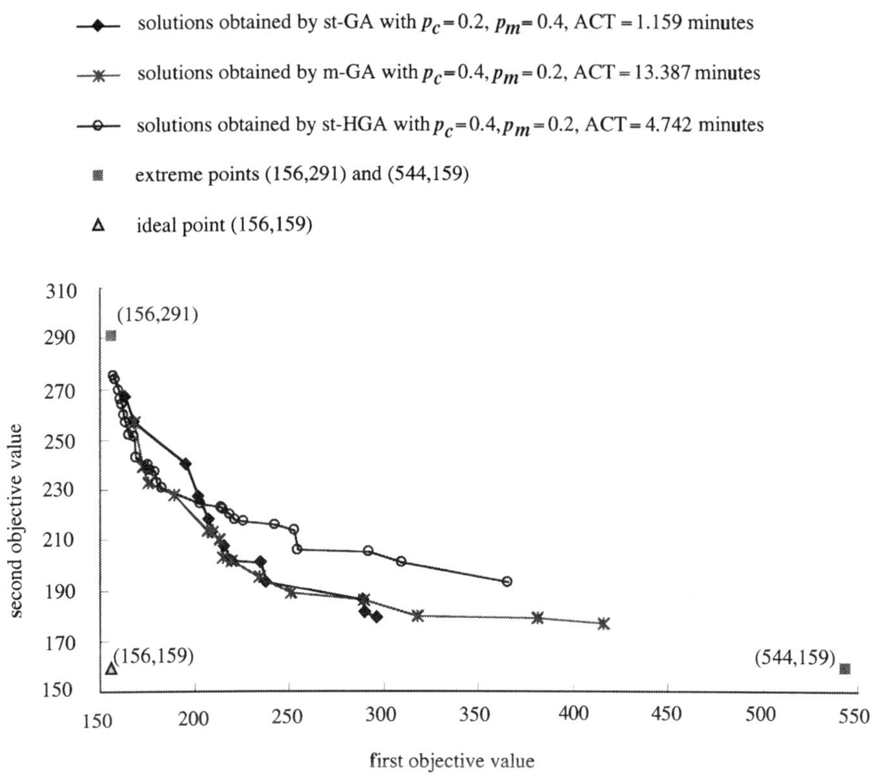

Figure 7.10. Comparison of Pareto solutions obtained with m-GA, st-GA, and st-HGA.

anism (p_c = 0.2 and p_m = 0.4). Figure 7.10 shows compares the m-GA, st-GA, and st-HGA results. The parameters p_c = 0.4 and p_m = 0.2 were used for the m-GA; the parameters p_c = 0.2 and p_m = 0.4 were used for the st-GA and st-HGA. All cases were run 20 times. From Figure 7.10, we can see that the results obtained by the st-GA are better than those by the m-GA in the sense of Pareto optimality because most of them are as good as those obtained by the m-GA. Therefore, we can conclude that they are closer to the real Pareto frontier than those by the m-GA. Furthermore, the solutions obtained by the st-HGA are better than those by the st-GA, in the sense of Pareto optimality. Obviously, the st-HGA is more effective than the st-GA in this multiple-objective problem.

Besides, the average computing times for the m-GA, st-GA, and the st-HGA were 13.387, 1.159, and 3.142 min, respectively. Obviously, the st-GA is more efficient than the m-GA and st-HGA with reduced computing time. It is because spanning tree-based encoding requires only $m + n - 2$ memory spaces for each chromosome, whereas matrix-based encoding requires $m \times n$ memory spaces to represent a chromosome. Since the st-HGA

TABLE 7.5. Comparison of the m-GA, st-GA, and st-HGA Approaches[a]

		Memory for a Solution		ACT (min.)			Parameter	
m	*n*	m-GA	st-GA (st-HGA)	m-GA	st-GA	st-HGA	*pop_size*	*max_gen*
3	4	12	5	0.236	0.225	0.302	30	1000
8	9	72	15	13.387	1.159	3.142	100	500
10	15	100	23	41.957	15.363	27.060	150	500

[a] *m*, Number of plants; *n*, number of warehouses; m-GA, matrix-based encoding genetic algorithm; st-GA, spanning tree-based encoding genetic algorithm; st-HGA, spanning tree-based hybridized genetic algorithm.

is a hybridized form of local improvement and the st-GA, it will take more computing time than the st-GA in the solution procedure. As a result of the increased memory, matrix-based encoding takes more time than the spanning tree-based encoding genetic algorithm for genetic operations. Therefore, for large-scale problems, the spanning tree-based encoding genetic algorithm will surely be more timesaving than the matrix-based encoding and hybridized genetic algorithms (see Table 7.5).

To demonstrate the effectiveness and efficiency of the spanning tree-based genetic algorithm for the multiobjective transportation problem, Gen and Li carried out numerical experiments and compared with it with the matrix-based genetic algorithm (m-GA), the spanning tree-based genetic algorithm (st-GA) and the spanning tree-based hybridized genetic algorithm (st-HGA). For the small-scale problem, there was no obvious difference on results. For the larger-scale problems, the spanning tree-based genetic algorithm can get the Pareto optimal solutions with much less CPU time than can the matrix-based genetic algorithm, and most of the results are much better than those obtained in the matrix-based genetic algorithm. Therefore, in the sense of Pareto optimality, the spanning tree-based GA approaches are more effective than the matrix-based genetic algorithm.

Recently, Jiménez and Verdegay proposed an evolutionary algorithm for solving interval solid transportation problem [723].

7.4 FIXED-CHARGE TRANSPORTATION PROBLEM

The fixed-charge transportation problem (fcTP) is an extension of the transportation problem. Many practical transportation and distribution problems, such as the minimum-cost network flow (transshipment) problem with a fixed charge for logistics, can be formulated as fixed-charge transportation problems. For instance, in a transportation problem, a fixed cost may be incurred for each shipment between a given plant and a given warehouse, and a plant or warehouse facility may result in a fixed amount on investment. The fcTP takes these fixed charges into account, so that the TP is a fcTP with equal fixed costs of zero for all routes. The fcTP is much more difficult to solve, due to the presence of fixed costs, which cause discontinuities in the objective function.

7.4.1 Mathematical Model

In the fixed-charge transportation problem, two types of costs are considered simultaneously when the best course of action is selection: (1) variable costs proportional to the activity level and (2) fixed costs. The fcTP seeks determination of a minimum-cost transportation plan for a homogeneous commodity from a number of plants to a number of warehouses. It requires specification of the level of supply at each plant, the amount of demand at each warehouse, and the transportation cost and fixed cost from each plant to each warehouse. The goal is to allocate the supply available at each plant so as to optimize a criterion while satisfying the demand at each warehouse. The usual objective function is to minimize the total variable cost and fixed costs from the allocation. It is one of the combinatorial problems involving constraints. This fixed-charge transportation problem with m plants and n warehouses can be formulated as follows:

$$\min \quad f(\boldsymbol{x}) = \sum_{i=1}^{m} \sum_{j=1}^{n} [f_{ij}(\boldsymbol{x}) + d_{ij} g_{ij}(\boldsymbol{x})] \tag{7.27}$$

$$\text{s.t.} \quad \sum_{j=1}^{n} x_{ij} \leq a_i, \qquad i = 1,2,\ldots,m \tag{7.28}$$

$$\sum_{i=1}^{m} x_{ij} \geq b_j, \qquad j = 1,2,\ldots,n \tag{7.29}$$

$$x_{ij} \geq 0, \qquad \forall i,j \tag{7.30}$$

$$\text{with} \quad g_{ij}(\boldsymbol{x}) = \begin{cases} 1, & \text{if } x_{ij} > 0 \\ 0, & \text{otherwise} \end{cases} \tag{7.31}$$

where $\boldsymbol{x} = [x_{ij}]$ is the unknown quantity to be transported on route (i,j), d_{ij} is fixed cost associated with route (i,j), a_i is number of units available at plant i, and b_j is number of units demanded at warehouse j, and where $f_{ij}(\boldsymbol{x})$ can be viewed as the objective function from the original transportation problem, which means that the total transportation cost for shipping per unit from plant i to warehouse j in which $f_{ij}(\boldsymbol{x}) = c_{ij}x_{ij}$ will be a cost function if it is linear.

7.4.2 Difficulty of fcTP Instances

The difficulty of a specific fcTP instance depends primarily on the number of m and n parameters. However, there are other properties of the problem structure that cause some instances to be harder to solve than others of the same size. Kennington and Unger [354] reported a strong correspondence between the difficulty of an instance and the F/C-ratio (of the fixed costs to the variable costs) of the optimal solution. Problems with an F/C-ratio between 3 and 10 seem to be the most difficult. Unfortunately, we cannot calculate this factor in advance. Hence Palekar, Karwan, and Zionts [498] suggest that the F/C-ratio be approximated by the difficulty ratio $\bar{d}(m + n - 1)/\bar{c}D$, with D being the sum of all demands, and $\bar{c}$ and $\bar{d}$ being the average of the variable and fixed charges, respectively. Experimental results indicated that the most difficult instances have ratios between 30 and 60 [498].

7.4.3 Solution Method for the fcTP

Many solution procedures have been proposed, including exact solution algorithms and heuristic methods for the fixed-charge transportation problem. Due to the excessive amount of computation time required, exact solution algorithms are not very useful when a problem reaches a certain level of complexity. Therefore, many heuristic methods have been proposed and developed. Hirsch and Dantzig [300] first studied the fixed-charge problem. They proved that the optimal solution occurs at one of the extreme points defined by the constraints of the problem.

The fixed-charge transportation problem is often formulated and solved as a mixed-integer network programming problem. Any general mixed-integer programming solution method can be used to solve the fcTP, such as the branch-and-bound method or the cutting-plane method. Because they do not take advantage of the special network structure of the fcTP, these methods are generally inefficient and computationally expensive [586].

Recently, for the fixed-charge transportation problem, Gottlieb and Paulmann [260] proposed a genetic algorithm with permutation representation based on a matrix, and a tabu search heuristic procedure was proposed by Sun et al. [599].

Since the fixed-charge transportation problem is an extension of the transportation problem, it has the same network structure as the solution characterized as a spanning tree. Therefore, the spanning tree-based genetic algorithm can be adapted to this problem. Experiments have shown the performance of the spanning tree-based genetic algorithm.

7.4.4 Implementation of the Genetic Algorithm

Representation. The Prüfer number, which is one means of node encoding for the tree, is used to encode a transportation tree. The criteria for the feasibility of a chromosome are used to check each chromosome (Prüfer number) generated.

Genetic Operators. As the genetic operator, one-point crossover is used. Inversion and displacement mutations are used in this genetic procedure.

TABLE 7.6. Total Supply (Demand) for Different Problem Sizes

Problem Size	Total Supply
5 × 10	5,000
10 × 10	10,000
10 × 20	15,000

Evaluation and Selection. The fixed-charge transportation problem differs from the traditional transportation problem only in the fixed costs incurred on the transportation routes. It indicates that the objective function value is calculated by adding the fixed cost to that corresponding edge of the transportation tree. Therefore, in the evaluation procedure of this problem, the calculation $f(T) \leftarrow f(T) + f_{ij}(x_{ij})$ of the TP is to be replaced with $f(T) \leftarrow f(T) + f_{ij}(x_{ij}) + d_{ij}$ for including the fixed charge.

A mixed strategy of $(\mu + \lambda)$-selection [567] and roulette wheel selection will be used that can enforce the best chromosomes into the next generation and can avoid premature convergence of the evolutionary process.

7.4.5 Numerical Examples

The spanning tree-based GA approach was implemented in C language and run on a HP 9000 Model 715/100 workstation under the UNIX operating system. The performance of the solution procedure is evaluated by both solution quality and CPU time. To compare this genetic algorithm with other GA approaches, the test problems were produced using the generator NETGEN by Klingman, Naoier, and Stutz [369] and modified by Barr, Glover, and Klingman [43].

The computational results for three problem sizes denoted by $m \times n$ are used; they are 5×10, 10×10, and 10×20. The total supply (demand) for different-sized problems are given in Table 7.6. The ranges of integer fixed costs for the different types of test problems are given in Table 7.7. The unit variable costs in all test problems are integers ranging from 3 through 8. Table 7.8 gives the sampling data for the 5×10 test problem with problem type A.

TABLE 7.7. Ranges of Integer Fixed Costs for Different Types of Test Problems

	Range of Fixed Costs	
Problem Type	Lower Limit	Upper Limit
A	50	200
B	100	400
C	200	800

TABLE 7.8. Sampling Data for the 5 × 10 Test Problem with Problem Type A

Plant, i	Transportation cost, c_{ij}										Supply, a_i
1	8	4	5	6	3	6	3	5	6	7	441
2	7	6	5	5	8	8	3	8	3	6	1267
3	7	8	4	4	3	8	6	5	6	6	1834
4	5	6	4	8	7	8	5	7	6	6	500
5	4	8	6	4	7	7	8	5	3	3	958
	Fixed charge, d_{ij}										
1	107	181	83	137	199	165	148	85	135	109	441
2	149	107	196	132	125	200	140	125	72	82	1267
3	112	184	178	89	84	82	175	63	199	151	1834
4	146	162	169	178	132	51	59	159	112	104	500
5	165	135	91	164	88	88	105	96	96	129	958
Demand, b_j	534	488	868	572	874	315	355	391	237	366	5000

The coded algorithm was run 10 times with the given parameters: mutation rate 0.4, crossover rate 0.2, maximum generation 2000, and different population sizes. For the different-sized problems, the given population sizes were 200 for the 5 × 10, 300 for the 10 × 10, and 400 for the 10 × 20. For the matrix-based GA, probabilities of the genetic operators were given as 0.2 for mutation and 0.4 for crossover, which were known as better parameter settings. Table 7.9 compares the results from the spanning tree-based genetic algorithm (st-GA) and the matrix-based genetic algorithm (m-GA) for the three test problems with problem types. To compare solutions, the solution quality was defined as follows:

$$\text{solution quality} = \frac{\text{best solution}}{\text{solution obtained}} \times 100$$

The best solution was the best of all the m-GA and st-GA solutions for a

TABLE 7.9. Comparison of Solution Qualities for the Two Genetic Algorithm

Problem		m-GA				st-GA			
Size	Type	Worst	Average	Best	Freq.	Worst	Average	Best	Freq.
5 × 10	A	90.80	95.06	98.73	0	98.49	99.12	100.00	2
	B	88.87	92.86	94.01	0	97.78	99.03	100.00	2
	C	85.41	86.99	90.82	0	98.31	99.83	100.00	9
10 × 10	A	99.62	99.86	100.00	4	99.42	99.77	100.00	2
	B	99.72	99.85	100.00	3	99.61	99.74	100.00	3
10 × 20	A	94.85	95.93	97.62	0	98.11	99.25	100.00	2

TABLE 7.10. Comparison of CPU Times for the Genetic Algorithm Approaches

Problem		Memory on a Solution		Average CPU Time (min)	
Size	Type	m-GA	st-GA	m-GA	st-GA
5 × 10	A	50	13	42.192	14.110
	B	50	13	42.192	12.160
	C	50	13	41.981	10.829
10 × 10	A	100	18	179.355	24.217
	B	100	18	178.047	24.383
10 × 20	A	200	28	675.819	471.758

given problem. The number of problems for which the best solution was found by a tree of genetic algorithms, out of a total of 10 trials, is given in the columns labeled "Freq."

As for the computational results shown in Table 7.9, the spanning tree-based genetic algorithm found the best solution in more cases than the matrix-based genetic algorithm and has a higher probability of evolving an optimal or near-optimal solution than m-GA on comparison of the solution quality for fcTP. The st-GA saved CPU time compared to that of the m-GA based on results given in Table 7.10. Certainly, if we increase the population size for m-GA, the optimal or near-optimal solution in encoding space may be found where the fixed costs have smaller integers (as the A type), but exceedingly longer CPU times will be taken. Figure 7.11 illustrates the average solution quality as a function of evolutionary process on the 5 × 10 problem of type A. It clearly shows that spanning tree-based encoding has a better performance than the matrix-based encoding for obtaining an optimal solution in the evolutionary process. Recently, Gen, Li, and Ida reported a spanning tree-based genetic algorithm for solving bicriteria fixed-charge transportation problems [234, 717].

7.5 CAPACITATED PLANT LOCATION PROBLEM

Plant location problems are concerned with the allocation of demands to plants. In some cases it is important that demands at site cannot be split between plants. In other cases, demands can be served by any available plant. Plant location models must reflect these different demand allocation policies and must then allocate demands (or fractions of the total demand in a region) to different plants. In many cases, demands will be allocated to the nearest plant; in other cases, doing so may not be optimal.

Plant location problems are actually a variation of the fixed-charge model. There are a number of receiving stations n, and demands at these destinations

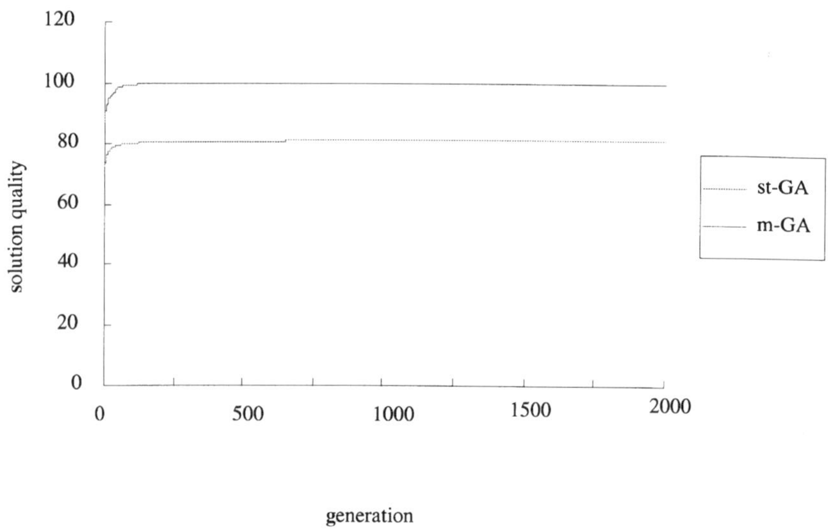

Figure 7.11. Evolutionary processes by two encoding genetic algorithms on the 5 × 10 fcTP.

are satisfied from m potential plants or warehouses. Usually, n is considerably larger than m. To satisfy demand, it is necessary to decide on the location and capacity of each plant. There is an installation (fixed) cost associated with the construction of each plant, and the objective is to minimize the total costs, including fixed costs and transportation costs, between origins and destinations. The problem reduces to an ordinary transportation model with the expection that a fixed-charge term d_i appears in the objective function if for any j, $x_{ij} > 0$ and vanishes otherwise, where x_{ij} is the amount transported from origin i to destination j. Clearly, for prespecified fixed charges, the problem is an ordinary transportation model.

7.5.1 Mathematical Model

The capacitated plant location problem (cPLP) is referred to as a fixed-charge problem to determine the locations of plants with the minimal total cost, including production, shipping costs, and fixed costs where the plants are located. In this case, m sources (or facility locations) produce a single commodity for n customers, each with a demand of b_j ($j = 1,...,n$) units. If a particular source i is opened (or facility is built), it has a fixed cost $d_i \geq 0$ and an associated production capacity $a_i > 0$. There is also a positive cost c_{ij} for shipping a unit from source i to customer j. The problem is to determine the locations of plants so that capacities are not exceeded and demands are met, all at a minimal total cost. The cPLP is the following mixed-integer program:

$$\text{cPLP:} \quad \min \quad z(\boldsymbol{x}) = \sum_{i=1}^{m} \sum_{j=1}^{n} c_{ij}x_{ij} + \sum_{i=1}^{m} d_i y_i \tag{7.32}$$

$$\text{s.t.} \quad \sum_{i=1}^{m} x_{ij} = b_j, \qquad j = 1,2,\dots,n \tag{7.33}$$

$$\sum_{j=1}^{n} x_{ij} \le a_i y_i, \qquad i = 1,2,\dots,m \tag{7.34}$$

$$x_{ij} \ge 0, \qquad \forall i,j \tag{7.35}$$

$$y_i = 0 \text{ or } 1, \qquad i = 1,2,\dots,m \tag{7.36}$$

The variables are x_{ij} and y_i, which represent the amount shipped from plant i to warehouse j and whether a plant is open (or located) ($y_i = 1$) or closed ($y_i = 0$), respectively.

The objective function (7.32) is the total shipping cost plus the total fixed cost. Note that d_i contributes to this sum only when $y_i = 1$ or plant i is open. Constraint (7.33) guarantees that each customer's demand is satisfied. Inequality (7.34) ensures that we do not ship from a plant that is not open, and it also restricts production from exceeding capacity.

When each plant can satisfy all customers' demand, it is possible to reformulate the capacitated plant location problem (7.32) to (7.36). Called as uncapacitated plant location problem (uPLP), it can be solved by inspection. To transform the problem, define g_{ij} to be the fraction of customer j's demand satisfied by plant i; that is,

$$g_{ij} = \frac{x_{ij}}{b_j}, \qquad \forall i,j \tag{7.37}$$

Before making direct use of equation (7.37), we first note that when $a_i \ge \sum_{j=1}^{n} b_j$ $(i = 1,\dots,m)$, inequality (7.34) is only necessary to ensure that shipments are not allowed from a closed plant. This allows us to replace constraint (7.34) by

$$\sum_{j=1}^{n} g_{ij} \le n y_i, \qquad i = 1,\dots,m \tag{7.38}$$

since $y_i = 0$ implies that $g_{ij} = 0$ $(i = 1,\dots,m)$ and thus no demand will be satisfied by plant i. Now, substituting $g_{ij}b_j$ for x_{ij} in expressions (7.32) to (7.34), and letting $t_{ij} = c_{ij}b_j$, we get the equivalent plant location problem:

$$\text{uPLP:} \quad \min \quad z(\boldsymbol{x}) = \sum_{i=1}^{m} \sum_{j=1}^{n} t_{ij} g_{ij} + \sum_{i=1}^{m} d_i y_i \tag{7.39}$$

$$\text{s.t.} \quad \sum_{i=1}^{m} g_{ij} = 1, \qquad j = 1,2,...,n \tag{7.40}$$

$$\sum_{j=1}^{n} g_{ij} \leq n y_i, \qquad i = 1,2,...,m \tag{7.41}$$

$$g_{ij} \geq 0, \qquad \forall i,j \tag{7.42}$$

$$y_i = 0 \text{ or } 1, \qquad i = 1,2,...,m \tag{7.43}$$

The variables are g_{ij} and y_i. This problem can be solved by inspection [171], with $y_i = 0$ or 1 replaced by $0 \leq y_i \leq 1$ ($i = 1,2,...,m$). Since the plants are uncapacitated, all demands of customer j will be assigned to the nearest open plant. The assignment variables g_{ij} will naturally assume integer values. The pin-backing model introduced in Section 2.3, belonging to the class of NP-hard problems, is a special case of the capacitated plant location models when coefficients $t_{ij} = 0$ for all i and j.

An important property of this type of problem is that the optimal solution must occur at an extreme point of the feasible space; that is, it must be associated with a feasible basic solution of a convex polyhedron. This means that the search for a global optimum can be restricted to considering the extreme points of a convex polyhedron only. This result is similar to that used with linear programming. However, because the objective function is concave, the associated algorithm is more complex.

There is a wide variety of algorithms and a rich literature for plant location problems. They vary primarily in the details of constructing the objective and constraint functions, but the basic idea remains unchanged. As in the fixed-charge problem, there are approximate and exact algorithms for plant problems [608].

The branch-and-bound algorithm for the uncapacitated plant location problem (uPLP) was investigated by Efroymson and Ray [171] and the branch-and-bound algorithm for the capacitated plant location problem was developed by Davis and Ray [148]. The efficiency of a branch-and-bound algorithm depends largely on how quickly the linear program can be solved. A heuristic technique with approximate solutions for cPLP was proposed by Sá [545]. Although heuristic methods generally produced good solutions, they are naturally very data dependent and will not work uniformly well.

7.5.2 Spanning Tree–Based Genetic Algorithm for Plant Location Problems

The spanning tree-based genetic algorithm for the capacitated plant location problem is the same as that of the fixed-charge transportation problem except that there is a different evaluation function in the evolutionary process.

TABLE 7.11. Capacities and Ranges of the Demands and Fixed Costs for a cPLP

Problem Size	Capacity of Plant	Demand	Fixed Cost
5 × 10	5,000	[100, 1000]	[200, 500]
10 × 15	10,000	[100, 2000]	[500, 1000]
10 × 20	10,000	[200, 2000]	[500, 1000]

Recently, Gen, Choi, and Tsujimura proposed a genetic algorithm based on two representation schemes for solving the capacitated plant location problem with single source constraints [715].

7.5.3 Numerical Examples

Experiments with spanning tree-based genetic algorithms were executed on a HP 9000 Model 715/100 workstation under the UNIX operating system. Solution quality and CPU time were used to evaluate the performance of the procedure. Computational results with three problem sizes denoted by $m \times n$ were used. Those are 5 × 10, 10 × 15, and 10 × 20. Each test problem was randomly generated. The ranges of the supplies, demands, and fixed costs are given in Table 7.11. The unit variable costs in all test problems are integers ranging from 3 through 10. Table 7.12 gives the sampling data generated randomly for the 5 × 10 test problem.

The implemented st-GA was run 10 times with given genetic parameters, mutation rate $p_m = 0.4$, and crossover rate $p_c = 0.2$. The maximum generation for the two problems 5 × 10 and 10 × 15 was taken as 2000, but for the problem 10 × 20 was 1000. The population sizes were given as 100, 150, and 150, respectively, for the 5 × 10, 10 × 15, and 10 × 20 test problems.

The computational results shown in Table 7.13 are compared with those of the matrix-based genetic algorithm (m-GA). The best solution, which might result from either genetic algorithm approach, was used as the basis for calculating the solution quality. The best solution has a percentage of 100. Other solutions have a percentage below 100. The solution quality is defined as follows:

TABLE 7.12. Sampling Data for the 5 × 10 Test Problem

Plant, i	Fixed cost, d_i	Delivery Cost, c_{ij}										Supply, a_i
1	304	10	9	8	7	4	6	4	7	8	4	1113
2	381	9	3	6	9	6	9	8	4	4	5	1247
3	459	3	5	8	7	3	10	10	8	5	4	902
4	292	6	6	3	10	8	7	4	4	6	3	912
5	241	10	10	9	4	8	5	3	10	4	6	826
	Demand, b_j	729	630	321	293	251	573	207	732	481	783	5000

TABLE 7.13. Comparison of Solution Qualities Between m-GA and st-GA

Problem Size	m-GA				st-GA			
	Worst	Average	Best	Freq.	Worst	Average	Best	Freq.
5 × 10	92.43	97.02	100.00	2	100.00	100.00	100.00	10
10 × 15	90.48	93.10	98.97	0	96.50	98.92	100.00	3
10 × 20	89.16	93.62	97.84	0	92.64	96.47	100.00	1

$$\text{solution quality} = \frac{\text{best solution}}{\text{solution obtained}} \times 100$$

The number of trials for which the best solution was found is given in the columns labeled "Freq."

From Table 7.13 we can see that st-GA found the best solution more times than the m-GA. In a comparison of the solution quality, the st-GA also has a higher probability to evolve the optimal solution or near-optimal solution than the m-GA. Table 7.14 shows that the st-GA used much less computational time than the m-GA for a small-scale problem. For large-scale problems, especially for the 10 × 20 test problem, the CPU time used by the st-GA was not greatly different from the computing time of the m-GA. This is because the st-GA requires more time to generate feasible chromosomes in the initialization procedure.

Figure 7.12 illustrates the average solution quality in evolutionary process on the problem 5 × 10 with sampling data. It clearly shown that spanning tree-based encoding is a better performance than matrix-based encoding for evolving an optimal solution in the evolutionary process.

Recently, Gen, Choi, and Tsujimura proposed a genetic algorithm based on two representation schemes for solving the capacitated plant location problem with single source constraints [715].

7.6 BICRITERIA TRANSPORTATION PROBLEM WITH FUZZY COEFFICIENTS

Real-world situations are often not as deterministic as those that we have been discussing. Precise mathematical models are not enough to tackle all

TABLE 7.14. Comparison of CPU Time Between m-GA and st-GA

Problem Size	Memory of a Chromosome		Average CPU Time (min.)	
	m-GA	st-GA	m-GA	st-GA
5 × 10	50	13	12.075	3.456
10 × 15	150	23	78.240	9.564
10 × 20	200	28	103.963	71.625

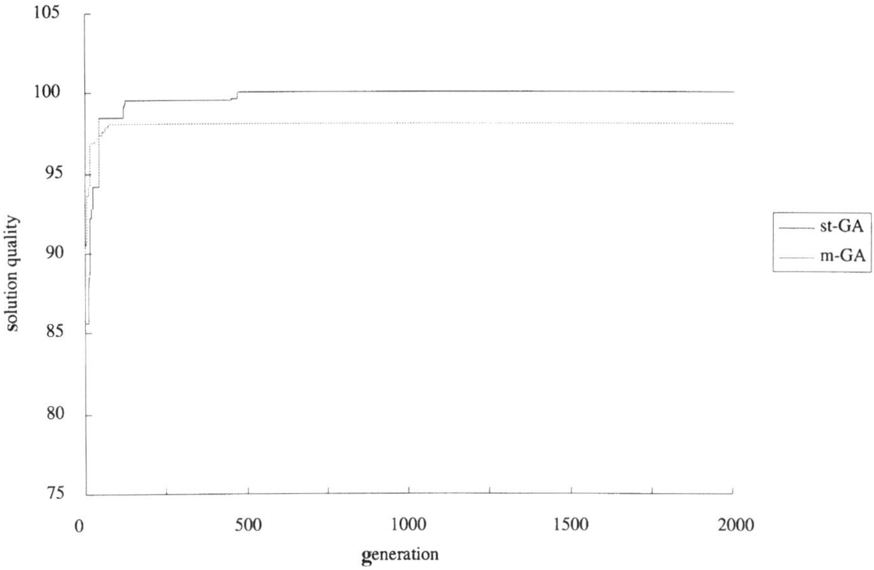

Figure 7.12. Evolutionary process on the 5 × 10 cPTP.

practical problems. To deal with imprecision and uncertainty, concepts and techniques of probability theory are usually employed. A common problem of these imprecise conditions is the difficulty in determining the proper values of model parameters. Fuzzy set theory has been applied to techniques of linear and nonlinear programming, integer programming, multicriteria decision making, and so on. It helps to improve oversimplified (crisp) models and provides more robust and flexible models for real-world complex systems. In a consideration of practical models, it is necessary to use fuzzy numbers to represent imprecise conditions. One way of handling such uncertainty in decision making is fuzzy mathematical programming. Kaufmann and Gupta [349] first examined a fuzzy transportation problem considering fuzzy coefficients. Gen, Li, and Ida [412, 414, 714] proposed an improved genetic algorithm for solving multiobjective solid transportation problems with fuzzy numbers. In this section we use a spanning tree-based GA to solve the bicriteria transportation problem (BTP) with fuzzy coefficients.

7.6.1 Problem Description

Frequently in transportation systems, the influence of traffic on a transportation system results in uncertainty as to some or all of the coefficients of objectives. For example, transportation costs, delivery times, and so on, may not be definitely known.

Consider the following two objectives: minimizing total transportation cost and minimizing total delivery time. Let $\tilde{c}^1_{ij}$ be the fuzzy data representing the transportation cost of shipping 1 unit from plant i to warehouse j, $\tilde{c}_{ij}^2$ be the fuzzy data representing the delivery time of shipping 1 unit of product from plant i to warehouse j, a_i be the number of units available at plant i, and b_j be the number of units demanded at warehouse j. This problem with m plants and n warehouses can be formulated as follows:

$$\min \quad \tilde{z}_1 = \sum_{i=1}^{m} \sum_{j=1}^{n} \tilde{c}_{ij}^1 x_{ij} \tag{7.44}$$

$$\min \quad \tilde{z}_2 = \sum_{i=1}^{m} \sum_{j=1}^{n} \tilde{c}_{ij}^2 x_{ij} \tag{7.45}$$

$$\text{s.t.} \quad \sum_{j=1}^{n} x_{ij} \leq a_i, \qquad i = 1,2,\ldots,m \tag{7.46}$$

$$\sum_{i=1}^{m} x_{ij} \geq b_j, \qquad j = 1,2,\ldots,n \tag{7.47}$$

$$x_{ij} \geq 0, \qquad \forall i,j \tag{7.48}$$

where x_{ij} is the unknown quantity to be transported from plant i to warehouse j.

7.6.2 Ranking Fuzzy Numbers

Liou and Wang's method is used to rank the fuzzy numbers [424]. This method ranks fuzzy numbers, which can be triangular or trapezoidal, using an integral value instead of a relative value. The left integral value is used to reflect the pessimistic viewpoint and the right integral value is used to reflect the optimistic viewpoint of the decision maker. A convex combination of right and left integral values, using an index of optimism, is called the total integral value.

Denote $I_L(\tilde{A})$ and $I_R(\tilde{A})$ as the left and right integral values for a triangular fuzzy number $\tilde{A} = (a_1,a_2,a_3)$. The total integral value $I_T^\alpha(\tilde{A})$ of the fuzzy number $\tilde{A}$ with a degree of optimism α of a decision maker is

$$\begin{aligned} I_T^\alpha(\tilde{A}) &= \alpha I_R(\tilde{A}) + (1 - \alpha) I_L(\tilde{A}) \\ &= \tfrac{1}{2}[\alpha a_1 + a_2 + (1 - \alpha) a_3] \end{aligned} \tag{7.49}$$

where $\alpha \in [0,1]$ is given. A large α indicates a higher degree of optimism [424].

When $\alpha = 0$, especially, the total integral value $I_T^0(\tilde{A})$ represents a pessimistic decision maker's viewpoint that $I_T^\alpha(\tilde{A})$ is equal to the left integral value of $\tilde{A}$ [i.e., $I_L(\tilde{A})$]. For an optimistic decision maker (i.e., $\alpha = 1$), the total integral value $I_T^1(\tilde{A})$ is equal to $I_R(\tilde{A})$. For a moderate decision maker (α = 0.5), the total integral value becomes $I_T^{0.5}(\tilde{A}) = \frac{1}{2}[I_R(\tilde{A}) + I_L(\tilde{A})]$, in which case the integral value is the same as with ordinary representation [349].

This total integral value of fuzzy numbers is used as the ranking function. For any fuzzy numbers $\tilde{A}_i$ and $\tilde{A}_j$, we have the following criteria for ranking fuzzy numbers:

1. If $I_T^\alpha(\tilde{A}_i) < I_T^\alpha(\tilde{A}_j)$, then $\tilde{A}_i < \tilde{A}_j$
2. If $I_T^\alpha(\tilde{A}_i) = I_T^\alpha(\tilde{A}_j)$, then $\tilde{A}_i = \tilde{A}_j$.
3. If $I_T^\alpha(\tilde{A}_i) > I_T^\alpha(\tilde{A}_j)$, then $\tilde{A}_i > \tilde{A}_j$.

For more details of the method, refer to [424].

In the fuzzy environment, we define the Pareto optimal solutions as follows:

Definition 7.1. A solution $\bar{\boldsymbol{x}} = [\bar{x}_{ij}]$ is said to be a Pareto optimal solution for fuzzy BTP if and only if there does not exist another $\boldsymbol{x} \in F$ such that

$$R(\tilde{z}_q(\boldsymbol{x})) \geq R(\tilde{z}_q(\bar{\boldsymbol{x}}))\ \forall q$$

$$R(\tilde{z}_p(\boldsymbol{x})) \neq R(\tilde{z}_p(\bar{\boldsymbol{x}}))\ \exists p.$$

where $R(\cdot)$ is called the ranking function and $\tilde{z}_q$ is the fuzzy number of the qth objective value to be minimized.

Pareto optimal solutions in the fuzzy environment are determined based on the ranked values of the fuzzy objectives. Several methods for ranking fuzzy numbers have been proposed [349, 424]. In this problem, the integral value of a fuzzy number is used as the ranking function.

7.6.3 Implementation of the Genetic Algorithm

A genetic algorithm approach based on a spanning tree is adopted for solving this problem. The basic implementation was the same as that given in Chapter 4. The major effort was given to handling fuzziness. In multicriteria optimization, we are interested in finding Pareto solutions. When the coefficients of objectives are represented with fuzzy numbers, the objective values become fuzzy numbers. Since a fuzzy number represents many possible real numbers, it is not easy to compare solutions to determine which is the Pareto solution. Fuzzy ranking techniques can help us to compare fuzzy numbers. In this approach, Pareto solutions are determined based on the ranked values of fuzzy objective functions, and genetic algorithms are used to search for Pareto solutions.

Representation. Spanning tree encoding with the Prüfer number is used to represent the candidate solution. The encoding and decoding procedures have been described in Section 7.2. The criterion of solution's feasibility designed in the spanning tree-based genetic algorithm is also employed.

Genetic Operators

Crossover. For simplicity, the one-point crossover is used.

Mutation. Inversion mutation and displacement mutation are used. The two mutation operators are randomly selected for mutation of a given chromosome.

Evaluation and Selection. In this approach, the evaluation procedure consists of two steps:

1. Convert a chromosome into a tree.
2. Calculate each objective function with fuzzy coefficients.

In objective functions the coefficients are represented by fuzzy numbers. The computations of objective functions are replaced with the following expressions:

$$\tilde{z}_1(T) \leftarrow \tilde{z}_1(T) + \tilde{c}_{ij}^1 x_{ij}$$

$$\tilde{z}_2(T) \leftarrow \tilde{z}_2(T) + \tilde{c}_{ij}^2 x_{ij}$$

The Pareto solutions are characterized as solutions to the multiobjective programming problem. Therefore, the module for obtaining Pareto solutions is added into genetic algorithms, which consist of the following two steps:

1. Evaluate chromosomes using objective functions.
2. Select Pareto solutions based on evaluation values.

Let E be a Pareto solution set generated up to the current iteration t. Then the procedure for Pareto solutions is given as follows:

Procedure: Pareto Solutions

Step 1. Set iteration $k = 1$, and $E = \{\phi\}$.

Step 2. If $k > i\_size$, then stop. Otherwise, go to step 3.

Step 3. Evaluate a chromosome T_k and obtain the solution vector $\tilde{z}_k = [\tilde{z}_1(T_k)\ \tilde{z}_2(T_k) \cdots \tilde{z}_Q(T_k)]$.

Step 4. Compare it with all Pareto solutions in E.

(4.1) If it is dominated by one Pareto solution based on the ranked value of solution vector $\tilde{z}_k$ using the ranking method of fuzzy numbers, go to step 5.

(4.2) If it dominates some Pareto solutions based on the ranked value of solution vector $\tilde{z}_k$ using the ranking method of fuzzy numbers, add it into E and delete solutions dominated by it.

(4.3) If it is a new Pareto solution and dominates none of exiting ones, simply add into E.

Step 5. Set $k = k + 1$ and go to step 2.

For the fitness of chromosomes in the evolutionary process, the following procedure is used to create the fitness function.

Fitness Function. The fitness function is derived in the following way:

1. Choose solution points that contain the minimum $I_T^\alpha(\tilde{z}_q^{\min})$ [or maximum $I_T^\alpha(\tilde{z}_q^{\max})$] that corresponds to each objective function value, compare with the stored solution points in the preceding generation, and select the best points to save again.

$$I_T^\alpha(\tilde{z}_q^{\min(t)}) = \min_k \{I_T^\alpha(\tilde{z}_q^{\min(t-1)}), I_T^\alpha(\tilde{z}_q^{(t)}(T_k)) \mid k = 1,2,...,i\_size\}, \qquad q = 1,2$$

$$I_T^\alpha(\tilde{z}_q^{\max(t)}) = \max_k \{I_T^\alpha(\tilde{z}_q^{\max(t-1)}), I_T^\alpha(\tilde{z}_q^{(t)}(T_k)) \mid k = 1,2,...,i\_size\}, \qquad q = 1,2$$

2. Solve the weights for each objective function and construct the fitness function.

$$\delta_q = I_T^\alpha(\tilde{z}_q^{\max(t)}) - I_T^\alpha(\tilde{z}_q^{\min(t)}), \qquad q = 1,2$$

$$\beta_q = \frac{\delta_q}{\sum_{q=1}^2 \delta_q}, \qquad q = 1,2$$

$$\text{eval}(T_k) = \sum_{q=1}^{2} \beta_q I_T^\alpha(\tilde{z}_q(T_k)), \qquad \forall k$$

Overall Procedure. Let $P(t)$ be a population of chromosomes, $C(t)$ be the chromosomes generated in current generation t, and $E(t)$ be the Pareto solution set generated up to current generation t. The overall procedure is as follows:

```
procedure: st-GA/b-fTP
begin
  t ← 0;
  initialize P(t);
  evaluate P(t) by the fitness function;
  determine E(t) by a module of Pareto solutions;
  while (not termination condition) do
  begin
   recombine P(t) to generate C(t);
```

TABLE 7.15. TFN Coefficients in the First Example with Three Plants and Four Warehouses

Objective, k	1: $\tilde{c}_{i1}^k$	2: $\tilde{c}_{i2}^k$	3: $\tilde{c}_{i3}^k$	4: $\tilde{c}_{i4}^k$	a_i
1	(1, 2, 3)	(1, 2, 3)	(5, 7, 9)	(6, 7, 9)	8
	(1, 1, 2)	(6, 9, 12)	(2, 3, 5)	(3, 4, 5)	19
	(6, 8, 10)	(7, 9, 10)	(2, 4, 6)	(5, 6, 8)	17
2	(3, 4, 5)	(2, 4, 6)	(1, 3, 5)	(2, 4, 6)	8
	(3, 5, 7)	(6, 8, 9)	(7, 9, 9)	(8, 9, 12)	9
	(4, 6, 8)	(1, 2, 3)	(4, 5, 6)	(1, 3, 3)	17
b_j	11	3	14	16	44

```
    evaluate C(t) by the fitness function;
    update E(t) by a module of Pareto solutions;
    select P(t + 1) from P(t) and C(t);
    t ← t + 1;
  end
end
```

7.6.4 Numerical Examples

Computer simulations were projected for the following problem, where its coefficients were described by the triangular fuzzy numbers (TFNs) given in Table 7.15.

The coded algorithm was run 30 times with the best mechanism (crossover rate $p_c = 0.2$ and mutation rate $p_m = 0.5$) and with three different degrees of optimism: $\alpha = 0$, 0.5, and 1. The results with population size 30 and maximum generation 1000 are shown in Figure 7.13. Results were found at the Pareto frontier. The Pareto solutions were obtained based on the integral value of each objective function value. The eight Pareto solutions obtained in the pessimistic case are given as follows:

$$\text{(1) solution matrix: } \begin{bmatrix} 5 & 3 & 0 & 0 \\ 6 & 0 & 0 & 13 \\ 0 & 0 & 14 & 3 \end{bmatrix}$$

Its corresponding objective values are $\tilde{z}_1 = (96, 148, 209)$, $\tilde{z}_2 = (202, 258, 334)$. The integral value of the objective value is $(I_T^0(\tilde{z}_1), I_T^0(\tilde{z}_2)) = (74.00, 129.00)$, and $P(T) = [6 \quad 1 \quad 4 \quad 2 \quad 2]$.

$$\text{(2) solution matrix: } \begin{bmatrix} 0 & 3 & 0 & 5 \\ 11 & 0 & 0 & 8 \\ 0 & 0 & 14 & 3 \end{bmatrix}$$

Its corresponding objective values are $\tilde{z}_1 = (111, 158, 224)$, $\tilde{z}_2 = (172, 238,$

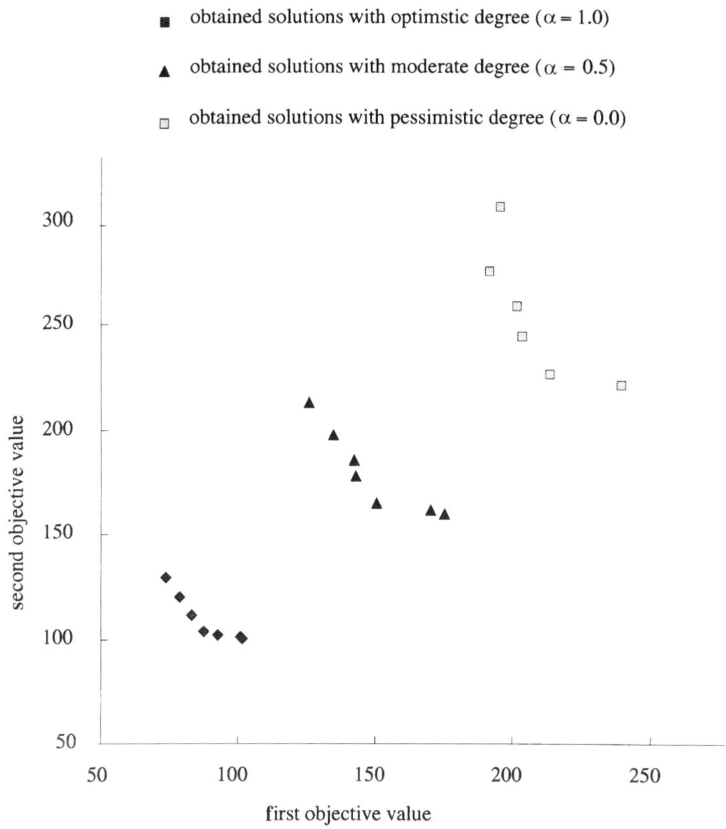

Figure 7.13. Solutions obtained with three degrees of optimism (σ) in the first example.

314). The integral value of the objective value is $(I_T^0(\tilde{z}_1), I_T^0(\tilde{z}_2)) = (79.00, 119.00)$, $P(T) = [6 \quad 2 \quad 1 \quad 6 \quad 2]$.

$$\text{(3) solution matrix:} \begin{bmatrix} 5 & 3 & 0 & 0 \\ 6 & 0 & 13 & 0 \\ 0 & 0 & 1 & 16 \end{bmatrix}$$

Its corresponding objective values are $\tilde{z}_1 = (122, 161, 235)$, $\tilde{z}_2 = (150, 232, 256)$. The integral value of objective value is $(I_T^0(\tilde{z}_1), I_T^0(\tilde{z}_2)) = (80.5, 116)$, $P(T) = [1 \quad 4 \quad 2 \quad 6 \quad 3]$.

$$\text{(4) solution matrix:} \begin{bmatrix} 0 & 3 & 0 & 5 \\ 11 & 0 & 8 & 0 \\ 0 & 0 & 6 & 11 \end{bmatrix}$$

Its corresponding objective values are $\tilde{z}_1 = (127, 166, 240)$, $\tilde{z}_2 = (140, 222,$

266). The integral value of objective value is $(T_T^0(\tilde{z}_1), I_T^0(\tilde{z}_2)) = (83, 111)$, $P(T) = [5 \quad 2 \quad 2 \quad 6 \quad 3]$.

$$\text{(5) solution matrix: } \begin{bmatrix} 0 & 3 & 5 & 0 \\ 11 & 0 & 8 & 0 \\ 0 & 0 & 1 & 16 \end{bmatrix}$$

Its corresponding objective values are $\tilde{z}_1 = (137, 176, 250)$ and $\tilde{z}_2 = (120, 207, 246)$. The integral value of the objective value is $(I_T^0(\tilde{z}_1), I_T^0(\tilde{z}_2)) = (88, 103.5)$, $P(T) = [7 \quad 2 \quad 1 \quad 6 \quad 2]$.

$$\text{(6) solution matrix: } \begin{bmatrix} 0 & 2 & 6 & 0 \\ 11 & 0 & 8 & 0 \\ 0 & 1 & 0 & 16 \end{bmatrix}$$

Its corresponding objective values are $\tilde{z}_1 = (146, 186, 260)$ and $\tilde{z}_2 = (116, 203, 242)$. The integral value of the objective value is $(I_T^0(\tilde{z}_1), I_T^0(\tilde{z}_2)) = (93, 101.5)$, and $P(T) = [2 \quad 6 \quad 1 \quad 5 \quad 3]$.

$$\text{(7) solution matrix: } \begin{bmatrix} 0 & 0 & 6 & 2 \\ 11 & 0 & 8 & 0 \\ 0 & 3 & 0 & 14 \end{bmatrix}$$

Its corresponding objective values are $\tilde{z}_1 = (160, 202, 276)$ and $\tilde{z}_2 = (116, 201, 242)$. The integral value of the objective value is $(I_T^0(\tilde{z}_1), I_T^0(\tilde{z}_2)) = (101, 100.5)$, and $P(T) = [5 \quad 2 \quad 2 \quad 6 \quad 1]$.

$$\text{(8) solution matrix: } \begin{bmatrix} 0 & 0 & 8 & 0 \\ 11 & 0 & 6 & 2 \\ 0 & 3 & 0 & 14 \end{bmatrix}$$

Its corresponding objective values are $\tilde{z}_1 = (160, 204, 276)$ and $\tilde{z}_2 = (116, 199, 246)$. The integral value of the objective value is $(I_T^0(\tilde{z}_1), I_T^0(\tilde{z}_2)) = (102, 99.5)$, and $P(T) = [4 \quad 3 \quad 4 \quad 1 \quad 1]$.

The second example has 6 plants and 9 warehouses. The coefficients of the two objectives are represented as the triangular fuzzy numbers given in Table 7.16. The genetic parameters were tested and selected as the best mechanism for the problem. Crossover rate $p_c = 0.2$, mutation rate $p_m = 0.5$, and the number of runs is 30 times. Figure 7.14 compares results for the three degrees of optimistic, moderate, and pessimistic. The computation results indicate that when the decision maker evaluates this transportation project based on the optimistic degree ($\alpha = 1$), the objective values are smallest for this problem. So the decision maker can obtain the range of objective values that he or she expected under imprecise conditions.

TABLE 7.16. TFN Coefficients in the Second Example with Six Destinations and Nine Warehouses

k	1: $\tilde{c}_{i1}^k$	2: $\tilde{c}_{i2}^k$	3: $\tilde{c}_{i3}^k$	4: $\tilde{c}_{i4}^k$	5: $\tilde{c}_{i5}^k$	6: $\tilde{c}_{i6}^k$	7: $\tilde{c}_{i7}^k$	8: $\tilde{c}_{i8}^k$	9: $\tilde{c}_{i9}^k$	a_i
1	(1, 1, 3)	(1, 2, 3)	(5, 7, 9)	(6, 7, 9)	(2, 4, 5)	(6, 7, 8)	(4, 5, 6)	(1, 3, 5)	(7, 8, 9)	8
	(1, 1, 2)	(6, 9, 12)	(2, 3, 5)	(3, 4, 5)	(7, 9, 10)	(6, 7, 8)	(2, 4, 6)	(3, 5, 8)	(5, 6, 7)	19
	(6, 8, 10)	(7, 9, 10)	(2, 4, 6)	(5, 6, 8)	(5, 6, 8)	(1, 4, 5)	(6, 9, 11)	(3, 5, 7)	(8, 9, 12)	17
	(5, 6, 7)	(1, 4, 5)	(8, 9, 11)	(3, 5, 7)	(3, 5, 7)	(6, 8, 10)	(7, 9, 9)	(8, 9, 12)	(9, 11, 13)	23
	(2, 4, 5)	(6, 7, 8)	(4, 5, 6)	(2, 3, 5)	(10, 13, 15)	(8, 9, 10)	(2, 5, 7)	(4, 7, 9)	(7, 8, 9)	10
	(7, 9, 11)	(3, 4, 6)	(2, 5, 7)	(4, 7, 9)	(7, 8, 9)	(5, 7, 9)	(9, 12, 15)	(3, 4, 6)	(5, 8, 10)	20
2	(3, 4, 5)	(2, 4, 6)	(1, 3, 5)	(2, 4, 6)	(10, 13, 15)	(8, 9, 10)	(2, 5, 7)	(4, 7, 9)	(7, 8, 9)	8
	(3, 5, 7)	(6, 8, 9)	(7, 9, 9)	(8, 9, 12)	(1, 4, 5)	(8, 9, 11)	(3, 5, 7)	(3, 5, 7)	(6, 8, 10)	19
	(4, 6, 8)	(1, 2, 3)	(4, 5, 6)	(1, 3, 3)	(6, 9, 12)	(2, 3, 5)	(8, 9, 10)	(2, 5, 7)	(4, 7, 9)	17
	(6, 8, 10)	(7, 9, 10)	(2, 4, 6)	(5, 6, 8)	(9, 12, 15)	(3, 4, 6)	(5, 8, 10)	(8, 9, 11)	(3, 5, 7)	23
	(5, 6, 7)	(1, 4, 5)	(8, 9, 11)	(3, 5, 7)	(5, 7, 9)	(9, 12, 15)	(3, 4, 6)	(5, 8, 10)	(3, 4, 6)	10
	(2, 4, 6)	(6, 7, 8)	(4, 5, 6)	(1, 3, 5)	(4, 7, 9)	(7, 8, 9)	(5, 7, 9)	(9, 12, 15)	(3, 4, 6)	20
b_j	11	3	14	16	8	6	10	20	9	

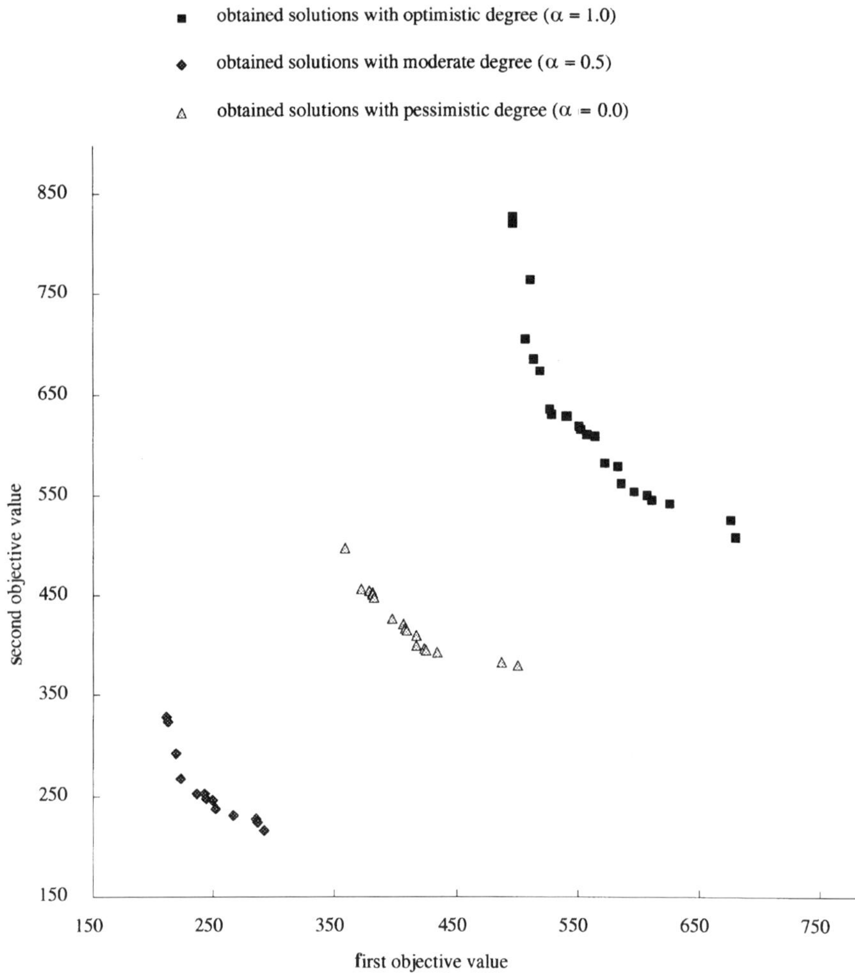

Figure 7.14. Solutions obtained with three degrees of optimism (α) in the second example.

8

NETWORK DESIGN AND ROUTING

8.1 INTRODUCTION

There has been an explosive growth in computer networks since the 1980s. Computer networks pervade our everyday life, from automated teller machines, to airline reservation systems, to electronic mail services, to electronic bulletin boards, to the Internet. There are many reasons for the explosive growth in computer networks. We are in an information age, and computer networks have become an integral part of the dissemination of information. Network design and routing are one of the important issues in the building and expansion of computer networks. Many ideas and methods have been proposed and tested in the past two decades. Recently, there has been increasing interest in applying genetic algorithms to problems related to computer networks [85, 89, 175, 210, 212, 380, 472, 474, 558, 632, 633]. In this chapter we explain how to solve some problems in network systems using genetic algorithms, including the shortest path problem, adaptive routing design of a centralized network, and a facility location problem on a network [228, 241, 474].

8.2 SHORTEST PATH PROBLEM

One of the most common problems encountered in analysis of networks is the shortest path problem: finding a path between two designated nodes having minimum total length or cost. This is a fundamental problem in many applications involving transportation, routing, and communications [528]. In many applications, however, several criteria are associated with traversing each edge of a network. For example, cost and time measures are both important in transportation networks, as are economic and ecological factors in highway construction. As a result, there has been recent interest in solving the bicriteria shortest path problem: finding paths that are efficient with respect to both criteria. There is usually no single path that gives a shortest path with respect to both criteria. Instead, a set of Pareto optimal paths or efficient paths are preferred. Several applications of the problem have been well documented in [293] and [656].

A number of algorithms have been proposed for finding efficient paths while considering bicriteria [125, 282, 293, 442, 656]. These methods have features common to Pareto solution–generating methods in general: intending to identify an entire efficient set or a substantial part of such a set. An empirical investigation of some of these methods was given by Smith and Shier [577].

The bicriteria shortest path problem is known as NP-hard [208]. The efficient set of paths may be very large, possibly exponential in size. Thus the computational effort required to solve it can increase exponentially with the problem size in the worst case [282]. While the tractability of the problem is of importance when solving large-scale problems, the issue concerning the size of the efficient set is important to a decision maker. Having to evaluate a large efficient set to select the best one poses a considerable cognitive burden on decision makers. Therefore, in such cases, obtaining the entire Pareto optimal set is of little interest. Murthy and Olson developed an interactive procedure to guide decision makers quickly to their preferred solutions [479].

Cheng and Gen proposed a compromise approach–based genetic algorithm to solve the bicriterion shortest path problem [109]. In contrast to the generating approach, the compromise approach identifies solutions that are closest to the ideal solution as determined by some measure of distance [686]. The difficult part of developing a genetic algorithm for this problem is how to encode a path in a graph into a chromosome. A priority-based encoding method is proposed that can potentially represent all possible paths in a graph. Besides, the capability of representing a path for an unplanar and undirected graph is one of the most significant features of the priority-based encoding method.

8.2.1 Problem Formulation

An undirected graph $G = (V,E)$ comprises a set of nodes $V = \{1,2,\ldots,n\}$ and a set of edges $E \in V \times V$ connecting nodes in V. Corresponding to each edge are two nonnegative numbers c_{ij}^1 and c_{ij}^2 representing the cost and distance, and others of interest, from node i to node j. A path from node i to node j is a sequence of edges (i,l), (l,m),..., (k,j) from E in which no node appears more than once. A path can also be represented equivalently as a sequence of nodes $(i,l,m,\ldots,k,j)$. For the example given in Figure 8.1, (1,4), (4,3), (3,5), (5,6) is a path from node 1 to node 6. The node representation is (1,4,3,5,6).

Let 1 denote the initial node and n denote the end node of the path. Let x_{ij} be an indicator variable defined as follows:

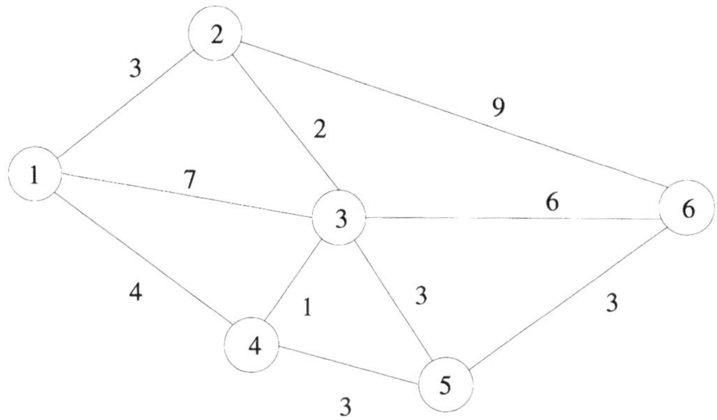

Figure 8.1. Simple undirected graph with six nodes and 10 edges.

$$x_{ij} = \begin{cases} 1, & \text{if edge } (i,j) \text{ is included in the path} \\ 0, & \text{otherwise} \end{cases}$$

The bicriteria shortest-path problem can be formulated as follows:

$$\min \quad z^1(\boldsymbol{x}) = \sum_i \sum_j c_{ij}^1 x_{ij} \tag{8.1}$$

$$\min \quad z^2(\boldsymbol{x}) = \sum_i \sum_j c_{ij}^2 x_{ij} \tag{8.2}$$

$$\text{s.t.} \quad \sum_j x_{ij} \leq 2, \qquad \forall i \in V \tag{8.3}$$

$$\sum_{j \neq k} x_{ij} \geq x_{ik}, \qquad \forall (i,k) \in E, \quad \forall i \in V \backslash \{1,n\} \tag{8.4}$$

$$\sum_j x_{1j} = \sum_j x_{jn} = 1, \qquad \forall i,j \in V \tag{8.5}$$

$$x_{ij} = x_{ji}, \qquad \forall (i,j) \in E \tag{8.6}$$

$$0 \leq x_{ij} \leq 1, \qquad \forall (i,j) \in E \tag{8.7}$$

where constraints (8.3) and (8.4) together imply that any node other than nodes 1 and n have either 0 or 2 nonzero incident edges. Constraint (8.5) makes nodes 1 and n the endpoints of the path.

8.2.2 Genetic Algorithm Approach

Priority-Based Encoding. How to encode a path for a graph is critical for developing a genetic algorithm to solve the shortest path problem. It is not easier to find a representation for this problem than for the traveling salesmen problem. Special difficulty arises because:

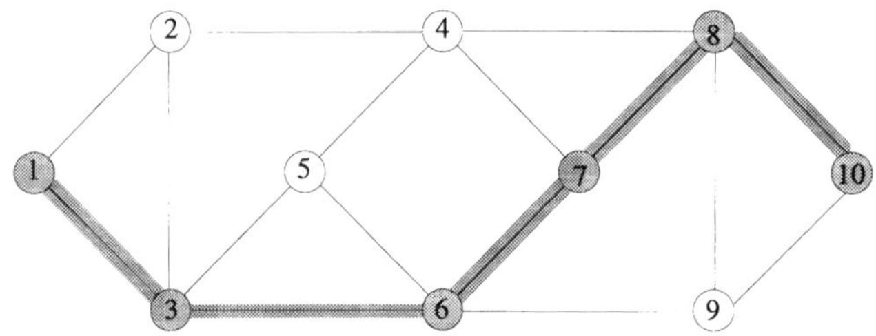

Figure 8.2. Simple undirected graph with 10 nodes and 16 edges.

1. A path contains a variable number of nodes, and the maximal number is $n - 1$ for an n-node graph.
2. A random sequence of edges usually does not correspond to a path.

To overcome such difficulties, Cheng and Gen adopted an indirect approach: Encode some guiding information for constructing a path in a chromosome but not the path itself [103, 221].

As we know, a gene in a chromosome is characterized by two factors: locus, the position of the gene within the structure of chromosome, and allele, the value the gene takes. The position of a gene is used to represent a node, and the value is used to represent the priority of the node for constructing a path among candidates. The encoding method is denoted as *priority-based encoding* [111]. The path corresponding to a given chromosome is generated by a sequential node appending procedure beginning with node 1 and terminating at node n. At each step, several nodes are usually available for consideration; only the node with the highest priority is added into the path.

Consider the undirected graph shown in Figure 8.2. Suppose that we are going to find a path from node 1 to node 10. In this instance the encoding is as shown in Figure 8.3. At the beginning, we try to find a node for the position next to node 1. Nodes 2 and 3 are eligible for the position, which can easily be fixed according to the adjacent relation among nodes. The priorities of the nodes are 3 and 4, respectively. Node 3 has the highest priority and is put into the path. The possible nodes following node 3 are nodes 2, 5, and 6. Because node 6 has the largest priority value, it is put into the path. Then we form the set of nodes available for the next position and select from among them the one with the highest priority. Repeat these steps until we obtain a complete path (1, 3, 6, 7, 8, 10).

position: node ID	1	2	3	4	5	6	7	8	9	10
value: priority	7	3	4	6	2	5	8	10	1	9

Figure 8.3. Example of priority-based encoding.

For an n-node problem, let Ω be a set containing integers from 1 up to n, that is, $\Omega = \{1,2,...,n\}$, and let p_i denote the priority for node i, which is a random integer exclusively from the set Ω. Priorities p_i of all nodes satisfy the following conditions:

$$p_i \neq p_j, \quad p_i,p_j \in \Omega, \qquad i \neq j, \quad i,j = 1,2,...,n \tag{8.8}$$

Then the priority-based encoding can be defined formally as follows:

$$[p_1,p_2,...,p_n]$$

Essentially, the mapping between encoding and path is many-to-one, which means that different chromosomes may produce an identical shortest path. But it is easy to prove that the probability of occurrence of many-to-one mapping is very low. Therefore, in most cases, there are no trivial genetic operations associated with the encoding. It is also easy to verify that any permutation of the encoding corresponds to a path, so that most existing genetic operators can easily be applied to the encoding. Also, any path has a corresponding encoding; therefore, any point in solution space is accessible for genetic search.

Path Growth Procedure. A path growth procedure is proposed to generate a path from an arbitrary chromosome. The basic idea of this procedure is to generate a path from initial node 1 to end node n by appending eligible edges into the path consecutively. At each step, several edges are usually available for consideration. The edge added to a partial path is always the edge incident to the node with the highest priority, which extends the partial path from the terminal node. The tricky part is, of course, finding a set of eligible edges. The following definitions and theorems will give you a better understanding of how to make such a set of eligible edges and how the path growth procedure works.

Let $G = (V,E)$ be a connected and undirected graph with a real-valued weight function defined on E. Let P_t^k be a partial path undergoing growth, which contains $k + 1$ nodes with terminal node t. The definition of an eligible edge is as follows:

Definition 8.1 (Eligible Edge). An edge is eligible for the path P_t^k if it can extend the path without forming a cycle with the edges in P_t^k.

Let $V_p \subseteq V$ be the set of nodes existing in the partial path P_t^k. Let $V_q = V - V_p$ be the complement set of V_p. Let $C(V_p,V_q) = \{(i,j) \mid i \in V_p, j \in V_q\}$ be the cut of the graph [71]. Let $T_t = \{(t,j) \mid j \in V\}$ be the set of edges incident with the terminal node t. Then we have the following property:

Property 8.1 (Eligible Edge Set). For a given partial path P_t^k, the set of eligible edges E_{qt} is given as follows:

$$E_{qt} = T_t \cap C(V_p, V_q) \tag{8.9}$$

Now we consider how to find the set of E_{qt}. For a connected and undirected graph $G = (V,E)$, the adjacency matrix A is the $n \times n$ matrix given by

$$a_{ij} = \begin{cases} 1, & \text{if } (i,j) \in E \\ 0, & \text{otherwise} \end{cases} \tag{8.10}$$

For a given partial path P_t^k, we can define a mesh matrix $M(k)$ as follows:

$$m_{ij} = \begin{cases} 0, & \text{if } i \in V \text{ and } \forall j \in V_p \\ 1, & \text{otherwise} \end{cases} \tag{8.11}$$

Then we can define a new adjacent matrix $A(k)$ according to a given partial path P_t^k as follows:

Definition 8.2 (Dynamic Adjacent Matrix). The dynamic adjacent matrix $A(k)$ with respect to a given partial path P_t^k is given by the following boolean intersection of matrix:

$$A(k) = A \cap M(k) \tag{8.12}$$

We will show that the dynamic adjacent matrix $A(k)$ removes all edges incident to the nodes in partial path P_t^k. In other words, it just contains the adjacent relation with respect to the eligible edges. Because such an adjacent relation varies as path P_t^k grows, we call it the dynamic adjacent matrix.

We further define a matrix $U(k)$ according to path P_t^k as follows:

$$u_{ij} = \begin{cases} 0, & \text{if } j = t \\ 1, & \text{otherwise} \end{cases} \tag{8.13}$$

Then we can have the following recursive equation of the mesh matrix and the dynamic adjacent matrix with respect to step k:

$$M(k) = M(k-1) \cap U(k), \ M(0) = [1]_{nxn} \tag{8.14}$$

$$\begin{aligned} A(k) &= A(k-1) \cap U(k) = A(k-1) \cap M(k) \\ &= A \cap M(k), \qquad A(0) = A \end{aligned} \tag{8.15}$$

Property 8.2 (Eligible Edge Set). Let $E(k) = \{(t,j) \mid a_{tj}(k) = 1, \forall j \in V\}$; we have

$$E(k) = E_{qt} \tag{8.16}$$

The set $E(k)$ can easily be obtained from $A(k)$, which takes recursive form and is easy to program. The basic idea of the path growth procedure is to extend path P_t^k by adding a new edge from set $E(k)$ at each step. We may take a risk that the path developed with this approach may suspend at a node; that is, we cannot reach final end node n from this node. Such a node is called a pendant node.

Definition 8.3 (Pendant Node). For a given partial path P_t^k with $t \neq n$, the terminal node t is a pendant node if all of its directly adjacent nodes belong to set V_p.

The following property gives the necessary condition for a pendant node.

Property 8.3 (Pendant Node). For a given partial path P_t^k, the terminal node t is a pendant node if and only if (1) $E(k) = \emptyset$ and (2) $t \neq n$.

In fact, there are some special nodes in V_q which are essentially unable to reach the final end node n. If one such node is appended to the partial path, it must mislead the path to a pendant node. The following definitions and theorems give us a means to distinguish such types of nodes from V_q.

Definition 8.4 (Dead Node). Given a partial path P_t^k with $t \neq n$, for a node $i \in V_q$, let $P(i)$ denote the set of all possible paths from node i to final end node n and V_{l_i} the set of nodes contained in a path l_i. Node i is a dead node if

$$V_p \cap V_{l_i} \neq \emptyset, \qquad \forall l_i \in P(i) \tag{8.17}$$

In other words, a node is a dead node if any of the possible paths from it to node n must contain at least one node in V_p.

The dead node can be identified by means of reachability matrix [534]. Let A be an adjacent matrix for a given graph $G = (V,E)$. The kth-order adjacent matrix of A is defined by the following boolean multiplication:

$$A^k = A^{k-1}A, \qquad A^0 = I \tag{8.18}$$

The kth-order adjacent matrix describes the relation between two nodes i and j that are not directly connected, but we can reach node j from i through a path of length $k - 1$. Node i is kth-order adjacent to node j if $a_{ij}^k = 1$.

Let $E_p = \{(i,j) \mid i \in V_p, j \in V\}$, which is the set of all edges incident to nodes in P_t^k. We can define the graph $G(k) = (V - V_p, E - E_p)$, which is a subgraph of graph G with removal of all nodes in V_p and all edges incident to V_p. The reachability matrix $R(k)$ for graph $G(k)$ is defined as follows:

Definition 8.5 (Reachability Matrix). The reachability matrix $R(k)$ with respect to graph $G(k)$ is given as follows:

$$R(k) = [A(k) + I]^{n-k-2}, \qquad k < n - 2 \tag{8.19}$$

where I is the identity matrix and $A(k)$ is the dynamic adjacent matrix with respect to path P_t^k.

The reachability matrix describes the relation between any two nodes when they are adjacent by at most $(n - k - 2)$th order. Then we have the following property for identifying a dead node.

Property 8.4 (Dead Node). For a given partial path P_t^k, a node i in set V_q is a dead node if and only if $r_{in}(k) = 0$.

Theoretically, we can check if a node is a dead node with Property 8.4 at each step as the procedure progresses and remove all dead nodes from set V. For a large problem, such checking carries a significant computational cost. We know that the evolutionary approach has the advantage that it allows infeasible chromosomes to enter the population pool and evolve with feasible chromosomes by means of a penalty mechanism. The infeasible chromosomes may provide some useful genes in the evolutionary process. So we do not check the dead nodes at each step, but instead, simply make a virtual connection between the terminal dead node and the final end node n with either a mild or a severe penalty. The following implementation of path growth procedure is based on this consideration:

Procedure: Path Growth

Step 1. Initialize. Let $k \leftarrow 0$, $V_p^k \leftarrow \{1\}$, and $A(k) \leftarrow A$, $t^k \leftarrow 1$.

Step 2. Perform the termination test. If $t^k = n$, go to step 9; otherwise, continue.

Step 3. Determine the eligible edge set. Make the edge set $E(k)$ according to dynamic incidence matrix $A(k)$.

Step 4. Perform the pendant node test. If $E(k) = \emptyset$, let $t^{k+1} \leftarrow n$, $V_p^{k+1} \leftarrow V_p^k \cup \{n\}$, give a penalty for virtual connection (t^k,n), and go to step 9; otherwise, continue.

Step 5. Extend the path. Select t^{k+1} from V_q^k subject to $(t^k,t^{k+1}) \in E(k)$ with the highest priority.

Step 6. Perform the mesh matrix update. Make a new mesh matrix $M(k + 1) \leftarrow M(k) \cap U(k + 1)$.

Step 7. Perform the dynamic incident matrix update. Let $A(k + 1) \leftarrow A(k) \cap M(k + 1)$.

Step 8. Perform the iteration index update. $k \leftarrow k + 1$ and go to step 2.

Step 9. Return the complete path.

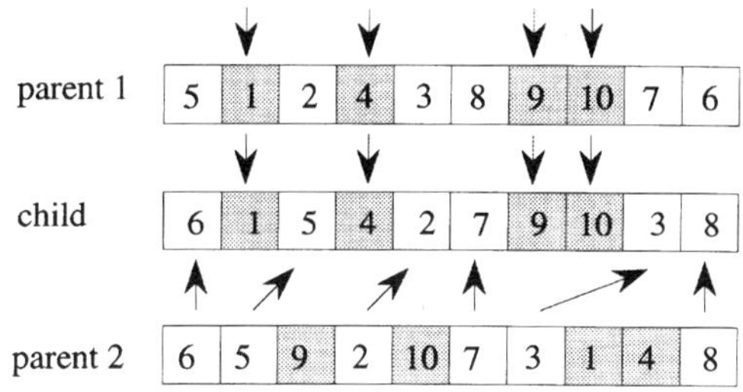

Figure 8.4. Position-based crossover operator.

Genetic Operators. The nature of the proposed encoding is a type of permutation representation. A number of recombination operators have been investigated for permutation representation during the past three decades. Here the position-based crossover operator proposed by Syswerda [604] was adopted. It can be viewed as a type of uniform crossover operator for integer permutation representation together with a repair procedure as shown in Figure 8.4. Essentially, it takes some genes from one parent at random and fills the vacuum position with genes from the other parent by a left-to-right scan. The swap mutation operator was used here, in which two positions are selected at random and their contents are swapped as shown in Figure 8.5.

During each iteration of a genetic algorithm, chromosomes are evaluated using a measure of fitness. There are three major steps included in the evaluation:

1. Convert the chromosome to a path.
2. Calculate the objective values.
3. Convert the objective values to fitness values.

The roulette wheel approach, a type of fitness-proportional selection, was adopted as the selection procedure. The elitist method was combined with this approach to preserve the best chromosome in the next generation and overcome stochastic errors of sampling. With the elitist selection, if the best individual in the current generation is not reproduced in the new generation, one individual is removed randomly from the new population and the best one is added to the new population.

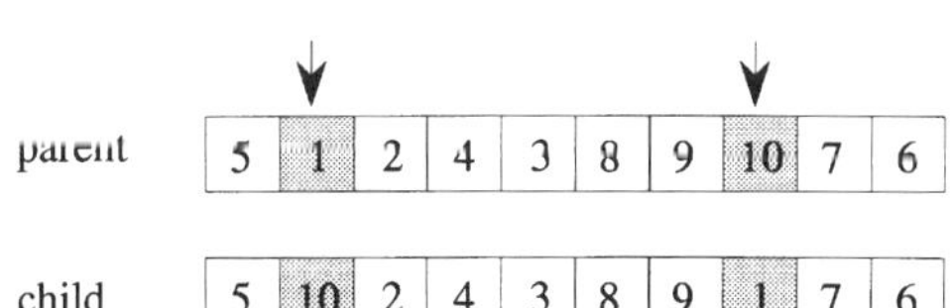

Figure 8.5. Mutation operator.

Compromise Approach–Based Fitness Assignment. As discussed in Sections 3.2 and 3.8, the compromise approach can be regarded as a type of mathematical formulation of goal-seeking behavior in terms of a distance function. The compromise approach identifies solutions that are closest to ideal as determined by following weighted L_p-norm:

$$r(z;p,w) = \|z - z^*\|_{p,w}$$
$$= \left[\sum_{j=1}^{2} w_j^p |z_j - z_j^*|^p \right]^{1/p}$$

where $z^* = (z_1^*, z_2^*)$ is the ideal solution to the problem. The parameter p is used to reflect the emphasis of decision makers. When set at $p = 1$, the sum of regrets of all objectives is emphasized; when set at $p = \infty$, the individual objective regret is emphasized.

For the bicriteria shortest path problem, the ideal point can be obtained easily by solving two single criterion problems using existing methods. For many complex problems, to find an ideal point is also difficult. To overcome the difficulty, the concept of a proxy ideal point can be suggested to replace the ideal point. The proxy ideal point is the ideal point corresponding to the current generation but not to a given problem. In other words, it is calculated in the partial solution space explored but not in the entire solution space. The proxy ideal point is easy to obtain in each generation. Along with evolutionary process, the proxy ideal point will gradually approximate the real ideal point. Let P denote the set of current population. The proxy ideal point $(z_{\min}^1, z_{\min}^2)$ can be calculated as follows:

$$z_{\min}^1 = \min\{z^1(\boldsymbol{x}) | \boldsymbol{x} \in P\} \tag{8.20}$$

$$z_{\min}^2 = \min\{z^2(\boldsymbol{x}) | \boldsymbol{x} \in P\} \tag{8.21}$$

Since the smaller the regret value, the better the individual, we have to convert the regret value into the fitness value to ensure that a fitter individual has a larger fitness value. Let $r(\boldsymbol{x})$ denote the regret value of individual $\boldsymbol{x}$, $r_{\max}$ the largest regret value, and $r_{\min}$ the smallest regret value in the current generation. The transformation is given as follows:

$$\text{eval}(\boldsymbol{x}) = \frac{r_{\max} - r(\boldsymbol{x}) + \gamma}{r_{\max} - r_{\min} + \gamma} \tag{8.22}$$

where γ is a positive real number that is usually restricted within the open interval (0,1). The purpose of using it is twofold: (1) to prevent equation (8.22) from zero division and (2) to make it possible to adjust the selection behavior from fitness-proportional selection to pure random selection.

The overall procedure is summarized as follows:

TABLE 8.1. Results of Random Test Problems

	Ideal Solution	$w_1 = 0.5$, $w_2 = 0.5$	$w_1 = 0.2$, $w_2 = 0.8$	$w_1 = 0.8$, $w_2 = 0.2$
L_1-norm	(861, 702)	(974, 754)	(1112, 712)	(930, 826)
L_2-norm	(861, 702)	(974, 754)	(974, 754)	(930, 826)
L_∞-norm	(861, 702)	(1243, 702)	(1243, 702)	(861, 1236)

Overall Procedure for the Genetic Algorithm

Step 0. Set genetic parameters and read in the data of a given instance.
Step 1. Generate the initial population randomly.
Step 2. Decode the chromosomes into paths.
Step 3. Calculate the objectives for each decoded individual.
Step 4. Calculate the proxy ideal point in current population.
Step 5. Calculate the regret value for each individual.
Step 6. Convert the regret value to the fitness value.
Step 7. Select the next generation using the roulette wheel method.
Step 8. If the maximal generation is reached, stop; otherwise, go to step 9.
Step 9. Produce offspring with crossover and mutation and then go back to step 2.

8.2.3 Numerical Examples

The primary computational experiments were conducted on a randomly generated test problem by Cheng and Gen [109]. It is a nonplanar and undirected graph with 100 nodes and 473 edges. Each edge is associated with two numbers: time and cost. The evolutionary environment is set as follows: population size = 40, maximum generation = 1000, crossover ratio = 0.4, and mutation ratio = 0.2.

The following Pareto solutions in one run were obtained: (861, 1236), (875, 1140), (878, 1115), (891, 1105), (894, 1080), (907, 931), (926, 904), (927, 851), (930, 826), (946, 824), (949, 799), (974, 754), (1096, 740), (1112, 712), and (1243, 702). The ideal solution and compromise solutions for various weights are given in Table 8.1, where the ideal solution is obtained by running the Floyd–Warshall algorithm twice, each time on one objective [134]. The regret functions used are the L_1-norm, L_2-norm, and L_∞-norm, respectively. The evolutionary process for a proxy ideal solution is depicted in Figure 8.6.

8.3 ADAPTIVE NETWORK ROUTING

A routing algorithm determines a route for a communication packet by employing topological information on networks and finds a route for each packet

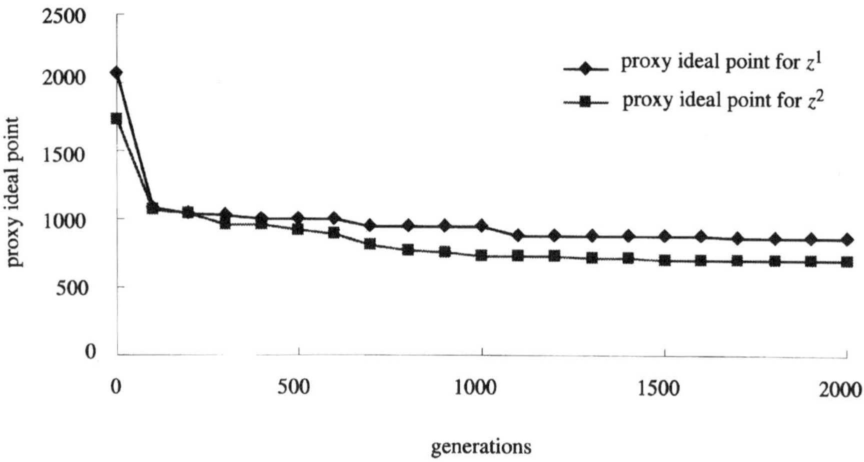

Figure 8.6. Evolutionary process of proxy ideal solution.

from its source node to its destination. Rapid advancement of the Internet has increased the importance of adaptive network routing algorithms. In the earlier history of the Internet, conventional vector-distance routing algorithms based on a hop count matrix were usually employed. There are several such types of routing algorithms, including the routing information protocol (RIP), shortest-path-first protocol (SPF), and others [474]. Most of these algorithms do not observe communication latency along the routes, which depend only on a hop count matrix to find the shortest routes. The RIP [292] is commonly used even now in small local area networks (LANs). The algorithm yields much communication overhead in larger networks because they periodically send broadcast messages that contain a routing table to all the nodes in the network, which greatly degrades the total performance of the network. To reduce communication overhead, routing algorithms based on link status information exchange, such as SPF [132], send broadcast messages that only contain information on link status. Based on a topological database generated from the information on link status, the algorithm calculates shortest paths employing Dijkstra's shortest path algorithm [164] in each node. This algorithm can reduce communication overhead because it broadcasts information only on the link status, not on routing tables in small network systems. However, this algorithm also has the defect that it leads to increased total overhead in larger networks.

The open shortest-path-first protocol (OSPF), which is based on the SPF, is a widely used network routing protocol for interior gateway protocol (IGP) [469]. For exterior gateway protocols, broader gateway protocols such as the BGP4 [507] have commonly been used on the Internet. The BGP4 employs a source routing strategy that holds complete paths to all destinations, which is not scalable either. The BGP4 (or other EGP) avoids inflation of routing tables by grouping a set of gateways as an autonomous system (AS) and assigning them AS numbers, which are limited from 0 to 65,535.

In recent years there have been some reports of genetic algorithms applied to solving network routing problems, which are maximum flow problems in which the flow is maximized by using global information on the network. Another application of genetic algorithms to routing in a network, given by Cox [136], solves a constrained optimization problem that allocates link bandwidth to each connection by employing information regarding the link bandwidth of the network. Carse, Fogarty, and Munro applied a fuzzy classifier system (FCS) to a distributed routing problem on packet-switching networks in which each packet is routed independently using a routing table [85]. This routing algorithm makes decisions as to whether a packet should take a direct route or an indirect route by using a FCS.

Munetomo, Takao, and Sato proposed an adaptive routing algorithm for a packet-switching network such as the Internet which attempts to minimize communication latency by observing route delays [474]. The routing algorithm maintains a limited number of alternative routes that are frequently used and also tries to balance load of links by distributing packets among the alternative routes in its routing table. Munetomo, Takao, and Sato employed a genetic algorithm to construct a routing table that is a population of strings, each of which represents a route.

8.3.1 Genetic-Based Adaptive Routing

The genetic-based adaptive routing (GAR) algorithm employs genetic path operators to create alternative routes in its routing table. The GAR algorithm is based on a source routing approach to effectively utilize the similarity among routes in a routing table. A source routing algorithm determines a complete route for a packet in its source node, so a packet needs to contain information on all the nodes along a route. The Internet protocol has a source routing option specifying whether or not a source routing is used [66, 132]. Data packets do not usually use this option and only a next hop is determined by looking up a routing table based on the shortest paths of the network. It is not obvious but does not seem difficult to create a routing table specifying only the next hop from the routes generated by the GAR algorithm. The important features of the GAR algorithm can be summarized as follows.

- The GAR algorithm is a source routing algorithm in which each route in a routing table contains all the nodes along the route.
- For each destination node to which packets are sent, there is a set of alternative routes.
- Each route has its weight value, which specifies the probability that the route will be selected from the alternative routes. This weight corresponds to a fitness value for the genetic operators.

TABLE 8.2. Example of a Routing Table[a]

Destination	Route	Frequency	Delay	Weight
2	(1 3 2)*	7,232	50	0.7
	(1 3 4 2)	2,254	60	0.2
	(1 3 4 5 2)	1,039	70	0.1
6	(1 8 6)*	20,983	100	0.4
	(1 10 11 6)	34,981	105	0.6
8	(1 8)*	30,452	40	0.9
	(1 7 8)	3,083	40	0.1

[a]Asterisks denote a default route.

- Weight values of routes are calculated from the communication latency observed by sending delay query packets.
- As an initial route, a default route is generated using Dijkstra's shortest path algorithm based on a hop count metric.
- Alternative routes in a routing table are generated by applying genetic operators such as mutation and crossover, which are invoked at a specified probability after every evaluation of weight values.
- To prevent overflow of a routing table, its size is reduced by applying selection operators. A route with the smallest weight value among the routes to the same destination is deleted. Moreover, all the routes to a destination are deleted when the smallest number of packets is sent to the destination among all the destinations in the routing table.

8.3.2 Representation of Chromosomes

A path (route) is encoded by listing a node from its source to its destination based on the topological database of a network. For example, a path from node 0 to node 9 is encoded into a list of nodes along the path: (0 12 5 8 2 9). If a path cannot be realized on the network, it cannot be encoded into a chromosome, which means that each step in a path must pass through a physical link in the network.

Table 8.2 is a routing table used in a routing algorithm. The table consists of five entries: destination, route, frequency, delay, and weight. The destination entry indicates the destination node of packets. For each destination there are a set of alternative routes. A route entry is a list of nodes along the route. The frequency entry specifies the number of packets sent to the destination by the route. The delay entry shows the value of communication latency of packets sent along the route. The weight entry of a route specifies the probability that a route will be selected when a packet is sent.

Routes marked with asterisks are the default routes that are the shortest paths in terms of hop-count metric. In initial status, the routing table is empty. When a packet is generated at a node, a default route is generated by Dijkstra's algorithm and is inserted in the routing table. After sending a specified number of packets along a route, a packet is sent to evaluate communication latency of the route. After receiving the packet's answer, genetic path operators are applied to general alternative routes at a specified probability.

8.3.3 Evaluation of Chromosomes

Evaluation of weight value (fitness value) of routes is based on the communication latency along the routes. A delay query message is sent at a specified interval to observe delay along the route. Using the delay value obtained, evaluation of the weight value of routes can be calculated as follows:

$$\text{eval}(v_k) = \frac{1/D_k}{\sum_{i \in S} 1/D_i} \tag{8.23}$$

where $D_i = d_i / \sum_{j \in S} d_j$; d_i is the delay for route i and S is a set of routes of the same destination. Equation (8.23) implies that a smaller value of d_i generates a larger evaluation value. By the evaluation method, a route with less communication latency is frequently employed in sending packets.

8.3.4 Genetic Operators

The path crossover operator exchanges subroutes between two chromosomes. The chromosomes should have the same source and destination nodes to apply the crossover. Crossing sites for the path crossover operator are limited to the nodes contained in both chromosomes. A node is randomly selected as a crossing site from the potential crossing sites and exchange subroutes. When applying the crossover operator to a pair of chromosomes V_1 and V_2, the operation proceeds according to the following procedure:

Procedure: Path Crossover Operator

Step 1. List a set of nodes N_c included in both V_1 and V_2 (excluding source and destination nodes) as potential crossing sites.

Step 2. Select a node i as a crossing site from the N_c.

Step 3. Cross over the chromosomes by exchanging all nodes after the crossing site i.

Figure 8.7 shows an overview of the operator applying for a pair of parents V_1 and V_2 from node 0 to node 20. Their potential crossing sites are nodes 7, 11, and 15. When we select node 11 as a crossing site, the new offspring are generated by exchanging the subroutes as shown in Figure 8.7.

When a common node in a pair of chromosomes does not exist, a crossing site is not able to select; therefore, it is impossible to perform the crossover

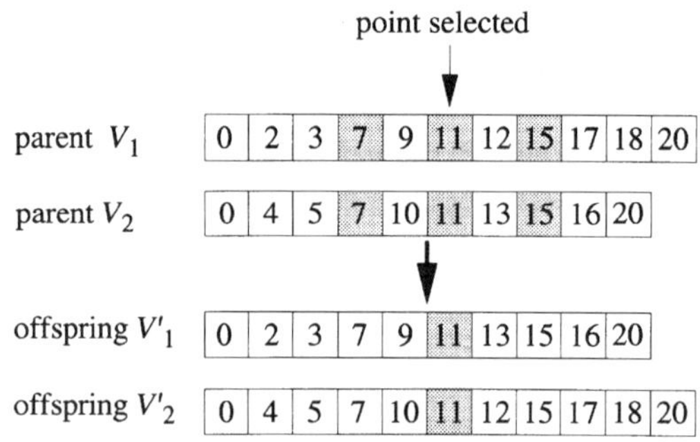

Figure 8.7. Example of crossover.

operator. A pair of parents without similarity is not worth crossing over because the operator may generate routes far from their parents, which means random regeneration of routes.

The path mutation operator generates an alternative chromosome from a chromosome. To perform a mutation, first, a node is randomly selected from the chromosome, which is called a mutation node. Then another node is randomly selected from the nodes directly connected to the mutation node. Finally, according to Dijkstra's shortest-path algorithm, an alternative route is generated by connecting the source node to the selected node and the selected node to the destination. The path mutation operator is described as follows:

Procedure: Path Mutation Operator

Step 1. Select a mutation node i randomly from all nodes in parent V.

Step 2. Select a node j from the neighbors of the mutation node i, where $j \in B$ and B is a set of neighbor nodes of mutation node i.

Step 3. Generate a shortest path r_l from the source node to j and another shortest path r_2 from j to the destination.

Step 4. If any duplication of nodes exists between r_l and r_2, discard the routes and do not perform a mutation. Otherwise, connect the routes to make up a mutated chromosome.

Figure 8.8 shows an example of a mutation operator. First, suppose that node 7 is selected as a mutation node. Second, node 8 is randomly selected from the neighbors of the mutation node. Third, by Dijkstra's shortest path algorithm, we connect the source node 0 and the selected node 8 to generate a subpath, r_1, and connect node 8 and the destination node 15 to generate another subpath, r_2. We finish the mutation by connecting the subroutes r_1 and r_2. An offspring V' isn't generated if any duplication of nodes exists between r_1 and r_2. This is because we need to avoid creating any loop in a route.

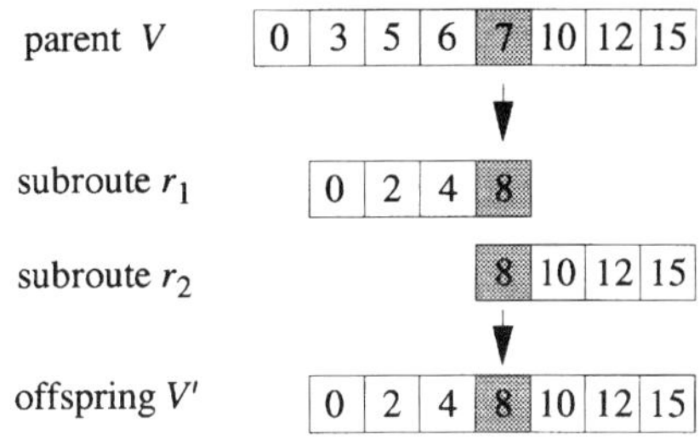

Figure 8.8. Example of mutation.

Overall GAR Algorithm. The details of Munetomo, Takai, and Sato's routing algorithm are described below [474]. Each node executes this algorithm independently to determine routes of packets. Each packet has entries such as *type, route,* and *next,* where *type* specifies type of packet, *route* is the route of the packet, and *next* indicates the next hop of the packet in its route. Types of packets are *DataPacket* (packet that contains data), *DelayRequest* (packet for requesting delay of a link), and *DelayAnswer* (packet for *DelayRequest* answer). *DelayRequest* and *DelayAnswer* packets have *DelayEntry,* which records the communication latency of the packets.

If the type of packet arriving at a node is *DataPacket,* the node forwards the packet according to its routing table. If a packet is created at a node, the node determines a route for the packet based on its routing table. If the routing table does not contain a route for the destination of the input packet, a default route is generated by employing Dijkstra's shortest path algorithm, which is to be inserted to the table. In initial state of the algorithm, the routing table does not contain any route. Generate routes only for destinations, to which communication packets are frequently sent.

At a specified interval of sending packets along a route, a *DelayRequest* packet is sent to observe communication latency along the route. If the packet arrived at its destination, a *DelayAnswer* packet is then sent back. When the answer packet is received, the communication latency of the route is obtained by calculating the average of the time from sending a *DelayRequest* packet to receiving a *DelayAnswer* packet.

```
Overall GAR Algorithm
begin
 initialize routing table;
  while not terminated do
  begin
   wait for a packet to be received or input from user;
   if packet.type = DataPacket then
   begin
    if the packet is sent from other node then
```

```
   begin
    if packet.destination = this node then
     receive packet;
    else
     send the packet to the node of packet.next;
   end
   else
   begin
    if routing table for the destination is empty then
    begin
     create a default route by using Dijkstra's
      algorithm;
     add the default route to the routing table;
     reset the frequency entry of the default route;
     packet.route ← default route;
     if the number of destination > limit then
      delete routes of a destination least frequently
       used;
    end
    else
    begin
     select a route from the routing table with
      roulette wheel selection;
     increment the frequency entry of the route
      selected;
     packet.route ← the route selected;
     if route.frequency mod EvaluationInterval = 0
      then
     begin
      packet.type ← DelayRequest;
      packet.route ← route;
      send the packet according to the route;
     end
    end
   end
  end
  if packet.type = DelayRequest then
  begin
   packet.DelayEntry ← CurrentTime -
    packet.CreateTime;
   packet.CreateTime ← CurrentTime;
   packet.type ← DelayAnswer;
   send packet to its source;
  end
  if packet.type ← DelayAnswer then
```

```
        begin
         packet.DelayEntry ← packet.DelayEntry + CurrentTime
          - packet.CreateTime;
         packet.delayentry ← packet.DelayEntry/2;
         change delay of the route based on
          packet.DelayEntry;
         if random < p_M then
         begin
          apply a mutation to a chromosome in the routing
           table;
          add the string created to the routing table;
          if table size > limit then
           delete a string of the lowest weight;
         end
         if random < p_C then
         begin
          apply a crossover to a pair of chromosomes in the
           routing table;
          add the string created to the routing table;
          if table size > limit then
           delete a string of the lowest weight;
         end
        end
      end
    end
```

After receiving a route delay, weight values of routes of the same destination are evaluated, according to equation (8.23). After every evaluation of weights, genetic operators are applied at a specified probability to create alternative routes in the routing table. If the size of the routing table exceeds a limit, the selection operation is performed to reduce its size. We have two types of selection operator:

1. *Local selection:* deletes a route with the smallest weight among routes of a same destination
2. *Global selection:* deletes all the routes of a destination of which the sum of frequencies is the smallest among all the destinations in the routing table

8.3.5 Numerical Examples

Munetomo, Takai, and Sato have done some experiments using a network simulator based on a discrete event simulation [474]. They also compared the GAR algorithm with conventional routing algorithms such as RIP and SPF.

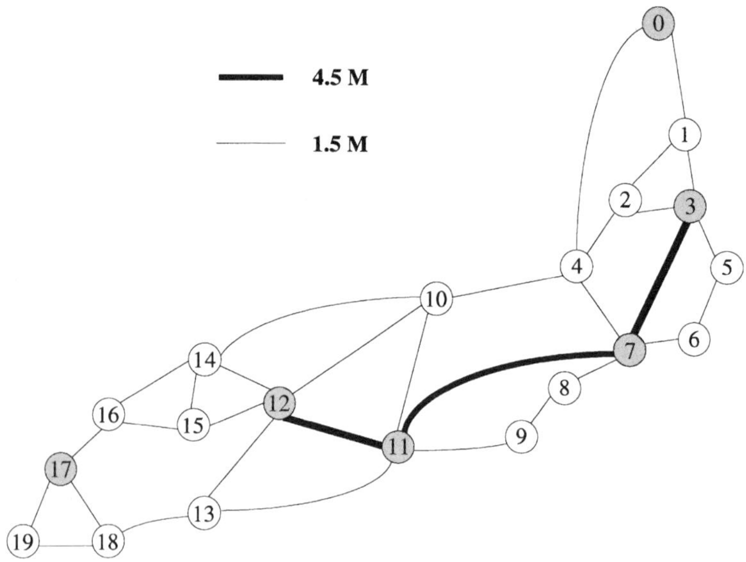

Figure 8.9. Sample network.

An example network is depicted in Figure 8.9, which is created from geographical information from major Japanese cities. The bandwidth of a link is represented by the thickness of the line. The thicker link is 4.5 Mbps, and the thinner link is 1.5 Mbps. The packets are randomly generated at a specified interval, which is exponentially distributed. The destination of a packet is randomly selected from the limited nodes, which are denoted by shaded circles.

Simulation conditions are given as follows: the fitness values for a chromosome (route) are evaluated at every tenth packet. The probability of applying a mutation operator is $p_m = 0.1$ and of applying a crossover operator is $p_c = 0.05$. The population size limit is 100. A simulation is performed for 3000 seconds and the following results are obtained after the simulation. To compare the mean response times for data packets, the mean value of time to arrive at destinations of packets from their creation on the source nodes, we can change frequency of creation of data packets by changing mean creation interval of data packets, which is exponentially distributed.

From the results we can see that the mean response time of packets is smallest when the GAR algorithm is employed for routing. SPF is slightly better than RIP. Especially for 2000-ms intervals of packet creation, which leads to heavily loaded links in the network, GAR achieves about 20% of mean response time compared with RIP and SPF. This means that packets sent by GAR arrive at their destinations five times faster than other algorithms.

Table 8.3 shows a routing table generated in node 0 after one simulation with 2200-ms arrival intervals for packet creation. For destination node 12,

TABLE 8.3. Routing Table in Node 0 After Experiment

Destination	Route	Delay	Weight
3	(0 1 3)	554	0.802636
	(0 4 2 3)	2253	0.197364
7	(0 4 7)	5052	1.000000
11	(0 4 7 11)	4423	0.533488
	(0 1 3 7 11)	5058	0.466512
12	(0 4 7 11 12)	2941	0.564116
	(0 4 10 12)	6210	0.267160
	(0 1 2 4 10 12)	9833	0.168724
17	(0 4 10 14 16 17)	2859	1.000000

the best route is (0 4 7 11 12). Another route (0 4 10 12) is the shortest path in the hop count metric but not the path with minimum delay because the links from node 7 to 11 and from 11 to 12 have more bandwidth than the other links. This result also shows that load balancing among alternative routes is realized by distributing packets probabilistically based on their weight values. Figures 8.10 to 8.12 show the load status of links during simulation. The width of each link represents log-scaled mean queue length for the link. Because the width is log-scaled, a thick link means highly overloaded. In the RIP result, a path from node (11–13–18) becomes extremely heavily loaded. On the other hand, a path from node (11–12–15–16–17–18), which is the alternative path (11–13–18), is not frequently used at all and

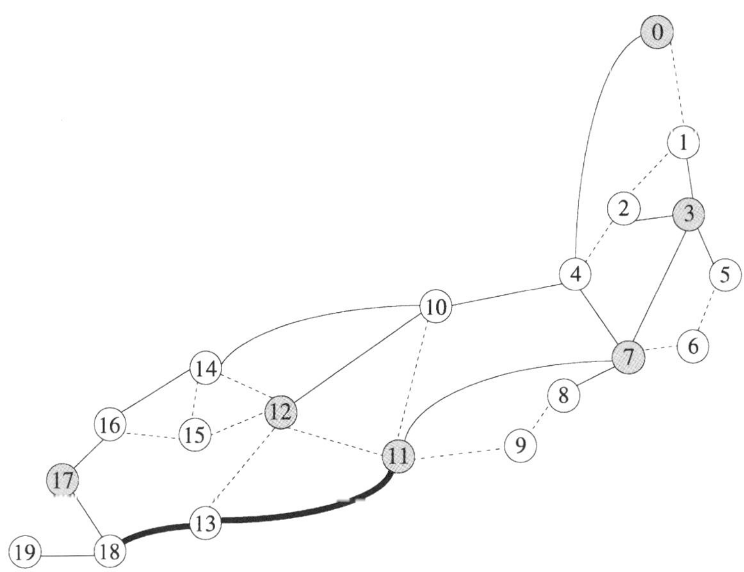

Figure 8.10. Load status of links (RIP).

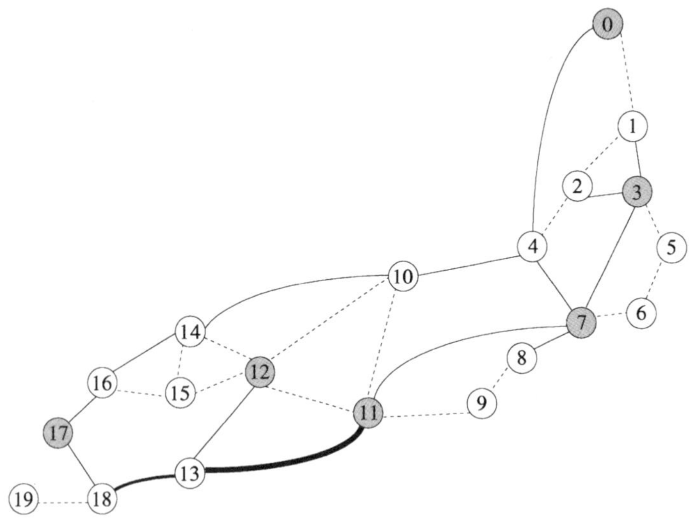

Figure 8.11. Load status of links (SPF).

links for the path become lightly loaded. By employing SPF, the load of the links is able to slightly reduce but the result is essentially the same as that of RIP. On the other hand, employing GAR algorithm, the load of links especially for heavily loaded ones is able to greatly reduce. This is not only because GAR algorithm calculates paths which minimizes communication latency but also because it distributes packets among alternative paths in the routing table, which leads to reduce load of links of heavily loaded. The

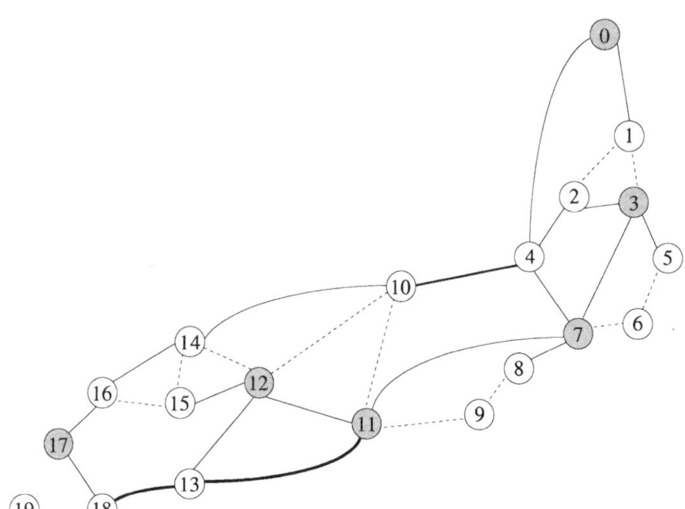

Figure 8.12. Load status of links (GAR).

results above show that the mean response time for packets can be made much smaller, especially in a heavily loaded network, as follows:

1. Take account of communication latency of routes.
2. Generate alternative routes employing genetic operators.
3. Distribute packets among the alternative routes so as to achieve load balancing among them.

8.4 CENTRALIZED NETWORK DESIGN

A centralized network is a network where all communication is to and from a single site [357]. In such networks, terminals are connected directly to the central site. Sometimes multipoint lines are used, where groups of terminals share a tree to the center and each multipoint line is linked to the central site by one link only. This means that optimal topology for this problem corresponds to a tree in a graph $G(V,E)$ with all but one of nodes in V corresponding to the terminals. The remaining node refers to the central site, and edges in E correspond to the feasible telecommunication wiring. Each subtree rooted in the central site corresponds to a multipoint line. Usually, the central site can handle, at most, a given fixed amount of information in communication. This, in turn, corresponds to restricting the maximum amount of information flowing in any link adjacent to the central site (which we will refer to as the root of the graph G) to that fixed amount. In the combinatorial optimization literature, this problem is known as the *capacitated minimum spanning tree problem.*

The problem has been shown to be NP-hard by Papadimitriou [502]. Much of the early works focused on heuristic approaches to find good feasible solutions. Among them are those by Chandy and Lo [94], Kershenbaum [356], and Elias and Ferguson [175]. The only full optimization algorithms that we are aware of are by Gavish [210] and Kershenbaum, Boorstyn, and Oppenheim [358], but their use is limited to problems involving up to 20 nodes. Gavish [211] also studied a new formulation and its several relaxation procedures for the capacitated minimum directed tree problem. Recently, this problem has elicited additional interest with the cutting plane algorithms of Gouveia [261] and Hall [277] and the branch-and-bound algorithm of Malik and Yu [434]. Zhou and Gen proposed a genetic algorithm approach to dealing with the problem [696, 699]. A tree-based encoding method was adopted to code candidate solutions of the problem.

8.4.1 Problem Formulation

Consider a complete, undirected graph $G = (V,E)$, and let $V = \{1,2,...,n\}$ be the set of nodes representing terminals. Denote the central site or "root" node as node 1, and let $E = \{(i,j)|i,j \in V\}$ be the set of edges representing all possible telecommunication wiring. For a subset of nodes S ($\subseteq V$), define $E(S) = \{(i,j)|i,j \in S\}$ as the edges whose endpoints are both in S. Define the following binary decision variables for all edges $(i,j) \in E$:

$$x_{ij} = \begin{cases} 1, & \text{if edge } (i,j) \text{ is selected} \\ 0, & \text{otherwise} \end{cases}$$

Let c_{ij} be the fixed cost with respect to edge (i,j) in the solution, and suppose that d_i represents the demand at each node $i \in V$, where by convention the demand of the root node $d_1 = 0$. Let $d(S)$, $S \subseteq V$ denote the sum of the demands of nodes of S. The subtree capacity is denoted as κ.

The centralized network design problem can be formulated as follows [210]:

$$\min \quad z = \sum_{i=1}^{n-1} \sum_{j=2}^{n} c_{ij} x_{ij} \tag{8.24}$$

$$\text{s.t.} \quad \sum_{i=1}^{n-1} \sum_{j=2}^{n} x_{ij} = 2(n-1) \tag{8.25}$$

$$\sum_{i \in S} \sum_{\substack{j \in S \\ j>1}} x_{ij} \le 2(|S| - \lambda(S)), \qquad S \subseteq V \backslash \{1\}, \quad |S| \ge 2 \tag{8.26}$$

$$\sum_{i \in U} \sum_{\substack{j \in U \\ j>1}} x_{ij} \le 2(|U| - 1), \qquad U \subset V, \quad |U| \ge 2, \quad \{1\} \in U \tag{8.27}$$

$$x_{ij} = 0 \text{ or } 1, \qquad i = 1,2,\ldots,n-1, \quad j = 2,3,\ldots,n \tag{8.28}$$

Equality (8.25) is true of all spanning trees: A tree with n nodes must have $n - 1$ edges. Inequality (8.27) is a standard inequality for spanning trees: If more than $|U| - 1$ edges connect the nodes of a subset U, the set U must contain a cycle. The parameter $\lambda(S)$ in inequality (8.26) refers to the *bin-packing number* of set S, namely, the number of bins of size κ needed to pack the nodes of items of size d_i for all $i \in S$. These constraints are similar to those for inequality (8.27), except that they reflect the capacity constraint: If the set S does not contain the root node, the nodes of S must be contained in at least $\lambda(S)$ different subtrees of the root.

When the demands of all nonroot nodes are 1, inequality (8.26) can be expressed more simply as

$$\sum_{i \in S} \sum_{\substack{j \in S \\ j>1}} x_{ij} \le |S| - \left\lceil \frac{|S|}{\kappa} \right\rceil, \qquad S \subseteq V \backslash \{1\}, \quad |S| \ge 2 \tag{8.29}$$

since items of unit size can always be packed in $\lceil S|/\kappa \rceil$ bins or subtrees.

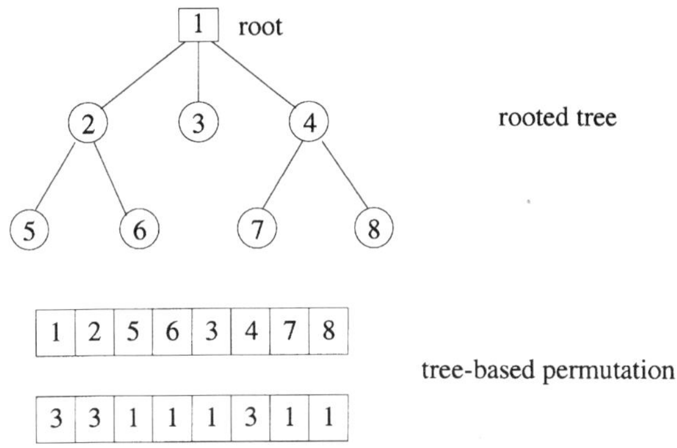

Figure 8.13. Rooted tree and its tree-based permutation.

Up to now, all heuristic algorithms for this problem are focused only on how to deal with the constraints to make the problem simpler to solve. In the cutting plane algorithms [261, 277] or branch-and-bound algorithm [434], the network topology of the problem is usually neglected. As a result, it leads to an exponential explosion of constraints.

8.4.2 Genetic Algorithm Approach

To solve centralized network design problems using genetic algorithms a tree-based permutation encoding method is adopted to encode the candidate solutions. This is illustrated in Figure 8.13.

Especially for the root node, its degree should take no less than the following value:

$$\left\lceil \frac{|V|}{\kappa} \right\rceil$$

which reflects the number of subtrees connected to the root node to satisfy the capacity constraint. For the initial population, each chromosome can be generated randomly except that node 1 should always be the first gene of the node dimension.

The same genetic operations as explained for the degree-constrained spanning tree problem in Section 2.5 are used here. It is necessary to modify infeasible solutions generated in the evolutionary process in the sense of violating the capacity constraint on the rooted node. Especially when the demands of all terminals are equal to 1, the problem is reduced to finding a rooted spanning tree in which each subtree of the root node contains at most κ nodes. Therefore, before evaluation, if there are such individuals whose subtrees violate the capacity constraint, the insertion mutation operation is used again to insert an extra branch on a subtree into another subtree with fewer nodes. The evaluation procedure is described as follows:

Step 1. Convert an individual into a tree solution according to the decoding procedure in Section 2.5 for the degree-based permutation encoding method.

Step 2. Calculate the total cost of solutions according to the objective function in problem (8.24) and take the reciprocal value of it as the fitness value of that individual.

Step 3. Repeat the procedure for all individuals.

The $(\mu + \lambda)$-selection strategy was adopted [30]. To avoid premature convergence of the evolutionary process, the selection strategy selects only μ different best individuals from μ parents and λ offspring. If no μ different individuals are available, the vacant population pool is filled with renewal individuals generated in the same way as for the initial population.

8.4.3 Numerical Example

Zhou and Gen tested their ideas on the instance given by Gavish [212]. The example consists of 16 nodes, unit traffic between each node and node 1, and capacity restriction $\kappa = 5$. The cost matrix is given in Table 8.4.

The parameters of the genetic algorithm were set as follows: population size = 200, mutation probabilities of mutation operations $p_m = 0.3$, and maximum generation = 500. Gavish used an augmented Lagrangian algorithm to solve this problem and got the optimal solution 8526 [212]. The same optimal solution is obtained using the genetic algorithm. Its corresponding topology of tree is illustrated in Figure 8.14.

8.5 COMPUTER NETWORK EXPANSION

The advent of low-cost computing devices has led to explosive growth in computer networks. There are several benefits from the distribution of computing across a network. One of the major advantages of computer networks over the centralized systems is their potential for improved system reliability. The reliability of a system depends not only on the reliability of its nodes and communication edges, but also on how nodes are connected by communication edges (i.e., the topology of a network). The network topology can be represented by an undirected graph $G = (N,E)$ with nodes N and edges E and can be characterized by network reliability, message delay, or network capacity. These performance characteristics depend on many properties of graphs, such as diameter, average distance, number of ports at each node (degree of a node), and number of edges: The higher the diameter, the larger the delay in the network; reliability increases with decreased diameter and average distance, and decreases with a decrease in the number of edges.

TABLE 8.4. Cost Matrix for the Numerical Example ($n = 16, \kappa = 5$)

$i \backslash j$	2	3	4	5	6	7	8	9	10	11	12	13	14	15	16
1	1616	1909	246	622	829	1006	2237	399	1717	632	1191	2116	824	1336	1519
2		2996	1419	2217	1213	2046	3753	1516	1180	1997	552	3622	2423	1367	862
3			1893	1543	1792	2785	1362	1667	3556	2332	2446	1248	1508	3233	3287
4				799	593	1188	2369	242	1670	857	962	2243	1004	1348	1425
5					1230	1253	1625	761	2301	801	1748	1509	206	1873	2119
6						1758	2597	480	1883	1449	663	2463	1420	1701	1573
7							2703	1399	1470	454	1849	2612	1350	960	1470
8								2238	3922	2304	3231	137	1442	3476	3743
9									1889	1029	1009	2108	959	1586	1628
10										1693	1437	3808	2480	511	340
11											1685	2206	909	1205	1603
12												3098	1952	1429	1100
13													1331	3368	3624
14														2038	2309
15															578

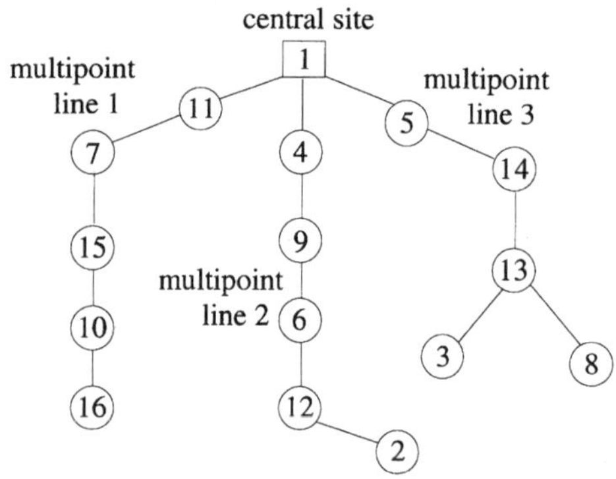

Figure 8.14. Optimal solution of the problem (cost = 8526).

Network expansion deals with an ever-growing need for more computing across a network. To meet this demand, the size of the network is expanded incrementally according to user requirements. In such a case, expansion is achieved by properly adding new communication edges and computer nodes such that reliability measures of the network are optimized within specified constraints.

8.5.1 Problem Description

Kumar, Pathak, and Gupta developed a genetic algorithm-based computer network expansion method to optimize a specified objective function (reliability measure) under a given set of network constraints [379, 380]. For the network expansion problem, the following assumptions are defined: Network topology is modeled by an undirected graph G, whose nodes represent processing elements and edges represent communication edges. A graph G does not have any self-loops. Failures of any edge are mutually statistically independent. An edge has two states, operational and faulty. The network is statistically coherent.

The network expansion problem maximizing computer network reliability under diameter and degree constraints can be formulated as follows:

$$\min \quad R(\boldsymbol{x})$$

$$\text{s.t.} \quad \sum_{j=1}^{n} x_{ij} \leq k_i, \qquad i = 1,\ldots,n$$

$$\max\{d_{ij} | i = 1,\ldots,n;\ j = 1,\ldots,n\} \leq D$$

where n is the number of notes and k_i indicates the degree of node i. If an edge exists between nodes i and j, $x_{ij} = 1$; otherwise, $x_{ij} = 0$. d_{ij} is a minimum number of hops between nodes i and j. D is the upper bound on the network diameter.

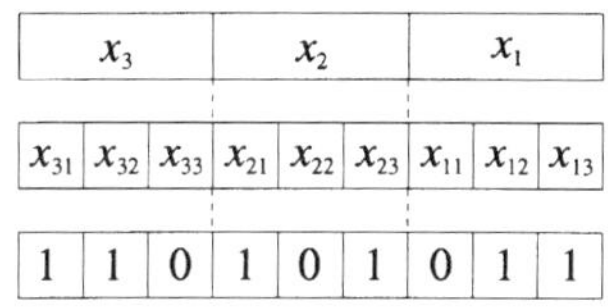

Figure 8.15. Chromosome for network expansion problem.

8.5.2 Kummar, Pathak, and Gupta's Approach

Representation and Initialization. To represent the connectivity of nodes, a coding scheme with binary string is used. The complete chromosome of network connectivity is divided into node fields, which are equal to the number of nodes in the expanded network after adding the new nodes to the existing network. One arrangement of connections for a three-node network, where one new node (x_3') is added to an existing two-node network whose two nodes are x_1 and x_2, is displayed in Figure 8.15. In general, if there are n nodes in the expanded network, and if the first $i - 1$ nodes represent the nodes in the old network, the remainder of the nodes are newly added to the network. The chromosome for this expanded network can be as shown in Figure 8.16. Usually, x_{ii}' and x_{ii} are zero because self-loops are not allowed. The initial population is absolutely randomly generated, not using any criterion to generate biased or adapted population at initialization.

Crossover. There are two types of genetic crossover operators: (1) SNFSO (simple node field swap operator) and (2) RNFSO (random node field swap operator). In SNFSO, crossover is performed at the boundaries of the node fields by randomly selecting two chromosomes for crossover from the preceding generation. Then, using a random number generator, an integer value generated in the range $(0, n - 1)$ is used as the crossover site. This operation produces two new chromosomes with information from their parents. Figure 8.17 shows the details of the crossover operation when the crossover site is the right boundary of the second decision varible, x_2.

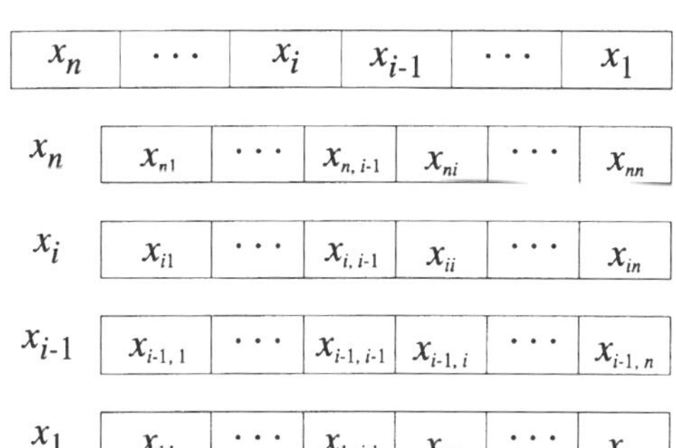

Figure 8.16. Generalized representation of chromosome for network expansion.

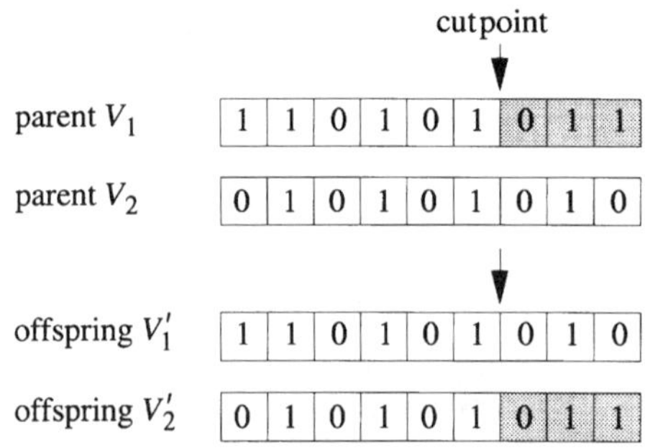

Figure 8.17. Simple node field swap operator (SNFSO).

The crossover operator sometimes generates a chromosome that does not represent valid network connectivity. For example, node field x_3 of child 1 shows that node 3 is connected to nodes 1 and 2, but x_1 indicates that node 1 is connected only to node 2. This anomaly is corrected by an adjustment operator, discussed later.

In RNFSO, the node fields are swapped randomly, one at a time, in the chromosome. Random numbers in the range $[0, n - 1]$ are generated for swapping a specified number of node fields. For example, if only one node field is to be swapped, a random number is generated to decide which field is to be swapped. Let the field to be swapped be 3; then RNFSO can be illustrated as shown in Figure 8.18. RNFSO can also create an anomaly in a chromosome as discussed for SNFSO. The anomaly is also corrected by the adjustment operator. For the network expansion problem, RNFSO, is preferred because it provides an unbiased and equal opportunity for each node to connect itself with other nodes, maintaining their own connectivity using an adjustment operator. But using SNFSO causes lowered-numbered nodes to become dominant in forcing the other nodes to change their connectivity through the adjustment operator.

Adjustment Operator. Since these crossover operators produce a discrepancy in a connectivity chromosome, the following simple procedure is executed:

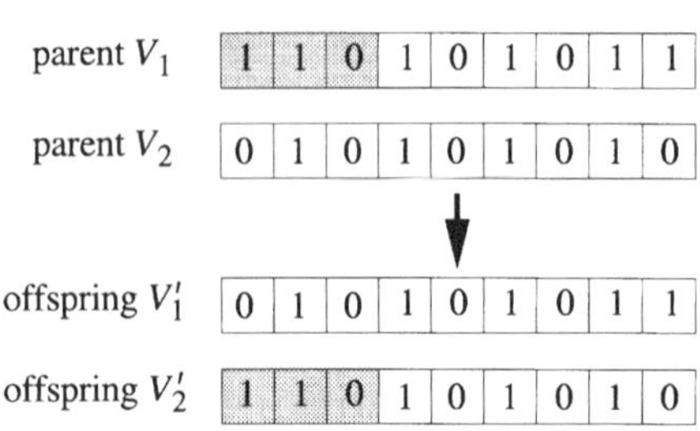

Figure 8.18. Random node field swap operator (RNFSO).

Procedure: Adjustment Operator for the Swapped Node Field (x_i)

Step 1. Let j be 1.

Step 2. If $x_{ij} = 1$, go to step 3; otherwise, go to step 4.

Step 3. If $x_{ji} = 0$, set $x_{ji} = 1$.

Step 4. If $x_{ji} = 1$, set $x_{ji} = 0$.

Step 5. Update j; that is, set $j = j + 1$. If $j > n$, then stop; otherwise, go to step 2.

This adjustment operation preserves the crossover characteristics by adjusting the chromosomes according to newly acquired fields.

Mutation. This operation is performed to improve the connectivity of a node field if it is appreciably reduced by the crossover operation. There are four steps in the process.

Step 1. Select at random a node field x_i from those nodes that are added to the existing network and have degree less than the upper bound.

Step 2. Select at random any other node x_k that is not connected to x_i and has degree less than its upper bound.

Step 3. If such a node cannot be found, go to step 1. If the number of nodes x_i tried for mutation exceeds the number of tries, stop.

Step 4. If proper values of x_i and x_k are found, create a new edge by setting bit x_{ik} in node field x_i to 1 and bit x_{ki} in node field x_k to 1. If the number of mutations done so far is smaller than $p_M \cdot pop\_size \cdot str\_len$ specified (where *str_len* is the chromosome length) go to step 1; else, stop.

Selection and Termination Rule. In Kumar, Pathak, and Gupta's method, each offspring generated after crossover, mutation, and adjustment is added to the new generation if it has a better objective function value than that of its parents. If the offspring is better than only one of the parents, select one randomly from the offspring and the better of the two parents. If the offspring is worse than both parents, either of the parents is selected at random for the next generation. This ensures that the best chromosome is always carried to the next generation, while the worse is not carried to succeeding generations. The execution of genetic algorithms terminates when the average and maximum fitness values of chromosomes in a generation are identical.

Fitness and Reliability Calculation. Since the objective function is a reliability function, the fitness value is determined by a reliability calculation. *Computer network reliability* is defined as the probability of each node communicating with all the other nodes in a given network. To compute network reliability, the following procedure was used:

Step 1. All the spanning trees of a network are computed and then made disjoint using an algorithm of Aggarawal and Rai [8]. The spanning tree

T_i of a network represented by graph G is defined as a connected subgraph of G that contains all the nodes of G. From the definition of a spanning tree, any T_i will connect all the nodes of G with $n - 1$ edges and hence represents the minimum interconnection required for providing communication between nodes.

Step 2. Using the edge reliabilities given in advance, compute the reliability of a spanning tree being operational, which in turn is used for network reliability computations.

Overall Algorithm

Step 1. Set the parameters. Set the population size (*pop_size*), the rate of crossover (p_c), the rate of mutation (p_m), the maximum generation (*max_gen*), and initialize the generation number $gen = 0$. Also input the number of nodes, degree of node, edge reliability, and upper gound on the network diameter (D).

Step 2. Randomly generate the initial population.

Step 3. Evaluate. Compute the network reliability for each chromosome in the population.

Step 4. Selection from the current population to form a mating pool.

Step 5. Perform crossover and mutation.

(5.1) Apply SNFSO or RNFSO to the chromosomes in the mating pool.

(5.2) Apply the adjustment operator to the new offspring.

(5.3) If the constraints are not satisfied, discard the chromosomes and go to step 6.

(5.4) Apply mutation operation.

(5.5) If $n < pop\_size - 1$, go to step 6, where n means the number of new offspring.

(5.6) If the fitness is better than that of the parent, add it to the new generation; otherwise, discard the chromosome.

(5.7) Go to step 6.

Step 6. Perform the terminating test.

(6.1) Set $gen = gen + 1$.

(6.2) If the terminating condition is not satisfied, return to step 3 for the next generation; otherwise, terminate.

8.5.3 Numerical Example

Four examples were tested by Kumar, Pathak, and Gupta [380]. The nodes are added such that network reliability is maximized while the diameter and degree constraints of the network are satisfied. The constraints for each case are shown in Table 8.5. The results are given in Table 8.6. The optimal result was obtained by a genetic algorithm in all cases.

TABLE 8.5. Parameters of Test Problems

Number of Nodes	k_i	$\max\{d_{ij}\}$
4 + 2	4	3
5 + 2	3	3
6 + 2	3	3
6 + 3	3 (i = 1,...,6) 2 (i = 7,8,9)	4

8.6 MULTISTAGE PROCESS PLANNING

Multistage process planning (MPP) problems are common among manufacturing systems. The problem, in general, provides a detailed description of manufacturing capabilities and requirements for transforming a raw stock of materials into a completed product through a multistage process. In modern manufacturing systems, the MPP can be classified into two types [95]: variant MPP and generative MPP. In the *variant MPP,* a family of components of similar manufacturing requirements are grouped together and are coded using a coding scheme. When a new component is required to be manufactured, its code is mapped onto one of these component families and the corresponding process plan is retrieved. In the *generative MPP,* the new process plan is generated automatically according to some decision logic. The decision logic interacts with some automatic manufacturing feature recognition mechanisms and the corresponding machining requirements are derived by utilizing optimization techniques. The generative MPP problem is now receiving more attention since it can accommodate both general and automated process planning models, which is particularly important in flexible manufacturing systems [387, 534].

The generative MPP problem lends itself naturally to optimization techniques designed to find the best process plan among numerous alternatives given a criterion such as minimum cost, minimum time, maximum quality, or with multiple criteria. The implicit enumeration of all these alternatives can be formulated as network flow, but with the increase in problem scale, has the difficulty known as dimension explosion. Even using notable shortest-path algorithms such as Dijkstra's [326], Floyd's [326], or out-of-kilter [660],

TABLE 8.6. Results of Test Problems

Nodes	$R(\boldsymbol{x})$
4 + 2	0.9993
5 + 2	0.9814
6 + 2	0.9906
6 + 3	0.9579

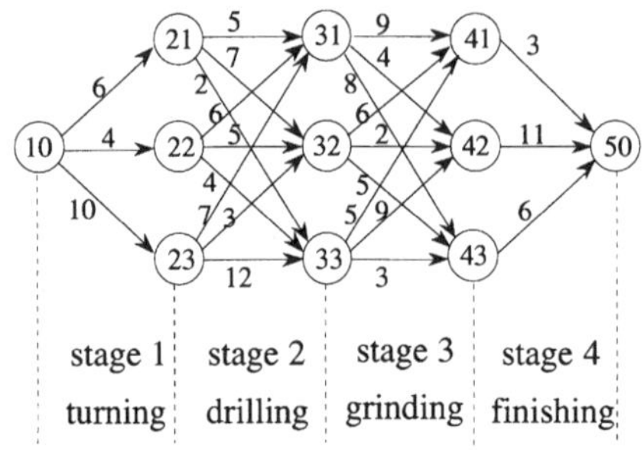

Figure 8.19. Flow network for a simple MPP problem.

it is still not easy to deal with a larger-scale MPP problem. For an MPP problem with multiple criteria, traditional techniques cannot be used. Although techniques based on goal programming were developed, they can only be illustrated for small networks [558, 578]. Zhou and Gen proposed a genetic algorithm to deal with the MPP problem [700].

8.6.1 Problem Description

The MPP system usually consists of a series of machining operations, such as turning, drilling, grinding, finishing, and so on, to transform a part into its final shape or product. The entire process can be divided into several stages. At each stage there is a set of similar manufacturing operations. The MPP problem is to find the optimal process planning among all possible alternatives given criteria such as minimum cost, minimum time, or maximum quality, or under several of these criteria. Figure 8.19 shows a simple MPP problem.

For an n-stage MPP problem, let s_k be a state at stage k, $D_k(s_k)$ be the set of possible states to be chosen at stage k, $k = 1,2,...,n$, x_k be the decision variable to determine which state to choose at stage k; obviously, $x_k \in D_k(s_k)$, $k = 1,2,...,n$. Then the MPP problem can be formulated as follows:

$$\min_{\substack{x_k \in D_k(s_k) \\ k=1,2,\ldots,n}} V(x_1,x_2,\ldots,x_n) = \sum_{k=1}^{n} v_k(x_k,x_x) \tag{8.30}$$

where $v_k(s_k,x_k)$ represents the criterion to determine x_k in state s_k at stage k, usually defined as a real number such as cost, time, or distance. Problem (8.30) can be rewritten as a dynamic recurrence expression [528]. If $f_k(x_k,s_k)$ represents optimal process planning from stage k to the last stage (n), the dynamic recurrence expression of problem (8.30) can be formulated as follows:

$$f_k(x_k,s_k) = \min_{x_k \in D_k(s_k)} \{v_k(s_k,x_k) + f_{k+1}(x_{k+1},s_{k+1})\}, \qquad k = 1,2,\ldots,n-1$$

Although this problem can be approached by the shortest-path method or by dynamic programming, with increased problem scale, many stages and states must be considered, which will greatly affect the efficiency of the procedure shortest path method or dynamic programming in getting the optimal solution. This is usually known as the dimension explosion.

If MPP problem (8.30) has multiple objective criteria, the problem has the following formulation:

$$\begin{aligned}
\min \quad & V_1(x_1,x_2,\ldots,x_n) = \sum_{k=1}^{n} v_k^1(s_k,x_k) \\
\min \quad & V_2(x_1,x_2,\ldots,x_n) = \sum_{k=1}^{n} v_k^2(s_k,x_k) \\
& \cdots \\
\min \quad & V_p(x_1,x_2,\ldots,x_n) = \sum_{k=1}^{n} v_k^p(s_k,x_k) \\
\text{s.t.} \quad & x_k \in D_k(s_k), \qquad k = 1,2,\ldots,n
\end{aligned} \tag{8.31}$$

where p is the number of objectives.

Obviously, it is difficult to transform problem (8.31) into its equivalent dynamic recurrence expression. For a small-scale problem, problem (8.31) can be approached by traditional multiple-criteria decision-making techniques such as goal programming. However, if the problem scale increases, it will become difficult to deal with even in the case of a single objective.

8.6.2 Genetic Algorithm Approach

Genetic Representation. For the MPP problem shown in Figure 8.19, the alternative states at each stage can be expressed by a series of integers to indicate node or state. If a state for an operation is chosen at one stage for process planning, its corresponding integer for that node or state can be assigned where the integer is within the number of possible states at that stage. Therefore, the MPP solution can be encoded concisely in a state permutation format by concatenating all the set states of stages as shown in Figure 8.20. This state permutation encoding has 1-to-1 mapping for the MPP problem. The probability of randomly producing process planning is definitely 1. It is also easy to decode and evaluate. Compared with a binary string encoding, state permutation encoding has two advantages:

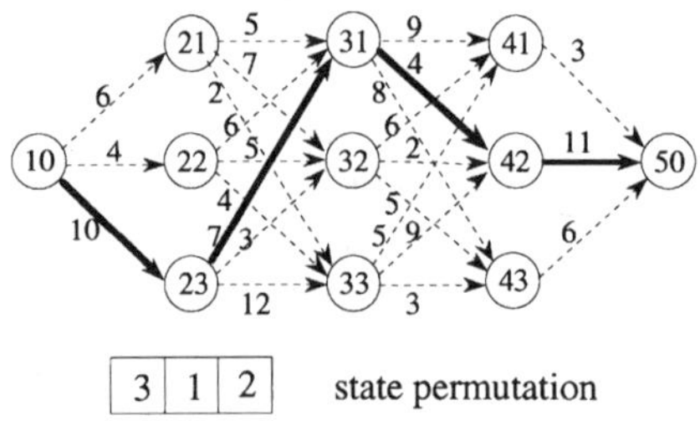

Figure 8.20. Process planning and its stage permutation encoding.

1. The encoding length is only $n - 1$ for an n-stage MPP problem, which can save much more computation memory when the problem scale gets larger.
2. The encoding is able to generate all feasible individuals in genetic operations such as crossover or mutation without modification.

As to the initial population for an n-stage MPP problem, each individual represents a permutation with $n - 1$ integers where the integers are generated randomly with the number of all possible states in the corresponding stage.

Genetic Operators. In Zhou and Gen's method, only the mutation operation was adopted because it is easy to hybrid the neighborhood search technique to produce more adapted offspring. This hybrid mutation operation provides a great chance to evolve an optimal solution. Figure 8.21 exemplifies this mutation operation with a neighborhood search technique. Suppose that a gene is at stage 3 and the number of possible states to be chosen is 4.

Evaluation. In the case of multiple criteria, the optimal solutions of problem (8.31) are a Pareto-optimal set. For an MPP problem with multiple objectives, a weighted-sum method was used to transform multiple objectives into a single objective and take its reciprocal as the fitness value:

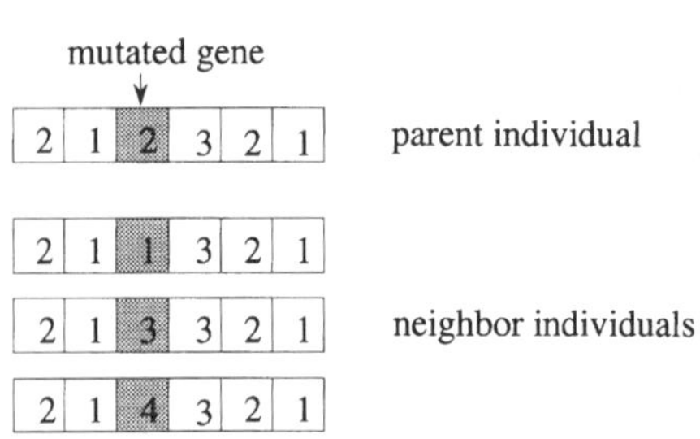

Figure 8.21. Mutation with neighborhood search.

TABLE 8.7. Comparison Among Different Approaches to the MPP Problem[a]

Problem	Number of Stages	Number of Nodes	CPU Time, OK (s)	CPU Time, SP (s) Min.	Ave.	%
1	7	24	3.01	0.03	0.04	100
2	7	27	2.84	0.03	0.04	100
3	8	38	4.31	0.06	0.09	100
4	9	37	4.25	0.06	0.12	100
5	10	47	4.48	0.09	0.27	100
6	11	53	4.58	0.34	0.45	100
7	12	63	4.69	0.32	0.48	100
8	13	72	5.08	0.61	1.03	100
9	14	79	5.19	0.49	1.01	90
10	15	89	5.35	0.97	1.28	80

[a]OK, out-of-kilter algorithm using SAS/OR program; SP, state permutation encoding by genetic algorithm, pop_size-1000; Min., the minimal CPU time in all 20 runs; Ave., the average CPU time in all 20 runs; %, the frequency to obtain the optimal solution in all 20 runs.

$$\text{eval}(\boldsymbol{x}) = \frac{1}{\sum_{i=1}^{p} \lambda_i V_i(\boldsymbol{x})} \tag{8.32}$$

where p is the number of objectives and λ_i $(i = 1,2,...,p)$ is the weight coefficient of objectives, which can be determined by the method introduced in Section 3.6.

8.6.3 Numerical Examples

To demonstrate the effectiveness and efficiency of the genetic algorithm on the MPP problem, 10 different MPP problems are randomly generated with the same scale as those given by Awadh, Sepehri, and Hawaleshka [23]. All results are given in Table 8.7. The results show that the genetic algorithm is more efficient than the traditional method in obtaining optimal solutions. When the problem scale is larger, the genetic algorithm has a great chance of obtaining the optimal solution.

Furthermore, an instance of the MPP problem with 15 stages and 89 nodes is generated. Two attributes are defined on each arc: processing time and processing cost. The weights are integers generated randomly and uniformly distributed over [1, 50] and [1, 100], respectively. Two objectives are to minimize the total processing time and minimize the total cost, respectively. The ideal point is (91, 159), and other two extreme points (Pareto optimal solutions) are (91, 686) and (402, 159). The results obtained by the genetic algorithm are plotted in Figure 8.22.

Compared with the enumeration of all possible real Pareto optimal solutions that takes 3840 of CPU time, 85% results obtained by the genetic al-

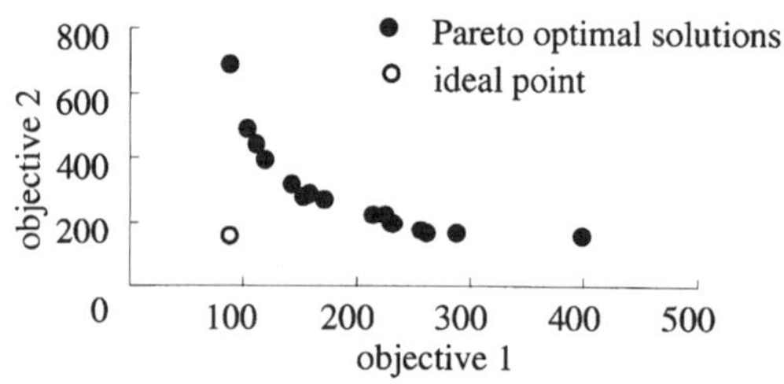

Figure 8.22. Multicriteria MPP problem.

gorithm are real Pareto optimal solutions, and it only takes 2.36 if CPU time on average to obtain the results. Also, Figure 8.22 shows that all the results form a Pareto frontier on the criteria space. It is evident that the adaptive hyperplane plays an important role in guiding the search toward the ideal point and enforces all chromosomes to the ideal point as close as possible in the evolutionary process.

8.7 *M/G/s* QUEUING FACILITY LOCATION ON A NETWORK

Facility location under uncertainty or stochastic facility location has been an active subject in the past 10 years [55, 75, 425]. Typical elements of the problem include the locations of customers, the presence or absence of each customer, the level of demand, the price of product, and the travel time or travel cost to the customer. Berman et al. examined the stochastic queue median (SQM) model on a network, which is for the optimal location of a single mobile server [75, 119]. In practice, a facility usually manages a number of mobile servers, not only one. It is much more reasonable and necessary to consider the case of multiple mobile serves. One may argue that it will be able to divide a large service area into several small regions so that we can require that one facility manages only one server within each small region. But the question is how to optimally design the service territories (see [56] and [57]). Furthermore, it should be noticed that to build up a new facility would cost much more money, time, and other resources than just to increase a number of mobile servers at an existing facility. These considerations motivate us to study SQM with multiple servers, which is denoted as the *M/G/s* SQM problem. Due to the intractabiity of the average waiting time in an *M/G/s* queue when $s \geq 2$, it is a difficult problem. One possible way is to do numerical simulation experiments to estimate the average response time of the system; however, it is hard to apply this method to large-scale networks. An efficient method for solving the *M/G/s* SQM problem is required to develop. The objective function of an *M/G/s* SQM problem is very complex and the location is constrained on the network. The *M/G/s* SQM problem is a multimodal optimization problem.

In this section we introduce Gong, Yamazaki, Gen, and Xu's method for the *M/G/s* SQM problem [256]. In their method a two-moment approximation is adopted to estimate the mean waiting time of a customer in the *M/G/s* queue and to build up an *M/G/s* SQM model for the problem. An evolutionary computation approach for obtaining a global or near-global optimal solution for the *M/G/s* SQM model is introduced. With this model and its solution approach, it is possible to solve *M/G/s* SQM on larger-scale networks efficiently.

8.7.1 Problem Description

The following notations of network will be used:

1. A network $G = (V,E)$ with node set $V = \{1,\ldots,n\}$ ($|V| = n$) and edge set $E = \{(i,j)|i,j \in V, i \neq j\}$ ($|E| = m$).
2. The length of an edge $(i,j) \in E$ is given and denoted by l_{ij}.
3. A point $P = (i,j{:}x) \in G$ is the point on the edge (i,j) and the distance between it and i is x.
4. $p(P_1,P_2)$ is the shortest path between the two points P_1 and P_2 on G, and $d(P_1,P_2)$ is the shortest distance i.e., the length of $p(P_1,P_2)$.

On a network G, demands are limited to the nodes of the network. At each node k, demands for service emerge according to a Poisson process with rate λ_k, independent of everything else. Let λ be the total generating rate of demands in the system and h_k be the weight of demands at node k ($k = 1,2,\ldots,n$). Then

$$\lambda = \sum_{k=1}^{n} \lambda_k \tag{8.33}$$

$$h_k = \frac{\lambda_k}{\lambda} \tag{8.34}$$

The facillity is located at a fixed point $(i,j{:}x)$ on the network and manages s mobile servers. The mobile servers are dispatched to serve demands and all of them should be in the facility when not busy in serving demands. If a emerged demand finds all servers busy, it is placed in a queue that is depleted in a first-in, first-out (FIFO) manner; otherwise, and idle server is dispatched to serve the demand immediately.

Travel occurs on the shortest path from the facility to a demand node at constant speed ξ. Further, travel from a demand node to the facility occurs at constant speed ξ/β, where β is a parameter to decide the return speed. The service of time of a demand is the sum of:

1. The travel time to the location of the demand from the facility of the server
2. The on-scene service time at the location of the demand
3. The travel time from the location of the demand to the facility of the server
4. The off-scene service time at the home location of the server

Assume that the net service times (on-scene plus off-scene) at node k form a renewal sequence, independent of everything else. This net service time is denoted by S_k and is assumed with finite first two moments $\overline{S_k}$ and $\overline{S_k^2}$. Then this system can be viewed as an $M/G'/s$ queue with FIFO discipline in which there are n classes of demands.

Let $W_q(i,j{:}x)$ be the waiting time of the qth emerged demand in the $M/G'/s$ and $T_q(i,j{:}x)$ be the travel time from the facility. Then the response time of the system to this demand, $R_q(i,j{:}x)$, can be written as

$$R_q(i,j{:}x) = W_q(i,j{:}x) + T_q(i,j{:}x) \tag{8.35}$$

Define the average response time of the system to a demand as follows:

$$\begin{aligned}\overline{R}(i,j{:}x) &= \lim_{N\to\infty} \frac{1}{N} \sum_{q=1}^{N} R_q(i,j{:}x) \\ &= \lim_{N\to\infty} \frac{1}{N} \sum_{q=1}^{N} W_q(i,jx) + \lim_{N\to\infty} \frac{1}{N} \sum_{q=1}^{N} T_q(i,j{:}x)\end{aligned} \tag{8.36}$$

Noting that the qth emerged demand belongs to a class of l to n, we have

$$\lim_{n\to\infty} \frac{1}{N} \sum_{q=1}^{N} T_q(i,j{:}x) = \sum_{k=1}^{n} h_k \frac{d(k,(i,j{:}x))}{\xi} \tag{8.37}$$

Thus equation (8.37) can be rewritten as

$$\overline{R}(i,j{:}x) = \overline{W}(M/G'(i,j{:}x)/s) + \sum_{k=1}^{n} h_k \frac{d(k,(i,j{:}x))}{\xi} \tag{8.38}$$

The goal, then, is to find the optimal location $(i,j{:}x)^*$ that minimizes $\overline{R}(i,j{:}x)$, or a nearly optimal location.

Approximations on $\overline{W}(M/G'(i,j{:}x/s)$. To obtain the optimal location $(i,j{:}x)^*$, we need to evaluate $\overline{W}(M/G'(i,j{:}x)$. Note that in equation (8.38), the classes of demands are neglected because of FIFO discipline. This means that $\overline{W}(M/G'(i,j{:}x)/s)$ can be evaluated by using an ordinary $M/G(i,j{:}x)/s$ queue, where

$G(i,j{:}x)$ denotes the distribution of service times of all demands when the facility is located at $(i,j{:}x)$. Then the distribution function $G(i,j{:}x)(t)$ is given by

$$G(i,j{:}x)(t) = \sum_{k=1}^{n} h_k G_k(i,j{:}x)(t) \tag{8.39}$$

where $G_k(i,j{:}x)(t)$ is the distribution function of the random variable $S_k + (\beta + 1)\,[d(k,(i,j{:}x))/\xi]$. $(i,j{:}x)$ of $G(i,j{:}x)$ is omitted below if there is no confusion. The first two moments of G, denoted by $\overline{G}$ and $\overline{G^2}$, are given as follows:

$$\overline{G} = \sum_{k=1}^{n} h_k(\beta + 1)\,\frac{d(k,(i,j{:}x))}{\xi} + \overline{S_k} \tag{8.40}$$

$$\overline{G^2} = \sum_{k=1}^{n} h_k\left((\beta + 1)\,\frac{d(k,(i,j{:}x))}{\xi}\right)^2 + 2(\beta + 1)\,\frac{d(k,(i,j{:}x))}{\xi}\,\overline{S_k} + \overline{S_k^2} \tag{8.41}$$

It is well known that for $s = 1$, $\overline{W}(M/G/1)$ is

$$\overline{W}(M/G/1) = \begin{cases} \dfrac{\lambda \overline{G^2}}{2(1-\rho)}, & \text{if } \rho < 1 \\ +\infty, & \text{otherwise} \end{cases} \tag{8.42}$$

where $\rho = \lambda\overline{G}$.

Unfortunately, there is no such explicit expression of $\overline{W}(M/G/s)$ for $s \geq 2$. Alternatively, may approximations on $\overline{W}(M/G/s)$ have been proposed (see [73] and [364]). Three approximations are described as follows:

1. Pollaczek–Khintchine [393]

$$\overline{W_a}(M/G/s)^{(\mathrm{P\&K})} = \frac{1 + c_G^2}{2}\,\overline{W}(M/M/s) \tag{8.43}$$

2. Bjorklund and Elldin [64]

$$\overline{W_a}(M/G/s)^{(\mathrm{B\&E})} = c_G^2\overline{W}(M/M/s) + (1 - c_G^2)\overline{W}(M/D/s) \tag{8.44}$$

3. Kimura [364]

$$\overline{W_a}(M/G/s)^{(\mathrm{Kimura})} = \frac{1 + c_G^2}{2c_G^2/\overline{W}(M/M/s) + (1 - c_G^2/\overline{W}(M/D/s)} \tag{8.45}$$

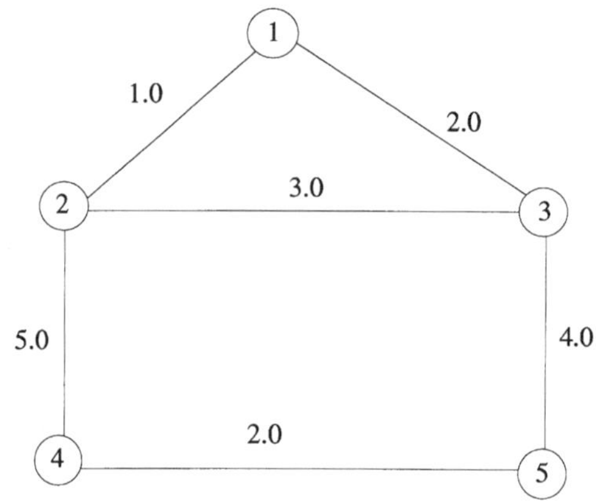

Figure 8.23. Simple five-node network problem.

where c_G^2 is the coefficient of variation of G and a represents approximations.

To determine which approximation is the best for the problem, Gong has carried out a number of numerical simulation experiments [255]. In his experiments, a network consisting of five nodes and six edges is used, which was introduced by Berman, Larson, and Chiu [55] (Figure 8.23). Assume an exponential distribution with man 1.0 for S_k, $k = 1,2,...,5$, $\xi = 1.0$, $\beta = 1.0$, $h_1 = 0.1$, $h_2 = 0.35$, $h_3 = 0.1$, $h_4 = 0.35$ and $h_5 = 0.1$. The range of parameters in the experiments includes s: 2 to 4 and λ: 0.02 to 0.48. To estimate the average response time at an arbitrary facility location $(i,j:x)$, each arc is equally divided into 100 intervals and the following experiments are performed to estimate the average response time when the facility is located at the beginning point of an interval. The estimate of average response time at each point and its 95% confidence interval are calculated from six independent runs, where a single run consists of 64,000 demands.

The three approximated average response times according to equations (8.43), (8.44), and (8.45) are calculated using the analytical formula on $\overline{W}(M/M/s)$ and an approximation on $\overline{W}(M/D/s)$ proposed by Kimura [365]. The results for the case of $s = 2$ and $\lambda = 0.12$ showed that the optimal location is on edge (2,4). His experiments show that (1) the results of the Pollaczek–Khinchine formula are always lower than the simulation results; (2) the results of the Bjorklund–Elldin formula are closer to the simulation results; and (3) the results of the Kimura formula fall almost within the 95% confidential interval of the simulation results. The experiments for other parameters as well as other arcs have yielded similar results. These suggest that $\overline{W}(M/G/s)^{(\text{Kimura})}$ is applicable to the problem. Table 8.8 shows that the optimal and best locations are obtained by simulation experiments and using $\overline{W}(M/G/s)^{(\text{Kimura})}$, respectively. Experiments on a network with five nodes and seven

TABLE 8.8. Best Locations by Kimura's Approximation and Scope of Optimal Locations

s	λ	Scope of Optimal Location After Experiments, $(i,j{:}x)^s$	Best Location by Kimura's Approximation, $(i,j{:}x)^b$
2	0.02	(2,4: 0.000)	(2,4: 0.000)
	0.12	(2,4: 0.65 to 0.75)	(2,4: 0.719)
	0.24	(2,4: 0.70 to 0.75)	(2,4: 0.725)
4	0.04	(2,4: 0.000)	(2,4: 0.000)
	0.24	(2,4: 0.000)	(2,4: 0.000)
	0.48	(2,4: 0.25 to 0.50)	(2,4: 0.298)

edges and a network with six nodes and eight edges were also carried out. The experiments also yielded results similar to those in Table 8.8.

With $\overline{W_a}(M/G/s)^{(\text{Kimura})}$, the problem can be formulated as

$$\min \quad \overline{W_a}(M/G(i,j{:}x)/s)^{(\text{Kimura})} + \sum_{k=1}^{n} h_k \frac{d(k,(i,j{:}x))}{\xi} \tag{8.46}$$

$$\text{s.t.} \quad (i,j{:}x) \in G \tag{8.47}$$

8.7.2 Evolutionary Computation Approach

The evolutionary computation approach is a population-based stochastic search method that evolves a population of candidate solutions to a given problem using operators inspired by natural genetic variation and natural selection. It uses very little information of the objective function (just the objective function value) and would converge to the global optimal solution if sufficient computing time is allowed (i.e., enough long generations to be evolved). In practice, an effective evolutionary computation approach should give a nearly global optimal solution within an acceptable computing time. Gong, Yamazaki, Gen, and Xu proposed an evolutionary computation approach (ECA) to find a nearly global optimal solution for an $M/G/s$ SQM model by incorporating the network structure into the evolutionary computation paradigm [247].

Representation of Chromosomes. Since a facility location is a point on the network, the point expression on the network is adopted:

$$F = (i,j{:}x) \tag{8.48}$$

This representation is very natural and simple; however, it requires the design of special evolutionary operators.

Fitness Function. For a given solution $F = (i,j{:}x)$, its corresponding objective function value in the $M/G/s$ SQM model is adopted as its fitness value.

$$\text{eval}(i,j{:}x) = \overline{W}(M/G/(i,j{:}x)/s)^{(\text{Kimura})} + \sum_{k=1}^{n} h_k \frac{d(k,(i,j{:}x))}{\xi} \tag{8.49}$$

Environment Parameters. The following parameters define the environment for the ECA in searching for a nearly global optimal solution:

pop_size = size of population

max_gen = maximal generations to be evolved

p_c = probability for crossover operator

p_m = probability for mutation operators

Initialization. ECA requires a group of initial candidate solutions to start searching. The initial solutions are generated as follows.

Step 1. Pick an edge $(i,j) \in E$ randomly.
Step 2. Take a random real number x in $[0,l_{ij}]$.
Step 3. Set $(i,j{:}x)$ as an initial solution.
Step 4. Repeat steps 1 to 3 until *pop_size* times.

Selection. Selection is to select the candidate individuals from the current generation to produce the next generation. The well-known rotating roulette wheel strategy together with the elitist strategy is adopted, which is regarded useful to accelerating convergence speed [219, 249, 455].

Genetic Operators

Crossover Operator. Let $P_1 = (i_1,j_1{:}x_1)$ and $P_2 = (i_2,j_2{:}x_2)$ be the two parents; then their children produced by crossover operator, $C_1 = (\bar{i}_1,\bar{j}_1{:}\bar{x}_1)$ and $C_2 = (\bar{i}_2,\bar{j}_2{:}\bar{x}_2)$ are two points on the shortest patch $p(P_1,P_2)$ corresponding to a random convex combination of them, as shown in Figure 8.24.

$$d(P_1,C_1) = rd(P_1,P_2) \tag{8.50}$$

$$d(P_1,C_2) = (1 - r)d(P_1,P_2) \tag{8.51}$$

where r is a random value and $r \in (0,1)$. It can be seen that this crossover operator can always ensure the feasibility of children.

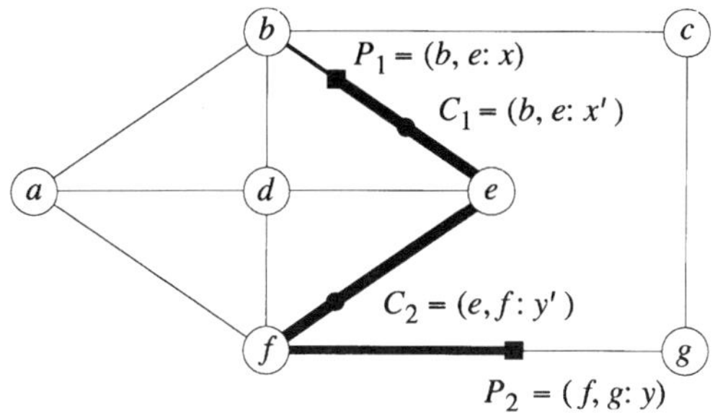

Figure 8.24. Crossover operator.

Mutation Operators. Let $P = (i,j{:}x)$ be the parent. Then the following three kinds of mutation operators are used.

(M_1) *ε-Neighborhood mutation:* to produce a child $C_1 = (i_1,j_1{:}x_1)$ randomly in the neighborhood of the parent, that is, $i_1 = i$, $j_1 = j$, and $x_1 \in (\max\{0, x - \delta\}, \min\{x + \delta, l_{ij})$ is a random number, where δ is a small positive number.

(M_2) *Local mutation:* to produce a child $C_2 = (i_2,j_2{:}x_2)$ randomly on the same edge (i,j), that is, $i_2 = i$, $j_2 = j$, and $x_2 \in (0,l_{ij})$ is a random number.

(M_3) *Edge mutation:* to produce a child $C_3 = (i_3,j_3{:}x_3)$ randomly on another edge $(i,k) \in E$, where k is a random integer between 1 and n, that is, $i_3 = i$, $j_3 = k$, and $x_3 \in (0,l_{ik})$ is a random number.

The mutation operators M_1, M_2, and M_3 are illustrated in Figure 8.25. All of them can ensure the feasibility of the child.

Overall Algorithm. The algorithm of the ECA for *M/G/s* SQM is summarized as follows.

Step 0. Initialize.

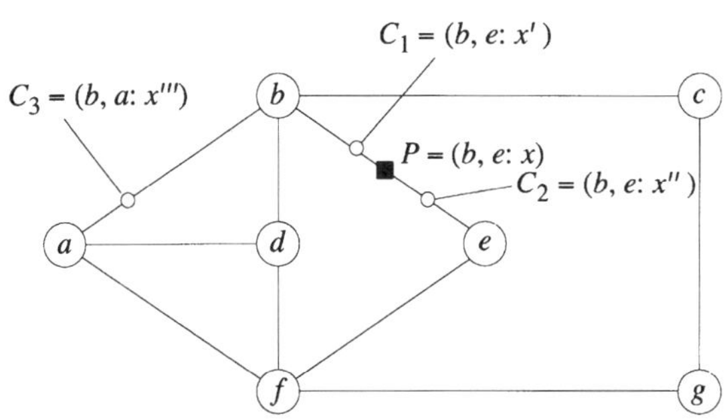

Figure 8.25. Mutation operator.

(0.1) Calculate the shortest-distance paths among all pairs of networks.

(0.2) Set environment parameters and $gen \leftarrow 0$.

(0.3) Produce an initial population.

(0.4) Calculate the fitness of individuals of the initial population.

Step 1. If $gen > max\_gen$, stop; otherwise, continue.

Step 2. Set $gen \leftarrow gen + 1$.

Step 3. Perform reproduction.

(3.1) Generate the roulette wheel of fitness of individuals in the current generation.

(3.2) Generate a random number $r \in (0,1)$.

(3.3) Copy an individual in the current generation as a member of the temporary population pool according to r and the roulette wheel.

(3.4) Repeat steps (3.2) and (3.3) until *pop_size* times.

Step 4. Perform crossover.

(4.1) Generate a random number $r \in (0,1)$.

(4.2) If $r < p_c$, continue; otherwise, go to step (4.6).

(4.3) Take two parents, P_1 and P_2, randomly from the temporary population pool.

(4.4) Produce two children, C_1 and C_2, according to the crossover operator.

(4.5) Replace parents P_1 and P_2 with children C_1 and C_2.

(4.6) Repeat steps (4.1) to (4.5) until *pop_size* times.

Step 5. Perform mutation.

(5.1) Generate a random number $r \in (0,1)$.

(5.2) If $r < p_m$, continue, otherwise, go to step (5.6).

(5.3) Take a parent P randomly from the temporary population pool.

(5.4) Produce a child C according to the mutation operators, which are applied alternatively (i.e., applying mutation operator in the order of $M_1 \rightarrow M_2 \rightarrow M_3 \rightarrow M_3$).

(5.5) Replace parent P with child C.

(5.6) Repeat steps (5.1) to (5.5) until *pop_size* times.

Step 6. Calculate the fitness of children in the temporary population pool.

Step 7. Apply the elitist strategy. Force the best individual of the current generation into the temporary population pool.

Step 8. Advance one generation.

(8.1) Replace the current generation with the temporary population.

(8.2) Go to step 1.

8.7.3 Numerical Examples

Gong, Yamazaki, Gen, and Xu tested their algorithm on the *M/G/s* SQM problems with a five-node network [256]. For a given group of environment

TABLE 8.9. Comparison of the Best Solutions by ECA and the Optimal Solutions

		Optimal		Best		
s	λ	$(i,j{:}x)^O$	$R(\cdot)^O$	$(i,j{:}x)^b$	$R(\cdot)^b$	Time (s)
1	0.01	(2,4: 0.000	3.230	(2,4: 0.000)	3.236	1.01
	0.06	(2,4: 1.614)	5.893	(2,4: 1.612)	5.946	1.02
	0.12	(2,4: 0.934)	23.525	(2,4: 0.913)	23.871	1.01
2	0.02	(2,4: 0.000)	2.877	(2,4: 0.000)	2.877	1.42
	0.12	(2,4: 0.65–0.75)	3.88–3.92	(2,4: 0.719)	3.895	1.42
	0.24	(2,4: 0.70–0.75)	12.0–12.6	(2,4: 0.725)	12.466	1.42
4	0.04	(2,4: 0.000)	2.850	(2,4: 0.000)	2.850	1.83
	0.24	(2,4: 0.000)	3.071	(2,4: 0.000)	3.071	1.84
	0.48	(2,4: 0.25–0.50)	6.9–7.2	(2,4: 0.298)	7.027	1.82

parameters, the program of ECA is executed 30 time to get an objective evaluation on its performance. To know whether or not the results given by ECA are really nearly global optimal solutions, when $s = 1$, the global optimal solutions of different generating rates are obtained by using Berman's pseudoenumerating method; when $s \geq 2$, the possible scopes of the global optimal solutions of different generating rates are estimated by using the results of numerical simulation experiments. The global optimal solutions (scopes) are given in Table 8.9, where $(\cdot)^O$ indicates the global optimal solution (scope), $(\cdot)^b$ indicates the best solution among 30 runs of ECA, and the time is the actual computing time in seconds.

The group of environment parameters for ECA is set as follows: $pop\_size = 20$, $max\_gen = 600$, $p_c = 0.4$, and $p_m = 0.2$. The results of the best run are given in Table 8.9, and the average performance of 30 runs is given in Table 8.10, where $(\cdot)^a$ indicates the average of 30 runs of ECA. From Table 8.9 we can see that ECA can always find the global or nearly global optimal solutions for all λ and s. These results show the global optimization ability of ECA. From Table 8.10 we see that all 30 runs of ECA for all given λ and

TABLE 8.10. Average Performance of 30 Runs on a Five-Node Network

s	λ	$(i,j{:}x)^a$	$R(\cdot)^a$	Variance	Computing Time (s)
1	0.01	(2,4: 0.000)	3.236	0.000	1.01
	0.06	(2,4: 1.612)	5.946	0.000	1.02
	0.12	(2,4: 0.913)	23.871	0.000	1.01
2	0.02	(2,4: 0.000)	2.877	0.000	1.42
	0.12	(2,4: 0.719)	3.895	0.000	1.42
	0.24	(2,4: 0.725)	12.466	0.000	1.42
4	0.04	(2,4: 0.000)	2.850	0.000	1.83
	0.24	(2,4: 0.000)	3.071	0.000	1.84
	0.48	(2,4: 0.298)	7.027	0.000	1.82

TABLE 8.11. Comparison of ECA and a Pseudoenumerating Method ($s = 1$)

	Optimal			Best		
λ	$(i,j{:}x)$	$\overline{R}(\cdot)$	Time (s)	$(i,j{:}x)$	$\overline{R}(\cdot)$	Time (s)
0.01	(45,9: 0.00)	7.06	42.64	(45,35: 0.00)	7.06	32.52
0.02	(23,43: 8,37)	9.00	42.64	(23,43: 8.37)	9.00	32.52
0.03	(23,43: 9.56)	11.66	42.64	(23,43: 9.54)	11.66	32.52
0.05	(23,43: 8.76)	22.05	42.64	(23,43: 8.76)	22.05	32.52
0.07	(45,9: 0.00)	65.87	42.64	(45,9: 0.00)	65.87	32.52

s converge to the global optimal or nearly optimal solutions (the variances of fitness values of 30 runs are all zero). This illustrates that ECA has a good convergence property.

To get the experience of the $M/G/s$ SQM model and ECA on large-scale network problems, a network with 64 nodes and 188 edges is randomly generated as follows:

1. Generate a complete graph with 64 nodes. The length of an edge (i,j) is a large constant 10,000 if $i = j$, otherwise is a random integer value in (10,80).
2. Pick out an arc randomly and delete it if this does not make the graph disconnected.
3. Repeat step 2 until only 188 edges remain.

The weights for the nodes of the network are assumed to be equal (i.e., $h_k = \frac{1}{64}$, and the first and second moments of service time of a demand are assumed to be $\overline{S_k} = \overline{S_k^2} = 1.0$ ($k = 1,2,...,64$), and the travel speed of the server $\xi = 10.0$ and $\beta = 1$.

The environment parameters for ECA are set as follows: $pop\_size = 40$, $max\_gen = 600$, $p_c = 0.4$, and $p_m = 0.2$. For the case of $s = 1$, the computing times for ECA and Berman's pseudoenumerating method for finding the optimal locations are compared in Table 8.11. For the case of $\lambda = 0.12$ and $s = 2$, the $M/G/s$ SQM model on the 64-node network can be solved using ECA. The computing results are: the best solution is (43,35 := 0.0), the best fitness (best average response time) is 18.041, and the computation time is 36.0 seconds. To check if this is a nearly global optimal solution, numerical simulation experiments are executed. Due to the huge amount of computation time [for a given point on edge (43, 35) the simulation experiments require 39.4 s, so if each edge is divided into 100 equal intervals and simulation experiments are carried out at each interval point on the entire network, it will cost 6.56×10^5 s, more than one week], only the experiments on edge (43, 35) are done.

From Table 8.10 it is shown that ECA used less computing time in finding a nearly optimal location than did a pseudoenumerating method on the 64-

node network. One thing should be noted: that the computing time used by ECA is usually more than the actual computing time to reach the nearly optimal location because 600 generations are a little longer to ensure the convergence of ECA. From the results, the approximated average response time also coincides with the simulation results, and the best location using ECA is the one suggested by the simulation results, although it requires only 35.52 sec. Since 64-node network is randomly generated, it can be expected that the proposed $M/G/S$ SQM model and ECA procedure will be applicable to general large-scale network problems. More details of experiment can be found in [255].

9 MANUFACTURING CELL DESIGN

9.1 INTRODUCTION

Group technology, established in 1959 by Mitrofanov, is now receiving considerable interest from academic researchers and practitioners. The concept is a method that identifies and exploits the similarity among the attributes of a set of objects. Cellular manufacturing is an application of group technology to organize cells that contain a set of machines to process a part family [81]. The important areas of group technology applications are classification and coding, process planning, part family and machine cell design, group technology layout, and group scheduling. In a recent automated manufacturing environment, cell configuration is a primary focus of research. Flexible manufacturing systems (FMSs), computer-integrated manufacturing (CIM), and just-in-time (JIT) are examples of automated cells. Wemmerlov and Johnson [658] summarized the reasons why firms established cellular manufacturing:

- To reduce throughput time
- To reduce work-in-process inventory
- To improve part/product quality
- To reduce response time to customer orders
- To reduce move times
- To increase manufacturing flexibility

Also, employee/worker benefits include increased worker flexibility, teamwork, reduced frustration, and improved status and job security [185]. The procedure of the machines and parts to form cells in group technology is called *manufacturing cell design* (MCD). The MCD problem has been proved to be an NP-hard problem [471]. To solve this problem, a number of methods have been developed during the past two decades. A number of attempts have recently been made to apply evolutionary search algorithms to the MCD problem. Simulated annealing [69, 116, 285, 421, 639], tabu search [598] and genetic algorithms [62, 274, 333, 334, 336, 350, 462, 463, 639] are very efficient evolutionary search approaches. Among them, genetic algorithms offer the best flexibility and extensibility for solving this complicated problem. In this chapter we review recent literature on solving the MCD problem using genetic algorithms and illustrate in detail typical implementations of genetic algorithms. In Section 9.2 we describe the basic concept of the MCD problem. In Section 9.3 we review traditional approaches to the MCD problem. Section 9.4 contains several approaches for the MCD problem using genetic algorithms. Finally, in Section 9.5 we discuss how to apply genetic algorithms to MCD problems with alternative process plans.

9.2 MANUFACTURING CELL DESIGN

The group technology concept differs from traditional job-shop arrangements. If the number of products being produced is small and the demand for these products is large (i.e., large lot sizes), a product layout is the most efficient way to mass produce the products. A product layout arranges a set of different machines in a linear flow and dedicates each line to a particular product, shown in Figure 9.1(a). Since a line is dedicated to one product and the machines are already arranged for the correct flow, setups are eliminated and work-in-progress (WIP), throughput time, and material handling are greatly reduced. Once the number of parts starts to increase and the demand for each part decreases (i.e., small to medium lot sizes), it is not cost-effective to dedicate a particular line to each product, owing to the large number of duplicate machines and additional space that would be needed. Therefore, a functional or job-shop layout is often used where similar machines and corresponding skilled personnel are located together in functional areas [Figure 9.1(b)]. The number of identical machines is reduced since machine utilization is increased by sharing demand across all products, not simply a single product. A functional layout offers the greatest flexibility in producing a large variety of parts (i.e., a large number of different part routings). New parts can be introduced without problems since a part can easily visit each of the necessary functional groups. Unlike the product layout, the machines would have to be rearranged and/or changed. The disadvantages of a typical job shop include the increased setup times, owing to the variety of parts that are produced; amount of material handling, since parts must now travel over the entire shop floor; scheduling difficulty, because of the number of parts and shared resources among these parts; and amount of WIP, because many different parts require the same machines, thus causing delays.

To overcome the disadvantages of a functional layout while maintaining routing flexibility for producing a variety of products in small to medium lot sizes, a layout based on the principles of group technology can be used. A

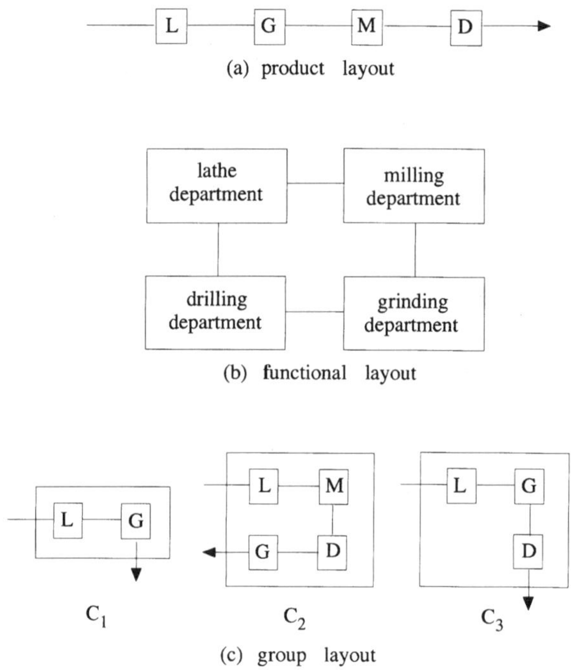

Figure 9.1. Material flow graph.

group layout can be considered a combination job-shop layout and product layout where a set of dissimilar machines are grouped together to process a set of similar parts [Figure 9.1(c)]. The machines are generally arranged in a linear or U-shaped layout. Therefore, a family of parts is dedicated to a very small job shop where one major flow pattern exists.

The MCD problem is an important step in the development and implementation of the Group layout and cellular manufacturing systems. A part family may consist of groups of parts requiring similar and sometimes identical operation processes, materials, and tools. The machines required to produce a family of parts are grouped to form a manufacturing cell. To illustrate the MCD problem, consider the machine part matrix shown in Figure 9.2(a). The (i,k)th element of the matrix is 1 if the jth part is processed by the kth machine; otherwise, it is 0.

After rearranging rows and columns in Figure 9.2(a), its final groups are presented in Figure 9.2(b). These final groups correspond to cells used to form a manufacturing system. Three machine cells, $C_1 = \{2,5\}, C_2 = \{3,4,6\}$, and $C_3 = \{1,7\}$, and corresponding part families, $F_1 = \{1,7\}, F_2 = \{3,4,6\}$, and $F_3 = \{2,5\}$, can be formed.

(a) part/machine matrix

machines \ parts	1	2	3	4	5	6	7
1		1			1		
2	1						1
3			1	1		1	
4			1	1		1	
5	1						1
6			1	1		1	
7		1			1		

(b) optimal clustering

machines \ parts	1	7	3	4	6	2	5
2	1	1					
5	1	1					
3			1	1	1		
4			1	1	1		
6			1	1	1		
1						1	1
7						1	1

Figure 9.2. Part machine matrix and optimal clustering.

9.3 TRADITIONAL MCD APPROACHES

The MCD problem can be classified as a combinatorial optimization problem because the part–machine incidence matrix contains n rows and m columns, offering $n!$ and $m!$ different permutations, respectively. Most of the methods for MCD problems have been developed on the basis of similarity coefficient methods, array-based methods, mathematical programming methods, graph and network methods, or other heuristic approaches. The objective in most of these methods is either to minimize intercell moves or to maximize cell independence considering shop floor factors such as production volume, cell size, alternative process plans, or machining times. However, most algorithms in the literature utilize only a part–machine incidence matrix that contains a prespecified single and fixed route. Relatively few researchers have addressed the issue of utilizing alternative operations, complete fixed alternative routings, or machine redundancy in determining cells and families. These fixed routings are often optimized to maximize machine utilization in a functional layout rather than utilizing the benefits of cellular manufacturing. A majority of the methods that include alternative operations and multiple routes utilize mathematical programming, limiting their usefulness for solving large-scale problems. Also, most of these techniques assume that demand is equal for all parts. However, when the demand among parts is usequal, which is most likely, the machine cells and part families formed using this assumption may be very inefficient. Although some methods offer generally superior results, no single technique has been shown to provide the best solution for a broad range of applications. These approaches typically involve single-criterion objective functions and do not permit an interchange of evaluation functions or the selective use of constraints. Also, most MCD algorithms cannot identify all naturally occurring clusters as well as finding solutions with a constrained number of clusters. A majority of the methods are unable to include other manufacturing information, such as part demand, alternative operations, and redundant machinery.

9.3.1 Similarity Coefficient Methods

A similarity coefficient method uses a similarity coefficient between each pair of objects and some specified threshold value to group them into cells. The threshold value is a similarity level at which two or more clusters are joined together. An example of similarity coefficient S_{ij} measure between two machines, i and j, is as follows [404]:

$$S_{ij} = \frac{N_{ij}}{N_i + N_j - N_{ij}} \tag{9.1}$$

where N_i is the number of parts visiting machine i and N_{ij} is the number of common parts visiting machines i and j. To illustrate the similarity coefficient of equation (9.1), consider the matrix in Figure 9.2(a). The similarity coefficients between each pair of machines are calculated as follows:

$$S_{13} = \frac{N_{13}}{N_1 + N_3 - N_{13}} = \frac{0}{2 + 3 - 0} = 0$$

$$S_{46} = \frac{N_{46}}{N_4 + N_6 - N_{46}} = \frac{3}{3 + 3 - 3} = 1$$

In the threshold value of similarity coefficient $S_{ij} = 1$, we can obtain two cells and corresponding two-part families as follows:

$$S_{25} = 1, \qquad S_{34} = S_{36} = S_{46} = 4, \qquad S_{17} = 1$$

$$S_{12} = S_{15} = S_{23} = S_{24} = S_{37} = 0$$

McAuley [446] was the first to use a single linkage cluster analysis (SLCA) method based on a similarity coefficient to form the machine-part groups. The coefficient is defined as "the number of parts visiting both machines divided by the number visiting either of the two machines." Subsequent modifications and an advanced version of the similarity measures have been proposed by Seifoddini and Wolfe [569] and Gupta and Seifoddini [273]. These methods present the user with a hierarchy of solutions from which the best alternative is chosen.

9.3.2 Array-Based Methods

Array-based methods use a machine–part incidence matrix as input and repeats to rearrange the matrix rows and columns to form final groups. McCormick et al. [447] proposed a quadratic assignment-based grouping method called the *bond energy algorithm* (BEA). They defined the bond energy, the sum of the bond strengths, between any two adjacent elements in a incidence matrix as their product. The objective is to maximizing the bond energy of the matrix, resulting in a block diagonal matrix, if one exists. Bhat and Haupt [61] have extended the basis BEA model to improve its computational efficiency. King [366, 367] presents an array-based method called rank order clustering (ROC). The method is described as follows:

Step 1. Assign binary weights to each row and column of the incidence matrix.

Step 2. Convert weights for each row and column of the matrix to their decimal equivalents.

Step 3. Rearrange the rows and columns alternately in decreasing order of magnitude until final machine–part groups emerge.

King and Nakornchai [368] propose an extended version of the basic ROC model to solve large problems. Subsequent improvements of ROC were developed by Chan and Milner [91], Chandrasekharan and Rajagopalan [93], and Kusiak and Chow [385].

9.3.3 Mathematical Programming Methods

One of the best known methods for manufacturing cell design is the mathematical programming model. Jensen [328] developed a dynamic programming model for solving the p-median model. The grouping measure of the model is the maximization of the total sum of similarity values for objects placed together in groups, under the condition that each object is assigned exactly once to a group, and assuming that the number of groups (p) is known. Kusiak and Heragu [388] provided a p-median model using integer programming for cell design and illustrated the importance of considering an alternative process plan. The p-median model, however, does not allow control of the cell size. An advanced p-median model has been proposed by Srinivasan, Narendran, and Mahadevan [582]. Ham and Han [278] proposed an integer goal programming model for part family formations in classification and coding. The objective of the model, which uses the absolute Minkowski distance, is to minimize these distances lexicographically. Kasilingam and Lashkari [348] proposed a nonlinear 0–1 integer programming model for manufacturing cell design considering alternative process plans. The objective of the model is to maximize the compatibility indices between machines and parts, which are defined based on tooling requirements and processing times. More recently, Adil, Ragamani, and Strong [6] developed a nonlinear integer programming model for manufacturing cell design considering alternative process plans. The model combines the evaluation procedure by considering the minimization of a weighted sum of the voids and intercell moves. A void indicates that a machine assigned to a cell is not required for the processing of a part in the cell. However, most of these methods have some limitations in solving large problems, owing to the computational complexity.

9.3.4 Graph and Network Methods

Rajagopalan and Batra [523] proposed a cell design method using a graph partitioning algorithm that utilizes cliques. Their algorithm uses a graph where the vertices represent machines and each edge represents parts processed by the two machines, representing the extremities of the edge. Lee and Garcia-Diaz [397] transformed the cell design problem into a network flow formulation and used a primal–dual algorithm developed by Bertsekas and Tseng [58] to solve for the cells. This algorithm is more efficient computationally than the p-median model of Kusiak and Heragu [388]. Other graph and network approaches include the minimum spanning tree of Ng [483], a heuristic graph partitioning approach by Askin and Chiu [21], and a network approach using a modified version of the Gomory–Hu algorithm by Vohra et al. [642]. More recently, Lee and Garcia-Diaz [398] proposed a three-phase network flow–based approach for minimizing intercellular part moves in cell design problems.

9.4 GENETIC ALGORITHM APPROACHES

9.4.1 Representation and Genetic Operators

To solve the cell design problem using genetic algorithms, the representation scheme determines how the problem is structured in a genetic algorithm as well as the genetic operators that can be used. Therefore, choosing an appropriate representation scheme is the first step in applying genetic algorithms to a cell design problem. A number of representation schemes for cell design problems have been presented: (1) cell number–based representation and (2) order-based representation.

Cell Number–Based Representation. A common integer programming formulation of the cell design problem consists of the following assignment of variable declarations and constraints, which ensure that each machine and part is assigned to only one cell and family, respectively:

$$x_{ic} = \begin{cases} 1, & \text{if machine } i \text{ is assigned to cell } c \\ 0, & \text{otherwise} \end{cases}$$

$$y_{jc} = \begin{cases} 1, & \text{if part } j \text{ is assigned to part family } c \\ 0, & \text{otherwise} \end{cases}$$

$$\sum_{c=1}^{k} x_{ic} = 1, \qquad i = 1,...,m$$

$$\sum_{c=1}^{k} y_{jc} = 1, \qquad j = 1,...,n$$

k = number of cells (families) specified

m = number of machines

n = number of parts

For an m machine/n parts problem with k cells, there are a total of $k(m + n)$ variables and $m + n$ constraints. As the number of parts and machines increases, the models become too large to be stored in memory or become computationally intractable [397]. To overcome these limitations, an integer programming model with the following set of variable declarations has been proposed independently by Joines et al. [333, 336], Venugopal and Narendran [639], and Moon, Kim, and Gen [463]:

$x_i = c$, machine i is assigned to cell c

$y_j = c$, part j is assigned to part family c

This model reduces the number of variables by a factor of k to $(m + n)$ and eliminates the constraints by employing a set notation. Each part and machine variable is equal to the number of its assigned family or cell. The value of a gene or chromosome in this scheme is that it represents the cell number, with the position of the gene corresponding to the machine number. An example of a chromosome representing a solution is as follows:

[3 1 1 2 3 1 2]

where the length of the chromosome depends on the number of machines considered in the process of production and each value of the gene indicates the cell number to which that machine has been assigned. We know from this that the number of machines and cell numbers considered in this problem were seven machines and three cells. Machines 2, 3, and 6 are assigned to cell 1; machines 4 and 7 are assigned to cell 2; and machines 1 and 5 are assigned to cell 3.

Order-Based Representation. Order-based representation is widely applied with genetic algorithms. Many representation schemes have been developed for order problems, but permutation representation is perhaps the most natural representation of a tour. Cheng et al. [101], Billo, Tate, and Bidanda, [63],

Kazerooni [350], and Kazerooni, Luong, and Abhary [351] applied it to cell design problems. For example, for a problem with seven machines (or parts) and three cells, a chromosome can be represented as follows:

$$[2\ 7\ 4 * 3\ 1\ 6 * 2\ 5]$$

In this chromosome, cell locations are identified with an asterisk (*) that designates the cutoff for the part numbers comprising the cell. The first cell contains machines 2, 7, and 4; the second cell contains machines 3, 1, and 6; and the third cell contains machines 2 and 5.

Crossover and mutation are the main genetic operators for cell design problem using genetic algorithms. In cell number–based representation, simple crossover operations such as one- or two-cut point crossover are used, whereas several crossover operators have been proposed to deal with order-based representation, such as partially mapped crossover (PMX), order crossover (OX), cycle crossover (CX), edge recombination crossover (ER), and so on [219]. Among them, the PMX operator is widely used in the order-based representation for cell design problems [333]. Mutation is performed as random perturbation. Each value of the chromosome is randomly selected to be mutated with the mutation rate. An example of a mutation operator is given as follows:

$$\begin{array}{c}\text{mutation point}\\ \downarrow\\ [3\ 7\ \underline{7}\ 1\ 6]\end{array}$$

In this chromosome the third gene is selected for mutation. The value of the gene is replaced by a random integer value with [1, 10]. If the value is 4, the offspring after mutation is

$$[3\ 7\ \underline{4}\ 1\ 6]$$

In the order-based representation, the mutation operator randomly selects two genes within a chromosome and exchanges their positions, or selects a gene at random and inserts it in a random position. Several mutation operators have been applied for order-based representation, such as inversion mutation, insert mutation, displacement mutation, reciprocal mutation, and so on.

Through empirical study, Joines [333] found the following five generic order-based operators to be well suited for the cellular formation problem.

1. Mutation operators
   (a) Single-point mutation
   (b) Swap mutation
   (c) Inversion mutation
2. Crossover operators
   (a) Partially mapped crossover (PMX)
   (b) Uniform order-based crossover (UOX)

9.4.2 Joines' Order-Based Approach

Joines [333] used two different order-based genetic algorithms to determine the best permutation of machines and parts that would maximize total bond energy. Given a particular row permutation (π) and column permutation (ρ), the total bond energy (TBE) can be defined by

$$\text{TBE}(\pi,\rho) = \frac{1}{2} \sum_{i=1}^{m} \sum_{j=1}^{n} a_{ij}(a_{i,j-1} + a_{i,j+1} + a_{i-1,j} + a_{i+1,j}) \qquad (9.2)$$

where

$$a_{0,j} = a_{m+1,j} = a_{i,0} = a_{i,n+1} = 0$$

Bonds are formed when a positive element (i.e., a part j needing an operation performed on a machine i) is arranged with other machines operating on the same part j [i.e., placed on either side of machine i (i.e., on top and bottom) in the sequence of the rows] or when other parts requiring operation on the same machine i are placed on either side of part j (i.e., to the left and right) in the sequence of columns. The maximum bond energy of any given element is four (i.e., a 1 would appear to the immediate left, right, top, and bottom of the element). Since the diagonals are not included in calculating the total bond energy (i.e., no dependency exists between columns and rows), the problem can be separated into a row bond energy (RBE) and a column bond energy (CBE), with the TBE being equal to the sum of the row and column bond energies. Therefore, the rows can be sequenced first and then the columns, without any loss of generality.

$$\max_{\pi} \text{RBE}(\pi) = \sum_{i=1}^{M+1} \sum_{j=1}^{N} a_{\pi(i),j} a_{\pi(i-1),j}$$

For the solution technique, the machines (rows) are clustered first to form machine cells and then the procedure is repeated for the parts (columns).

TABLE 9.1. Parameters Used in Order-Based Experiments

Parameter	Value or Number of Operations
Single-point mutation	4
Swap mutation	4
Inversion mutation	4
Partially mapped crossovers	6
Uniform order-based crossover	6
Probability of selecting the best individual, q	0.1
Maximum number of generations, T	10,000

Genetic Operators. Order-based operators that work well for TSP may not be effective for cell formation problems. Order-based crossover (OX) has been shown to be better than partially mapped crossover (PMX) and cyclic crossover (CX). However, the OX operator is not suited for the clustering problem because a TSP solution allows a salesman to start in any city and requires only that all the cities be visited and that the salesman return to the city of origin. Therefore, the two chromosomes [1 4 0 5 2 3 6] and [2 3 6 1 4 0 5] are considered equivalent. The OX operator uses this phenomenon to develop new solutions. However, in a clustering problem, the last element in the individual is not considered adjacent to the first element as in a TSP. Therefore, the PMX crossover operator is used instead of the OX operator.

Computational Experience. The order-based genetic algorithm will be applied to two manufacturing cell data sets from the literature. The genetic algorithm employs a ranking scheme based on the normalized geometric distribution (see Section 9.4.4), an elitist model, and the order-based genetic operators of the preceding section. The genetic algorithm parameters in Table 9.1 were used for both data sets. The number of genetic operations given in Table 9.1 are applied deterministically (e.g., single-point mutation would applied to four randomly chosen parents). The order-based genetic algorithm clustering tool is applied to two test data sets. The first data set, taken from King and Nakornchai [368], is a 16×43 problem that contains two bottleneck machines (i.e., machines 5 and 7; Figure 9.3). The solution is shown in Figure 9.4. The second data set, a 20×35 matrix, came from Burbidge [80]. The original data set and order-based genetic algorithm solution can be seen in figure 9.5 and 9.6. The cell boundaries were determined from visual inspection. The order-based genetic algorithm approach was able to find the best solution in terms of total bond energy. However, the cell designer must inspect the final output visually to obtain the actual members of each cell. This was not difficult except in the King data set, where it was difficult to determine to which cell the bottleneck machines should be assigned. Finally, the algorithm does not necessarily produce solutions with cells formed along the

	0	1	2	3	4	5	6	7	8	9	10	11	12	13	14	15	16	17	18	19	20	21	22	23	24	25	26	27	28	29	30	31	32	33	34	35	36	37	38	39	40	41	42
0:																																				1					1		
1:		1								1																		1				1					1	1		1		1	
2:							1										1																	1	1	1							
3:					1				1					1					1		1		1						1														
4:					1			1	1					1	1	1			1		1		1						1				1								1		1
5:	1	1				1	1	1				1	1	1			1		1				1									1	1	1			1		1	1		1	1
6:	1												1												1																		
7:	1	1	1					1	1		1	1			1				1	1	1		1	1			1	1			1						1	1			1		1
8:		1		1						1								1										1				1					1	1		1		1	
9:	1											1	1												1	1					1								1				
10:			1						1											1				1			1			1													
11:											1											1		1			1			1													
12:			1																					1																			
13:		1				1											1																		1								
14:					1									1					1		1												1								1		1
15:		1					1			1								1														1					1	1				1	

Figure 9.3. Original part–machine incidence matrix for the 16 × 43 King problem.

	21	29	23	26	10	14	15	35	34	16	33	6	31	41	36	1	39	27	37	9	17	3	28	4	13	18	22	7	32	42	40	20	8	2	19	30	25	24	12	0	11	38	5
2:								1	1	1	1	1																															
13:									1	1						1																											1
0:														1	1																												
1:													1	1	1	1	1	1	1	1																							
8:													1	1	1	1	1	1	1	1	1	1																					
15:												1	1	1	1	1			1	1	1																						
5:										1	1	1	1	1	1	1	1								1	1	1	1	1	1									1	1	1	1	1
9:																																				1	1	1	1	1	1	1	
6:																																						1	1	1			
12:			1																															1									
11:	1	1	1	1	1																																						
10:		1	1	1																													1	1	1								
7:			1	1	1	1									1	1		1	1							1	1	1		1	1	1	1	1	1	1				1	1		
4:						1	1																1	1	1	1	1	1	1	1	1	1	1										
3:																							1	1	1	1	1					1	1										
14:																								1	1	1			1	1	1	1											

Figure 9.4. Order-based genetic algorithm solution to the King data set with a total bond energy = 154.

	0	1	2	3	4	5	6	7	8	9	10	11	12	13	14	15	16	17	18	19	20	21	22	23	24	25	26	27	28	29	30	31	32	33	34
0:	1		1																	1			1		1										
1:		1					1			1		1	1					1						1			1				1				
2:	1		1		1										1		1												1		1				
3:		1					1					1	1											1			1								
4:								1						1		1																		1	
5:								1						1		1			1			1				1								1	
6:	1		1		1										1		1			1			1												
7:	1		1		1										1		1			1			1		1				1						
8:								1						1					1			1				1									
9:								1						1		1			1			1				1									
10:				1		1			1		1										1							1		1		1			1
11:				1		1			1		1										1												1		
12:		1										1	1											1											
13:		1					1			1		1	1					1						1			1				1				
14:				1		1			1		1										1							1		1					
15:				1		1			1		1										1							1		1		1			
16:	1		1		1										1		1			1			1		1				1						
17:		1								1		1	1					1						1							1				
18:				1					1		1										1							1		1		1			
19:								1						1					1				1			1									

Figure 9.5. Original part–machine incidence matrix for the 20 × 35 Burbidge problem.

	34	31	5	20	3	10	8	27	29	6	26	11	1	23	12	9	30	17	21	18	25	7	13	15	33	32	28	14	16	4	0	2	22	19	24
11:			1	1	1	1	1																			1									
14:			1	1	1	1	1	1	1																										
10:	1	1	1	1	1	1	1	1	1																										
15:		1	1	1	1	1	1	1	1																										
18:		1		1	1	1	1	1	1																										
19:																				1	1	1	1										1		
8:																			1	1	1	1	1												
9:																			1	1	1	1	1	1											
5:																			1	1	1	1	1	1	1										
4:																						1	1	1	1										
0:																															1	1	1	1	1
6:																												1	1	1	1	1	1	1	
7:																											1	1	1	1	1	1	1	1	1
16:																											1	1	1	1	1	1	1	1	1
2:																	1										1	1	1	1	1	1			
17:												1	1	1	1	1	1	1																	
13:										1	1	1	1	1	1	1	1	1																	
1:										1	1	1	1	1	1	1	1	1																	
3:										1	1	1	1	1	1																				
12:												1	1	1	1																				

Figure 9.6. Order-based genetic algorithm solution to the 20 × 35 Burbidge problem with a total bond energy = 211.

TABLE 9.2. Process Plans and Production Volumes for Parts

Part	Process Plan	Production Volume
1	$M_1 \rightarrow M_3 \rightarrow M_4$	5
2	$M_2 \rightarrow M_5 \rightarrow M_1$	10
3	$M_1 \rightarrow M_2$	2
4	$M_1 \rightarrow M_5 \rightarrow M_2$	7

diagonal, due to the fact that the machines are grouped into cells separately from parts into families. Even though this order-based genetic algorithm worked well for all data sets, it does not group cells and families simultaneously, and it requires visual inspection of the output in the determination of cells and families.

9.4.3 Moon and Kim's Approach

Mathematical programming approaches for the clustering problem are nonlinear or linear integer programming problems [69, 386, 388]. These formulations suffer from three critical limitations. First, because of the resulting nonlinear form of the objective function, most approaches do not concurrently group machines into cells and parts into families [69]. Second, the number of machine cells must be specified a priori, which affects the grouping process and potentially obscures natural cell formations in the data. Third, since the variables are constrained to integer values, most of these models are computationally intractable for realistically sized problems [397]. However, a common integer programming formulation of the cell design problem can reduce the size of the variables using a genetic algorithm with cell number–based representation. Moon and Kim [463] used a genetic algorithm to solve an integer programming formulation model for the design of machine cells which maximizes the parts flow between machines within the same cell. For this, a set of closely related machines should be included in the same cell. The computation of parts flow between machines is based on process plans, production volumes, material handling time, and available transfer device capacity per trip.

Problem Statement. Parts that are not similar in shape may require a similar manufacturing process. Since similar processes are required for all family members, a machine cell can be built to manufacture the family. From the process plan sheets, a list of parts and their process plans and production volumes are shown in Table 9.2. The relationship between machines can be established in terms of arc weights representing material flow between machines when using the data of Table 9.2 as input. The basic concept used in the computation of material flow between machines is based on the process

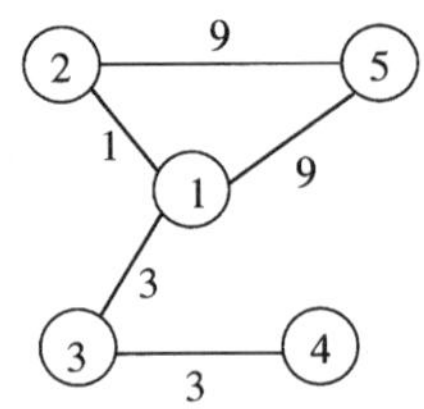

Figure 9.7. Material flow graph.

plans, production volumes, and available capacity of transfer device per trip. For example, the available capacity of transfer device per trip for production is 2 units; then the material flow graph can be represented as shown in Figure 9.7. In this graph, the nodes and arcs represent machines and their relationships defined as the value of total parts flow between machines, respectively. Designing the machine cells essentially means recognizing and using the relationship between machines for the purpose of minimizing intercell moves and maximizing the total number of parts processed within cells concurrently. Thus a set of closely related machines should be included in the same cell. In this process, the cell size, defined as a number of machines assigned to each cell, and the number of cells are determined in the design process.

Integer Programming Model. A part or part family is processed within a cell with minimum interaction with other cells. This is clearly equivalent to maximizing the total number of parts processed within cells. The model proposed is based on the objective of maximizing the total number of parts processed within cells. For the machine cell design the following preconditions are adopted.

Assumptions

1. Part families are formed by the classification and coding.
2. Each part has a predetermined process plan from the process plan sheets.

Notation

i and j are machine indexes ($i,j = 1,2,\ldots,N_m$).
k is a part index ($k = 1,2,3,\ldots,N_p$).
c is a cell index ($c = 1,2,3,\ldots,N_c$).
pv_k is the production volume of part k.
cd_k is available transfer units per trip for part k using a transfer device.
U_c is the upper limit of cell size.
nt_{ijk} is the number of trips made by part k between machines i and j:

$$nt_{ijk} = \left\lceil \frac{pv}{cd_k} \right\rceil$$

where $\lceil w \rceil$ indicates the smallest integer value greater than or equal to w.

Variable

$$x_{ic} = \begin{cases} 1, & \text{if machine } i \text{ is assigned to cell } c \\ 0, & \text{otherwise} \end{cases}$$

The overall integer programming model can be formulated as follows:

$$\max \quad z = \frac{1}{2} \sum_{c=1}^{N_c} \sum_{i=1}^{N_m} \sum_{j=1}^{N_m} \sum_{k=1}^{N_p} nt_{ijk} x_{ic} x_{jc} \tag{9.3}$$

$$\text{s.t.} \quad \sum_{c=1}^{N_c} x_{ic} = 1, \qquad \forall i \tag{9.4}$$

$$1 \le \sum_{i=1}^{N_m} x_{ic} \le U_c, \qquad \forall c \tag{9.5}$$

$$x_{ic} = \{0,1\}, \qquad \forall i,c \tag{9.6}$$

The objective function (9.3) focuses on maximizing the total moves of parts processed within cells as a result of minimizing the intercell moves. Constraints (9.4) guarantee that all machines are allocated to a cell. Constraints (9.5) ensure that the number of machines assigned to each cell does not exceed the upper limit of cell size. Constraints (9.6) indicate the binary variables.

Genetic Algorithm

Representation and Initialization. The cell number–based representation scheme is applied in this problem. An example of the chromosome to represent a solution is as follows:

[2 1 3 1 2 3 2]

where the length of the chromosome depends on the number of machines considered in the process of production and each gene value indicates the cell number to which that machine has been assigned. From this, we know that the number of machines and cell numbers considered in this problem were seven machines and three cells. Machines 2 and 4 are assigned to cell 1; machines 1, 5, and 7 are assigned to cell 2; and machines 3 and 6 are assigned to cell 3. The second step in the genetic algorithm is to initialize the population of chromosomes. The initialization process can be executed with either a randomly generated or a well-adapted population. The random approach is used to initialize the population in this model.

Evaluation. Each chromosome is evaluated using some measure of fitness. A fitness value is computed for each chromosome in the population and the objective is to find a chromosome with the maximum fitness value. We use the objective function defined in the integer programming model for the fitness function. For the cell design problem, the fitness is simply equal to the value of the objective function as follows:

$$\text{eval}(st_i) = z(st_i), \qquad i = 1,2,\ldots,pop\_size \tag{9.7}$$

By the way, the genetic algorithm may yield illegal chromosomes in the sense of violating constraint (9.5). To handle illegal chromosomes, a repair technique is developed as follows:

```
Repairing Procedure
begin
for i = 1 to pop_size
 repeat
   check the U_c and N_c in each chromosome;
   if (not feasible) then
    begin
    select machines having overflow cell number;
    replace current cell numbers by new cell numbers
     generated randomly;
    update data:
   end
 until (U_c and N_c become feasible)
end
```

Genetic Operators. To create the next generation, a new set of chromosomes called *offspring* is formed by the execution of genetic operators such as crossover and mutation. Two-cutpoint crossover is used here. For example, chromosomes st_1 and st_2 were selected for crossover as follows, and the positions of the cutpoint were randomly selected after the second and fifth genes:

$$st_1 = [\underline{3\ \ 1}\ |\ 1\ \ 2\ \ 3\ |\ \underline{1\ \ 2}]$$

$$st_2 - [\underline{2\ \ 3}\ |\ 2\ \ 1\ \ 1\ |\ \underline{2\ \ 1}]$$

The offspring obtained by exchanging the left and right parts of their parents would be as follows:

$$op_1 = [2\ \ 3\ |\ 1\ \ 2\ \ 3\ |\ 2\ \ 1]$$

$$op_2 = [3\ \ 1\ |\ 2\ \ 1\ \ 1\ |\ 1\ \ 2]$$

The mutation is performed after crossover as random perturbation. It selects

TABLE 9.3. Data for a Problem of Ten Parts and Eight Machines

Part	Process Plan	Production Volume
1	$M_1 \to M_3 \to M_4$	1
2	$M_2 \to M_5 \to M_1$	1
3	$M_1 \to M_2$	3
4	$M_1 \to M_5 \to M_2$	1
5	$M_3 \to M_8 \to M_6$	1
6	$M_4 \to M_6 \to M_8 \to M_3$	4
7	$M_6 \to M_7 \to M_2$	2
8	$M_5 \to M_2$	1
9	$M_6 \to M_7 \to M_8$	1
10	$M_7 \to M_8 \to M_2$	2

a gene randomly and replaces it in a cell number generated within $[1, N_c]$. If the fifth gene of the chromosome st_1 is selected for mutation and the random number generated is 1, the chromosome after mutation would be

$$op_3 = [3 \;\; 1 \;\; 1 \;\; 2 \;\; \underline{1} \;\; 1 \;\; 2]$$

Selection. Selection strategy is concerned with the problem of how to select chromosomes from population space. It may create a new population for the next generation based on either all parents and offspring or part of them. A mixed selection strategy based on roulette wheel and elitist selection is adopted for cell design problems.

Numerical Example. Numerical examples have been provided to verify the approach proposed in previous sections. The first example considered has eight machines and 10 parts. The process plan and production volume for each part are given in Table 9.3. We assumed that the available capacity of transfer device per trip for production is 1 unit. The total parts flow between machines is calculated in Table 9.4. It is desired to find three cells with cell size equal to $U_c = 3$ or 4. To solve the problem using the proposed genetic algorithm, the genetic parameters are set as follows: $max\_gen = 100$, $pop\_size = 20$, $p_c = [0.5, 0.7]$, $p_m = [0.1, 0.3]$, and run size $= 10$. In this setting, the best crossover and mutation probabilities are $p_c = 0.5$ and $p_m = 0.1$. For each case we have obtained the same result over all run times. For the case of three cells with cell size equal to $U_c = 4$, the best chromosome is obtained as [1 1 2 3 1 2 2 2], which implies that machines 1, 2, and 5 are assigned to cell 1, machines 3, 6, 7, and 8 to cell 2, and machine 4 to cell 3. The fitness value is 24. The results are reported in Table 9.5. For the large example, we selected a problem with 46 parts and 18 machines. It is desired to find four or five cells with cell size equal to $U_c = 5$ or 6. The

TABLE 9.4. Total Parts Flow Between Machines

	M_1	M_2	M_3	M_4	M_5	M_6	M_7	M_8
M_1	0	3	1	0	2	0	0	0
M_2		0	0	0	3	0	2	2
M_3			0	1	0	0	0	5
M_4				0	0	4	0	0
M_5					0	0	0	0
M_6						0	3	5
M_7							0	3
M_8								0

production volume of each part is 1 constantly. The parts flow between machines can be calculated as shown in Table 9.6. The genetic parameters are set as follows: $max\_gen = 300$, $pop\_size = 100$, $p_c = 0.5$, and $p_m = 0.1$. In this numerical experiment, we have run the genetic algorithm five times and obtained a best solution for each case. The best solution was obtained in the case of four cells with a five-cell size. The maximum parts flow within cells is 118. The results are summarized in Table 9.7.

From the experimental results, the approach proposed can be applied to solving complex and large problems for cell design efficiently.

9.4.4 Joines' Integer Programming Formulation Approach

To overcome limitations with order-based genetic algorithms as well as other techniques, Joines [332, 333, 336] developed a solution technique using genetic algorithms to solve the cell design problem formulated by integer programming model.

Chromosome Representation. For any genetic algorithm, a chromosome representation is needed to describe each individual in the population of interest. The representation scheme determines how the problem is structured in the

TABLE 9.5. Results of Cell Design

Number of cells = 3 Cell size = 4	Number of cells = 3 Cell size = 3	
Cell 1: M_1 M_2 M_5 Cell 2: M_3 M_6 M_7 M_8 Cell 3: M_4	Alternative 1: Cell 1: M_4 M_6 M_7 Cell 2: M_1 M_2 M_5 Cell 3: M_3 M_8	Alternative 2: Cell 1: M_1 M_2 M_5 Cell 2: M_3 M_4 Cell 3: M_6 M_7 M_8
Fitness value = 24	Fitness value = 20	Fitness Value = 20
Maximum parts flow: 34		

TABLE 9.6. Parts Flow Between Machines for Gas Turbine Model

	M_1	M_2	M_3	M_4	M_5	M_6	M_7	M_8	M_9	M_{10}	M_{11}	M_{12}	M_{13}	M_{14}	M_{15}	M_{16}	M_{17}	M_{18}
M_1													11					
M_2								1	32						3			
M_3										3				1				
M_4						1	1											
M_5									1		16	8			16	23		
M_6							2		1									
M_7																		
M_8																		
M_9													4		6			
M_{10}																		
M_{11}																		
M_{12}																	8	
M_{13}																		
M_{14}																		
M_{15}																		
M_{16}																		
M_{17}																		8
M_{18}																		

TABLE 9.7. Summary of Results

Number of Cells	Cell Size	Machine Cell	Fitness Value	Maximum Parts Flow
4	5	Cell 1: $M_1M_3M_{10}M_{13}M_{14}$ Cell 2: $M_2M_5M_9M_{15}M_{16}$ Cell 3: $M_4M_6M_7M_8$ Cell 4: $M_{11}M_{12}M_{17}M_{18}$	116	146
	6	Cell 1: $M_1M_2M_8M_9M_{11}M_{13}$ Cell 2: M_4 Cell 3: $M_3M_4M_6M_7M_{10}$ Cell 4: $M_5M_{12}M_{15}M_{16}M_{17}M_{18}$	118	
5	4	Cell 1: $M_5M_{12}M_{15}M_{16}$ Cell 2: $M_3M_{10}M_{11}$ Cell 3: $M_4M_6M_7M_8$ Cell 4: $M_1M_2M_9M_{13}$ Cell 5: $M_{14}M_{17}M_{18}$	109	
	5	Cell 1: $M_3M_8M_{10}$ Cell 2: $M_1M_6M_7M_{11}M_{13}$ Cell 3: M_{14} Cell 4: $M_4M_{12}M_{17}M_{18}$ Cell 5: $M_2M_5M_9M_{15}M_{16}$	113	

genetic algorithm and also determines the genetic operators that are used. Each individual or chromosome is made up of a sequence of genes from a certain alphabet. An alphabet could consist of binary digits (0 and 1), floating-point numbers, integers, symbols (i.e., A, B, C, D), matrices, and so on. In the representation of [336], the first m variables represent the machines and the last n variables are associated with the parts. Therefore, each chromosome is a vector of $m + n$ integer variables in the range of 1 to the maximum number of cells or families (k_{max}):

$$\text{chromosome} \rightarrow (\underbrace{x_1, x_2, \ldots, x_m}_{\text{machines}}, \underbrace{y_1, y_2, \ldots, y_n}_{\text{parts}})$$

Later, we will see that this representation can be extended to include the route (or process plan) variables as described in Section 9.5.2.

Selection Function. The selection of individuals to produce successive generations plays an extremely important role in a genetic algorithm. A probabilistic selection is performed based on the chromosome's fitness such that the better chromosomes have an increased chance of being selected. However, all the individuals in the population have a chance of being selected to reproduce into the next generation. A chromosome in the population can be selected more than once, with all chromosomes in the population having a chance of being selected to reproduce into the next generation. There are several schemes for the selection process: roulette wheel selection and its extensions, scaling techniques, tournament, elitist models, and ranking methods [219, 249, 455].

A common selection approach assigns a probability of selection, P_j, to each chromosome j based on its fitness value. A series of N random numbers is generated and compared against the cumulative probability, $C_i = \sum_{j=1}^{i} P_j$, of the population. The appropriate individual, i, is selected and copied into the new population if $C_{i-1} < U(0,1) \leq C_i$. Ranking methods only require the evaluation function to map the solutions to a totally ordered set, thus allowing for minimization and negativity. Ranking methods assign P_i based on the rank of solution i when all solutions are sorted. Normalized geometric ranking [339] defines P_i for each chromosome by

$$P[\text{selecting the } i\text{th chromosome}] = q'(1 - q)^{r-1} \tag{9.8}$$

where q is the probability of selecting the best chromosome; r the rank of the chromosome, where 1 is the best; P the population size; and

$$q' = \frac{q}{1 - (1 - q)^P}$$

Joines [332, 333, 336] used a normalized geometric ranking scheme and employed an elitist model in which the best individual from the preceding generation is always included in the current generation.

Genetic Operators. Genetic operators provide the basic search mechanism for genetic algorithms. The operators are used to create new solutions based on existing solutions in the population. There are two basic types of operators: crossover and mutation. Mutation operators tend to make small random changes in one parent to form one child in an attempt to explore all regions of the state space. Crossover operators combine information from two parents to form two offspring such that the two children contain a "likeness" (a set of building blocks) from each parent. The application of these two basic types of operators and their derivatives depends on the chromosome representation used.

Six float operators described by Michalewicz [455] were modified to work with the integer representation: *uniform mutation, multiuniform mutation, nonuniform mutation, multi-nonuniform mutation, boundary mutation, simple crossover, and arithmetic crossover* [336]. Two problem-specific genetic operators (*cell-swap crossover* and *cell-two-point crossover*) based on the proposed cell formation representation were also developed and shown to enhance the genetic algorithm's performance [336]. Let $\overline{X} = (x_1, x_2, \ldots, x_n)$ and $\overline{Y} = (y_1, y_2, \ldots, y_n)$ be two n-dimensional integer row vectors denoting individuals (parents) from the population. Each of these operators was used in the experiments described by Michalewicz and is discussed briefly below. Let a_i and b_i be the lower and upper bound, respectively, for each variable i. For the cell formation problem (i.e., the machine and part variables), the lower bound is 0 and the upper bound is equal to $k_{\max}$, the maximum number of cells and families, and P_j, the number of routes for part j for the machine–part variables and route variables, respectively. These operators are applied to the population a discrete number of times (see Table 9.8). Parents are randomly selected from the population to undergo the various operations.

Uniform Mutation. Randomly select one variable, j, and set it equal to a truncated uniform random number, $\lfloor U(a_i,b_i) \rfloor$, where $\lfloor x \rfloor$ is the largest integer less than or equal to x.

$$x_i' = \begin{cases} \lfloor U(a_i,b_i) \rfloor, & \text{if } i = j \\ x_i, & \text{otherwise} \end{cases} \tag{9.9}$$

Multiuniform Mutation. Apply equation (9.9) to all the variables in the parent.

Nonuniform Mutation. Randomly select one variable, j, and set it equal to a nonuniform random number based on equation (9.10). The new variable is equal to the old variable plus or minus a random displacement.

$$x_i' = \begin{cases} \lceil x_i + (b_i - x_i) f(G) \rceil, & \text{if } r_1 < 0.5 \\ \lfloor x_i - (x_i + a_i) f(G) \rfloor, & \text{if } r_1 \geq 0.5 \\ x_i, & \text{otherwise} \end{cases} \tag{9.10}$$

where

$$f(G) = \left[r_2 \left(1 - \frac{G}{G_{\max}} \right) \right]^b \tag{9.11}$$

r_1 and r_2 are uniform random numbers between 0 and 1, G is the current generation, $G_{\max}$ the maximum number of generations, b a shape parameter, and $\lceil x \rceil$ the smallest integer greater than or equal to x.

Multi-nonuniform Mutation. Apply equation (9.10) to all the variables in the parent.

Boundary Mutation. Randomly select one variable, j, and set it equal to either its lower or upper bound, where $r = U(0,1)$.

$$x_i' = \begin{cases} a_i, & \text{if } i = j, \quad r < 0.5 \\ b_i, & \text{if } i = j, \quad r \geq 0.5 \\ x_i, & \text{otherwise} \end{cases} \tag{9.12}$$

Simple Crossover. Generate a random number r from a discrete uniform distribution from 2 to $(m + n - 1)$ and create two new individuals ($\overline{X}'$ and $\overline{Y}'$) according to

$$x_i' = \begin{cases} x_i, & \text{if } i < r \\ y_i, & \text{otherwise} \end{cases} \tag{9.13}$$

$$y_i' = \begin{cases} y_i, & \text{if } i < r \\ x_i, & \text{otherwise} \end{cases} \tag{9.14}$$

Arithmetic Crossover. Arithmetic crossover produces two complementary linear combinations of the parents, where $r = U(0,1)$.

$$\overline{X}' = r\overline{X} + (1 - r)\overline{Y} \tag{9.15}$$

$$\overline{Y}' = (1 - r)\overline{X} + r\overline{Y} \tag{9.16}$$

To achieve the necessary integer representation of the variables, the following is performed:

$$\overline{X}' = (\langle ax_1 + by_1\rangle, \langle ax_2 + by_2\rangle, \langle ax_3 + by_3\rangle, \langle ax_4 + by_4\rangle, \langle ax_5 + by_5\rangle) \tag{9.17}$$

where

$$\langle ax_i + by_i\rangle = \begin{cases} \lceil ax_i + by_i \rceil, & \text{if } x_i > y_i \\ \lfloor ax_i + by_i \rfloor, & \text{otherwise} \end{cases}$$

Cell-Swap Crossover. The previous genetic operators work for any integer programming formulation. The next two operators work only with the cell formation representation. Let $\overline{A} = (x_1,x_2,\ldots,x_m,y_1,y_2,\ldots,y_n)$ and $B = (w_1,w_2,\ldots,w_m,z_1,z_2,\ldots,z_n)$ be two $(m + n)$-dimensional cell formation individuals (parents). Create two new individuals $\overline{A}'$ and $\overline{B}'$:

$$a'_i = \begin{cases} x_i, & \text{if } i < m \\ z_i, & \text{otherwise} \end{cases} \tag{9.18}$$

$$b'_i = \begin{cases} w_i, & \text{if } i < um \\ y_i, & \text{otherwise} \end{cases} \tag{9.19}$$

Cell-Two-Point Crossover. Generate two random numbers, r_1 and r_2, from a discrete uniform distribution from 2 to $(m - 1)$ and $(m + 2)$ to $(m + n - 1)$, respectively, and create two new individuals ($\overline{A}'$ and $\overline{B}'$) according to

$$a'_i = \begin{cases} x_i, & \text{if } i < r_1 \\ w_i, & \text{if } r_1 < i \leq m \\ y_i, & \text{if } m < i < r_2 \\ z_i, & \text{otherwise} \end{cases} \tag{9.20}$$

$$b'_i = \begin{cases} w_i, & \text{if } i < r_1 \\ x_i, & \text{if } r_1 < i \leq m \\ z_i, & \text{if } m < i < r_2 \\ y_i, & \text{otherwise} \end{cases} \tag{9.21}$$

Joines [336] demonstrated that employing these additional operators improved both the computational efficiency and the solution quality.

Initialization of Population. The genetic algorithm must be provided with an initial population. The most common method is to randomly generate solutions for the entire population. However, since genetic algorithms can iteratively improve existing solutions (i.e., solutions from other heuristics and/or current practices), the beginning population can be seeded with potentially good solutions, with the remainder of the population being randomly generated solutions. The experiments described by Jaines used a random initial population with common random numbers across all methods for each replication.

Termination Criteria. The genetic algorithm moves from generation to generation selecting and reproducing parents until a termination criterion is met. The most frequently used stopping criterion is a specified maximum number of generations. Another termination strategy involves population convergence criteria. In general, genetic algorithms will force much of the entire population to converge to a single solution. When the sum of the deviations among individuals becomes smaller than a specified threshold, the algorithm can be terminated. The algorithm can also be terminated due to a lack of improvement in the best solution over a specified number of generations. Alternatively, a target value for the evaluation measure can be established based on an arbitrarily "acceptable" threshold. Several strategies can be used in conjunction with each other.

Evaluation Measure. The *grouping efficacy* measure was used as the evaluation criterion to test the integer-based genetic algorithm on several data sets from the literature. Grouping efficacy was chosen as the initial evaluation measure because it has been used frequently in the literature and results are available for comparison. It generates block diagonal cell formation and practically feasible solutions [381, 483]. Grouping efficacy seeks to minimize the number of exceptional elements and the number of voids (zeros) in the diagonal blocks. Exceptional elements represent intercell movements of parts, which reduce the effectiveness of cellular manufacturing. More specifically, grouping efficacy attempts to minimize the number of voids plus the number of exceptional elements divided by the total operational zone. The *operational zone* is defined by the number of operations (exceptional elements plus operations along the diagonal) plus the number of voids. Grouping efficacy has a value of 1 when there are no exceptional elements and no voids and a value of zero is the number of exceptional elements equals the total number of operations. Formally, grouping efficacy (Γ) is defined as

$$\max \Gamma = \frac{1 - \psi}{1 + \phi} = \frac{1 - e_o/e}{1 + e_v/e} = \frac{e - e_o}{e + e_v}, \tag{9.22}$$

where e is the number of operations in the data matrix, e_v the number of voids in the diagonal blocks, and e_o the number of exceptional elements.

Grouping efficacy cannot be used as an objective function in classical integer programming formulations because of its nonlinearity [336]. Using the classical 0–1 assignment variable declarations and constraints from Section 9.4.3 and the definition of grouping efficacy (Γ), the nonlinear closed form of grouping efficacy is

$$\Gamma = \frac{\sum_{l=1}^{k_{\max}} \sum_{j=1}^{n} \sum_{i=1}^{m} y_{jl} x_{il} a_{ij}}{\sum_{j=1}^{n} \sum_{i=1}^{m} a_{ij} + \sum_{l=1}^{k_{\max}} \left[\left(\sum_{j=1}^{n} y_{jl} \right) \left(\sum_{i=1}^{m} x_{il} \right) \right] - \sum_{l=1}^{k_{\max}} \sum_{j=1}^{n} \sum_{i=1}^{m} y_{jl} x_{il} a_{ij}}$$

Experiments Using the Integer-Based Genetic Algorithm. The integer-based genetic algorithm was tested on 17 data sets from the literature to study its effectiveness as a clustering tool. In all experiments, the genetic algorithm employed a ranking scheme based on the normalized geometric distribution, an elitist model, and the seven integer-based genetic operators described previously. Joines [336] selected the values for the genetic algorithm parameters in Table 9.8 after extensive experimentation and were shown to perform well with all the data sets. The values associated with various genetic operators indicate the number of parents modified by that operator within each gener-

TABLE 9.8. Parameters Used in the Genetic Algorithm Experiments

Parameter	Value
Number of boundary mutation oeprators	4
Number of uniform mutation operators	4
Number of nonuniform mutation operators	4
Number of multi-nonuniform mutation operators	8
Number of swap crossover opertors	6
Number of multipoint crossover operators	6
Number of arithmetical crossover operators	6
Maximum permissible number of cells k_{max}	k^L
Probability of selecting the best individual, q	0.08
Maximum number of generations, G_{max}	5000
Population size	80

ation. For data sets taken from the literature, the maximum number of permissible cells, k_{max}, was initially set equal to the best known number of cells, k^L, as determined by other cell formation algorithms. Because grouping efficacy was chosen as the criterion for comparing individuals, the solutions forced clusters to form along the diagonal block, grouping parts and machines simultaneously. Figure 9.5 is the 20 × 35 (part–machine) incidence matrix of Brubidge [80] used in the order-based experiments in Section 9.4.2. The solution to this data set contains exceptional elements and is displayed in Figure 9.8. The next data set, taken from King and Nakornchai [368], is a 16 × 43 problem containing two bottleneck machines and is shown with its solution in Figures 9.3 and 9.9. Notice that no visual inspection of the output is necessary to determine the machine cell and part family composition. The cells (families) are determined explicitly by the genetic algorithm as shown in the tabular area of the solution figures.

9.4.5. Other Approaches

Utilizing order-based genetic algorithms to solve the problem usually requires visual inspection of the part–machine incidence matrix to determine the cell–family composition, which may be very difficult in large, complicated data sets [333, 335]. Billo, Tate, and Bidanda [62, 63] used an order-based genetic algorithm to minimize the number of cells and maximize the similarity of parts. However, their algorithm modified the representation to include cell dividers (i.e., the sequence was broken up to determine cell compositions automatically). It was shown to outperform a hierarchical clustering algorithm.

Daskin [139] used a genetic algorithm to determine a good starting machine permutation. Machines were assigned to groups based on the order of the machines in the permutation, which minimized the combined capital cost

family/cell number	parts	machines
0	0, 2, 6, 7, 16	0, 2 , 4, 14, 16, 19, 22, 24,28
1	10, 11,14, 15, 18	3, 5, 8, 10, 20, 27, 29, 31, 32, 34
2	1, 3, 12, 13, 17	1, 6, 9, 11, 12, 17, 23, 26, 30
3	4, 5, 8, 9, 19	7, 13, 15, 18, 21, 25, 33

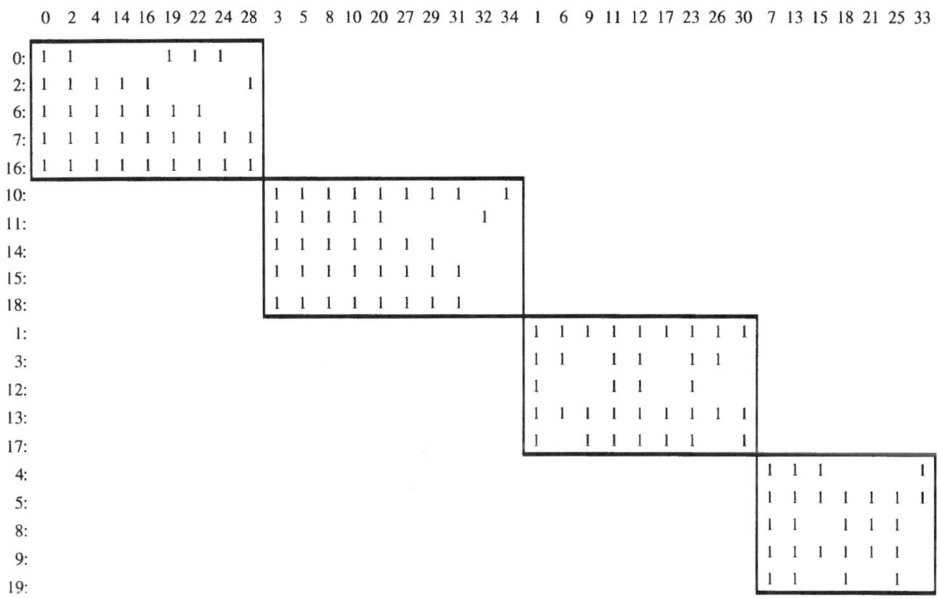

Figure 9.8. Integer genetic algorithm solution for the 20 × 35 problem of Burbidge [80] with $\Gamma = 0.75751$.

of the machines in all groups and the material handling cost of moving parts between cells. Only the machine cells were formed.

Kazerooni, Luong, and Abhary [350, 351] developed two new similarity coefficients for machines and parts that included production volume and process sequences. Similar to the approach in Section 9.4.2, two independent order-based genetic algorithms were used to determine the best sequence of machines and parts separately that maximized each similarity.

Venugopal and Narendran [640] used a genetic algorithm to solve a multiobjective integer programming formulation of the cell formation problem which minimizes the volume of intercell moves and the total within-cell load variation. This algorithm utilized only the uniform mutation and simple crossover used by Joines in Section 9.4.4 and determined only the machine cell composition.

family/cell number	parts	machines
0	3, 4, 7, 14	4, 7, 8, 13, 14, 15, 18, 20, 22, 28, 30, 32, 40, 42
1	0, 1, 8, 15	1, 3, 9, 17, 27, 31, 36, 37, 39, 41
2	10, 11, 12	2, 10, 19, 21, 23, 26, 29
3	2, 5, 6, 9, 13	0, 5, 6, 11, 12, 16, 24, 25, 33, 34, 35, 38

	0	2	4	14	16	19	22	24	28	3	5	8	10	20	27	29	31	32	34	1	6	9	11	12	17	23	26	30	7	13	15	18	21	25	33
0:	1	1				1	1	1																											
2:	1	1	1	1	1				1																										
6:	1	1	1	1	1	1	1																												
7:	1	1	1	1	1	1	1	1	1																										
16:	1	1	1	1	1	1	1	1	1																										
10:										1	1	1	1	1	1	1	1		1																
11:										1	1	1	1	1				1																	
14:										1	1	1	1	1	1	1																			
15:										1	1	1	1	1	1	1	1																		
18:										1	1	1	1	1	1	1	1																		
1:																				1	1	1	1	1	1	1	1	1							
3:																				1	1		1	1		1	1								
12:																				1			1	1		1									
13:																				1	1	1	1	1	1	1	1	1							
17:																				1		1	1	1	1	1		1							
4:																													1	1	1				1
5:																													1	1	1	1	1	1	1
8:																													1	1		1	1	1	
9:																													1	1	1	1	1	1	
19:																													1	1		1		1	

Figure 9.9. Integer genetic algorithm solution for the 16×43 problem taken from King [366] with $\Gamma = 0.4926$.

9.5 CELL DESIGN WITH ALTERNATIVE PROCESS PLANS

Most algorithms in the literature utilize only the part–machine incidence matrix, which contains only a prespecified single fixed routing for each part. Relatively few researchers have addressed the issue of utilizing alternative operations, alternative routings, or machine redundancy in determining cells and families. These single fixed routings are often optimized to maximize machine utilization in a functional layout rather than the benefits of cellular manufacturing. The existence of functionally similar machines or workcenters is not considered in a functional layout because parts can be routed or schedules to the next available machine [21]. These functionally identical machines lead to alternative operations or routings that can be used to obtain better cell design and independence.

The p-median algorithm, a mathematical programming approach, was extended by Kusiak and Cho [384] to include complete fixed alternative routes, while Shtub [574] used a generalized assignment problem to solve a similar problem in an attempt to obtain better cell independence. Nagi, Harbalakis, and Proth [479] developed a two-phase approach that minimizes intercell traffic in the manufacturing system while considering part demand, capacity considerations, multiple fixed part routings, and multiple functionally identical machines. Kang and Wemmerlov [345] developed a similarity index heuristic that determines machine cells first and then part families by considering multiple routes specified in terms of both operations and machines. This procedure also considers capacity when combining operations into routes. However, most heuristic methods lack the flexibility to form cells using a variety of evaluation measures or to handle constraints adequately. A majority of the methods that include alternative operations and multiple routes utilize mathematical programming, limiting their usefulness for solving large cell formation problems.

To move toward a more satisfactory algorithmic result, other manufacturing information, such as setup time requirements, tooling and crew requirements, machine capacity, alternative routings, machine costs, intercell transfers, intracell and intercell layout, and reduced machine utilization needs to be included.

The integer programming approach to cell design developed by Joines [336] (see Section 9.4.4) allows the system designer to substitute various types of evaluation functions, permitting alternative designs to be generated and reviewed quickly. Joines et al. [332, 335–338] have demonstrated the flexibility and extensibility of the genetic algorithm approach by incorporating new representations and other manufacturing information (i.e., part sequencing or routing information).

9.5.1 Incorporation of Alternative Operations and Machine Redundancy

One objective is to form completely independent cells, since the benefits of cellular manufacturing (reduced WIP, throughput time, etc.) are reduced when intercell movement exists. Most techniques try to minimize intercell movement as represented by the number of exceptional elements. However, the number of exceptional elements may not accurately reflect the level of intercell movement required. For example, if intercell movement occurs in the middle of the operational sequence, two (not one) intercell movements will be required. The operational sequence can be extremely valuable in identifying the effect of intercell traffic and backtracking. To eliminate exceptional elements, the machine causing intercell movement can be duplicated or the parts routing sequence can be changed by subcontracting the part, redesigning the part, or using an alternative routing or operation. Routings are often generated to maximize machine utilization rather than the effectiveness of group technology. Therefore, a fixed relationship between a part and a particular set of machines can be constrained since there may exist alternative machines for a specific operation which may lead to greater cell independence [345].

Chan and Milner [91] and King [366] eliminate exceptional elements by duplicating machines interactively and re-solving the problem. Exceptional elements are interdependent because actions used to eliminate one element may affect other elements in the incidence matrix. Therefore, deciding to duplicate a machine early in the cell design process may not lead to the best solution. Others have suggested interactive sessions after the initial cell formation to eliminate the cells by redesigning the part or assigning the operation to another machine. Neither of these is satisfactory since real problems are often large and complicated.

Most cell design algorithms use a binary part–machine matrix, A, consisting of elements $a_{ij} = 1$ if part j requires processing on machine i, otherwise $a_{ij} = 0$. This format leads to fixed routing sequences with no machine substitutions. Several researchers have proposed the following representation, where the operational sequence is embedded in the matrix [332, 479]:

$$a_{ij} = \begin{cases} k, & k\text{th operation of part } j \text{ is required on machine } i \\ 0, & \text{no processing on machine } i \end{cases}$$

Using this variable definition, alternative operations can be incorporated into the part–machine incidence matrix by specifying that several machines can perform the kth operation of a particular part (see Figure 9.12). Because the evaluation function is independent of the decision rules, the genetic algorithm provides the flexibility to interchange various objective functions without changing the algorithm. The evaluation function must take into account only the new part–machine incidence matrix representation. Specifying alternative operations has the advantage of allowing the algorithm to determine the most appropriate routing sequences in the context of minimizing intercell flows. It also allows one to determine the true effect of intercell movement, backtracking, and sequence dependent setup times since the actual sequence of operations is specified.

Alternative Operations Integer Programming Model. The ability of a genetic algorithm to incorporate alternative operations will be demonstrated using the evaluation function used in Section 9.4.4. The chromosomal representation employed by Joines [332, 336] in Section 9.4.4 does not change, but calculation of the grouping efficacy has to take allow for alternative operations.

For the alternative operations problem to be modeled as a nonlinear integer programming problem, the following new variable definitions and constraint sets need to be added to the grouping efficacy as defined in [336]. In this model, a particular operation may be performed by more than one machine. Therefore, the new variable (o_{jkl}) determines which operation is selected. The first set of constraints ensures that operation k of part j is assigned to exactly one cell. The second set of constraints makes sure that at least one machine that can perform operation k has been assigned to the same cell.

$$o_{jkl} = \begin{cases} 1, & \text{if operation } k \text{ of part } j \text{ is assigned to cell } l \\ 0, & \text{otherwise} \end{cases}$$

$$\sum_{l=1}^{k_{\max}} o_{jkl} = 1, \qquad j = 1,\ldots,n, \quad k = 1,\ldots,|O_j|$$

$$o_{jkl} \leq \sum_{i \in Mkj} x_{il}, \qquad l = 1,\ldots,k_{\max}, \quad j = 1,\ldots,n, \quad k = 1,\ldots,|O_j|$$

where O_j is the set of operations for part j and M_{jk} is the set of machines that can perform operation k for part j. Using these new variables with the assignment variables defined previously, x_{il} and y_{jl}, grouping efficacy (Γ) can now be defined as follows:

$$\Gamma = \frac{\sum_{l=1}^{k_{\max}} \sum_{j=1}^{n} \sum_{k=1}^{|O_j|} y_{jl} o_{jkl}}{\sum_{j=1}^{n} |O_j| + \sum_{l=1}^{k_{\max}} \left[\left(\sum_{j=1}^{n} y_{jl} \right) \left(\sum_{i=1}^{m} x_{il} \right) \right] - \sum_{i=1}^{k_{\max}} \sum_{j=1}^{n} \sum_{k=1}^{|O_j|} y_{jl} o_{jkl}}.$$

Computational Experience. Notice how much easier the model is using set notation. Only the evaluation function has to decode the chromosome. An example taken from Nagi, Harhalakis, and Proth [479] is used to demonstrate the genetic algorithm's ability to use this new incidence matrix representation that permits alternative operation substitution. The problem consists of 20 different parts and 20 workcenters. Workcenters 5 and 6 are interchangeable, as are workcenters 17, 18, and 19 for all parts. In comparison, the Nagi problem was modified by removing all duplicate workcenters (machines 6, 18, and 19), and a standard binary part–machine incidence matrix was created as shown in Figure 9.10. The genetic algorithm requires 136.25 generations

machines \ parts	0	1	2	3	4	5	6	7	8	9	10	11	12	13	14	15	16	17	18	19
0:		1	1	1	1															
1:						1		1	1	1										
2:											1	1	1							
3:																		1	1	1
4:						1	1			1										
5:	1	1	1	1		1	1	1	1	1										
7:											1	1	1							
8:	1		1		1															
9:														1	1	1	1			
10:											1		1							
11:	1	1	1	1	1															
12:																		1	1	
13:														1		1	1			
14:																		1	1	1
15:							1	1	1	1										
16:														1	1		1			
17:					1		1		1		1	1	1	1	1	1				

Figure 9.10. Original matrix without machine redundancy.

on average for 10 replications to solve the problem. The solution, having a Γ of 0.7308 and 10 exceptional elements, is shown in Figure 9.11. Further evaluation using the operation sequence information in the part–machine matrix shown in Figure 9.12 reveals that the 10 exceptional elements actually generate 13 intercell movements. This illustrates the limitation of fixed binary representations of the routings in accurately measuring intercell flows. Next, the genetic algorithm was used to solve the problem with the inclusion of redundant machines along with the operational sequences (Figure 9.12). The algorithm required an average of 143.2 generations to reach a final Γ value of 0.7952 over 10 replications. The solution (Figure 9.13) contains only one exceptional element. The fourth operation of part 4 cannot be processed inside its cell (cell 4), and one intercell movement to either of three cells (0, 1, or 3) is required. Thus an improved cell formation was achieved by the genetic algorithm through the consideration of alternative machines for selected operations and including redundant machines.

9.5.2. Incorporation of Alternative Routings

Even though the model from Section 9.5.1 can generate cells using dynamic routes, the formulation suffers from two distinct disadvantages. First, it forms

family/cell number	parts	machines
0	5, 6, 7, 8,	1, 4, 5, 15
1	10, 11, 12	2, 7, 10, 17
2	0, 1, 2, 3, 4	0, 8, 11
3	13, 14, 15, 16	9, 13, 16
4	17, 18, 19	3, 12, 14

parts

machines	5	6	7	8	9	10	11	12	0	1	2	3	4	13	14	15	16	17	18	19
1:	1		1	1	1															
4:	1	1			1															
5:	1	1	1	1	1				1	1	1	1								
15:		1	1	1	1															
2:						1	1	1												
7:						1	1	1												
10:						1		1												
17:		1		1		1	1	1					1	1	1	1				
0:										1	1	1	1							
8:									1		1		1							
11:									1	1	1	1	1							
9:														1	1	1	1			
13:														1		1	1			
16:														1	1		1			
3:																		1	1	1
12:																		1	1	
14:																		1	1	1

Figure 9.11. Solution to Nagi's problem without machine redundancy.

cells considering alternative operations in which preferred sets of machines cannot be specified and preserved through the clustering process. This can be important for manufacturers who possess a mix of manual machines and flexible computerized numerically controlled (CNC) technology. To illustrate, consider a part with two valid machining sequences. One sequence employs several manually operated machines, and the second sequence might consist of a single CNC workstation. The genetic algorithm formulation of Section 9.3 considering alternative operations would allow the CNC workstation to be substituted for any of the individual machines in the manual sequence. In practice it would be unlikely that a part that can be produced entirely on a CNC workstation would visit this workstation for only one operation while the other operations are performed on manually operated machines. It is more likely that the part would be machined entirely on the CNC workstation or produced exclusively using the manually operated machines. A second disadvantage of this model over complete alternative fixed routings is the ability not to assign priorities to certain routes. This would allow the cell designer

parts

machines	0	1	2	3	4	5	6	7	8	9	10	11	12	13	14	15	16	17	18	19
0:		3	1	1	2															
1:						3		3	2	1										
2:											2	1	3							
3:																		2	1	2
4:						2	1			3										
5:	3	1	4	3		1	4	2	4	4										
6:	3	1	4	3		1	4	2	4	4										
7:											1	2	2							
8:	2		2		1															
9:														1	3	2	3			
10:											3		1							
11:	1	2	3	2	3															
12:																		1	2	
13:														3		3	2			
14:																		3	3	1
15:							2	1	1	2										
16:														4	2		1			
17:					4		3		3		4	3	4	2	1	1				
18:					4		3		3		4	3	4	2	1	1				
19:					4		3		3		4	3	4	2	1	1				

Figure 9.12. Original part–machine operation incidence matrix. (From Nagi et al. [479].)

to produce cells that balance workloads, generate minimum tooling costs, take full advantage of labor skills, and so on.

Fixed Alternative Routings Integer Programming Model. A generalized cell formation problem that can consider complete fixed alternative routings which overcomes these two disadvantages can be formulated from the classic integer programming formulation [336] by adding constraints to ensure that each part is assigned one and only one route. However, it is easily shown to be NP-hard since the original problem is NP-hard [639, 640]. The classical formulation must incorporate the following additional variable declarations and constraints to handle fixed routings:

family/cell number	parts	machines
0	10, 11, 12	2, 7, 10, 17
1	5, 6, 7, 8, 9	1, 4, 5, 15, 18
2	17, 18, 19	3, 12, 14
3	13, 14, 15, 16	9, 13, 16, 17
4	0, 1, 2, 3, 4	0, 6, 8, 11

machines \ parts	10	11	12	5	6	7	8	9	17	18	19	13	14	15	16	0	1	2	3	4
2:	2	1	3																	
7:	1	2	2																	
10:	3		1																	
17:	4	3	4																	
1:				3		3	2	1												4
4:				2	1			3												
5:				1	4	2	4	4												
15:					2	1	1	2												
18:					3		3													
3:									2	1	2									4
12:									1	2										
14:									3	3	1									
9:												1	3	2	3					
13:												3		3	2					
16:												4	2		1					
19:												2	1	1						4
0:																	3	1	1	2
6:																3	1	4	3	
8:																2		2		1
11:																1	2	3	2	3

Figure 9.13. Solution to the operation incidence matrix. (From Nagi et al. [479].)

$$z_{jh} = \begin{cases} 1, & \text{if part } j \text{ uses route } h \\ 0, & \text{otherwise} \end{cases} \tag{9.23}$$

$$\sum_{l=1}^{p_j} z_{jh} = 1, \qquad j = 1,\ldots,m \tag{9.24}$$

where p_j is the number of alternative routes for part j.

The constraints ensure that each part is assigned only one routing and the nonlinear integer formulation [336] can easily be updated as shown below.

Using these new variables with the previously defined assignment variables, x_{il} and y_{jl}, grouping efficacy (Γ) can now be defined as follows:

$$\Gamma = \frac{\sum_{l=1}^{k_{\max}} \sum_{i=1}^{m} \sum_{j=1}^{n} \sum_{h=1}^{p_j} x_{il} y_{jl} z_{jh} a_{ijh}}{\sum_{i=1}^{m} \sum_{j=1}^{n} \sum_{h=1}^{p_j} z_{jh} a_{ijh} + \sum_{l=1}^{k_{\max}} \left[\left(\sum_{j=1}^{n} y_{jl} \right) \left(\sum_{i=1}^{m} x_{il} \right) \right] - \sum_{l=1}^{k_{\max}} \sum_{i=1}^{m} \sum_{j=1}^{n} \sum_{h=1}^{p_j} x_{il} y_{jl} z_{jh} a_{ijh}}$$

The genetic algorithm approach is extended by developing an integer programming-based genetic algorithm with a new set of variable declarations that allow evaluation of fixed alternative routings. Specifically, let $z_j = h$ if part j uses route h. If there is only a single routing for part j, this variable is unnecessary and is not be included in the programming model. Notice that the number of variables has grown by at most a factor of n. With this new variable definition, a new chromosome representation can be developed easily by adding the z variable component to individuals in the population:

$$\text{individual} \rightarrow (\underbrace{x_1,x_2,\ldots,x_m}_{\text{machines}},\ \underbrace{y_1,y_2,\ldots,y_n}_{\text{parts}},\ \underbrace{z_i,z_2\ldots,z_n}_{\text{route}})$$

Modified Grouping Efficacy Example and Algorithm. Again, because of the flexibility of the genetic algorithm, the algorithm remains unchanged and only the evaluation function is modified to accommodate this new representation. Next we provide an algorithm explaining grouping efficacy and an example of decoding an individual and the computing grouping efficacy.

Computational Experience. The new formulation of the genetic algorithm is also demonstrated on the Nagi problem. Figure 9.14 shows the original part–machine incidence matrix with the alternative route for each part. Notice that parts 16 to 19 have only one process plan. The genetic algorithm then determines the appropriate values for only 56 different variables (20 machine, 20 part, and 16 route assignments). As in the preceding model, there is only one exceptional element and intercell movement (part 4, shown in the solution in Figure 9.15). The same value Γ of 0.7952 was obtained after an average of 231.6 generations. For this problem, identical cell configurations are obtained because the complete alternative routings matrix (Figure 9.14) can be reduced to the operational sequence matrix in Figure 9.12. This would not be possible if the number of operations per route were different. For example, if part 1 could be produced on machines 5, 8, and 11 or machines 6 and 9, these two alternative routes cannot be reduced to the form of the operation sequences matrix in Figure 9.3. Therefore, different cell configurations are possible.

part:	P0		P1		P2		P3		P4			P5		P6						P7		P8					
route:	1	2	1	2	1	2	1	2	1	2	3	1	2	1	2	3	4	5	6	1	2	1	2	3	4	5	6
0:			1	1	1	1	1	1	1	1	1																
1:												1	1							1	1	1	1	1	1	1	1
2:																											
3:																											
4:												1	1	1	1	1	1	1	1								
5:	1		1		1		1					1					1	1	1		1		1		1		1
6:		1		1		1		1					1	1	1	1				1		1		1		1	
7:																											
8:	1	1			1	1			1	1	1																
9:																											
10:																											
11:	1	1	1	1	1	1	1	1	1	1	1																
12:																											
13:																											
14:														1	1	1	1	1	1	1	1	1	1	1	1	1	1
15:																											
16:										1					1			1						1	1		
17:									1					1			1					1	1				
18:											1					1			1							1	1
19:																											

part:	P9		P10			P11			P12			P13			P14			P15			P16	P17	P18	P19
route:	1	2	1	2	3	1	2	3	1	2	3	1	2	3	1	2	3	1	2	3	1	1	1	1
0:																								
1:	1	1																						
2:			1	1	1	1	1	1	1	1	1													
3:																						1	1	1
4:	1	1																						
5:	1																							
6:		1																						
7:			1	1	1	1	1	1	1	1	1													
8:																								
9:												1	1	1	1	1	1	1	1	1	1			
10:			1	1	1				1	1	1													
11:																						1	1	
12:												1	1	1				1	1	1	1			
13:																						1	1	1
14:	1	1																						
15:												1	1	1	1	1	1				1			
16:			1			1			1			1			1			1						
17:				1			1			1			1			1			1					
18:					1			1			1			1			1			1				
19:																								

(rows: machine)

Figure 9.14. Original matrix for the Nagi problem with complete alternative routings.

family/cell number	parts	machines
0	13, 14, 15, 16	9, 13, 16, 17
1		3, 12, 14
2	5, 6, 7, 8, 9	1, 4, 5, 15, 19
3	0, 1, 2, 3, 4	0, 6, 8, 11
4	10, 11, 12	2, 7, 10, 18

Part:	0	1	2	3	4	5	6	7	8	9
Route:	2	2	2	2	3	1	6	2	6	1

part:	10	11	12	13	14	15	16	17	18	19
route:	2	2	2	1	1	1	–	–	–	–

parts

machines	13	14	15	16	17	18	19	5	6	7	8	9	0	1	2	3	4	10	11	12
9:	1	1	1	1																
13:	1		1	1																
16:	1	1		1																
17:	1	1	1	1																
3:					1	1	1													
12:					1	1														
14:					1	1	1													
1:								1		1	1	1								
4:								1	1			1								
5:								1	1	1	1	1								
15:									1	1	1	1								
19:									1		1						1			
0:														1	1	1	1			
6:													1	1	1	1				
8:													1		1		1			
11:													1	1	1	1	1			
2:																		1	1	1
7:																		1	1	1
10:																		1		1
18:																		1	1	1

Figure 9.15. Solution for the Nagi problem with complete alternative routings.

Complete Alternative Routings with Machine Redundancy. Even though consideration of complete alternative routings offers the ability to specify a preferred set of machines or to assign priorities to certain routes, it can also be limiting. The operational sequences model offers advantages by allowing machine redundancy as well as determination of the true effect of intercell movement. These two models can be combined into one representation by allowing the complete alternative routings to contain their operation sequences as well.

Again, because of the flexibility of the genetic algorithm, the algorithm remains unchanged and only the evaluation function is modified to accommodate a combination of the operation sequences with complete alternative routings. This formulation of the genetic algorithm is demonstrated on a modified version of the Nagi problem. Figure 9.16 shows the original part–machine incidence matrix with the alternative routes for each part, where each alternative route now contains the operation sequences. Some routes contain machine redundancies (i.e., certain operations of a route can be performed on several different machines). Again, parts 16 to 19 have only one route. As in Section 9.5.2, the genetic algorithm determines the appropriate values for only 56 variables. This model was able to eliminate the exceptional element as shown in Figure 9.17, as well as to obtain a slightly higher value Γ of 0.7975 in an average of 227.7 generations. Notice that the genetic algorithm was able to determine the most appropriate route to use as well as to dynamically build the operational sequence.

By combining machine redundancy with the complete fixed alternative routes, this model offers distinct advantages over previous models. Whereas the model of the preceding section can handle machine redundancy through the use of separate routes, the approach in this section is more computationally and state-space efficient. Take the same example of a part that can be manufactured at a single CNC workstation or with a set of manual machines. As stated before, one does not want the genetic algorithm to assign certain operations to the CNC workstation and the remaining operations to manual machines. Therefore, a preferred set of machines needs to be specified. However, several manual machines may be substituted for one another or a set of interchangeable multipurpose CNC machines assigned. The genetic algorithm has the capability of utilizing the most appropriate CNC workstation or set of manual machines within the most appropriate complete alternative routing. Also, this model, unlike the complete alternative routings model, is able to determine the true effect of intercell movement.

9.5.3 Moon, Gen, and Kim's Approach to Independent Cells

Exceptional elements (i.e., intercell movements) diminish the effectiveness of manufacturing cells. Therefore, most models try to minimize the number of exceptional elements. In many cases, elimination of all exceptional elements is not possible under the current conditions (parts, machines, process plans, etc.) Moon and Kim [462] consider a more complicated mathematical model that utilizes alternative fixed routing similar to Joines [332, 336] but produces completely independent cells by duplicating additional machines if necessary. The objective is to minimize the production costs plus the cost of duplicating additional machines.

Problem Statement. In this section, a 0–1 integer programming model is developed to determine the process plan for parts, part family, and machine cell simultaneously. It incorporates the process plans and other manufacturing factors.

part:	P0		P1		P2		P3		P4			P5		P6				P7		P8				P9		P10		P11		P12		P13		P14		P15		P16	P17	P18	P19
route:	1	2	1	2	1	2	1	2	1	2	3	1	2	1	2	3	4	1	2	1	2	3	4	1	2	1	2	1	2	1	2	1	2	1	2	1	2	1	1	1	1
0:			3			1		1			2				1										1																
1:													3					1	3				2			1				1											
2:		1			2		3									1						1					2		1		3	1									
3:					1				2																	2		2											2	1	2
4:													2				1								3					2				3	4						
5:	3		1					3					1		2		4		2		2		4		4	3				3											
6:	3		1			4		3					1		2		4		2		2		4		4	3				3											
7:					1											2						2					1		2		2	3									
8:	2					2					1																														
9:										2				1						2				1									1		3		2	3			
10:																											3				1	4									
11:	1		2			3		2			3										1																				
12:				2			2											2		1																			1	2	
13:										1		1		2										2									3				3	2			
14:					3		1		1									3																					3	3	1
15:																	2		1				1		2			1						2		2					
16:												2		3						3				3									4		2			1			
17:		2			3					3	4	3				3	3					3	3				4		3	4	4	2	2	1	1	1	1				
18:		2			3					3	4	3				3	3					3	3				4		3	4	4	2	2	1	1	1	1				
19:		2			3					3	4	3				3	3					3	3				4		3	4	4	2	2	1	1	1	1				

(Rows: machines)

Figure 9.16. Original matrix with complete alternative routings and machine redundancy.

family/cell number		parts	machines
family/cell	0	7, 8, 12	1, 4, 5, 15, 19
	1	0, 3	0, 6, 8, 11
	2	1, 17, 18, 19	3, 12, 14
	4	4, 5, 9, 13, 14, 15, 16	9, 13, 16, 18
	3	2, 6, 10, 11	2, 7, 10, 17

part:	0	1	2	3	4	5	6	7	8	9
route:	1	1	1	1	2	1	3	2	4	2

part:	10	11	12	13	14	15	16	17	18	19
route:	2	2	2	2	2	2	—	—	—	—

machines \ parts	7	8	12	0	3	1	17	18	19	4	5	9	13	14	15	16	2	6	10	11
1:	3	2	1																	
4:			2																	
5:	2	4	3																	
15:	1	1																		
19:		3	4																	
0:					1															
6:				3	3															
8:				2																
11:				1	2															
3:						1	2	1	2											
12:						2	1	2												
14:						3	3	3	1											
9:										2		1	1	3	2	3				
13:										1	1	2	3		3	2				
16:											2	3	4	2		1				
18:										3	3		2	1	1					
2:																	2	1	2	1
7:																	1	2	1	2
10:																			3	
17:																	3	3	4	3

Figure 9.17. Solution matrix with complete alternative routings and machine redundancy.

Assumptions

1. Alternative process plans for each part are known.
2. The number of cells for configuration and upper limits on the number of parts in each cell are known.
3. Only one same type of machine exists within each cell.

Notation

i is a part index ($i = 1,2,...,np$).

j is a process plan index ($j = 1,2,...,npp_i$).
k is a machine index ($k = 1,2,...,nm$).
c is a cell index ($c = 1,2,...,nc$).
p_{ij} is the process plan j for part i.
ms_{ij} is the set of machines for part i when the process plan j is selected.
pm_{ij} is the cardinality of ms_{ij}.
pc_{ijk} is the processing cost of machine k for part i under process plan j.
mc_k is the cost of machine type k.
U_c is the upper limit on the number of parts in cell c.
um_k is the number of copies of machine type k.
cam_k is the capacity of machine type k.
pv_i is the production volumes for part i.

$$x_{ijc} = \begin{cases} 1, & \text{if part } i \text{ produced under process plan } j \text{ belongs to cell } c \\ 0, & \text{otherwise} \end{cases}$$

$$y_{kc} = \begin{cases} 1, & \text{if machine type } k \text{ belongs to cell } c \\ 0, & \text{otherwise} \end{cases}$$

The objective function of the model is to minimize the sum of the machining and duplication costs. The overall model is formulated as follows:

$$\min \quad \text{TC} = \sum_i \sum_j \sum_k \sum_c pv_i pc_{ijk} x_{ijc} + \sum_c \sum_k mc_k y_{kc} \tag{9.25}$$

$$\text{s. t.} \quad \sum_j \sum_k x_{ijc} = 1, \qquad \forall i \tag{9.26}$$

$$\sum_i \sum_j x_{ijc} \leq U_c, \qquad \forall c \tag{9.27}$$

$$\sum_i \sum_j pv_i pc_{ijk}\, x_{ijc} \leq cam_k, \qquad \forall (k,c) \tag{9.28}$$

$$\sum_{k \in msij} y_{kc} \geq pm_{ij} x_{ijc}. \qquad \forall (i,j,c) \tag{9.29}$$

$$x_{ijc}, y_{kc} = \{0,1\}, \qquad \forall (i,j,k,c) \tag{9.30}$$

Constraint (9.26) ensures that only one process plan is selected for each part and that a part belongs to only one cell. Constraint (9.27) imposes the restriction on the maximum number of parts in a cell to have easy planning and control. Constraint (9.28) implies that total processing costs for a machine type are less than or equal to one's available capacity. Constraint (9.29) represents that all machines needed for the production of a part under a selected process plan ought to be assigned to the same cell. Constraint (9.30) indicates the 0–1 integer variables.

Genetic Algorithms The model described above has a few limitations. Because the variables are constrained to integer values, the model is difficult to solve for a large number of parts and machines, due to the computational complexity. Also, this model does not offer cell designers the flexibility to change objective functions and constraints. In this section an approach using genetic algorithms is developed to overcome these drawbacks and to solve the simultaneous process plans selection and independent manufacturing cell design problem.

Representation and Initialization. Representation is the first step in applying genetic algorithm to cell design problem. In the machine cell design problem, each gene represents a process plan number and cell number for each part. The length of the chromosome represents the number of parts considered in the problem. Let the number of parts be *np;* then a chromosome can be represented as follows:

$$S_i = [m_1 \ m_2 \ m_3 \ \cdots \ m_k \ \cdots \ m_{np}]$$

where S_i is the ith chromosome and m_k is the kth gene within the S_i. In this chromosome, let the number of process plans for each part be npp_k and the upper limit on the number of cells be cn and the range of m_k be $[1, npp_k x cn]$. If d is a m_k from S_i, the process plan number for part i and the cell number can be calculated as follows:

$$\begin{aligned} j &\leftarrow \frac{d-1}{cn+1} \\ c &\leftarrow (d-1) \bmod cn + 1 \end{aligned} \tag{9.31}$$

By equations (9.31), the S_i can be represented as follows:

$$SS_i = [(j_1,c_1), (j_2,c_2), (j_3,c_3), \ldots, (j_k,c_k), \ldots, (j_{np},c_{np})]$$

For instance, let $np = 5$, $npp_1 = 2$, $npp_2 = 3$, $npp_3 = 2$, $npp_4 = 1$, $npp_5 = 2$, and $cn = 4$. Then we have the following chromosome:

$$S_1 = [5 \ 2 \ 7 \ 1 \ 6]$$

By equations (9.31), S_1 calculates the process plan and cell numbers for each part. The resulting values for the chromosome above are as follows:

$$SS_1 = [(2,1) \ (1,2) \ (2,3) \ (1,1) \ (2,2)]$$

The second step in the genetic algorithm is to initialize the population of chromosomes. In this model, the chromosome is generated randomly.

Evaluation. In many optimization problems, the objective function is more naturally stated as the minimization of some cost function. The objective function for the cell design model is formulated to minimize the sum of the machining and duplication costs. The fitness value is calculated based on the original objective function. The fitness function for chromosome i is defined as follows:

$$fit\_fun(i) = \text{TC} \tag{9.32}$$

However, the solution space for the cell design problem contains two parts: a feasible area and an infeasible area. For handling infeasible chromosomes, we use the regenerating technique. This technique has two steps for infeasible solutions. If a chromosome is not included in the feasible area, discard it and generate a new chromosome.

Crossover and Mutation. Crossover is aimed at exchanging bit strings between two parent chromosomes. The crossover used in this model is a one-cutpoint method. For example, the parent chromosomes are randomly selected and the cutpoint is then randomly selected at position 2 as follows:

crossover point

$$\downarrow$$

$$[5\ \ 2\ \ \underline{7\ \ 1\ \ 6}]$$

$$[3\ \ 7\ \ \underline{2\ \ 2\ \ 5}]$$

and offspring are formed by exchanging the end parts of their parents as follows:

$$[5\ \ 2\ \ \underline{2\ \ 2\ \ 5}]$$

$$[3\ \ 7\ \ \underline{7\ \ 1\ \ 6}]$$

Mutation is performed as a random perturbation. Each value with the chromosome is randomly selected to be mutated with the mutation rate. The mutation operator for the cell design problem is designed to perform random exchange. For a selected gene m_k, it will be replaced by a random integer with [1, $npp_k xcn$]. An example is given as follows:

mutation point

$$\downarrow$$

$$[3\ \ 7\ \ \underline{7}\ \ 1\ \ 6]$$

In this chromosome, the third gene is selected for mutation. The value of the

TABLE 9.9. Data of the Machine–Part Processing[a]

Part	npp	M: 1	2	3	4	5	6	7	8	9	10	pv_i
1	1	2		3							2	80
2	1					2				3	1	80
	2	4		5		4				7		
	3	6		5		7				5		
3	1	6				5		6				80
	2					3		3				
4	1			4		5					7	80
	2		4			7					6	
	3					3				4	3	
5	1			4	2					5	2	80
6	1	5	6				4		8			80
	2	2	4				2			4		
7	1			5	7					3		80
		2500	2300	2000	2200	2000	2500	2500	2000	2000	2000	

[a]npp: number of plans, M: number of machines.

gene is replaced by a random integer value within [1,12]. If the value is 4, the offspring after mutation is

$$[3\ \ 7\ \ 4\ \ 1\ \ 6]$$

Selection. Deterministic selection strategy is adopted as the selection strategy i.e., sort of parents and offspring on ascending order, delete all duplicate individuals and select the first *pop_size* chromosomes as a new population.

Numerical Examples. To demonstrate the efficiency of the approach proposed, we considered a manufacturing system with 10 machines, 7 parts, 3 cells, and 200 duplication costs for each machine, and the maximum number of parts in each cell is restricted to three. A set of process plans, production volumes for each part, process cost per unit, and machine capacity is given in Table 9.9. The genetic parameters for the example problem are as follows: $p_c = 0.6$, $p_m = 0.2$, $max\_gen = 1000$, and $pop\_size = 100$. The computational experiments are performed on a personal computer with a Pentium 75 chip under a Windows 95 operating system environment. For this setting we ran our genetic algorithm five times and obtained two alternative solutions. The results are listed in Table 9.10. Also, we solved the 0–1 integer programming model using a LINDO/PC package. The optimal solution over five runs is 8120, and the average computational time is 340 sec. For evaluation of the effectiveness of the genetic algorithm–based approach, three different-sized

TABLE 9.10. Optimal Solution

Solution	Cell Number	Part (Plan)	Machine	Fitness Value	CPU Time (s)
Alternative solution 1	1	1(1)5(1)7(1)	1 3 4 9 10	8120	247
	2	2(1)3(2)4(3)	5 7 9 10		
	3	6(2)	1 2 6 9		
Alternative solution 2	1	3(2)	5 7		
	2	2(1)4(3)6(2)	1 2 5 6 9 10		
	3	1(1)5(1)7(1)	1 3 4 9 10		

TABLE 9.11. Comparison of Results

Problem Size, $m \times p$	CPU Time of GA (s)	CPU Time of LINDO	Optimal Value
10×10	293	440	10,720
10×15	362	549	13,780
10×20	637	Cannot solve	18,438

problems are adopted. The test results are shown in Table 9.11. In this result the genetic algorithm–based approach is able to find an optimal solution within a satisfactory execution time. However, the LINDO program does not solve the third problem because the number of constraints and the number of integer variables get larger.

9.6 DESIGNING INDEPENDENT CELLS

In this section, potential application of genetic algorithms to independent cells is discussed. Independent cells are self-sufficient in the sense that products are completed within a cell without having to visit any other cell. This problem is observed in many labor-intensive manufacturing cells where lightweight, inexpensive, and usually small machines and equipment are used. Since machines are inexpensive, machine duplication does not constitute a major concern as in machine-intensive cells. One of the main reasons for having independent cells is the simplified planning and scheduling process. As the scheduling task gets simpler, the probability of implementing schedules also increases. Furthermore, it is not unusual to have multiple cells assigned to a family due to high production volume.

Besides labor-intensive cells, independent cells might be the only cell type allowed in some regulated environments where products are not allowed to leave the cell and share machines in other cells due to traceability requirements. This also prevents products from mixing with each other, which is very critical in pharmaceutical and medical device industries. As a result, identifying product families becomes more important in this environment. The cellular design in independent cells environment can therefore be considered as a sequential process where first, product families are determined, and then, corresponding independent cells are formed.

9.6.1 Family Formation for Minimizing Number of Machine Types

In this section the objective is to form product families so that the total number of machine types is minimized.

Chromosome Representation. Süer et al. [597] proposed a family number–based representation to address this problem.. The family number–based representation ensures that each part is assigned to a family. X_i indicates the family number to which product i is assigned. The number of families allowed is a design issue. The number of families (f) may be decided considering the number of desirable products in a family or organizational concern. An example of the chromosome to represent a solution is shown as follows:

$$[1\ 2\ 3\ 1\ 2\ 3]$$

where the length of chromosome indicates the number of products to be considered in cellular design. Example shows that six parts are assigned to three families. Parts 1 and 4 are assigned to family 1, parts 2 and 5 to family 2, and parts 3 and 6 to family 3.

The probability that a part can be assigned to any family is equal and it can easily be determined as $1/f$. The assignment of parts to families is made by drawing random numbers and determining corresponding family numbers. However, there is a possibility that some families may never have parts assigned to them. This will lead to fewer families than planned originally. Moreover, since fewer families will result in a lower total number of machine types, it will incorrectly look favorable in terms of fitness value. Therefore, there is a need to modify the procedure to avoid this problem. This requires the last $(f - 1)$ positions to be filled cautiously so that at least one part is assigned to each family.

Fitness Function. The fitness function used in their study is the total number of machine types needed. The first step is to find the number of machine types needed for each family. This is determined by considering the union of the set of machines for all products assigned to family k:

$$\mathrm{SM}_k = \cup_{i \in \mathrm{SP}_k} \mathrm{SO}_i \tag{9.33}$$

where SO_i is the set of machine types for product i, SP_k the set of products included in family k, and SM_k the set of machine types needed for family k.

Having determined the set of machine types needed for each family, one can easily find the number of machine types for each family (NM_k) from SM_k. The next step is to find the total number of machine types for all families (FF):

$$FF = \sum_{k=1}^{f} NM_k \tag{9.34}$$

Selection. In their study they have used a roulette wheel selection procedure. Each chromosome is assigned a reproduction probability based on its fitness value. Since the objective is of minimization type, the lower the fitness function, the higher the reproduction probability is supposed to be. Therefore, there is a need to apply a transformation function. One possible alternative is to sum all fitness values first (9.35). Then it is feasible to determine an adjusted fitness value for every chromosome by dividing the total fitness value by corresponding fitness value (9.36). Finally, a reproduction probability is calculated based on the adjusted fitness value as shown in equation (9.37).

$$TFF = \sum_{i=1} FF_i \tag{9.35}$$

$$AF_i = \frac{TFF}{FF_i} \tag{9.36}$$

$$P_i = \frac{AF_i}{\Sigma_{i=1}^{S} AF_i} \tag{9.37}$$

where FF_i is the fitness function value for chromosome i, TFF the total fitness function value, AF_i the adjusted fitness function value for chromosome i, P_i the reproduction probability for chromosome i, and S is the number of chromosomes.

Crossover Operator. A single cut-point crossover strategy is adapted. The cut point is selected randomly and then the genes after the cut point in both parents are swapped, thus generating offspring. Assume that in the following example, the cut point is after the fourth gene. The last two genes of the first parent are replaced by the last two genes of the second parent, and vice versa.

parent 1	[1 2 3 1 $\underline{3\ 1}$]
parent 2	[2 1 3 2 $\underline{1\ 2}$]
offspring 1	[1 2 3 1 $\underline{1\ 2}$]
offspring 2	[2 1 3 2 $\underline{3\ 1}$]

However, the feasibility check has to be performed since there is the possibility that a chromosome may not have at least one part assigned to each family as a result of crossover. In that case, a randomly selected part is assigned back to the empty family.

Mutation Operator. The mutation operator is applied to each gene independently until a gene mutates. As soon as a gene mutates, the application ceases. When a gene is selected for mutation, the position for the product remains unchanged. The only change is the family assignment. In the example problem considered below, mutation is applied to the fourth gene and product 4 is assigned to family 3. In this problem, a mutation operator is quite desirable. The main reason is that the chromosomes in the initial population are expected to have an equal number of parts for each family (since a probability of $1/f$ has been used to assign parts to families). This initial grouping may not form the most desirable families because the families should be formed based on processing similarity, not based on an equal number of parts. Both crossover and mutation operators are expected to correct this problem. Once again, the feasibility check has to be performed since the only part in a family might be selected for mutation, thus creating an empty family. In that case, a randomly selected part is assigned back to the empty family.

$$[1\ 2\ 3\ \underline{1}\ 2\ 3] \quad \text{original chromosome}$$

$$[1\ 2\ 3\ \underline{3}\ 2\ 3] \quad \text{mutated chromosome}$$

Crossover and Mutation Strategies. Süer et al. [594] used the following crossover and mutation strategies in their study:

1. *Crossover and mutation* ($C + M$). Reproduced chromosomes are paired and they all go through a crossover operation. Following crossover, each gene is tested independently until a gene mutates. As soon as a gene mutates, no other genes are tested for mutation.
2. *Crossover only* (*C-O*). This is similar to the previous strategy except that mutation is skipped.
3. *Mutation only* (*M-O*). Reproduced chromosomes only go through a mutation operator. However, in this case the mutation is considered for each gene with increasing probability. The mutation probability for the first gene is $1/n$, for the second gene is $1/(n - 1)$, for the third gene is $1/(n - 2)$, and so on.
4. *Mutation and crossover* (M_x, C_y). x of the reporduced chromosomes go through a mutation operator and y chromosomes go through crossover. The mutation strategy adapted is in the form of increasing probability as explained in *M-O* strategy.

TABLE 9.12. Machine–Part Incident Matrix

	Product									
Machine	P_1	P_2	P_3	P_4	P_5	P_6	P_7	P_8	P_9	P_{10}
M_1	1			1	1			1		
M_2				1	1		1	1	1	1
M_3	1		1	1		1	1	1	1	1
M_4	1	1	1	1	1	1	1		1	
M_5		1				1				
M_6		1	1			1				

Numerical Example. The product–machine incidence matrix given in Table 9.12 is used to illustrate how the procedure works. In the matrix, $M_{ij} = 1$ indicates that part j has an operation on machine i; a zero entry indicates that it does not.

Assume that a population consists of the following chromosomes:

$$V_1 = [1\ 2\ 1\ 3\ 1\ 2\ 2\ 3\ 1\ 3]$$

$$V_2 = [1\ 2\ 3\ 3\ 1\ 3\ 2\ 2\ 1\ 2]$$

$$V_3 = [2\ 3\ 1\ 3\ 2\ 1\ 2\ 2\ 1\ 3]$$

$$V_4 = [3\ 1\ 2\ 2\ 2\ 1\ 3\ 1\ 3\ 2]$$

The details of the computations are presented for the first chromosome. There are three randomly formed families. Family 1 consists of products 1, 3, 5, and 9, family 2 of products 2, 6, and 7, and family 3 of products 4, 8, and 10. The results for families are summarized below:

$$\mathrm{SP}_1 = \{P_1, P_3, P_5, P_9\}$$

$$\mathrm{SP}_2 = \{P_2, P_6, P_7\}$$

$$\mathrm{SP}_3 = \{P_4, P_8, P_{10}\}$$

The set of operations can easily be formed by using the incidence matrix given in Table 9.12. For example, part 1 needs machines 1, 3, and 4. The sets of machines for the parts in the first family are summarized as follows:

$$\mathrm{SO}_1 = \{M_1, M_3, M_4\}$$

$$\mathrm{SO}_3 = \{M_3, M_4, M_6\}$$

$$\mathrm{SO}_5 = \{M_1, M_2, M_4\}$$

$$\mathrm{SO}_9 = \{M_2, M_3, M_4\}$$

The set of machines for family 1 is determined as follows:

$$\begin{aligned} SM_1 &= \{\{M_1, M_3, M_4\} \cup \{M_3, M_4, M_6\} \cup \{M_1, M_2, M_4\} \\ &\quad \cup \{M_2, M_3, M_4\}\} \\ &= \{M_1, M_2, M_3, M_4, M_6\} \end{aligned}$$

Now the number of machine types needed for family 1 can easily be determined ($NM_1 = 5$).

By repeating this procedure for families 2 and 3, the following results are obtained:

$$SM_2 = \{M_2, M_3, M_4, M_5, M_6\}, \qquad NM_2 = 5$$

$$SM_3 = \{M_1, M_2, M_3, M_4\}, \qquad NM_3 = 4$$

Fitness function value for chromosome 1 is calculated as

$$FF_1 = 5 + 5 + 4 = 14$$

In a similar fashion, the values of the fitness function are obtained as 16, 15, and 15 for chromosomes 2, 3, and 4, respectively. Having determined fitness function values, the next step is to calculate reproduction probability for every chromosome as shown in the steps below:

$$TFF = 14 + 16 + 15 + 15 = 60$$

$$AF_1 = \frac{60}{14} = 4.28, \qquad AF_2 = 3.75, \qquad AF_3 = 4.0, \qquad AF_4 = 4.0$$

$$P_1 = \frac{4.30}{16.08} = 0.268, \qquad P_2 = 0.234, \qquad P_3 = 0.249, \qquad P_4 = 0.249$$

Alternative Solutions. The example problem introduced in the preceding section is solved. A population size of 25 is chosen and the number of generations is set to 500. The best fitness value obtained is 10. Three solutions provide the same result since they correspond to the same family and cell definitions. The results are summarized in Table 9.13. Frequency measures how many times the optimal solution was obtained during the run.

9.6.2 Determining Number of Families

In some cases it may be difficult to establish the number of families in advance. In that case, the designer will have to run a genetic algorithm for various f values.

TABLE 9.13. Alternative Solutions for Three-Family Problem

Solution	Fitness Value	Frequency	Chromosome	Families	Machine in the Cells
1	10	1	3223323331	(10)	[2,3]
				(6,3,2)	[3,4,5,6]
				(9,8,7,5,4,1)	[1,2,3,4]
2	10	1	1331131112	(9,8,7,5,4,1)	[1,2,3,4]
				(10)	[2,3]
				(6,3,2)	[3,4,5,6]
3	10	1	3113313332	(6,3,2)	[3,4,5,6]
				(10)	[2,3]
				(9,8,7,5,4,1)	[1,2,3,4]

Different Number of Families. In this section the results presented by Süer et al. [595] are discussed. In their analysis they also tried $f = 2$ and $f = 4$ for the same example problem. The results are summarized in Tables 9.14 and 9.15, respectively.

However, there is not enough information to decide whether a two-, three-, or four-family solution is better. Furthermore, there are alternative solutions within each family category as well. Even though alternative solutions within a family category resulted in the same fitness value, it is difficult to assess which alternative is better. As a result, the number of machines had to be calculated to compare the alternative solutions. Processing times and demand information are needed to proceed with the computations.

General Methodology. The general methodology to determine the optimal number of families and the best alternative is summarized in the following steps:

TABLE 9.14. Results for Two-Family Solution

Solution	Fitness Value	Frequency	Chromosome	Families	Machine in the Cells
1	8	15	1221121111	(10,9,8,7,5,4,1)	[1,2,3,4]
				(6,3,2)	[3,4,5,6]
2	8	9	2112212222	(6,3,2)	[3,4,5,6]
				(10,9,8,7,6,5,4,3,2,1)	[1,2,3,4]
3	8	7	2222222221	(10)	[2,3]
				(9,8,7,6,5,4,3,2,1)	[1,2,3,4,5,6]
4	8	22	1111111112	(9,8,7,6,5,4,3,2,1)	[1,2,3,4,5,6]
				(10)	[2,3]

TABLE 9.15. Results for Four-Family Solution

Solution	Fitness Value	Frequency	Chromosome	Families	Machine in the Cells
1	13	1	433434142	(8)	[1,2,3]
				(10)	[2,3]
				(6,3,2)	[3,4,5,6]
				(9,7,5,4,1)	[1,2,3,4]
2	13	1	3113313234	(6,3,2)	[3,4,5,6]
				(8)	[2,3]
				(9,7,5,4,1)	[1,2,3,4]
				(10)	[2,3]

1. Set f = MIN (minimum acceptable number of families).
2. Apply evolutionary programming to assign n products to f families such that the total number of machine types will be minimized.
3. Determine the number of machines, M_f.
4. Increase f by 1 ($f = f + 1$).
5. If $f \leq$ MAX (maximum allowable number of families), go to step 2. Otherwise, continue with step 6.
6. The alternative (g) is the final configuration where $M_g = \min\{M_f\}$. MIN $\leq f \leq$ MAX.

Number of Machines: Bottleneck Machine-Based Approach. In this section the bottleneck machine-based approach is used to determine the number of machines Süer et al. [595]. This approach is feasible when unit transfers are used between machines. Due to material handling devices, and so on, a machine remains idle if it is not used by a product. Another reason for using bottleneck-based computation is the single product restriction in a cell at all times, as in the pharmaceutical industry. Obviously, this approach requires more machines, due to increased machine idle time. It is assumed that there will be only one machine of each required type in a cell. The production rate of a product is calculated based on the maximum processing time of its operations:

$$t_{i,\max} = \max_{j \in O_i}\{t_{ij}\} \tag{9.38}$$

where t_{ij} is the processing time of operation j of product i, O_i the set of operations for product i, and $t_{i,\max}$ the maximum operation time for product i.

The total time required to meet the demand for all products in a family is obtained by considering the demand and production rates:

TABLE 9.16. Processing Time and Product Demand

	Product									
Machine	P_1	P_2	P_3	P_4	P_5	P_6	P_7	P_8	P_9	P_{10}
M_1	0.15			0.25	0.15			0.25		
M_2				0.20	0.10		0.15	0.20	0.30	0.15
M_3	0.18		0.25	0.30		0.25	0.20	0.14	0.25	0.20
M_4	0.20	0.10	0.20	0.28	0.11	0.20	0.15		0.20	
M_5		0.20				0.30				
M_6		0.12	0.15			0.15				
Annual demand	7k	4k	10k	6k	8k	6k	2.5k	2k	3k	2.5k
Production rate (h)	5	5	4	3.33	6.66	3.33	5	4	3.33	5
Time (h) required	1400	800	2500	1800	1200	1800	500	500	900	500

$$\text{TR}_k = \sum_{i \in SP_k} \frac{d_i}{t_{i,\max}} \tag{9.39}$$

where d_i is the annual demand for product i and TR_k is the total time required by family k.

The number of cells required for family k (NC_k) is calculated by dividing the total time required by the available hours in a year per cell (A):

$$\text{NC}_k = \frac{\text{TR}_k}{\text{A}} \tag{9.40}$$

Next, the total number of machines needed (MN_k) for a family is determined by multiplying the number of cells by the number of machine types in a cell:

$$\text{MN}_k = \text{NC}_k \times \text{MN}_k \tag{9.41}$$

Finally, the total number of machines needed for a configuration is obtained by summing the number of machines needed for all families:

$$M_f = \sum_{i=1}^{f} \text{MN}_i \tag{9.42}$$

Numerical Example. The previous example is extended by including processing times and demand information as given in Table 9.16. The maximum unit processing time for product 1 is 0.20 h on Machine 4. Therefore, the maximum output rate for product 1 from the cell will be limited to machine 4, and thus it will be 5 units/h. If the annual demand for product 1 is 7000

TABLE 9.17. Results for Two-Family Solution with Fitness Value 8

Solution	Number of Machines	Chromosome	Families	Machine in the Cells	Cells
1	28	1221121111	(10,9,8,7,5,4,1)	[1,2,3,4]	4
			(6,3,2)	[3,4,5,6]	3
2	28	2112212222	(6,3,2)	[3,4,5,6]	3
			(10,9,8,7,5,4,1)	[1,2,3,4]	4
3	38	2222222221	(10)	[2,3]	1
			(9,8,7,6,5,4,3,2,1)	[1,2,3,4,5,6]	6
4	38	1111111112	(9,8,7,6,5,4,3,2,1)	[1,2,3,4,5,6]	6
			(10)	[2,3]	1

units, that implies that product 1 will require 1400 h of the cell. Hours required by each product have been computed and they are also included in Table 9.16. The results obtained for two-, three-, and four-family solutions are summarized in Tables 9.17, 9.18, and 9.19, respectively. As one can observe from the tables, the number machines were 30 for all alternatives in the three-family category (identical solutions), 29 machines in the four-family category (identical solution). However, the results varied significantly in the two-family category. Some alternatives (two identical) required 28 machines, whereas two other alternatives required 38 machines. In many cases, several identical cells were required to meet the demand.

The products in the first family for solution 1 in Table 9.17 are P_1, P_4, P_5, P_7, P_8, P_9, and P_{10}. TR_1 is calculated by summing the time required for all parts in the first family:

$$TR_1 = 1400 + 1800 + 1200 + 500 + 500 + 900 + 500 = 6800 \text{ h}$$

Assuming that a cell is available 2000 h/yr, it is easy to determine that four cells are needed to meet the demand:

TABLE 9.18. Results for Three-Family Solution with Fitness Value 10

Solution	Number of Machines	Chromosome	Families	Machine in the Cells	Cells
1	30	3223323331	(10)	[3,4]	1
			(6,3,2)	[3,4,5,6]	3
			(9,8,7,5,4,1)	[1,2,3,4]	4
2	30	1331131112	(9,8,7,5,4,1)	[1,2,3,4]	4
			(10)	[3,4]	1
			(6,3,2)	[3,4,5,6]	3
3	30	3113313332	(6,3,2)	[3,4,5,6]	3
			(10)	[2,3]	1
			(9,8,7,5,4,1)	[1,2,3,4]	4

TABLE 9.19. Results for Four-Family Solution with Fitness Value 13

Solution	Number of Machines	Chromosome	Families	Machine in the Cells	Cells
1	29	4334434142	(8)	[1,2,3]	1
			(10)	[2,3]	1
			(6,3,2)	[3,4,5,6]	3
			(9,7,5,4,1)	[1,2,3,4]	3
2	29	3113313234	6,3,2)	[3,4,5,6]	3
			(8)	[1,2,3]	1
			(9,7,5,4,1)	[1,2,3,4]	3
			(10)	[2,3]	1

$$NC_1 = \left\lceil \frac{6800}{2000} \right\rceil = 4 \text{ cells}$$

where $\lceil x \rceil$ indicates the smallest integer larger than x. Hence the total number of machines needed are computed by multiplying the number of cells by the number of machines in a cell.

$$MN_1 = NC_1 \times NM_1 = 4 \times 4 = 16 \text{ machines}$$

Similarly, TR_2, NC_2, and MN_2 values are calculated as 5100 h, 3 cells, and 12 machines. As a result, the total number of machines needed if two families are formed is determined as

$$M_2 = 16 + 12 = 28 \text{ machines}$$

Best Solution. The best solution is to group the parts in two families since that configuration results in the minimum number of machines:

$$M_g = \min\{28, 30, 29\} = 28 \text{ machines}, \qquad \text{MIN} \leq f \leq \text{MAX}$$

9.6.3 Minimizing Number of Machines

Süer et al. [596] also used directly the total number of machines needed as the fitness function. All other operators were exactly the same and 10 runs were made. The number of machine computation was kept the same. The same example problem was run with a population size of 25 and 500 generations. Better results are obtained by including the number of machines directly as the fitness function. The total number of machines required were lowered to 26 in many cases, as shown in Tables 9.20 to 9.22. Their study also concluded that the m25, cm25, c10m15, m15c10, and m20c5 strategies produced exactly the same results and outperformed both the m5c20 and c25

TABLE 9.20. Solution for m25, cm25, c10m15, m15c10, and m20c5 Strategies

	Number of Families				
Results	2	3	4	5	6
Minimum	27	26	26	26	26
Maximum	27	26	26	26	26
Average	27	26	26	26	26
Best solution	0	10	10	10	10

strategies. The number of families didn't affect the results except the two-family configuration in most cases.

9.6.4 Other Considerations

In designing independent cells, other issues, such as machine weights or machine costs, can also be included in the fitness function. Obviously, a bottleneck-based number of machine computations is not the only way to compute the number of machines required. Another possibility is to compute machine requirements based on individual processing times on the machines required, as opposed to those based on the bottleneck machine. Undoubtedly, this would require the same number of machines or fewer. Finally, one could also focus on designing remainder cells. The remainder family is treated like any other family except that at least one of each machine type is placed in the remainder cell, since by definition it is expected to handle any product and any process. Süer et al. [593] discussed this problem briefly in an earlier paper.

TABLE 9.21. Solution for c25 Strategy

	Number of Families				
Results	2	3	4	5	6
Minimum	27	26	27	26	27
Maximum	32	32	29	29	30
Average	30	28.5	28.2	27.4	28.2
Best solutions	0	1	0	1	0

TABLE 9.22. Solution for m5c20 Strategy

	Number of Families				
Results	2	3	4	5	6
Minimum	27	26	27	26	26
Maximum	27	27	26	27	26
Average	27	26.4	26	26.2	26
Best solutions	0	6	10	8	10

REFERENCES

[1] Abido, M. A., Intelligent techniques approach to power system identification and control, Ph.D. dissertation, University of Petroleum and Minerals (Saudi Arabia), 1997.

[2] Abuali, F. N., Using determinant and cycle basis schemes in genetic algorithms for graph and network applications, Ph.D. dissertation, University of Tulsa, 1995.

[3] Abuali, F., R. Wainwright, and D. Schoenefeld, A new encoding scheme for spanning trees applied to the probabilistic minimum spanning tree problem, in Eshelman [177], pp. 470–475.

[4] Adams, J., E. Balas, and D. Zawack, The shifting bottleneck procedure for job shop scheduling, *International Journal of Flexible Manufacturing Systems,* vol. 34, no. 3, pp. 391–401, 1988.

[5] Adar, N., Allocation and scheduling on multi-computers using genetic algorithms, Ph.D. dissertation, Lehigh University, 1994.

[6] Adil, G. K., D. Ragamani, and D. Strong, Cell design considering alternative routings, *Int. Journal of Production Research,* vol. 34, no. 5, pp. 1361–1380, 1996.

[7] Aggarwal, K. K., Y. C. Chopra, and J. S. Bajwa, Topological layout of lips for optimizing the overall reliability in a computer communication system, *Microelectronics and Reliability,* vol. 22, no. 3, pp. 347–351, 1982.

[8] Aggarwal, K. K. and S. Rai, Reliability evaluation of computer-communication networks, *IEEE Transactions on Reliability,* vol. 30, no. 1, pp. 32–35, 1981.

[9] Ahuja, R. K., T. L. Magnanti, and J. B. Orlin, *Network Flows: Theory, Algorithms, and Applications,* Prentice Hall, Upper Saddle River, NJ, 1993.

[10] Al-harkan, I. M., One merging sequencing and scheduling theory with genetic algorithms to solve stochastic job shop problems, Ph.D. dissertation, University of Oklahoma, 1997.

[11] Alander, J., *An Indexed Bibliography of Genetic Algorithms: 1957–1993,* Art of CAD Ltd., Espoo, Finland, 1994.

[12] Alvarez-Valdés, R. and J. Tamarit, Heuristic algorithms for resource constrained project scheduling: a review and an empirical analysis, in Slowinski, R. and J. Weglarz, editors, *Advances in Project Scheduling,* pp. 113–134, Elsevier Science Publishers, Amsterdam, 1989.

[13] Anderson, C., K. Jones, and J. Ryan, A two-dimensional genetic algorithm for the ising problem, *Complex Systems,* vol. 5, pp. 327–333, 1991.

[14] Aneja, Y. and K. Nair, Bicriteria transportation problem, *Management Science,* vol. 25, pp. 73–78, 1978.

[15] Ann, B.-H. and J.-H. Hyun, Single facility multi-class job scheduling, *Computers and Operations Research,* vol. 17, pp. 265–272, 1989.

[16] Annaiyappa, P. V., Critical analysis of genetic algorithms for global optimization, Ph.D. dissertation, New Mexico State University, 1991.

[17] Applegate, D. and W. Cook, A computational study of the job shop scheduling problem, *ORSA Journal of Computing,* vol. 3, no. 2, pp. 149–156, 1991.

[18] Arabeyre, T., J. Faernley, F. Steiger, and W. Teather, The airline crew scheduling problem, *Transportation Science,* vol. 3, no. 2, pp. 140–163, 1969.

[19] Areibi, S., Toward optimal circuit layout using advanced search techniques, Ph.D. dissertation, University of Waterloo (Canada), 1995.

[20] Arguelles, D., Hybrid artificial neutral network/genetic algorithm approach to the on-line optimization of electrical power systems, Ph.D. dissertation, George Washington University, 1996.

[21] Askin, R. and K. Chiu, A graph partitioning procedure for machine assignment and cell formation, *Int. Journal of Production Research,* vol. 28, no. 8, pp. 1555–1572, 1990.

[22] Atiqullah, M. M. and S. S. Rao, Reliability optimization of communication networks using simulated annaeling, *Microelectronics and Reliability,* vol. 33, no. 9, pp. 1303–1319, 1993.

[23] Awadh, B., N. Sepehri, and O. Hawaleshka, A computer-aided process planning model based on genetic alogorithms, *Computers and Operations Research,* vol. 22, no. 8, pp. 841–856, 1995.

[24] Awadth, B., Manufacturing process planning model using genetic algorithms, Ph.D. dissertation, University of Manitoba, 1994.

[25] Bäck, T., GENEsYs 1.0 software distribution and installation notes, Technical report, Systems Analysis Research Group, LSXI, Department of Computer Science, University of Dortmund, Germany, 1992.

[26] Bäck, T., Optimal mutation rates in genetic search, in Forrest [204], pp. 2–9.

[27] Bäck, T., Selective pressure in evolutionary algorithms: a characterization of selection mechanisms, in Fogel [197], pp. 57–62.

[28] Bäck, T., *Evolutionary Algorithms in Theory and Practice,* Oxford University Press, New York, 1996.

[29] Bäck, T., editor, *Proceedings of the 7th International Conference on Genetic Algorithms,* Morgan Kaufmann Publishers, San Francisco, 1997.

[30] Bäck, T. and F. Hoffmeister, Extended selection mechanisms in genetic algorithms, in Belew and Booker [51], pp. 92–99.

[31] Bäcker, T., F. Hoffmeister, and H. Schwefel, A survey of evolution strategy, in Belew and Booker [51], pp. 2–9.

[32] Bäck, T. and H. Schwefel, An overview of evolution algorithms for parameter optimizations, *Evolutionary Computation,* vol. 1, no. 1, pp. 1–23, 1993.

[33] Bäck, T. and H. Schwefel, Evolutionary computation: a survey, in Fogel [195], pp. 20–28.

[34] Bakalis, O. L., Encoding the invariant measure of iterated affine transforms with a genetic algorithm, Ph.D. dissertation, University of New Mexico, 1996.

[35] Baker, J., Reducing bias and inefficiency in the selection algorithm, in Grefenstette [266], pp. 14–21.

[36] Baker, J., Adaptive selection methods for genetic algorithms, in Grefenstette [267], pp. 100–111.

[37] Baker, K., *Introduction to Sequencing and Scheduling,* Wiley, New York, 1974.

[38] Balas, E. and A. Ho, Set covering algorithms using cutting planes, heuristics, and subgradient optimization: a computational study, *Mathematical Programming Studies,* vol. 20, pp. 37–60, 1980.

[39] Balinski, M., Integer programming: methods, uses and computation, *Management Science,* vol. 12, no. 3, pp. 253–313, 1965.

[40] Balinski, M. and R. Quandt, On an integer program for a delivery problem, *Operations Research,* vol. 12, no. 2, pp. 300–304, 1964.

[41] Ball, M. and R. M. Van Slyke, Backtracking algorithms for network reliability analysis, *Annals of Discrete Mathematics,* vol. 1, pp. 49–64, 1977.

[42] Bangalore, S. S., Data analysis stragegies for qualitative and quantitative determination of organic compounds by Fourier transform infrared spectroscopy, Ph.D. dissertation, Ohio State University, 1996.

[43] Barr, R., R. Glover, and D. Klingman, A new optimization method for large scale fixed charge transportation problems, *Operations Research,* vol. 29, no. 3, pp. 448–463, 1981.

[44] Bazaraa, M., J. Jarvis, and H. Sherali, *Linear Programming and Network Flows,* 2nd edition, Wiley, New York, 1990.

[45] Bazaraa, M., H. Sherali, and C. Shetty, *Nonlinaer Programming: Theory and Algorithms,* 2nd edition, Wiley, New York, 1993.

[46] Bean, J. C., A. Lagrangian algorithm for the multiple choice integer program, *Operations Research,* vol. 32, pp. 1185–1193, 1984.

[47] Bean, J., Genetic algorithms and random keys for sequencing and optimization, *ORSA Journal of Computing,* vol. 6, no. 2, pp. 154–160, 1994.

[48] Beasley, J. E., OR-library: distributing test problems by electronic mail, *Journal of the Operations Research Society,* vol. 41, no. 11, pp. 1069–1072, 1990.

[49] Beasley, J. E. and P. C. Chu, A genetic algorithm for the set covering problem, *Europaen Journal of Operational Research,* vol. 94, pp. 392–404, 1996.

[50] Beasley, J. E. and K. Jornsten, Enhancing an algorithm for set covering problems, *Europaen Journal of Operational Research,* vol. 58, pp. 293–300, 1992.

[51] Belew, R. and L. Booker, editors, *Proceedings of the 4th International Conference on Genetic Algorithms,* Morgan Kaufman Publishers, San Francisco, 1991.

[52] Bellman, R. and L. Zadeh, Decision making in a fuzzy environment, *Management Science,* vol. 17, pp. 141–146, 1970.

[53] Bellmore, M., Set covering and involutional bases, *Management Science,* vol. 18, no. 3, pp. 194–206, 1971.

[54] Bellmore, M., H. Greenberg, and J. Jarvis, Multi-commodity networks, *Management Science,* vol. 16, no. 6, pp. 427–433, 1970.

[55] Berman, O., R. C. Larson, and S. S. Chiu, Optimal server location on a network operating as an $M/G/1$ queue, *Operations Research,* vol. 33, no. 4, pp. 746–771, 1985.

[56] Berman, O. and B. LeBlance, Location–relocation of N mobile facilities on a stochastic network, *Transportation Science,* vol. 21, pp. 207–216, 1984.

[57] Berman, O. and R. R. Mandowsky, Location–allocation on congested network, *European Journal of Operational Research,* vol. 26, pp. 238–250, 1986.

[58] Bertsekas, D. and P. Tseng, Relaxation methods for minimum cost oridinary and generalized network flow problems, *Operations Research,* vol. 36, no. 1, pp. 93–114, 1988.

[59] Bertismas, D., The probabilistic minimum spanning tree problem, *Networks,* vol. 20, pp. 245–275, 1990.

[60] Besson, N. W., Evaluation of the genetic algorithm as a computational tool in protein NMR, Ph.D. dissertation, Harvard University, 1995.

[61] Bhat, M. V. and A. Haupt, An efficient clustering algorithm, *IEEE Transactions on Systems, Man and Cybernetics,* vol. 6, no. 1, pp. 61–64, 1976.

[62] Billo, R., D. Tate, and B. Bidanda, A genetic cluster algorithm for the machine-component grouping problem, Technical report, University of Pittsburgh, 1995.

[63] Billo, R. E., D. Tate, and B. Bidanda, Comparison of a genetic algorithm and clustyer analysis for the cell formation problem: a case study, in *Proceedings of the 3rd Industrial Engineering Research Conference,* pp. 543–548, Atlanta, GA, 1994.

[64] Bjorklund, M. and A. Elldin, A practical method of calculation for certain types of complex common control systs, *Ericsson Technics,* vol. 20, pp. 3–75, 1964.

[65] Bjorndal, M., A Caprara, P. Cowling, F. Croce, H. Lourence, F. Malucelli, A. Orman, D. Pisinger, C. Rego, and J. Salazar, Some thoughts on combinatorial optimization, *European Journal of Operational Research,* vol. 83, pp. 253–270, 1995.

[66] Black, U., *TCP/IP and Related Protocols,* McGraw-Hill, New York, 1995.

[67] Blackstone, J., D. Phillips, and G. Hogg, A state-of-the-art survey of dispatching rules for manufacturing job shop operations, *Int. Journal of Production Research,* vol. 20, pp. 27–45, 1982.

[68] Blom, G., *Probability and Statistics: Theory and Applications,* Discrete Applied Mathematics, 1996.

[69] Boctor, F., A linear formulation of the machine-part cell formation problem, *Int. Journal of Production Research,* vol. 29, no. 2, pp. 343–356, 1991.

[70] Bollobas, B., *Graph Theory: An Introductory Course,* Springer-Verlag, New York, 1979.

[71] Booker, L., Improving search in genetic algorithms, in Davis, L., editor, *Genetic Algorithms and Simulated Annealing,* Morgan Kaufmann Publishers, San Francisco, 1987.

[72] Bowden, R. O., Genetic algorithm–based machine learning applied to the dynamic routing of discrete parts, Ph.D. dissertation, Mississippi State University, 1992.

[73] Boxman, O. J., J. W. Cohen, and N. Huffels, Approximations of mean waiting time in an $M/G/s$ queuing systems, *Operations Research,* vol. 27, pp. 1115–1127, 1979.

[74] Brady, R., Optimization strategies gleaned from biological evolution, *Nature,* vol. 317, no. 31, pp. 804–806, 1985.

[75] Brandeau, M. and S. Chiu, An overview of representative problems in location research, *Management Science,* vol. 35, pp. 645–673, 1989.

[76] Brindle, A., Genetic algorithms for function optimization, Ph.D. dissertation, University of Alberta–Edmondton, 1981.

[77] Brown, E. C., Using the facility location problem to explore operator policies and constraint-handling methods for genetic algorithms, Ph.D. dissertation, University of Virginia, 1996.

[78] Brown, W. G., Optimal rate concept acquisition using version spaces and genetic algorithms, Ph.D. dissertations, Wayne State University, 1994.

[79] Bui, T. N. and B. R. Moon, On multi-dimensional encoding/crossover, in Eshelman [177], pp. 49–56.

[80] Burbidge, J. L., An introduction of group technology, in *Seminar on Group Technology,* Turin, Italy, 1969.

[81] Burbidge, J. L., *The Introduction of Group Technology,* Halsted Press, division of Wiley, New York, 1975.

[82] Busacker, R. and T. Saaty, *Finite Graphs and Networks,* Martinus Nijhoff, New York, 1965.

[83] Cancela, H. and M. E. Khadiri, A recursive variance-reduction algorithm for estimating communication-network reliability, *IEEE Transactions on Reliability,* vol. 44, no. 4, pp. 595–602, 1995.

[84] Canpolate, N., Optimization of seasonal irrigation scheduling by genetic algorithms, Ph.D. dissertation, Oregon State University, 1997.

[85] Carse, B., T. C. Fogarty, and A. Munro, Adaptive distributed routing using evolutionary fuzzy control, in Belew and Booker [51], pp. 389–396.

[86] Caskey, K. R., Genetic algorithms and neutral networks applied to manufacturing scheduling, Ph.D. dissertation, University of Washington, 1993.

[87] Cavaretta, M. J., Cultural algorithms and real-valued function optimization, Ph.D. dissertation, Wayne State University, 1995.

[88] Cedeño, W., the multiniche crowding genetic algorithm: analysis and applications, Ph.D. dissertation, University of California–Davis, 1995.

[89] Ceneño, W. and V. R. Vemuri, Database design with genetic algorithms, in Dasgupta, D. and Z. Michalewicz, editors, *Multiple Criteria Decision Making: Theory and Applications,* pp. 189–206, Springer-Verlag, Heidelberg, 1997.

[90] Chaer, W. S., Mixture-of-experts approach to adaptive estimation, Ph.D. dissertation, University of Texas–Austin, 1996.

[91] Chan, H. M. and D. A. Milner, Direct clustering algorithm for group formation in cellular manufacture, *OMEGA: International Journal of Management Science,* vol. 1, no. 1, pp. 65–74, 1982.

[92] Chanas, S., Fuzzy programming in multiobjective linear programming: a parametric approach, *Fuzzy Sets and Systems,* vol. 29, pp. 303–313, 1989.

[93] Chandrasekharan, M. P. and R. Rajagopalan, An ideal seed non-hierarchical clustering algorithm for cellular manufacturing, *Int. Journal of Production Research,* vol. 24, no. 2, pp. 451–464, 1986.

[94] Chandy, K. M. and T. Lo, The capacitated minimum spanning tree, *Networks,* vol. 3, pp. 173–182, 1973.

[95] Chang, T. C. and R. A. Wysk, *An Introduction to Automated Process Planning Systems,* Prentice Hall, Upper Saddle River, NJ, 1985.

[96] Chankong, A. and M. Miller, A model for optimal programming of railway freight train movements, *Management Science,* vol. 3, no. 1, pp. 74–92, 1956.

[97] Changkong, V. and Y. Y. Haimes, *Multiobjective Decision Making Theory and Methodology,* North-Holland, New York, 1983.

[98] Charnes, A. and W. W. Cooper, *Management Models and Industrial Applications of Linear Programming,* Wiley, New York, 1961.

[99] Chbat, N. W., Direct-learning neutral and fuzzy control of two unknown dynamic systems using Powell optimization and genetic algorithms, Ph.D. dissertation, Columbia University, 1996.

[100] Cheng, C., Study on optimal design of fuzzy facility layout, master's thesis, Ashikaga Institute of Technology, March 1995.

[101] Cheng, C. H., Y. P. Gupta, W. H. Lee, and K. F. Wong, A TSP-based heuristic for forming machine groups and part families, *Int. Journal of Production Research,* vol. 28, no. 5, pp. 1325–1337, 1998.

[102] Cheng, M. Y., Control design and robustness measurement for biped locomotion, Ph.D. dissertation, University of Missouri-Columbia, 1996.

[103] Cheng, R., Study on genetic algorithm–based optimal scheduling techniques, Ph.D. dissertation, Tokyo Institute of Technology, 1997.

[104] Cheng, R. and M. Gen, A hybrid genetic algorithm for the parallel machine minmax weighted absolute lateness problems, *Engineering Design and Automation* (forthcoming).

[105] Cheng, R. and M. Gen, Evolution program for resource constrained project scheduling problem, in Fogel [197], pp. 736–741.

[106] Cheng, R. and M. Gen, Resource constrained project scheduling problem using genetic algorithms, *International Journal of Intelligent Automation and Soft Computing,* vol. 3, no. 3, pp. 273–286, 1997.

[107] Cheng, R. and M. Gen, An adaptive superplane approach for multiple objective optimization problems, Technical report, Ashikaga Institute of Technology, 1998.

[108] Cheng, R. and M. Gen, Loop layout design problem in flexible manufacturing system using genetic algorithm, *Computers and Industrial Engineering,* vol. 34, no. 1, pp. 53–61, 1998.

[109] Cheng, R. and M. Gen, Compromise approach-based genetic algorithms for bicriterion shortest path problems, Technical report, Ashikaga Institute of Technology, 1998.

[110] Cheng, R. and M. Gen, An evolution program for the resource constrained project scheduling problem, *Computer Integrated Manufacturing,* vol. 11, no. 3, pp. 274–287, 1998.

[111] Cheng, R. and M. Gen, A priority based encoding and shortest path problem, Technical report, Ashikaga Institute of Technology, 1998.

[112] Cheng, R. and M. Gen, A survey of genetic multiobjective optimizations, Technical report, Ashikaga Institute of Technology, 1998.

[113] Cheng, R., M. Gen, and T. Tozawa, Genetic algorithms for multi-row machine layout problem, *Engineering Design and Automation* (forthcoming).

[114] Cheng, R., M. Gen, and Y. Tsujimura, A tutorial survey of job-shop scheduling problems using genetic algorithms: I. Representation, *Computers and Industrial Engineering,* vol. 30, pp. 983–997, 1996.

[115] Cheng, R., M. Gen, and Y. Tsujimura, A tutorial survey of job-shop scheduling problems using genetic algorithms: II. Hybrid genetic search strategies, *Computers and Industrial Engineering,* vol. 36, no. 2, 1999.

[116] Cheng, S. J. and C. S. Cheng, A neural network-based cell formation algorithm in cellular manufacturing, *Int. Journal of Production Research,* vol. 33, no. 2, pp. 293–318, 1995.

[117] Cheng, T. and X. Cai, On the complexity of completion time variance minimization problem, Working paper, University of Western Australia–Nedlands, 1990.

[118] Chintrakulchai, P., High-performance fractal image compression, Ph.D. dissertation, Dartmouth College, 1995.

[119] Chiu, S. S., O. Berman, and R. C. Larson, Locating a mobile server queueing facility on a tree network, *Management Science,* vol. 31, no. 6, pp. 764–772, 1985.

[120] Chopra, Y. C., B. S. Sohi, R. K. Tiwari, and K. K. Aggarwal, Network topology for maximizing the terminal reliability in a computer communication network, *Microelectronics and Reliability,* vol. 24, pp. 911–913, 1984.

[121] Christofides, N., R. Alvarez-Valdés, and J. Tamarit, Project scheduling with resource constraint: a branch bound approach, *European Journal of Operational Research,* vol. 29, pp. 262–273, 1987.

[122] Chu, T., Some problems in fuzzy decision making, Ph.D. dissertation, University of Texas–Arlington, 1993.

[123] Chunduru, R. K., Global and hybrid optimization in geophysical inversion, Ph.D. dissertation, University of Texas–Austin, 1998.

[124] Chung, W. S., Analysis of the effects of different representation schemes on genetic algorithms, Ph.D. dissertations, University of South Florida, 1995.

[125] Climaco, J. and E. Martins, A bicriterion shortest path algorithm, *European Journal of Operational Research,* vol. 11, pp. 399–404, 1982.

[126] Cobham, A., R. Fridshal, and J. North, An application of linear programming to the minimization of Boolean functions, Report RC 472, IBM Research Center, 1961.

[127] Cohoon, P. and W. Paris, Genetic placement, in *Proceedings of the IEEE International Conference on Computer-Aided Design,* pp. 422–425, 1986.

[128] Coit, D. W. and A. E. Smith, Solving the redundancy allocation problem using a combined neural network/genetic algorithm approach, *Computers and Operations Research,* vol. 23, no. 6, pp. 515–526, 1996.

[129] Coit, D., Optimization of reliability design problems considering uncertainty in component reliability and time-to-failure, Ph.D. dissertation, University of Pittsburgh, 1996.

[130] Coit, D. W. and A. E. Smith, Redundancy allocation to maximize a lower precentile of the system time-to-failure distribution, *IEEE Transactions on Reliability,* vol. 47, no. 1, pp. 79–87, 1998.

[131] Collins, R. J., Studies in artificial evolution, Ph.D. dissertation, University of California–Los Angeles, 1992.

[132] Comer, D. E., *Internetworking with TCP/IP,* vol. 1: *Principles, Protocols, and Architecture,* 3rd edition, Prentice Hall, Upper Saddle River, NJ, 1995.

[133] Corcoran, A. L., Techniques for reducing the disruption of superior building blocks in genetic algorithms, Ph.D. dissertation, University of Tulsa, 1994.

[134] Cormen, T. H., C. E. Leiserson, and R. L. Rivest, *Introduction to Algorithms,* MIT Press, Cambridge, MA, 1990.

[135] Cowgill, M. C., Monte Carlo validation of two genetic clustering algorithms, Ph.D. dissertation, Virginia Polytechnic Institute and State University, 1993.

[136] Cox, L. A., Dynamic anticipatory routing in circuit-switched telecommunications networks, in Davis [147], pp. 124–143.

[137] Crawford, K. D., Role of recombination in genetic algorithms for fixed-length subset problems, Ph.D. dissertation, University of Tulsa, 1996.

[138] Crossley, W. A., Using genetic algorithms as an automated methodology for conceptual design of rotorcraft, Ph.D. dissertation, Arizona State University, 1995.

[139] Daskin, M. S., An overview of recent research on assigning products to groups for group technology production problems, Technical report, Northwestern University, 1991.

[140] Davern, J. J., Architecture for job shop scheduling with genetic algorithms, Ph.D. dissertation, University of Central Florida, 1994.

[141] Davidor, Y., A genetic algorithm applied to robot trajectory generation, in Davis [147], pp. 923–932.

[142] Davidor, Y., H. Schwefel, and R. Männer, editors, *Parallel Problem Solving from Nature: PPSN III,* Springer-Verlag, Berlin, 1994.

[143] Davis, E. and G. Heidorn, An algorithm for optimal project scheduling under multiple resources constraints, *Management Science,* vol. 17, pp. B803–B816, 1971.

[144] Davis, L., Applying adaptive algorithms to epistatic domains, *in Proceedings of the International Joint Conference on Artificial Intelligence,* pp. 162–164, 1985.

[145] Davis, L., Job shop scheduling with genetic algorithm, in Grefenstette [266], pp. 136–140.

[146] Davis, L., Adapting operator probabilities in genetic algorithm, in Schaffer [564], pp. 61–69.

[147] Davis, L., editor, *Handbook of Genetic Algorithms,* Van Nostrand Reinhold, New York, 1991.

[148] Davis, P. and T. Ray, A branch-bound algorithm for the capacitated facilities location problem, *Naval Research Logistics Quarterly,* vol. 16, pp. 331–344, 1969.

[149] Dawkins, R., *The Selfish Gene,* Oxford University Press, Oxford, 1976.

[150] Day, R., On optimal extraction from a multiple file data storage system, *Operations Research,* vol. 13, no. 3, pp. 482–494, 1965.

[151] Deb, K., Genetic algorithms in multimodel function optimization, M.S. dissertation, University of Alabama, 1989.

[152] DeChaine, M.D., Stochastic fuel management optimization using genetic algorithms and heuristic rules, Ph.D. dissertation, Pennsylvania State University, 1995.

[153] Deeter, D. L. and A. E. Smith, Heuristic optimization of network design considering all-terminal reliability, in *Proceedings of the Annual Reliability and Maintainability Symposium,* pp. 194–199, 1997.

[154] Deeter, D. L. and A. E. Smith, Economic design of reliable network, *IIE Transactions,* vol. 30, pp. 1161–1174, 1998.

[155] Demmeyer, F. and S. Voss, Dynamic tabulist management using the reverse elimination method, *Annals of Operations Research,* vol. 41, pp. 31–46, 1993.

[156] Dengiz, B., F. Altiparmak, and A. E. Smith, A genetic algorithm approach to optimal topology design of all terminal networks, in *Intelligent Engineering Systems Through Artificial Neural Networks,* vol. 5, pp. 405–410, ASME Press, New York, 1995.

[157] Dengiz, B., F. Altiparmak, and A. E. Smith, Efficient optimization of all-terminal reliable networks using evolutionary approach, *IEEE Transactions on Reliability,* vol. 46, no. 1, pp. 18–26, 1997.

[158] Dengiz, B., F. Altiparmak, and A. E. Smith, Local search genetic algorithm for optimal design of reliable networks, *IEEE Transactions on Evolutionary Computation,* vol. 1, no. 3, pp. 179–188, 1997.

[159] Dengiz, B., F. Altiparmak, and A. E. Smith, Local search genetic algorithm for optimization of highly reliable communications networks, in Bäck [29], pp. 650–657.

[160] Deo, N. and N. Kumar, Computation of constrained spanning trees, in Pardalos, P. M., editor, *Network Optimization,* pp. 194–233, Springer-Verlag, Berlin, 1997.

[161] Dev, K., *Optimization for Engineering Design: Algorithms and Examples,* Prentice-Hall, New Delhi, 1995.

[162] Dhingra, A. K., Optimal apportionment of reliability and redundancy in a series system under multiple objectives, *IEEE Transactions on Reliability,* vol. 41, no. 4, pp. 576–582, 1992.

[163] Dighe, R., Human pattern nesting strategies in a genetic algorithms framework, Ph.D. dissertation, Massachusetts Institute of Technology, 1996.

[164] Dijkstra, E., A note on two problems in connection with graphs, *Numerische Mathematick,* vol. 1, pp. 269–271, 1959.

[165] Dorndorf, U. and E. Pesch, Evolution based learning in a job shop scheduling environment, *Computers and Operations Research,* vol. 22, no. 1, pp. 25–40, 1995.

[166] Dozier, G. V., Constraint processing using adaptive microevolutionary/systematic hill climbing, Ph.D. dissertation, North Carolina State University, 1995.

[167] Drexl, A. and J. Gruenewald, Nonpreemptive multi-mode resource constrained project scheduling, *IIE Transactions,* vol. 25, pp. 74–81, 1993.

[168] Dudzinski, K. and S. Walukiewicz, Exact methods for the knapsack problem and its generalizations, *European Journal of Operational Research,* vol. 28, pp. 3–21, 1987.

[169] Eberlein, M. V., GA heuristic generically has hyperbolic fixed points, Ph.D. dissertation, University of Tennessee, 1996.

[170] Edirisinghe, C. D., Optimization of radiotherapy planning using genetic algorithms, Ph.D. dissertation, State University of New York–Buffalo, 1996.

[171] Efroymson, M. and T. Ray, A branch–bound algorithm for plant location, *Operations Research,* vol. 14, pp. 361–368, 1966.

[172] Egan, M. A., Validity-guided robust fuzzy clustering methods for characterizing clusters in noisy data, Ph.D. dissertation, Rensselaer Polytechnic Institute, 1997.

[173] Eichler, W. R. M., On the development and interpretation of parameter manifolds for biophysically robust compartmental models of CA3 hippocampal neurons, Ph.D. dissertation, University of Minnesota, 1996.

[174] Elbaum, R. and M. Sidi, Topological design of local-area networks using genetic algorithms, *IEEE/ACM Transactions on Networking,* vol. 4, no. 5, pp. 766–778, 1996.

[175] Elias, D. and M. Ferguson, Topological design of multipoint teleprocessing networks, *IEEE Transactions on Communications,* vol. 22, pp. 1753–1762, 1974.

[176] Elketroussi, M., Relapse from tobacco smoking cessation: mathematical and computer microsimulation modeling including parameter optimization with genetic algorithms, Ph.D. dissertation, University of Minnesota, 1993.

[177] Eshelman, L. J., editor, *Proceedings of the 6th International Conference on Genetic Algorithms,* Morgan Kaufmannn Publishers, San Francisco, 1995.

[178] Eshelman, L., K. Mathias, and J. Schaffer, Convergence controlled variation, in Belew, R. and M. Vose, editors, *Foundations of Genetic Algorithms,* vol. 4, Morgan Kaufmann Publishers, San Francisco, 1997.

[179] Eshelman, L., K. Mathias, and J. Schaffer, Crossover operator biases: exploiting the population distributions, in Bäck [29], pp. 354–361.

[180] Eshelman, L. and J. Schaffer, Real-coded genetic algorithms and interval-schemata, in Whitley, L., editor, *Foundations of Genetic Algorithms,* vol. 2, pp. 187–202, Morgan Kaufmann Publishers, San Francisco, 1993.

[181] Falkenauer, E., A new representation and operators for genetic algorithms applied to grouping problems, *Evolutionary Computation,* vol. 2, no. 2, pp. 123–144, 1994.

[182] Falkenauer, E., Tapping the full power of genetic algorithms through suitable representation and local optimization: application to bin packing, in *Evolutionary Algorithms in Management Applications,* pp. 167–182, Springer-Verlag, Berlin, 1995.

[183] Falkenauer, E. and S. Bouffoix, A genetic algorithm for job shops, in *Proceedings of the IEEE International Conference on Robotics and Automation,* pp. 824–829, 1991.

[184] Falkenauer, E. and A. Delchambre, A genetic algorithm for bin packing and line balancing, in *Proceedings of the IEEE International Conference on Robotics and Automation,* pp. 1186–1192, 1992.

[185] Fazakerley, G. M., A research report on the human aspects of group technology and cellular manufacture, *Int. Journal of Production Research,* vol. 14, no. 1, pp. 123–134, 1976.

[186] Fernandes, L. M. and L. Gouveia, Minimal spanning trees with a constraint on the number of leaves, *European Journal of Operational Research,* vol. 104, pp. 250–261, 1998.

[187] Fetterlof, P. C. and G. Anandalingam, Optimal design of LAN–WAN internetworks: an approach using simulated annealing, *Annals of Operations Research,* vol. 36, pp. 275–298, 1992.

[188] Filho, A. A., Optimizing hydrocarbon field development using a genetic algorithm–based approach, Ph.D. dissertation, Stanford University, 1997.

[189] Fisher, M. L. and P. Kedia, Optimal solution of set covering/partitioning problems using dual heuristics, *Management Science,* vol. 36, pp. 674–688, 1990.

[190] Fishman, G. S., A comparison of four Monte Carlo methods for estimating the probability of s-t connectedness, *IEEE Transactions on Reliability,* vol. R-35, no. 2, pp. 145–155, 1986.

[191] Fishman, G. S., A Monte Carlo sampling plan for estimating network reliability, *Annals of Operations Research,* vol. 34, pp. 581–594, 1986.

[192] Fluerent, C., *Algorithmes génétiques hybrides pour l'optimisation combinatoire,* Ph.D. dissertation, University of Saskatchewan, 1994.

[193] Forgarty, T., Varying the probability of mutation in the genetic algorithm, in Schaffer [564], pp. 104–109.

[194] Fogarty, T., editor, *Evolutionary Computing,* Springer-Verlag, Berlin, 1994.

[195] Fogel, D., Editor, *Proceedings of the IEEE International Conference on Evolutionary Computation,* IEEE Press, Piscataway, NJ, 1996.

[196] Fogel, D., editor, *Proceedings of the IEEE International Conference on Evolutionary Computation,* IEEE Press, Piscataway, NJ, 1998.

[197] Fogel, D., editor, *Proceedings of the First IEEE Conference on Evolutionary Computation,* IEEE Press, Piscataway, NJ, 1994.

[198] Fogel, D., *Evolutionary Computation: Toward a New Philosophy of Machine Intelligence,* IEEE Press, Piscataway, NJ, 1995.

[199] Fogel, D. and J. Atmar, Comparing genetic operators with Gaussian mutations in simulated evolutionary process using linear systems, *Biological Cybernetics,* vol. 63, pp. 111–114, 1990.

[200] Fogel, L., A. Owens, and M. Walsh, *Artificial Intelligence Through Simulated Evolution,* Wiley, New York, 1966.

[201] Fonseca, C. and P. Fleming, Genetic algorithms for multiobjective optimization: formulation, discussion and generalization, in Forrest [204], pp. 416–423.

[202] Fonseca, C. and P. Fleming, An overview of evolutionary algorithms in multiobjective optimization, *Evolutionary Computation,* vol. 3, no. 1, pp. 1–16, 1995.

[203] Forrest, S., Documentation for prisoners dilemma and norms programs that use the genetic algorithm, Ph.D. dissertation, University of Michigan–Ann Arbor, 1985.

[204] Forrest, S., editor, *Proceedings of the 5th International Conference on Genetic Algorithms,* Morgan Kaufmann Publishers, San Francisco, 1993.

[205] Fourman, M., Compaction of symbolic layout using genetic algorithms, in Grefenstette [266], pp. 141–153.

[206] Freeman, D. W., Genetic algorithms: a new technique for solving the neutron spectrum unfolding problem, Ph.D. dissertation, University of Missouri–Rolla, 1998.

[207] Fujimoto, Y., Study on parallel processing architecture of neural networks and genetic algorithms, Ph.D. dissertation, Osaka University, 1993.

[208] Garey, M. and D. Johnson, *Computers and Intractability: A Guide to the Theory of NP-Completeness,* W.H. Freeman, New York, 1979.

[209] Garfinkel, R., Optimal political districting, Ph.D. dissertation, Johns Hopkins University, 1968.

[210] Gavish, B., Topological design of centralized computer networks: formulation and algorithms, *Networks,* vol. 12, pp. 355–377, 1982.

[211] Gavish, B., Formulation and algorithms for the capacitated minimal directed tree problem. *Journal of the Association for Computing Machinery,* vol. 30, no. 1, pp. 118–132, 1983.

[212] Gavish, B., Augmented Lagrangean based algorithms for centralized network design, *IEEE Transactions on Communications,* vol. 33, pp. 1247–1257, 1985.

[213] Gen, M. Reliability optimizaton by 0–1 programming for a system with several failure models, *IEEE Transactions on Reliability,* vol. 24, pp. 206–210, 1975; also in Rai, S. and D. P. Agrawal, editors, *Distributed Computing Network Reliability,* pp. 252–256, IEEE Computer Society Press, Los Alamitos, CA, 1990.

[214] Gen, M. and R. Cheng, Interval programming using genetic algorithms, in Gen, M. and Y. Tsujimura, editors, *Evolutionary Computations and Intelligent Systems,* Gordon and Breach, New York (forthcoming).

[215] Gen, M. and R. Cheng, Optimal design of system reliability under uncertainty using interval programming and genetic algorithm. Technical report, ISE94-6, Intelligent Systems Engineering Laboratory, Ashikaga Institute of Technology, 1994.

[216] Gen, M. and R. Cheng, Interval programming using genetic algorithms, in Jamshidi, M., M. Fathi, and F. Pierrot, editors, *Intelligent Automation and Control,* vol. 4, pp. 243–248, TSI Press, Albuquerque, NM, 1996.

[217] Gen, M. and R. Cheng, Optimal design of system reliability using interval programming and genetic algorithms, *Computers and Industrial Engineering,* vol. 31, no. 1–2, pp. 237–240, 1996.

[218] Gen, M. and R. Cheng, A survey of penalty techniques in genetic algorithms, in Fogel [195], pp. 804–809.

[219] Gen, M. and R. Cheng, *Genetic Algorithms and Engineering Design,* Wiley, New York, 1997.

[220] Gen, M., R. Cheng, M. Sasaki, and Y. Jin, Multiple-choice knapsack problem using genetic algorithms, In Parsaei, H., M. Gen, Y. Tsujimura, and J. Wong, editors, *Advances in Engineering Design and Automation Research II,* Integrated Technology Systems, Maui, HI, 1998.

[221] Gen, M., R. Cheng, and D. Wang, Genetic algorithms for solving shortest path problems, in Porto [512], pp. 401–406.

[222] Gen, M. and K. Ida, *Goal Programming Using Turbo C,* Harcourt Brace Japan, Tokyo, 1993 (in Japanese).

[223] Gen, M., K. Ida, and J. R. Kim, A spanning tree-based genetic algorithm for bicriteria topological network design, in Fogel [196], pp. 15–20.

[224] Gen, M., K. Ida, and J. R. Kim, System reliability optimization with fuzzy goals using genetic algorithm, *Journal of the Japan Society of Fuzzy Theory and Systems,* vol. 10, no. 2, pp. 356–365, 1998.

[225] Gen, M., K. Ida, J. R. Kim, and J. U. Lee, Bicriteria network design using spanning tree-based genetic algorithm, in *Proceedings of the 3rd International Symposium on Artificial Life and Robotics,* vol. 1, pp. 43–46, 1998.

[226] Gen, M., K. Ida, and Y. Z. Li, Bicriteria transportation problem by hybrid genetic algorithm, *Computers and Industrial Engineering,* vol. 35, no. 1–2, pp. 363–366, 1998.

[227] Gen, M. and J. R. Kim, Bicriteria reliability design using hybrid genetic algorithm, in *Proceedings of the 6th European Congress on Intelligent Techniques and Soft Computing,* vol. 1, pp. 407–412, 1998.

[228] Gen, M. and J. R. Kim, GA-based optimal network design, in Dagli, C. H., M. Akay, A. L. Buczak, and B. R. Fernandez, eds., *Intelligent Engineering Systems, Through Artificial Neural Networks,* vol. 8, pp. 247–252, ASME Press, New York, 1998.

[229] Gen, M. and J. R. Kim, GA-based optimisation of reliability design, in Bengley, P., *Evolutionary Design by Computers,* Chapter 8, pp. 191–218, Morgan Kaufmann, San Francisco, 1999.

[230] Gen, M., J. R. Kim, and Y. Li, Hybrid genetic algorithm for solving bicriteria reliability design problems, Technical report, Ashikaga Institute of Technology, 1998.

[231] Gen, M. and T. Kobayashi, editors, *Proceedings of the 16th International Conference on Computers and Industrial Engineering,* Ashikaga, Japan, 1994.

[232] Gen, M. and Y. Li, Hybrid genetic algorithms for transportation problem in logistics, *Journal of Logistics* (forthcoming).

[233] Gen, M. and Y. Li, Solving multi-objective transportation problem by spanning tree-based genetic algorithm, in Parmee, I., editor, *Adaptive Computing in Design and Manufacture,* pp. 95–108, Springer-Verlag, New York, 1998.

[234] Gen, M. and Y. Li, Spanning tree-based genetic algorithm for bicriteria fixed charge transportation problem, in *Proceedings of the Congress on Evolutionary Computation,* pp. 2265–2271 Washington, DC, 1999.

[235] Gen, M. and B. Liu, Evolution algorithm for optimal capacity expansion, *Journal of the Operations Research Society of Japan,* vol. 40, no. 1, pp. 1–9, 1997.

[236] Gen, M. and B. Liu, A genetic algorithm for nonlinear goal programming, *Evolutionary Optimization,* vol. 1, no. 1, 1999.

[237] Gen, M., B. Liu, and K. Ida, Evolution program for deterministic and stochastic optimizations, *European Journal of Operational Research,* vol. 94, no. 3, pp. 618–625, 1996.

[238] Gen, M., B. Liu, K. Ida, and D. Zheng, Evolutin program for constrained nonlinear optimization, in Gen and Kobayashi [231], pp. 576–579.

[239] Gen, M. and Y. Tsujimura, editors, *Genetic Algorithms and Intelligent Systems,* Gordon and Breach, New York, (forthcoming).

[240] Gen, M., Y. Tsujimura, and E. Kubota, Solving job-shop scheduling problem using genetic algorithms, in Gen and Kobayashi [231], pp. 576–579.

[241] Gen, M., G. Zhou, and J. R. Kim, Genetic algorithms for solving network design problems: state-of-the-art survey, *Evolutionary Optimization,* vol. 1, no. 3, 1999.

[242] Gen, M., G. Zhou, and M. Takayama, Matrix-based genetic algorithm approach on bicriteria minimum spanning tree problem with interval coefficients, *Journal of the Japan Society for Fuzzy Theory and Systems,* vol. 10, no. 6, pp. 1144–1153, 1998 (in Japanese).

[243] Geoffrion, A. M., Lagrangian relaxation for integer programming, *Mathematical Programming,* pp. 82–114, 1974.

[244] Giddens, T. D., Determination of invariant parameters and operators of the traditional genetic algorithm using an adaptive genetic algorithm generator, Ph.D. dissertation, Texas Tech University, 1994.

[245] Gillies, A., Machine learning procedures for generating image domain feature detectors, Ph.D. dissertation, University of Michigan–Ann Arbor, 1985.

[246] Glover, F., M. Lee, and J. Ryan, Least-cost network topology design for a new service: an application of a tabu search, *Annuals of Operations Research,* vol. 33, pp. 351–362, 1991.

[247] Gold, F. M., Application of the genetic algorithm to the kinematic design of turbine blade fixtures, Ph.D. dissertation, Worcester Polytechnic Institute, 1998.

[248] Goldberg, D. and J. Richardson, Genetic algorithms with sharing for multimodal function optimization, in Grefenstette [267], pp. 41–49.

[249] Goldberg, D., *Genetic Algorithms in Search, Optimization and Machine Learning,* Addison-Wesley, Reading, MA, 1989.

[250] Goldberg, D. and K. Deb, A comparative analysis of selection schemes used in genetic algorithms, in Rawlins [529], pp. 69–93.

[251] Goldberg, D., B. Korb, and K. Deb, Messy genetic algorithms: motivation, analysis, and first results, *Complex Systems,* vol. 3, pp. 493–530, 1989.

[252] Goldberg, D. and R. Lingle, Alleles, loci and the traveling salesman problem, in Grefenstette [266], pp. 154–159.

[253] Golden, J. B., Adaptive approximation and optimization of transform functions, Ph.D. dissertation, Vanderbilt University, 1995.

[254] Gong, D., G. Yamazaki, M. Gen, and W. Xu, Hybrid evolutionary method for capacitated location–allocation problem, *Engineering Design and Automation,* vol. 3, no. 2, pp. 179–190, 1997.

[255] Gong, D., Evolutionary computation and artificual neural networks for optimization problems and their applications, Ph.D. dissertation, Tokyo Metropolitan Institute of Technology, 1998.

[256] Gong, D., G. Yamazaki, M. Gen, and W. Xu, A genetic method for one-dimensional machine location problems, *International Journal of Production Economics,* vol. 60–61, pp. 337–342, 1999.

[257] González Hernández, L. F. and D. W. Corne, Evolutionary divide and conquer for the set-covering problem, in Fogarty, T., editor, *Evolutionary Computing,* pp. 198–208, Springer-Verlag, Berlin, 1996.

[258] Gordon, V. S., Exploitable parallelism in genetic algorithms, Ph.D. dissertation, Colorado State University, 1994.

[259] Gordon, V. and D. Whitley, Serial and parallel genetic algorithms as functions optimizers, in Forrest [204], pp. 177–183.

[260] Gottlieb, J. and L. Paulmann, Genetic algorithms for the fixed charge transportation problems, in *Proceedings of the IEEE International Conference Evolutionary Computation,* pp. 330–335, Anchorage, 1998.

[261] Gouveia, L., A $2n$ constraint formulation for the capacitated minimal spanning tree problem, *Operations Research,* vol. 43, no. 1, pp. 130–141, 1995.

[262] Graham, R. and P. Hell, On the history of the minimum spanning tree problem, *Annals of the History of Computing,* vol. 7, pp. 43–57, 1985.

[263] Grand, S. M., Application of the genetic algorithm to protein tertiary structure prediction, Ph.D. dissertation, Pennsylvania State University, 1993.

[264] Graybeal, M. E., Pest and resistance management simulation model of the Colorado potato beetle on potatoes, Ph.D. dissertation, Virginia Commonwealth University, 1998.

[265] Grefenstette, J. J., A user's guide to GENESIS, Technical report, Computer Science Department, Vanderbilt University, 1983.

[266] Grefenstette, J. J., editor, *Proceedings of the First International Conference on Genetic Algorithms,* Lawrence Erlbaum Associates, Hillsdale, N.J., 1985.

[267] Grefenstette, J. J., editor, *Proceedings of the 2nd International Conference on Genetic Algorithms,* Lawrence Erlbaum Associates, Hillsdale, N.J., 1987.

[268] Grefenstette, J. J., Optimization of control parameters for genetic algorithms, *IEEE Transactions on Systems, Man and Cybernetics,* vol. 16, pp. 122–128, 1986.

[269] Grefenstette, J. J., Lamarkian learning in multi-agent environment, in Belew and Booker [51], pp. 303–310.

[270] Grefenstette, J. J. and J. Baker, How genetic algorithms work: a critical look at implicit parallelism, in Schaffer [564], pp. 20–27.

[271] Guan, J., Applications of genetic algorithms in groundwater quality management, Ph.D. dissertation, Georgia Institute of Technology, 1998.

[272] Guo, Z., Nuclear power plant fault diagnostics and thermal performance studies using neural networks and genetyic algorithms, Ph.D. dissertation, University of Tennessee, 1992.

[273] Gupta, T. and H. Seifoddini, Production data based similarity coefficient for machine-component grouping decisions in the design of a cellular manufacturing system, *Int. Journal of Production Research,* vol. 28, no. 7, pp. 1247–1269, 1990.

[274] Gupta, Y., M. Gupta, A. Kumar, and C. Sundram, Minimizing total intercell and intracell moves in cellular manufacturing: a genetic algorithm approach, *Computer Integrated Manufacturing,* vol. 8, no. 2, pp. 92–101, 1995.

[275] Hachino, T., Study on identification of continuous time-delay systems, and lower dimensionizing of models by genetic algorithms, Ph.D. dissertation, Kyushu Institute of Technology, 1995 (in Japanese).

[276] Hajda, P., Genetic algorithms for control of wastewater conveyance systems, Ph.D. dissertation, Marquette University, 1997.

[277] Hall, L., Experience with a cutting plane algorithm for the capacitated spanning tree problem, *INFORMS Journal on Computing,* vol. 8, no. 3, pp. 219–234, 1996.

[278] Ham, I. V. and C. Han, Multiobjective cluster analysis for part family formations, *OMEGA: International Journal of Management Science,* vol. 5, no. 4, pp. 223–229, 1986.

[279] Han, S. S., Modeling and optimization of plasma-enhanced chemical vapor deposition using neural networks and genetic algorithms, Ph.D. dissertation, Georgia Institute of Technology, 1996.

[280] Hanagandi, V., Process control, optimization, and modeling of chemical systems using genetic algorithms and neutral networks, Ph.D. dissertation, Texas A&M University, 1995.

[281] Hancock, P., An empirical comparison of selection methods in evolutionary algorithms, in Fogarty, T., editor, *Evolutionary Computing,* pp. 80–95, Springer-Verlag, Berlin, 1994.

[282] Hansen, P.., Bicriterion path problems, in Fandel, G. and T. Gal, editors, *Multiple Criteria Decision Making: Theory and Applications,* pp. 109–127, Springer-Verlag, Heidelberg, 1980.

[283] Hanzhong, C. and L. Dongkui, A new algorithm for computing the reliability of complex networks by the cut method, *Microelectronics and Reliability,* vol. 34, no. 1, pp. 175–177, 1994.

[284] Harche, F. and G. L. Thompson, The column subtraction algorithm: an exact method for solving weighted set covering, packing and partitioning problems, *Computers and Operations Research,* vol. 21, pp. 689–705, 1994.

[285] Harhalakis, G., R. Nagi, and J. M. Proth, An efficient heuristic in manufacturing cell formation for group technology applications, *Int. Journal of Production Research,* vol. 28, no. 1, pp. 185–198, 1990.

[286] Harik, G., Learning gene linkage to efficiently solve problems of bounded difficulty using genetic algorithms, Ph.D. dissertation, University of Michigan, 1997.

[287] Hart, W. E., Adaptive global optimization with local search, Ph.D. dissertation, University of California–San Diego, 1994.

[288] Hase, K., Simulation on mutual adaptation of a body form and biped walking by nerve muscle skelation model and genetic algorithms, Ph.D. dissertation, Keio University, 1996.

[289] Hashimoto, Y., Study of neural network and genetic algorithms on an application to production planning problems, Ph.D. dissertation, Ritsumeikan University, 1995.

[290] Haupt, R., A survey of priority-rule based scheduling problems, *OR Spectrum,* vol. 11, pp. 3–16, 1989.

[291] Healy, W. C., Multiple choice programming, operations research, *Computers and Operations Research,* vol. 12, pp. 122–138, 1964.

[292] Hedrick, C., *RFC-1058: Routing Information Protocol,* Network Working Group, New York, 1988.

[293] Henig, M., The shortest path problem with two objective functions, *European Journal of Operational Research,* vol. 25, pp. 281–291, 1986.

[294] Herrera, F. and M. Lozano, Adaptation of genetic algorithm parameters based on fuzzy logic controllers, in Herrera, F. and J. Verdegay, editors, *Genetic Algorithms and Soft Computing,* pp. 95–125, Physica-Verlag, 1996.

[295] Herrera, F., E. Viedma, and M. Lozano, Fuzzy tools to improve genetic algorithms, in Zimmermann, H., editor, *Proceedings of the 2nd European Congress on Intelligent Techniques and Soft Computing,* pp. 1532–1539, Aachen, Germany, 1994.

[296] Hesser, J. and R. Männer, Towards an optimal mutation probability for genetic algorithms, in Schwefel and Männer [568], pp. 23–32.

[297] Hightower, R. R., Computational aspects of antibody gene families, Ph.D. dissertation, University of New Mexico, 1996.

[298] Hinterding, R., Mapping, order-independent genes and the knapsack problem, in Fogel [197], pp. 13–17.

[299] Hinterding, R., Z. Michalewicz, and A. Eiben, Adaptation in evolutionary computation: a survey, in Porto [512], pp. 65–69.

[300] Hirsch, W. M. and G. B. Dantzig, The fixed charge problem, *Naval Research Logistics Quarterly,* vol. 15, pp. 413–424, 1968.

[301] Ho, L. L., Simulation of condensed-phase physical, chemical, and biological systems on massively parallel computers, Ph.D. dissertation, Yale University, 1997.

[302] Hocaogle, C., Multipath planning using evolutionary computation with specification, Ph.D. dissertation, Rensselaer Polytechnic Institute, 1997.

[303] Holland, J., *Adaptation in Natural and Artificial Systems,* University of Michigan Press, Ann Arbor, MI, 1975; MIT Press, Cambridge, MA, 1992.

[304] Holsapple, C., V. Jacob, R. Pakath, and J. Zaveri, A genetics-based hybrid scheduler for generating static schedules in flexible manufacturing contexts, *IEEE Transactions on Systems, Man and Cybernetics,* vol. 23, pp. 953–971, 1993.

[305] Horn, J., Multicriterion decision making, in Bäck, T., D. Fogel, and Z. Michalewicz, editors, *Handbook of Evolutionary Computation,* pp. 517–522, Oxford University Press, New York, 1997.

[306] Horn, J., N. Nafpliotis, and D. Goldberg, A niched Pareto genetic algorithm for multiobjective optimization, in Fogel [197], pp. 82–87.

[307] Hoschei, G. C., Tabu search/genetic algorithm hybrid heuristic for solving a master production scheduling problem with sequence-dependent changeover times, Ph.D. dissertation, Colorado School of Mines, 1995.

[308] Hou, E. S., H. Ren and N. Ansari, *Dynamic, Genetic, and Chaotic Programming: Efficient Multiprocessor Scheduling Based on Genetic Algorithms,* 1992.

[309] Houck, C. R., Meta-heuristics for manufacturing problems, Ph.D. dissertation, North Carolina State University, 1996.

[310] Huang, R. J., Detection strategies for face recognition using learning and evolution, Ph.D. dissertation, George Mason University, 1998.

[311] Huang, X., Data-driven description of reservoir petrophysical properties, Ph.D. dissertation, University of Tulsa, 1995.

[312] Hughell, D. A., Simulated adaptive management for timber and wildlife under uncertainty, Ph.D. dissertation, North Carolina State University, 1997.

[313] Hwang, C. and K. Yoon, *Multiple Attribute Decision Making: Methods and Applications,* Springer-Verlage, Berlin, 1981.

[314] Igata, S., Study on estimation of short-time rainy areas, using optimized neural networks by genetic algorithms, Ph.D. dissertation, Muroran Institute of Technology, 1995 (in Japanese).

[315] Ignizio, J., *Linear Programming in Single and Multiple-Objective Systems,* Prentice Hall, Upper Saddle River, NJ, 1982.

[316] Ijiri, Y., *Management Goals and Accounting for Control,* North-Holland, Amsterdam, 1965.

[317] Ikonen, I. T., Genetic algorithm for a three-dimensional non-convex bin packing problem, Ph.D. dissertation, University of Louisville, 1998.

[318] Inoue, H., Study on an automatic generation method of fuzzy rules by genetic algorithms, Ph.D. dissertation, Ritsumeikan University, 1997.

[319] Ishibuchi, H. and T. Murata, A multiobjective genetic local search algorithm and its application to flowshop scheduling, *IEEE Transactions on Systems, Man and Cybernetics,* vol. 28, no. 3, pp. 392–403, 1998.

[320] Ishii, H., H. Shiode, and T. Nishida, Stochastic spanning tree problem, *Discrete Applied Mathematics,* vol. 3, pp. 263–273, 1981.

[321] Jacobs, L. W. and M. J. Brusco, A simulated annealing based heuristic for the set-covering problem, Technical report, Operations Management and Information Systems Department, North Illinois University, 1993.

[322] Jain, L. C. and K. Jain, editors, *Proceedings of the 1998 2nd International Conference on Knowledge-Based Intelligent Electronic Systems,* IEEE Press, Piscataway, NJ, 1998.

[323] Jallas, E., Improved model-based decision support by modeling cotton variability and using evolutionary algorithms, Ph.D. dissertation, Mississippi State University, 1998.

[324] Jan. R. H., Design of reliable networks, *Computers and Operations Research,* vol. 20, no. 1, pp. 25–34, 1993.

[325] Jan, R. H., F. J. Hwang, and S. T. Chen, Topological optimization of a communication network subject to a reliability constraint, *IEEE Transactions on Reliability,* vol. 42, no. 1, pp. 63–70, 1994.

[326] Jensen, A. P. and J. W. Barnes, *Network Flow Programming,* Wiley, New York, 1980.

[327] Jensen, E. D., Topological structural design using genetic algorithms, Ph.D. dissertation, Purdue University, 1992.

[328] Jenson, R. E., A dynamic programming algorithm for cluster analysis, *Operations Research,* vol. 12, pp. 1034–1057, 1969.

[329] Jeon, Y. C., Genetic algorithm–based non-linear modeling and its application to control and mechanical diagnostics, Ph.D. dissertation, Columbia University, 1994.

[330] Jin, K., Study on optimal neural network design and its application using genetic algorithms, Ph.D. dissertation, Hokkaido University, 1996.

[331] Johnson, D. S., A. Demers, J. D. Ullman, M. R. Garey, and R. L. Graham, Worst-case performance bounds for simple one-dimensional packing algorithms, *Journal of SIAM,* vol. 3, pp. 299–325, 1974.

[332] Joines, J. A., Hybrid genetic search for manufacturing cell design, Ph.D. dissertation, North Carolina State Univesity, 1996.

[333] Joines, J. A., Manufacturing cell design using genetic algorithms, M.S. dissertation, North Carolina State University, 1993.

[334] Joines, J. A., C. T. Culbreth, and R. E. King, Manufacturing cell design: an integer programming model employing genetic algorithms, Technical Report NCSU-IE 93-16, North Carolina State University, 1993.

[335] Joines, J. A., C. T. Culbreth, and R. E. King, A genetic algorithm based integer program for manufacturing cell design, *Proceedings of the International Conference on Flexible Automation and Integrated Manufacturing,* Stuttgart, Germany, 1995.

[336] Joines, J. A., C. T. Culbreth, and R. E. King, Manufacturing cell design: an integer programming model employing genetics, *IIE Transactions,* vol. 28, no. 1, pp. 69–85, 1996.

[337] Joines, J. A., R. E. King, and C. T. Culbreth, Utilizing a hybrid genetic search for manufacturing cell design, *IIE Transactiuons* (in review).

[338] Joines, J. A., R. E. King, and C. T. Culbreth, *Cell formation using genetic algorithms,* in Suresh, N. C. and J. M. Kay, editors, *Group Technology and Cellular Manufacturing,* pp. 185–204, Kluwer Academic Publishers, Norwell, MA, 1998.

[339] Joines, J. and C. Houck, On the use of non-stationary penalty functions to solve nonlinear constrained optimization problems with GAs, in Fogel [197], pp. 579–584.

[340] Jones, M. H., The Hereford Ranch algorithm: an improvement in genetic algorithms using selective breeding, Ph.D. dissertation, Utah State University, 1995.

[341] Julstrom, B., What have you done for me lately? Adapting operator probabilitics in a steady-state genetic algorithm, in Eshelman [177], pp. 81–87.

[342] Kaiser, T., Dynamic load distributions for adaptive computations on MIMD machines usign hybrid genetic algorithms, Ph.D. dissertation, University of New Mexico, 1997.

[343] Kamat, S. J. and M. W. Riley, Determination of reliability using event-based Monte Carlo simulation, *IEEE Transactions on Reliability,* vol. 24, no. 1, pp. 73–75, 1975.

[344] Kang, M. H., Learning evasive maneuvers using evolutionary algorithms and neural networks, Ph.D. dissertation, Mississippi State University, 1997.

[345] Kang, S. and U. Wemmerlov, A work load-oriented heuristic methodology for manufacturing cell formation allowing reallocation of operations, *European Journal of Operational Research,* vol. 69, no. 3, pp. 292–311, 1993.

[346] Kargupta, H., Search, polynomial complexity, and the fast messy genetic algorithm, Ph.D. dissertation, University of Illinois at Urbana-Champaign, 1996.

[347] Kasahara, H. and S. Narita, Parallel processing of robot-arm control computation on a multimicroprocessor system, *IEEE Journal of Robotics and Automation,* vol. 1, no. 2, pp. 104–113, 1985.

[348] Kasilingam, J. A. and R. S. Lashkari, Cell formation in the presence of alternative process plans in FMS, *Production Planning and Control,* vol. 2, no. 2, pp. 135–141, 1991.

[349] Kaufmann, A. and M. Gupta, *Fuzzy Mathematical Models in Engineering and Management Science,* 2nd edition, North-Holland, Amsterdam, 1991.

[350] Kazerooni, M., Cell formation using genetic algorithms, in *Proceedings of the International Conference on Flexible Automation and Integrated Manufacturing,* Stuttgart, Germany, 1995.

[351] Kazerooni, M., L. Luong, and K. Abhary, Cell formation using genetic algorithms, *Proceedings of the International Conference on Flexible Automation and Integrated Manufacturing,* vol. 3, no. 3–4, pp. 283–301, 1996.

[352] Keeney, R. L. and H. Raiffa, *Decisions with Multiple Objectives: Preferences and Value Tradeoffs,* Wile, New York, 1976.

[353] Kennedy, S., Five ways to a smarter genetic algorithm, *AI Expert,* pp. 35–38, 1993.

[354] Kennington, J. L. and V. E. Unger, A new branch and bound algorithm for the fixed charge transportation problem, *Management Science,* vol. 22, no. 10, pp. 1116–1126, 1976.

[355] Kermani, B. G., On using artificial neural networks and genetic algorithms to optimize performance of an electric noise, Ph.D. dissertation, North Carolina State University, 1996.

[356] Kershenbaum, A., Computing capacitated minimal spanning trees efficiently, *Networks,* vol. 4, pp. 299–310, 1974.

[357] Kershenbaum, A., *Telecommunications Network Design Algorithms,* McGraw-Hill, New York, 1993.

[358] Kershenbaum, A., R. R. Boorstyn, and R. Oppenheim, Centralized teleprocessing network design, *Networks,* vol. 13, pp. 279–293, 1983.

[359] Kershenbaum, A. and R. V. Slyke, Recursive analysis of network reliability, *Networks,* vol. 3, pp. 81–94, 1973.

[360] Khuri, S., T. Bäck, and J. Heitkötter, The zero/one multiple knapsack problem and genetic algorithms, in *Proceedings of the ACM Symposium on Applied Computation,* 1994.

[361] Kim, J. R., M. Gen, and K. Ida, Bicriteria network design using spanning tree-based genetic algorithm, *Artificial Life and Robotics,* 1999 (forthcoming).

[362] Kim, J. R., M. Gen, and K. Ida, A spanning tree-based genetic algorithm for reliable multiplexed network topology design, in *Proceedings of the 4th International Symposium on Artificial Life and Robotics,* vol. 2, pp. 460–463, 1999.

[363] Kim, J. R. and M. Gen, A GA for bicriteria communication network topology design, *Engineering Valuation and Cost Analysis,* (forthcoming).

[364] Kimura, T., A two-moment approximation for the mean waiting time in the $GI/G/s$ queue, *Management Science,* vol. 32, no. 6, pp. 751–763, 1986.

[365] Kimura, T., Approximations for multi-server queues: system interpolations, *Queuing Systems,* vol. 17, pp. 347–382, 1994.

[366] King, J. R., Machine-component grouping formation in group technology, *OMEGA: International Journal of Management Science,* vol. 8, no. 2, pp. 193–199, 1980.

[367] King, J. R., Machine-component grouping in production flow analysis: an approach using a rank order clustering algorithm, *Int. Journal of Production Research,* vol. 18, no. 2, pp. 213–232, 1980.

[368] King, J. R. and V. Nakornchai, Machine-component group formation in group technology: review and extension, *Int. Journal of Production Research,* vol. 20, no. 2, pp. 117–133, 1982.

[369] Klingman, D., A. Naoier, and J. Stutz, A program for generating large scale capacitated assignment, transportation and minimum cost flow networks, *Management Science,* vol. 20, no. 5, pp. 814–822, 1974.

[370] Kobayashi, S., Foundations of genetic algorithms and its applications, *Communications of ORSJ,* vol. 45, pp. 256–261, 1993 (in Japanese).

[371] Kobayashi, S., I. Ono, and M. Yamamura, An efficient genetic algorithm for job shop scheduling problems, in Eshelman [177], pp. 506–511.

[372] Koh, S. J. and C. Y. Lee, A tabu search for the survivable fiber optic communication network design, *Computers and Industrial Engineering,* vol. 28, no. 4, pp. 689–700, 1995.

[373] Kommu, V., Enhanced genetic algorithms in constrained search spaces with emphases on parallel environments, Ph.D. dissertation, University of Iowa, 1993.

[374] Kou, S., Decision support for irrigated project planning using a genetic algorithm, Ph.D. dissertation, University of Oklahoma, 1997.

[375] Koza, J. R., *Genetic Programming,* MIT Press, Cambridge, MA, 1992.

[376] Kruska, J. Jr., On the shortest spanning subtree of a graph and the traveling salesman problem, *Proceedings of the ACM,* vol. 7, no. 1, pp. 48–50, 1956.

[377] Kubota, N. and T. Fukuda, Schema representation in virus-evolutionary genetic algorithm for knapsack problem, in Fogel [196], pp. 834–838.

[378] Kumamoto, H., K. Tanaka, and K. Inoue, Efficient evaluation of system reliability by Monte Carlo method, *IEEE Transactions on Reliability,* vol. 26, no. 5, pp. 311–315, 1977.

[379] Kumar, A. R. and Y. P. Gupta, Genetic-algorithm-based reliability optimization for computer network expansion, *IEEE Transactions on Reliability,* vol. 44, no. 1, pp. 63–72, 1995.

[380] Kumar, A. R., M. Pathak, Y. P. Gupta, and H. R. Parsaei, a genetic algorithm for distributed system topology design, *Computers and Industrial Engineering,* vol. 28, no. 3, pp. 659–670, 1995.

[381] Kumar, K. R. and M. P. Chandrasekharan, Grouping efficacy: a quantitative criterion for goodness of block diagonal forms of binary matrices in group technology, *Int. Journal of Production Reseach,* vol. 28, no. 2, pp. 233–243, 1990.

[382] Kuo, W., V. R. Prasad, F. A. Tillman, and C. L. Hwang, *Fundamentals and Applications of Reliability Optimization* (forthcoming).

[383] Kursawa, F., A variant of evolution strategies for vector optimization, in Schwefel and Männer [568], pp. 193–197.

[384] Kusiak, A. and M. Cho, Similarity coefficient algorithms for solving the group technology problem, *Int. Journal of Production Research,* vol. 30, no. 11, pp. 2633–2646, 1992.

[385] Kusiak, A. and W. Chow, Efficient solving of the group technology problem, *OMEGA: International Journal of Management Science,* vol. 6, no. 2, pp. 117–124, 1987.

[386] Kusiak, A. and W. Chow, Decomposition of manufacturing systems, *IEEE Journal of Robotics and Automation,* vol. 4, no. 5, pp. 457–471, 1988.

[387] Kusiak, A. and G. Finke, Selection of process plans in automated manufacturing systems, *IEEE Journal of Robotics and Automation,* vol. 4, no. 4, pp. 397–402, 1988.

[388] Kusiak, A. and S. Heragu, The facility layout problem, *European Journal of Operational Research,* vol. 29, pp. 229–251, 1987.

[389] Labordere, H., Partitioning algorithm in mixed and pseudo mixed integer programming, Ph.D. dissertation, Rensselaer Polytechnic Institute, 1969.

[390] Lai, Y. and C. Hwang, *Fuzzy Mathematical Programming,* Springer-Verlag, Berlin, 1992.

[391] Larson, C. B., Vibration testing by design: excitation and sensor placement using genetic algorithms, Ph.D. dissertation, University of Florida, 1996.

[392] Lazio, T. J. W., Genetic algorithms, pulsar planets, and ionized interstellar microturbulence, Ph.D. dissertation, Cornell University, 1997.

[393] Lee, A. M. and P. A. Longton, Queuing process associated with airline passenger check-in, *Operations Research Quarterly,* vol. 10, pp. 56–71, 1957.

[394] Lee, B., Three new algorithms for exact D-optimal design problems, Ph.D. dissertation, Ohio State University, 1993.

[395] Lee, C., Fuzzy logic in control systems: fuzzy logic controller, part I, *IEEE Transactions on Systems, Man and Cybernetics,* vol. 20, pp. 404–418, 1990.

[396] Lee, C., Fuzzy logic in control systems: fuzzy logic controller, part II, *IEEE Transactions on Systems, Man and Cybernetics,* vol. 20, pp. 419–435, 1990.

[397] Lee, H. and A. Garcia-Diaz, A network flow approach to solve clustering problems in group technology, *Int. Journal of Production Research,* vol. 31, no. 3, pp. 603–612, 1993.

[398] Lee, H. and A. Garcia-Diaz, Network flow procedures for the analysis of cellular manufacturing systems, *IIE Transactions,* vol. 28, pp. 333–345, 1996.

[399] Lee, J., Tolerance optimization using genetic algorithm and approximated simulation, Ph.D. dissertation, University of Michigan, 1992.

[400] Lee, J., Genetic algorithms in multidisciplinary design of low-vibration rotors, Ph.D. dissertation, Rensselaer Polytechnic Institute, 1996.

[401] Lee, M. A., Automatic design and adaptation of fuzzy systems and genetic algorithms using soft computing techniques, Ph.D. dissertation, University of California–Davis, 1994.

[402] Lee, M. and H. Takagi, Dynamic control of genetic algorithm using fuzzy logic techniques, in Forrest [204], pp. 76–83.

[403] Lee, S., *Goal Programming for Decision Analysis,* Auerbach Publishers, Philadelphia, 1972.

[404] Leskowsky, L., L. Logan, and Vannelli, Modern production management systems, in Kusiak, A., editor, *Group Technology Decision Aides in a Expert System for Plant Layout,* pp. 561–585, North-Holland, Amsterdam, 1987.

[405] Levin, A., Fleet routing and scheduling problems for air transportation systems, Ph.D. dissertation, Massachusetts Institute of Technology, 1969.

[406] Levine, D. M., Parallel genetic algorithm for the set partitioning problem, Ph.D. dissertation, Illinois Institute of Technology, 1994.

[407] Li, D., Multiobjective optimum design of static and seismic-resistant structures with genetic algorithms, fuzzy logic, and game theory, Ph.D. dissertation, University of Missouri–Rolla, 1998.

[408] Li, J., On an algorithm for solving fuzzy linear systems, *Fuzzy Sets and Systems,* vol. 61, pp. 369–371, 1994.

[409] Li, J., Oil tanker market modeling, analysis, and forecasting using neural networks, fuzzy logic, and genetic algorithms, Ph.D. dissertation, University of Michigan, 1997.

[410] Li, Y. X., Multi-criteria optimization techniques and its applications to engineering design problems, M.S. dissertation, Ashikga Institute of Technology, 1996.

[411] Li, Y. Z., Study on hybridized genetic algorithms for production distribution planning problems, Ph.D. dissertation, Ashikaga Institute of Technology, 1999.

[412] Li, Y. Z., and M. Gen, Spanning tree-based genetic algorithm for bicriteria transportation problem with fuzzy coefficients, *Australian Journal of Intelligent Information Processing Systems,* vol. 4, no. 3, pp. 220–229, 1998.

[413] Li, Y. Z. and M. Gen, Spanning tree-based genetic algorithm for solving bicriteria transportation problem, *Journal of the Japan Society for Fuzzy Theory and Systems,* vol. 10, no. 5, pp. 888–898, 1998.

[414] Li, Y. Z., M. Gen, and K. Ida, Improved genetic algorithm for solving multiobjective solid transportation problem with fuzzy numbers, *Japanese Journal of Fuzzy Theory and Systems,* vol. 9, no. 2, pp. 239–250, 1997.

[415] Li, Y. H., Genetic algorithm–based design of multiplierless two-dimensional state-space digital filters, Ph.D. dissertation, Tohoku University, 1996.

[416] Lin, S. C., Genetic algorithm–based scheduling system for dynamic job shop scheduling problem, Ph.D. dissertation, Michigan State University, 1997.

[417] Lin, W., High-quality tour hybrid genetic schemes for TSP optimization problems, Ph.D. dissertation, State University of New York–Binghamton, 1995.

[418] Liu, B., *Uncertain Programming,* Wiley, New York, 1999.

[419] Liu, B. and A. O. Esogbue, *Decision Criteria and Optimal Inventory Processes,* Kluwer Academic Publishers, Boston, 1999.

[420] Liu, C., M. Dai, X. Y. Wu, and W. K. Chen, A network overall reliability algorithm, in *Proceedings of Neural, Parallel and Scientific Computations,* pp. 287–292, Atlanta, GA, 1995.

[421] Liu, C. and J. Wu, Machine cell formation: using the simulated annealing algorithm, *Computer Integrated Manufacturing,* vol. 6, no. 6, pp. 335–349, 1993.

[422] Liu, D. G., Application of multiobjective genetic algorithms to control systems design, Ph.D. dissertation, Tohoku University, 1996.

[423] Loh, J. D., Automated discovery of self-replicating structures in cellular space automata models, Ph.D. dissertation, University of Maryland, 1996.

[424] Liou, T. S. and M. J. Wang, Ranking fuzzy numbers with integral value, *Fuzzy Sets and Systems,* vol. 50, pp. 247–255, 1992.

[425] Louveaux, F. V., Stochastic location analysis, *Location Science,* vol. 1, no. 2, pp. 127–154, 1993.

[426] Lu, B. L., Study of genetic algorithms on a application to industrial design, Ph.D. dissertation, Muroran Institute of Technology, 1996.

[427] Ludvig, J., J. Hesser, and R. Manner, Tackling the representation problem by stochastic averaging, in Bäck [29], pp. 196–203.

[428] Luenberger, D., *Linear and Nonlinear Programming,* 2nd edition, Addison-Wesley, Reading, MA, 1984.

[429] Luhandjula, M. K., Fuzzy optimization: an appraisal, *Fuzzy Sets and Systems,* vol. 30, pp. 257–282, 1989.

[430] Luhandjula, M., Linear programming under randomness and fuzziness, *Fuzzy Sets and Systems,* vol. 10, pp. 45–55, 1983.

[431] Luhandjula, M., On possibilistic linear programming, *Fuzzy Sets and Systems,* vol. 18, pp. 15–30, 1986.

[432] Mahfoud, S., Niching methods for genetic algorithms, Ph.D. dissertation, University of Illinois at Urbana-Champaign, 1995.

[433] Maifeld, T. T., Genetic-based unit commitment algorithm, Ph.D. dissertation, Iowa State University, 1997.

[434] Malik, K. and G. Yu, A branch and bound algorithm for the capacitated minimum spanning tree problem, *Networks,* vol. 23, pp. 525–532, 1993.

[435] Mandaltsis, D. and J. M. Kontolen, Overall reliability determination of computer networks with hierarchical routing strategies, *Microelectronics and Reliability,* vol. 27, no. 1, pp. 129–143, 1987.

[436] Mantawy, A. H., Unit commitment by artificial intelligence techniques, Ph.D. dissertation, King Fahd University of Petroleum and Minerals (Saudia Arabia), 1997.

[437] Marchelya, J. C., Enforcing pollution control laws: Stackleberg–Markov game models for improving efficiency and effectiveness, Ph.D. dissertation, Johns Hopkins University, 1997.

[438] Maria, A., Genetic algorithm for multimodel continuous optimization problems, Ph.D. dissertation, University of Oklahoma, 1995.

[439] Martello, S. and P. Toth, The 0–1 knapsack problem, in Christofides, N., A. Mingozzi, and C. Sandi, editors, *Combinatorial Optimization,* pp. 237–279, Wiley, Chichester, West Sussex, England, 1979.

[440] Martello, S. and P. Toth, Algorithms for knapsack problems, *Annals of Discrete Mathematics,* vol. 31, pp. 213–257, 1987.

[441] Martello, S. and P. Toth, *Knapsack Problems: Algorithms and Computer Implementations,* Wiley, Chichester, West Sussex, England, 1990.

[442] Martins, E., On a multicriteria shortest path problem, *European Journal of Operational Research,* vol. 16, pp. 236–245, 1984.

[443] Masters, B. P., Evolutionary design of controlled structures, Ph.D. dissertation, Massachusetts Institute of Technology, 1997.

[444] Matsui, K., Expression of genetic algorithms to macroadaptation degree and its application to picture processing systems, Ph.D. dissertation, Tokyo Institute of Technology, 1996.

[445] Matthew, R. M., Applications and limitations of genetic algorithms for the optimization of multivariate calibration techniques, Ph.D. dissertation, Clemson University, 1998.

[446] McAuley, J., Machine grouping for efficient production, *Production Engineering,* vol. 51, no. 2, pp. 53–57, 1972.

[447] McCormick, W. T., P. J. Schweitzer, and T. W. White, Problem decomposition and data reorganization by a cluster technique, *Operations Research,* vol. 20, no. 5, pp. 993–1009, 1972.

[448] Meddin, M., Genetic algorithms: a Markov chain and detail balance approach, Ph.D. dissertation, Georgia Institute of Technology, 1995.

[449] Memon, G. Q., Optimization methods for real-time traffic control, Ph.D. dissertation, University of Pittsburgh, 1995.

[450] Merkle, L. D., Analysis of linkage-friendly genetic algorithms, Ph.D. dissertations, Air Force Institute of Technology, 1996.

[451] Michael, F., Form-finding analysis of vibrating towered shells of revolution as an inverse eigenproblem and as a genetic algorithm optimization problem, Ph.D. dissertation, University of Tokyo, 1996.

[452] Michalewicz, Z., Genetic algorithms, numerical optimization, and constraints, Eshelman [170], pp. 151–158.

[453] Michalewicz, Z., A survey of constraint handling techniques in evolutionary computation methods, In McDonnell, J., R. Reynolds, and D. Fogel, editors, *Proceedings of the 4th Annual Conference on Evolutionary Programming,* pp. 135–155, MIT Press, Cambridge, MA, 1995.

[454] Michalewicz, Z., Evolutionary computation: practical issues, in Fogel [195], pp. 30–39.

[455] Michalewicz, Z., *Genetic Algorithm + Data Structure = Evolution Programs,* 3rd edition, Springer-Verlag, New York, 1996.

[456] Michalewicz, Z., T. Logan, and S. Swaminathan, Evolutionary operators for continuous convex parameter spaces, in Sebald, A. and L. Fogel, editors, *Proceedings of the 3rd Annual Conference on Evolutionary Programming,* pp. 84–97, World Scientific Publishing, River Edge, NJ, 1994.

[457] Michalewicz, Z., G. A. Vignaux, and M. Hobbs, A non-standard genetic algorithm for the nonlinear transportation problem, *ORSA Journal of Computing,* vol. 3, no. 4, pp. 307–316, 1991.

[458] Mock, K. J., Intelligent information filtering via hybrid techniques: hill climbing, case-based reasoning, index patterns, and genetic algorithms, Ph.D. dissertation, University of California–Davis, 1996.

[459] Modi, A. K. and I. A. Karimi, Design of multiproduct batch processes with finite intermediate storage, *Computers and Chemical Engineering,* vol. 13, pp. 127–139, 1989.

[460] Monem, M. J., Performance evaluation and optimization of irrigation canal system using genetic algorithm, Ph.D., dissertation, University of Calgary, 1996.

[461] Monma, C. L. and C. N. Potts, On the complexity of scheduling with batch setup times, *Operations Research,* vol 37, pp. 798–804, 1989.

[462] Moon, B. R. and C. K. Kim, A two-dimensional embedding of graphs for genetic algorithms, in Bäck [29], pp. 204–211.

[463] Moon, C. and C. K. Kim, and M. Gen, Genetic algorithm for maximizing the parts flow within cells in manufacturing cell design, *Computers and Industrial Engineering,* vol. 2, pp. 1730–1733, 1999.

[464] Moon, C. and M. Gen, A genetic algorithm–based approach for design of independent manufacturing cells, *International Journal of Production Economics,* vol. 60–61, pp. 421–426, 1999.

[465] Moon, C., M. Gen, and A. Suer, A genetic algorithm–based model for minimizing additional capital investment in manufacturing cell design, *Engineering Valuation and Cost Analysis,* vol. 2, pp. 133–142, 1999.

[466] Moon, Y., High-performance simulation-based optimization environment for large-scale systems, Ph.D. dissertation, University of Arizona, 1996.

[467] Morikawa, K., Study on scheduling using genetic algorithms, Ph.D. dissertation, Nagoya University, 1996.

[468] Moscato, P. and M. Norman, A memetic approach for the travelling salesman problem: implementation of a computational ecology for combinatorial optimization on message-passing systems, in *Proceedings of the International Conference on Parallel Computing and Transputer Applications,* Amsterdam, 1992.

[469] Moy, J., *RFC-1131: The OSPF Specification,* 2nd edition, Network Working Group, New York, 1989.

[470] Mühlenbein, H., M. Schlouter, and O. Kraemer, Evolution algorithms in combinatorial optimization, *Parallel Computing,* vol. 7, pp. 65–85, 1988.

[471] Mulvey, J. and H. Crowder, Cluster analysis: an application of Lagrangian relaxation, *Management Science,* vol. 25, pp. 329–340, 1979.

[472] Munakata, T. and D. J. Hashier, A genetic algorithm applied to the maximum flow problem, in Forrest [204], pp. 488–493.

[473] Munemoto, M., Study on disperse control–type dynamic load balance using genetic algorithms, Ph.D. dissertation, Hokkaido University, 1995.

[474] Munetomo, M., Y. Takai, and Y. Sato, An adaptive network routing algorithm employing path genetic operators, in Bäck [29], pp. 643–649.

[475] Murata, T., Genetic algorithms for multiobjective optimization, Ph.D. dissertation, University of Osaka Prefecture, 1996.

[476] Murata, T., H. Ishibuchi, and H. Tanaka, Multiobjective genetic algorithm and its application to flowshop scheduling, *Computers and Industrial Engineering,* vol. 30, no. 4, pp. 957–968, 1996.

[477] Murthy, J. and D. Olson, An interactive procedure using domination cones for bicriterion shortest path problems, *European Journal of Operational Research,* vol. 72, pp. 417–431, 1994.

[478] Murty, K., *Linear and Combinatorial Programming,* Wiley, New York, 1976.

[479] Nagi, R., G. Harhalakis, and J. M. Proth, Multiple routings and capacity considerations in group technology applications, *Int. Journal of Production Research,* vol. 28, no. 12, pp. 2243–2257, 1990.

[480] Nakamura, M., Study on protocol design and combinatorial optimization in resource assignment problems on the basis of net theory and genetic algorithms, Ph.D. dissertation, Osaka University, 1995.

[481] Nakanishi, Y., Optimization of a structure phase by homology theory and genetic algorithms, Ph.D. dissertation, University of Tokyo, 1994.

[482] Narula, S. and C. Ho, Degree-constrained minimum spanning tree, *Computers and Operations Research,* vol. 7, pp. 239–249, 1980.

[483] Ng, S., Worst-case analysis of an algorithm for cellular manufacturing, *European Journal of Operational Research,* vol. 69, no. 3, pp. 384–398, 1993.

[484] Niesse, J. A., Structural optimization of atomic and molecular microclusters using a genetic algorithm in real-valued space-fixed coordinates, Ph.D. dissertation, University of New Hampshire, 1998.

[485] Nims, J. W., Contingency ranking for voltage stability using a genetic algorithm, Ph.D. dissertation, University of Alabama, 1995.

[486] Noga, A. J., Performance analysis and simulation of a class of numerical demodulation techniques for large-deviation FM signals, Ph.D. dissertation, Syracuse University, 1995.

[487] Nolte, B. J., Global optimization algorithms for seismic inversion, Ph.D. dissertation, University of Hawaii, 1995.

[488] Oei, C. K., D. E. Goldberg, and S. J. Chang, Tournament selection, niching, and the preservatin of diversity, Technical report, Illinois Genetic Algorithms Laboratory, University of Illinois at Urbana-Champaign, 1991.

[489] Oh, K., On-line optimization of substrates feeding in a hybrid cell culture using an artificial neural network and a genetic algorithm, Ph.D. dissertation, Osaka University, 1996.

[490] Oliver, I., D. Smith, and J. Holland, A study of permutation crossover operators on the traveling salesman problem, in Grefenstette [267], pp. 224–230.

[491] Olsen, A., Penalty functions and the knapsack problem, in Fogel [197], pp. 554–558.

[492] Ono, I. and S. Kobayashi, A real-coded genetic algorithm for function optimization using unimodal normal distribution crossover, in Bäck [29], pp. 246–253.

[493] Ono, I., M. Yamamura, and S. Kobayashi, A genetic algorithm for job-based order crossover, in Fogel [195], pp. 547–552.

[494] Orvosh, D. and L. Davis, Using a genetic algorithm to optimize problems with feasibility constraints, in Fogel [197], pp. 548–552.

[495] Osyczka, A. and S. Kundu, A new method to solve generalized multicriterion optimization problems using genetic algorithm, *Structural Optimization,* vol. 10, no. 2, pp. 94–99, 1995.

[496] Osyczka, A. and S. Kundu, A modified distance method for multicriterion optimization using genetic algorithm, *Computers and Industrial Engineering,* vol. 30, no. 4, pp. 871–882, 1996.

[497] Palazolo, P. J., Use of genetic algorithms in bounded search for design of biological nitrification/denitrification waste treatment systems, Ph.D. dissertation, George Mason University, 1997.

[498] Palekar, U. S., M. H. Karwan, and S. Zionts, Approaches for solving fixed charge transportation problem, Ph.D. dissertation, State University of New York–Buffalo, 1986.

[499] Palmer, C. C., Approach to a problem in network design using genetic algorithms, Ph.D. dissertation, Polytechnic University, 1994.

[500] Palmer, C. C. and A. Kershenbaum, An approach to a problem in network design usign genetic algorithms, *Networks,* vol. 26, pp. 151–163, 1995.

[501] Panwalkar, S. and W. Iskander, A survey of scheduling rules, *Operations Research,* vol. 25, pp. 45–61, 1977.

[502] Papadimitriou, C. H., The complexity of the capacitated tree problem, *Networks,* vol. 8, pp. 217–230, 1978.

[503] Pareto, V., *Manuale di Econòmica Polìttica,* Società Editrice Libràia, Milan, Italy, 1906; translated into English by A. S. Schwier, as *Manual of Political Economy,* Macmillan, New York, 1971.

[504] Parrill, A. L., Applications of artificial intelligence in drug design, Ph.D. dissertation, University of Arizona, 1996.

[505] Patel, H. C., Genetic algorithms: a tool for quantitative structure-activity relationship analyses, Ph.D. dissertation, University of Illinois–Chicago, 1996.

[506] Patterson, J. and G. Roth, Scheduling a project under multiple resource constraints: a 0–1 programming approach, *AIIE Transactions,* vol. 8, pp. 449–455, 1976.

[507] Peterson, L. L. and B. S. Davie, *Computer Networks: A Systems Approach,* Morgan Kaufmann Publishers, San Francisco, 1996.

[508] Pheatt, C. B., Construction of D-optimal experimental designs using genetic algorithms, Ph.D. dissertation, Illinois Institute of Technology, 1995.

[509] Picardo, L., Framework for the analysis of evolutionary algorithms, Ph.D. dissertation, Case Western Reserve University, 1996.

[510] Pierre, S., M. Hyppolite, J. Bourjolly, and O. Dioume, Topological design of computer communication networks using simulated annealing, *Engineering Applications of Artificial Intelligence,* vol. 8, no. 1, pp. 61–69, 1995.

[511] Pinon, E., Investigation of the applicability of genetic algorithms to spacecraft trajectory optimization, Ph.D. dissertation, University of Texas–Austin, 1995.

[512] Porto, B., editor, *Proceedings of the IEEE International Conference on Evolutionary Computation,* IEEE Press, Piscataway, NJ, 1997.

[513] Potter, M. A., Design and analysis of a computational model of cooperative coevolution, Ph.D. dissertation, George Mason University, 1997.

[514] Potts, C. N. and L. V. Wassenhove, Integrating scheduling with batch and lot-sizing: a review of algorithms and complexity, *Journal of the Operations Research Society,* vol. 43, pp. 395–406, 1992.

[515] Prim, R., Shortest connection networks and some generalizations, *Journal of Bell Systems Technology,* vol. 36, pp. 1389–1401, 1957.

[516] Pritsker, A., L. Waters, and P. Wolfe, Multiproject scheduling with limited resources: a 0–1 approach, *Management Science,* vol. 16, pp. 93–108, 1069.

[517] Prüfer, H., Neuer Beweis eines Satzes über Permutationen, *Archiv fuer Mathematische und Physik,* vol. 27, pp. 742–744, 1918.

[518] Psaraftis, H. N., A dynamic programming approach for sequencing groups of identical jobs, *Operations Research,* vol. 28, pp. 1347–1359, 1980.

[519] Qiao, Z., Y. Zhang, and G. Wang, On fuzzy random linear programming, *Fuzzy Sets and Systems,* vol. 65, pp. 31–49, 1994.

[520] Radcliffe, N. and P. Surry, Formal memetic algorithms, in Fogarty [194], pp. 1–16.

[521] Rai, S. and D. Agrawal, editors, *Distributed Computing Network Reliability,* IEEE Computer Society Presss, Los Alamitos, CA, 1990.

[522] Raidl, G. R., An improved genetic algorithm for the multiconstrained 0–1 knapsack problem, in Fogel [196], pp. 207–211.

[523] Rajagopalan, R. and J. L. Batra, Design of cellular production systems: a graph-theoretic approach, *Int. Journal of Production Research,* vol. 13, no. 6, pp. 567–579, 1975.

[524] Rajyabaquse, R. A., Fuzzy logic application of genetic algorithm techniques for adaptive control of nonlinear systems, Ph.D. dissertation, University of Tokyo, 1998.

[525] Ramakumar, R., *Engineering Reliability: Fundamentals and Applications,* Prentice Hall, Upper Saddle River, NJ, 1993.

[526] Rao, S. S., *Engineering Optimization: Theory and Practice,* Wiley, New York, 1996.

[527] Rasheed, K. M., GADO: a genetic algorithm for continuous design optimization, Ph.D. dissertation, University of New Jersey–New Brunswick, 1998.

[528] Ravindran, A., D. T. Phillips, and J. J. Solberg, *Operations Research: Principles and Practice,* Wiley, New York, 1987.

[529] Rawlins, G., editor, *Foundations of Genetic Algorithms,* Morgan Kaufmann Publishers, San Francisco, 1991.

[530] Rechenberg, I., *Evolutionstrategie: Optimieriung technischer Systeme nach Prinzipien der biologischen Evolution,* Frommann-Holzboog, Stuttgart, Germany, 1973.

[531] Reeves, C., Diversity and diversification in genetic algorithms: some connections with tabu search, in Albrecht, R., C. Reeves, and N. Steele, editors, *Artificial Neural Nets and Genetic Algorithms,* pp. 344–351, Springer-Verlag, New York, 1993.

[532] Reklaitis, G. V., A. Ravindran, and K. M. Ragsdell, *Engineering Optimization: Methods and Applications,* Wiley, New York, 1983.

[533] Renders, J. and H. Bersini, Hybridizing genetic algorithms with hill-climbing methods for global optimization: two possible ways, in Fogel [1977], pp. 312–317.

[534] Requicha, A. G. and J. Vandenbrande, Automated systems for process planning and part programming, in *Artificial Intelligence: Implication for CIM,* pp. 301–326, Kempston, Bedfordshire, England, 1988.

[535] Revelle, C., D. Marks, and J. Liebman, An analysis of private and public sector location models, *Management Science,* vol. 16, no. 11, pp. 692–707, 1970.

[536] Rho, S., Distributed database design: allocation of data and operations to nodes in distributed database systems, Ph.D. dissertation, University of Minnesota, 1995.

[537] Richardson, J., M. Palmer, G. Liepins, and M. Hilliard, Some guidelines for genetic algorithms with penalty functions, in Schaffer [564], pp. 191–197.

[538] Rocha, L., Evidence sets and contextual genetic algorithms: exploring uncertainty, context, and embodiment in cognitive and biological systems, Ph.D. dissertation, State University of New York–Binghamton, 1997.

[539] Rogers, L. L., Optimal groundwater remediation using artificial neural networks and a genetic algorithm, Ph.D. dissertation, Stanford University, 1992.

[540] Root, J., An application of symbolic logic to a selection problem, *Operations Research,* vol. 12, no. 4, pp. 519–526, 1964.

[541] Ros, J. P., Learning Boolean functions with genetic algorithms: a PAC analysis, Ph.D. dissertation, University of Pittsburgh, 1992.

[542] Rosin, C., Coevolutionary search among adversaries, Ph.D. dissertation, University of California–San Diego, 1997.

[543] Rankin, R. P., Consideration of rapidly converging genetic algorithms designed for application to problems with expensive evaluation functions, Ph.D. dissertation, University of Missouri–Rolla, 1993.

[544] Rubin, J., A technique for the solution of massive set-covering problems with applications to airline crew scheduling, *Transportation Science,* vol. 4, pp. 34–48, 1973.

[545] Sá, G., Branch and bound approximate solutions to the capacitated plant location problem, *Operations Research,* vol. 17, no. 6, pp. 1006–1016, 1969.

[546] Sage, A. P., *Systems Engineering,* Wiley, New York, 1992.

[547] Sakawa, M., *Fuzzy Sets and Interactive Multiobjective Optimization,* Plenum Press, New York, 1993.

[548] Sakawa, M., H. Ishii, and I. Nishizaki, *Soft Optimization,* Asakura Publishing, Tokyo, 1995 (in Japanese).

[549] Sakawa, M., K. Kato, H. Sunada, and T. Shibano, Fuzzy programming for multiobjective 0–1 programming problems through revised genetic algorithms, *European Journal of Operational Research,* vol. 8, pp. 149–158, 1997.

[550] Sakawa, M., I. Nishizaki, and M. Hitaka, Interactive fuzzy programming for multilevel 0–1 programming problems through genetic algorithms, *European Journal of Operational Research,* vol. 114, pp. 580–588, 1999.

[551] Sakawa, M. and T. Shibano, Interactive fuzzy programming for multiobjective 0–1 programming problems through genetic algorithms with double strings, in Da Ruan, editor, *Fuzzy Logic Foundations and Industrial Applications,* pp. 111–128, Kluwer Academic Publishers, Norwell, MA, 1996.

[552] Sakawa, M. and T. Shibano, Multiobjective fuzzy satisficing methods for 0–1 knapsack problems through gentic algorithm, in W. Pedrycz, editor, *Fuzzy Evolutionary Computation,* pp. 157–177, Kluwer Academic Publishers, Norwell, MA, 1997.

[553] Sakawa, M. and H. Yano, An interactive fuzzy satisficing method using augmented minimax problems and its application to environmental systems, *IEEE Transactions on Systems, Man and Cybernetics,* vol. 15, pp. 720–728, 1985.

[554] Salkin, C. A. and D. De Kluyver, The knapsack problem: a survey, *Naval Research Logistics Quarterly,* vol. 22, pp. 127–144, 1975.

[555] Salkin, H. and J. Saha, Set covering: uses, algorithms and results, Technical report 272, Department of Operations Research, Case Western Reserve University, 1973.

[556] Salkin, H. and K. Mathur, *Foundations of Integer Programming,* North-Holland, New York, 1989.

[557] Salverson, M., The assembly line balancing problem, *Journal of Industrial Engineering,* vol. 6, no. 1, pp. 18–25, 1955.

[558] Sancho, N. G., A multi-objective routing problem, *Engineering Optimization,* vol. 10, pp. 71–76, 1986.

[559] Sarma, J. A., Analysis of decentralized and spatially distributed genetic algorithms, Ph.D. dissertation, University of New Hampshire, 1998.

[560] Sasaki, M., Studies on solution methods for fuzzy reliability design problems by hybrid genetic algorithms, Ph.D. dissertation, Hakkaido University, 1999 (in Japanese).

[561] Sato, H., TA proposal and analysis on the alternation of generation models of genetic algorithms, Ph.D. dissertation, Tokyo Institute of Technology, 1996.

[562] Savelsbergh, M., Local search for routing problem with time windows, *Annals of Operations Research,* vol. 4, no. 23, pp. 285–305, 1985.

[563] Schaffer, J., Multiple objective optimization with vector evaluated genetic algorithms, in Grefenstette [266], pp. 93–100.

[564] Schaffer, J., editor, *Proceedings of the 3rd International Conference on Genetic Algorithms,* Morgan Kaufmann Publishers, San Francisco, 1989.

[565] Schoenauer, M. and Z. Michalewicz, Boundary operators for constrained parameter optimization problems, in Bäck [29], pp. 322–329.

[566] Schwefel, H., *Numerical Optimization of Computer Models,* Wiley, Chichester, West Sussex, England, 1981.

[567] Schwefel, H., *Evolution and Optimum Seeking,* Wiley, New York, 1995.

[568] Schwefel, H. and R. Männer, editors, *Parallel Problem Solving from Nature,* Sringer-Verlag, New York, 1990.

[569] Seifoddini, H. and P. M. Wolfe, Application of the similarity coefficient method in group technology, *IIE Transactions,* vol. 18, no. 3, pp. 271–277, 1986.

[570] Sen, S., Minimal cost set covering using probabilistic methods, in *Proceedings of the 1993 ACM/SIGAPP Symposium on Applied Computing,* pp. 157–164, 1993.

[571] Sendhoff, B., M. Kreuts, and W. Seelen, A condition for the genotype-phenotype mapping: casualty, in Back [29], pp. 354–361.

[572] Shaefer, C., The ARGOT strategy: adaptive representation genetic optimizer technique, in Grefenstette [267], pp. 50–55.

[573] Shahooker, K. D., Application of the genetic algorithm for computer-aided design of VLSI layout, Ph.D. dissertation, University of Michigan, 1994.

[574] Shtub, A., Modeling group technology cell formation as a generalized assignment problem, *Int. Journal of Production Research,* vol. 27, no. 5, pp. 775–782, 1989.

[575] Shu, L., Impact of data structures on the performance of genetic algorithm–based learning, Ph.D. dissertation, University of Alberta, 1992.

[576] Skiena, S., *Implementing Discrete Mathematics Combinatorics and Graph Theory with Mathematics,* Addison-Wesley, Reading, MA, 1990.

[577] Smith, J. B. and D. Shier, An empirical investigation of some bicriterion shortest path algorithms, *European Journal of Operational Research,* vol. 43, pp. 216–224, 1989.

[578] Sniedovich, M., A multi-objective routing problem revisited, *Engineering Optimization,* vol. 13, pp. 99–108, 1988.

[579] Sollin, M., Le Trace de canalisation, in Berge, C. and A. Ghouilla-Houri, editors, *Programming, Games, and Transportation Networks,* Wiley, New York, 1965.

[580] Spears, W. M., Role of imitation and recombination in evolutionary algorithms, Ph.D. dissertation, George Mason University, 1998.

[581] Srinivas, N. and K. Deb, Multiobjective optimization using nondominated sorting in genetic algorithms, *Evolutionary Computation,* vol. 2, no. 3, pp. 221–248, 1995.

[582] Srinivasan, G., T. Narendran, and B. Mahadevan, An assignment model for the part-families problem in group technology, *Int. Journal of Production Research,* vol. 28, no. 1, pp. 145–152, 1990.

[583] Stadler, W., A survey of multicriteria optimization or the vector maximization problem: I. 1776–1960, *Journal of Optimization Theory and Applications,* vol. 69, pp. 1–52, 1979.

[584] Stadler, W., A comprehensive bibliography on multicriteria decision making and related areas, Technical report, University of California–Berkeley, 1981.

[585] Starns, G. K., Optimal synthesis of a planar four-bar mechanism with prescribed timing using generalized reduced gradient, simulated annealing and genetic algorithms, Ph.D. dissertation, Iowa State University, 1996.

[586] Steinberg, D. I., The fixed charged problem, *Naval Research Logistics Quarterly,* vol. 17, no. 2, pp. 217–236, 1970.

[587] Steinmann, H. and R. Schwinn, Computational experience with a zero–one programming problem, *Operations Research,* vol. 17, no. 5, pp. 917–920, 1969.

[588] Steuer, R. E., *Multiple Criteria Optimization: Theory, Computation, and Application,* Wiley, New York, 1986.

[589] Stinson, J., E. Davis, and B. Khumawala, Multiple resource constrained scheduling using branch and bound, *AIIE Transactions,* vol. 10, pp. 252–259, 1978.

[590] Storer, R., S. Wu, and R. Vaccari, New search spaces for sequencing problems with application to job shop scheduling, *Management Sciences,* vol. 38, no. 10, pp. 1495–1510, 1992.

[591] Streifel, R. J., Application of computational intelligence to electromechanical systems, Ph.D. dissertation, University of Washington, 1996.

[592] Stumpf, J. D., Studies on enhanced operator-oriented genetic algorithms, Ph.D. dissertation, Marquette University, 1996.

[593] Süer, G., Evolutionary programming for designing manufacturing cells, *Proceedings of the IEEE International Conference on Evolutionary Computation,* pp. 379–384, 1997.

[594] Süer, G., R. Vazquez, Y. Pena, and M. Cortes, Evolutionary programming for part family formation, in Gen, M. et al eds. *Proceedings of the First Japan–USA Joint Workshops on Intelligent Manufacturing Systems,* Ashikaga, pp. 29–41, 1998.

[595] Süer, G., R. Vazquez, Y. Pena, and M. Cores, Cellular design by using evolutionary programming as a tool, in Katai, O., M. Gen, et al eds. *Proceedings of the 2nd Japan–Australia Joint Workshops on Intelligent Evolutionary Systems,* Kyoto, pp. 379–384, 1998.

[596] Süer, G., R. Vazquez, and Y. Pena, Evolutionary programming in cellular design, Presented at INFORMS Spring Meetings, Cincinnati, 1999.

[597] Süer, G., M. Gen, and C. Moon, Evolutionary programming for designing and controlling manufacturing cells, in *Evolutionary Computations and Intelligent Systems,* Gordon and Breach New York (forthcoming).

[598] Sun, D. L. and R. Batta, Cell formation using tabu search, *Computers and Industrial Engineering,* vol. 28, no. 4, pp. 485–494, 1995.

[599] Sun, M., J. E. Aronson, P. G. McKeown, and D. Drinka, A tabu search heuristic procedure for the fixed charge transportation problem, *European Journal of Operational Research,* vol. 106, pp. 441–456, 1998.

[600] Sun, R., Evolving population-based search algorithms through thermodynamic operation: dynamic system design and integration, Ph.D. dissertation, University of Maryland, 1995.

[601] Sweeney, D. J. and R. A. Murphy, Branch-and-bound methods for multi-item scheduling, *Operations Research,* vol. 29, pp. 853–864, 1981.

[602] Swets, D. L., Self-organizing hierarchical optimal subspace learning and inference framework for view-based object recognition and image retrieval, Ph.D. dissertation, West Virginia University, 1996.

[603] Syswerda, G., Uniform crossover in genetic algorithms, in Schaffer [564], pp. 2–9.

[604] Syswerda, G., Scheduling optimization using genetic algorithms, in Davis [147], pp. 332–349.

[605] Tadikonda, S. K., Automated image segmentation and interpretation using genetic algorithms and semantic nets, Ph.D. dissertation, University of Iowa, 1993.

[606] Tagami, T., Study on application method of genetic algorithms for combinatorial optimization problems, Ph.D. dissertation, University of Tokushima, 1995.

[607] Taguchi, T., K. Ida, and M. Gen, Method for solving nonlinear goal programming with interval coefficients using genetic algorithms, *Computers and Industrial Engineering,* vol. 33, no. 1–4, pp. 579–600, 1997.

[608] Taha, H. A., *Integer Programming,* Academic Press, San Diego, CA, 1975.

[609] Takagi, T., Study on an automatic structure method using a fuzzy reasoning controller with neural networks and genetic algorithms, Ph.D. dissertation, Tokai University, 1994.

[610] Takashi, K., Study of evolutional mechanisms of cooperative problem solving: a hybrid genetic algorithm for multialgorithm problems, Ph.D. dissertation, Keio University, 1995.

[611] Talbot, F. and J. Patterson, An efficient integer programming algorithm with network cuts for solving resource-constraints scheduling problems, *Management Science,* vol. 24, pp. 1163–1174, 1978.

[612] Tamaki, H., H. Kita, and S. Kobayashi, Multiobjective optimization by genetic algorithms: a review, in Fogel [195], pp. 517–522.

[613] Tanaka, H. and K. Asia, Fuzzy solution in fuzzy linear programming, *IEEE Transactions on Systems, Man and Cybernetics,* vol. 1, pp. 325–328, 1984.

[614] Tanenbaum, A. S., editor, *Computer Networks,* Prentice Hall, Upper Saddle River, 1996.

[615] Tang, A., Genetic algorithms for optimal operation of soil aquifer treatment systems, Ph.D. dissertation, Arizona State University, 1997.

[616] Tang, J. and D. Wang, An interactive approach based on a GA for a type of quadratic programming problem with fuzzy objective and resources, *Computers and Operations Research,* vol. 24, pp. 413–422, 1997.

[617] Tay, E. H., Automated generation and analysis of dynamic system designs, Ph.D. dissertation, Massachusetts Institute of Technology, 1995.

[618] Terano, T., K. Asai, and M. Sugeno, *Applied Fuzzy Systems,* Academic Press, London, 1994.

[619] Thierens, D. and D. Goldberg, Convergence models of genetic algorithm selection schemes, in Davidor et al. [142], pp. 119–129.

[620] Tillman, F., C. Hwang, and W. Kuo, *Optimization of Systems Reliability,* Marcel Dekker, New York, 1980.

[621] Timothy, S. J., Optimization of sequencing problems using genetic algorithms, Ph.D. dissertation, Colorado State University, 1993.

[622] Tiwari, R. N., S. Dharmar, and J. R. Rao, Priority structure in fuzzy goal programming, *Fuzzy Sets and Systems,* vol. 19, pp. 251–259, 1986.

[623] Tom, A. and A. Zilinskas, *Global Optimization,* Springer-Verlag, New York, 1989.

[624] Tomasz, O., Nonlinear adaptive filtering: the genetic algorithm approach, Ph.D. dissertation, Politechnika Warszawska, 1995.

[625] Tompkins, G. H., Optimization of qualitative variables in discrete event simulation models, Ph.D. dissertation, Kansas State University, 1995.

[626] Toregas, C., The location of emergency service facilities, *Operations Research,* vol. 19, no. 6, pp. 1363–1373, 1971.

[627] Tsujimura, Y. and M. Gen, Genetic algorithms for solving multiprocessor scheduling problems, in Yao, X., J. H. Kim, and T. Furuhashi, editors, *Simulated Evolution and Learning,* pp. 106–115, Springer-Verlag, Heidelberg, 1997.

[628] Tsujimura, Y. and M. Gen, Entropy-based genetic algorithm for solving TSP, in Jain and Jain [322], pp. 285–290.

[629] Tsujimura, Y., M. Gen, and R. W. Cheng, Improved genetic algorithms for job-shop scheduling problems, *Engineering Design and Automation,* vol. 3, no. 2, pp. 133–144, 1997.

[630] Tsujimura, Y., M. Gen, R. Cheng, and T. Momota, Comparative studies on encoding methods of GA for open shop scheduling, *Australian Journal of Intelligent Information Processing Systems,* vol. 4, no. 3–4, pp. 214–219, 1997.

[631] Tsujimura, Y., M. Gen, and D. Zheng, GA-based approach for fuzzy multiple-item inventory control, *Engineering Valuation and Cost Analysis,* vol. 2, pp. 151–157, 1999.

[632] Tsujimura, Y., J. H. Kim, and M. Gen, GA-based design of star-ring LAN topology, in Hwang, H. and N. Tawara, editors, *Proceedings of the First Korea–Japan Joint Conference on Industrial Engineering and Management,* pp. 206–210, Taejon, 1998.

[633] Tsuijimura, Y., S. B. Ko, and M. Gen, Data distribution considered communication flow in network using genetic algorithm, in Jain and Jain [322], pp. 264–271.

[634] Tsujimura, Y. and M. Gen, Parts loading scheduling in a flexible forging machine using an advanced genetic algorithm, *Journal of Intelligent Manufacturing,* vol. 10, no. 2, pp. 149–159, 1999.

[635] Tuson, A. and P. Ross, Cost based operator rate adaptation: an investigation, in Voigt, H., W. Ebelling, I. Rechenberg, and H. Schwefel, editors, *Proceedings of the 4th International Conference on Parallel Problem Solving from Nature,* pp. 461–469, Springer-Verlag, Berlin, 1996.

[636] Valenta, J., Capital equipment decisions: a model for optimal systems interfacing, Master's thesis, Massachusetts Institute of Technology, 1969.

[637] Velthuizen, R. P., Feature extraction with genetic algorithms for fuzzy clustering, Ph.D. dissertation, University of South Florida, 1996.

[638] Venetsanopoulos, A. N. and I. Singh, Topological optimization of communication networks subject to reliability constraints, *Problems of Control and Information Theory,* vol. 15, pp. 63–78, 1986.

[639] Venugopal, V. and T. T. Narendran, Cell formation in manufacturing systems through simulated annealing: an experimental evaluation, *European Journal of Operational Research,* vol. 63, no. 3, pp. 409–422, 1992.

[640] Venugopal, V. and T. T. Narendran, A genetic algorithm approach to the machine-component grouping problem with multiple objectives, *Computers and Industrial Engineering,* vol. 22, no. 4, pp. 469–480, 1992.

[641] Vignaux, G. A. and Z. Michalewicz, A genetic algorithm for the linear transportation problem, *IEEE Transactions on Systems, Man and Cybernetics,* vol. 21, pp. 445–452, 1991.

[642] Vohra, T., D. Chen, J. Chang, and H. Chen, A network approach to cell formation in cellular manufacturing, *Int. Journal of Production Research,* vol. 28, no. 11, pp. 2075–2084, 1990.

[643] Voicu, L. I., Multiresolution aspects in genetic algorithms, Ph.D. dissertation, University of Central Florida, 1998.

[644] Volgenant, A., A Lagrangean approach to the degree-constrained minimum spanning tree problem, *European Journal of Operational Research,* vol. 39, pp. 325–331, 1989.

[645] Walker, W., Using the set-covering problem to assign fire companies to fire houses, *European Journal of Operational Research,* vol. 22, pp. 275–277, 1974.

[646] Wall, M. B., Genetic algorithm for resource-constrained scheduling, Ph.D. dissertation, Massachusetts Institute of Technology, 1996.

[647] Walters, G. A. and D. K. Smith, Evolutionary design algorithm for optimal layout of tree networks, *Engineering Optimization,* vol. 24, pp. 261–281, 1995.

[648] Wang, D., A simulated annealing approach for scheduling fundamental runs of grouped jobs, *Computers and Operations Research,* (submitted).

[649] Wang, D., An inexact approach for linear programming with fuzzy objective and resource, *Fuzzy Sets and Systems,* vol. 24, pp. 261–281, 1995.

[650] Wang, D., Modelling and optimization for a type of fuzzy nonlinear programming problems in manufacture systems, *Proceedings of the IEEE Conference on Decision and Control,* vol. 4, pp. 4401–4405, 1996.

[651] Wang, D., R. Cheng, and M. Gen, Scheduling grouped jobs on single machine with genetic algorithm, in Gen, M. and Y. Tsujimura, editors, *Evolutionary Computations and Intelligent Systems,* Gordon and Breach, New York (forthcoming).

[652] Wang, L., Genetic algorithm–based approach to subtask matching and scheduling in heterogeneous computing environments and a comparative study of parallel genetic algorithms, Ph.D. dissertation, Purdue University, 1997.

[653] Wang, M. Y., Scheduling in flowshops with reentrant flows and pallet requirements, Ph.D. dissertation, University of Toronto, 1995.

[654] Wang, P., Matrix genome encoding for genetic algorithms, Ph.D. dissertation, Polytechnic University, 1996.

[655] Wang, P. Y., G. S. Wang, and Z. G. Hu, Speeding up the search process of genetic algorithms by fuzzy logic, in Zimmermann [708], pp. 665–671.

[656] Warburton, A., Approximation of Pareto optima in multiple objective shortest path problems, *Operations Research,* vol. 35, pp. 70–79, 1987.

[657] Webster, S. and K. Baker, Scheduling groups of jobs on a single machine, *Operations Research,* vol. 43, pp. 692–703, 1995.

[658] Wemmerlov, U. and D. J. Johnson, Cellular manufacturing at 46 user plants: implementation experiences and performance improvements, *Int. Journal of Production Research,* vol. 35, no. 1, pp. 29–49, 1997.

[659] Wetzel, A., Evaluation of the effectiveness of genetic algorithms in combinatorial optimization, Technical report, University of Pittsburgh, 1983.

[660] Whatley, J. K., editor, *SAS/OR User's Guide: Version 5. Netflow Procedure,* SAS Institute, Inc., Cary, NC, 1985.

[661] White, D. W., GANNet: a genetic algorithm for searching topology and weight spaces in neural network design—the first step in finding a neural network solution, Ph.D. dissertation, Univesity of Maryland, 1993.

[662] Whitley, D., GENITOR: a different genetic algorithm, in *Proceedings of the Rocky Mountain Conference on Artificial Intelligence,* Denver, 1989.

[663] Whitley, D., V. Gordan, and K. Mathias, Lamarckian evolution, the Baldwin effect and function optimization, in Davidor et al. [142], pp. 6–15.

[664] Wilkov, R. S., Design of computer networks based on a new reliability measure, in *Symposium on Computer-Communications Networks and Teletraffic,* pp. 371–384, 1972.

[665] Wong, H., Performance analysis for genetic algorithms, Ph.D. dissertation, New Jersey Institute of Technology, 1996.

[666] Wozniak, S. J., Knowledge evolution in active database systems via fuzzy constraint management, Ph.D. dissertation, George Mason University, 1998.

[667] Wright, A., Genetic algorithms for real parameter optimization, in Rawlins [512], chapter 4.

[668] Wright, T., Genetic algorithm approach to scheduling resources for a space power system, Ph.D. dissertation, Case Western Reserve University, 1994.

[669] Wu, A. S., Noncoding DNA and floating building blocks for the genetic algorithm, Ph.D. dissertation, University of Michigan, 1995.

[670] Ziao, Y. L., Computer-assisted drug design: genetic algorithms and structures of molecular clusters of aromatic hydrocarbons and actinomycin D-deoxyguanosine, Ph.D. dissertation, University of Louisville, 1994.

[671] Xie, M. C., Properties and characteristics of genetic algorithms as a problem-solving method, Ph.D. dissertation, Fukui University, 1996.

[672] Xie, Z., Genetic algorithms: the method of regularization and their application to hypocenter location, Ph.D. dissertation, Texas A&M University, 1997.

[673] Xu, H. and G. Vukovich, Fuzzy evolutionary algorithms and automatic robot trajectory generation, in Fogel [197], pp. 595–600.

[674] Xu, Q., Genetic algorithms searching capability and their application for optimization of a model reference fuzzy adaptive controller for linear systems with time delay, Ph.D. dissertation, Musashi Institute of Technology, 1996.

[675] Xu, W., Quadratic minimum spanning tree problems and related topics, Ph.D. dissertation, University of Maryland, 1984.

[676] Yamada, T. and R. Nakano, A genetic algorithm applicable to large-scale job-shop problems, in Männer, R. and B. Manderick, editors, *Proceedings of the 2nd International Conference on parallel Problem Solving from Nature,* pp. 281–290, Elsevier Science Publishers, New York, 1992.

[677] Yang, X. and M. Gen, Evolution program for bicriteria transportation problem, in Gen and Kobayashi [231], pp. 451–454.

[678] Yang, X. H., Study on signature verification using a genetic algorithm method, Ph.D. dissertation, Nagoya University, 1995.

[679] Yao, L., Parameter estimation for nonlinear systems, Ph.D. dissertation, University of Wisconsin-Madison, 1992.

[680] Yazenin, A. V., Fuzzy and stochastic programming, *Fuzzy Sets and Systems,* vol. 22, pp. 171–180, 1987.

[681] Ye, J., Fuzzy modeling of nonlinear systems and eugenics-based genetic algorithms, Ph.D. dissertation, Okayama University, 1995.

[682] Yeh, M. S., J. S. Lin, and W. C. Yeh, A new Monte Carlo method for estimating network reliability, in Gen and Kobayashi [231], pp. 723–726.

[683] Yokota, D., Study on solving system reliability optimization problems with interval data by genetic algorithms, Ph.D. dissertation, Meiji University, 1996 (in Japanese).

[684] Yokota, T., M. Gen, and Y. Li, Genetic algorithms for nonlinear mixed integer programming problems and its applications, *Computers and Industrial Engineering,* vol. 30, no. 4, pp. 905–917, 1996.

[685] Yokota, T. and M. Gen, Optimal interval design for system reliability with incomplete FDS by means of improved genetic algorithms, *Electronics and Communications in Japan,* vol. 81, no. 1, pp. 84–94, 1998.

[686] Yu, P., Multiple criteria decision making: five basic concepts, in Nemhauser, G., A. R., Kan, and M. Todd, editors, *Handbooks in Operations Research and Management Science, Optimization,* vol. 1, chapter 10, pp. 663–699, Elsevier Science Publishers, New York, 1989.

[687] Yu, T. Crystal deformation due to a dipping fault in an elastic gravitational layer overlying a viscoelastic gravitational half-space, Ph.D. dissertation, University of Colorado–Boulder, 1995.

[688] Yue, K. K. and D. J. Lilja, Designing multiprocessor scheduling algorithms using a distributed genetic algorithms system, in Dasgupta, D. and Z. Michalawicz, editors, *Evolutionary Algorithms in Engineering Applications,* pp. 207–222, Springer-Verlag, Heidelberg, 1997.

[689] Yunker, J. M., Optimization of simulation models by genetic algorithms: a comparative study, Ph.D. dissertation, Virginia Polytechnic Institute and State University, 1993.

[690] Zadeh, L., Optimality and non-scalar-valued performance criteria, *IEEE Transactions on Automatic Control,* vol. 8, no 59, 1963.

[691] Zadeh, L., Fuzzy sets as a basis for a theory of possibility, *Fuzzy Sets and Systems,* vol. 1, pp. 3–28, 1978.

[692] Zeng, X. and B. Rabenasolo, A fuzzy logic based design for adaptive genetic algorithms, in Zimmermann [707], pp. 660–664.

[693] Zheng, D., M. Gen, and R. Cheng, Multiobjective optimization using genetic algorithims, *Engineering Valuation and Cost Analysis,* vol. 2, pp. 303–310, 1999.

[694] Zhou, C. G., Study on the convergence of genetic algorithms and its application to structural determination of feed forward neural networks, Ph.D. dissertation, Saitama University, 1997.

[695] Zhou, G., Study on constrained spanning tree problems with genetic algorithms, Ph.D. dissertation, Ashikaga Institute of Technology, 1999.

[696] Zhou, G. and M. Gen, Evolutionary computation on extended minimum spanning tree problems, Technical report, Intelligent Systems Engg. Lab., Ashikaga Institute of Technology, 1999.

[697] Zhou, G. and M. Gen, A new tree encoding for the degree-constrained spanning tree problem, *Soft Computing* (forthcoming).

[698] Zhou, G. and M. Gen, The genetic algorithms approach to the multicriteria minimum spanning tree problem, in Kim, J. H., X. Yao, and T. Furuhashi, editors, *Proceedings of the First Asia-Pacific Conference on Simulated Evolution and Learning,* pp. 387–394, Taejon, 1996.

[699] Zhou, G. and M. Gen, Approach to degree-constrained minimum spanning tree problem using genetic algorithm, *Engineering Design and Automation,* vol. 3, no. 2, pp. 157–165, 1997.

[700] Zhou, G. and M. Gen, Evolutionary computation on multicriteria production process planning problem, in Porto [512], pp. 419–424.

[701] Zhou, G. and M. Gen, A note on genetic algorithm approach to the degree-constrained spanning tree problems, *Networks,* vol. 30, pp. 105–109, 1997.

[702] Zhou, G. and M. Gen, An effective genetic algorithm approach to the quadratic minimum spanning tree problem, *Computers and Operations Research,* vol. 25, no. 3, pp. 229–247, 1998.

[703] Zhou, G. and M. Gen, Genetic algorithm approach on multi-criteria minimum spanning tree problem, *European Journal of Operational Research,* vol. 114, pp. 141–152, 1999.

[704] Zhuang, Z., Ergonomic evaluation of assistive devices and manual methods for resident-handling tasks in nursing homes, Ph.D. dissertation, West Virginia University, 1995.

[705] Zimmermann, H. J., Fuzzy programming and linear programming with several objective functions, *Fuzzy Sets and Systems,* vol. 1, no. 1, pp. 45–55, 1978.

[706] Zimmerman, H., Description and optimization of fuzzy system, *International Journal of General System,* vol. 2, pp. 209–215, 1976.

[707] Zimmermann, H., *Fuzzy Set Theory and Its Applications,* Kluwer-Nijhoff, Norwell, MA, 1985.

[708] Zimmerman, H., editor, *Proceedings of the 5th European Congress on Intelligent Techniques and Soft Computing,* Aachen, Germany, 1997.

[709] Bierwirth, C. and D. C. Mattfeld, Production scheduling and rescheduling with genetic algorithms. *Evolutionary Computation,* vol. 7, no. 1, pp. 1–17, 1999.

[710] Bilchev, G., Evolutionary metaphors for the bin packing problem. *Evolutionary Programming,* pp. 333–341, 1995.

[711] Chen, T., A hybrid intelligent system for process modeling and control using a neural network and a genetic algorithm, PhD dissertation, The University of Iowa, 1997.

[712] Deb, K. and M. Goyal, A flexible optimization procedure for mechanical component design based on genetic adaptive search, *Trans. of the ASME J. of Mechanical Design,* vol. 120, 1998.

[713] Deb, K., Multi-objective genetic algorithms: problem difficulties and construction of test problems. *Evolutionary Computation,* vol. 7, no. 3, pp. 205–230, 1999.

[714] Gen, M., K. Ida, and Y. Z. Li, Solving bicriteria solid transportation problem with fuzzy numbers by genetic algorithm, *Computer and Industrial Engineering,* vol. 29, pp. 537–543, 1995.

[715] Gen, M, J. O. Choi, Y. Tsujimura, Genetic algorithm for the capacitated plant location problem with single source constraints, *Proc. of 7th European Congress on Intelligent Techniques & Soft Computing, Session CD-7, Aachen, 1999.*

[716] Gen, M., Y. Z. Li, and K. Ida, Solving multi-objective transportation problem by spanning tree-based genetic algorithm, *IEICE Trans. Fundamentals,* vol. E82-A, 1999 (forthcoming).

[717] Gen, M., Y. Z. Li, and K. Ida, Spanning tree-based genetic algorithm for bicriteria fixed charge transportation problem, *Journal of Japan Society of Fuzzy Theory and Systems,* 2000 (forthcoming).

[718] Gen, M., Y. Tsujimura, and Y. X. Li, Fuzzy assembly line balancing using genetic algorithms, *Computer and Industrial Engineering,* vol. 31, no. 3–4, pp. 631–634, 1996.

[719] Hadj-Alouane, A. B. and J. C. Bean, A genetic algorithm for the multiple-choice integer program. *Oper. Res.,* vol. 45, no. 1, pp. 92–101, 1997.

[720] Hadj-Alouane, A. B. and J. C. Bean, and K. G. Murty, A hybrid genetic/optimization algorithm for a task allocation problem, *Journal of Scheduling,* vol. 2, pp. 189–201, 1999.

[721] Hsieh, Y., T. Chen, and D. Bricker, Genetic algorithms for reliablity design problems, *Microelectronics and Reliability,* vol. 38, no. 10, pp. 1599–1605, 1998.

[722] Hung, Y. F., C. C. Shin, and C. P. Chen, Evolutionary algorithms for production planning problem with setup decisions, *Journal of the Operational Research Society,* vol. 50, pp. 857–866, 1999.

[723] Jim é nez, F. and J. L. Verdegay, An evolutionary algorithm for interval solid transportation problems, *Evolutionary Computation,* vol. 7, no. 1, pp. 103–107, 1999.

[724] Kim, Y. K., Y. J. Kim, and Y. H. Kim, Genetic algirithms for assembly line balancing with various objectives, *Computer and Industrial Engineering,* vol. 30, no. 3, pp. 397–409, 1996.

[725] Levitin, G., A. Lisnianski, H. Ben-Haim, and D. Elmakis, Redundancy optimization for series-parallel multistate systems, *IEEE Transaction on Reliability,* vol. 47, no. 2, pp. 165–172, 1998.

[726] Levitin, G., Sh. Mazal-Tov, and D. Elmakis, Genetic algorithm for open-loop distribution system design, *Electric Power Systems Research,* vol. 32, pp. 81–87, 1995.

[727] Özdamar, L., A genetic algorithm approach to a general category project scheduling problem, *IEEE Transaction on Systems, Man, and Cybernetics,* vol. 29, no. 1, pp. 44–59, 1999.

[728] Mühlenbein, H. and D. Schlierkamp-Voosen, Predictive models for the breeder genetic algorithm I, continuous parameter optimization, Evolutionary Computation, vol. 1, pp. 25–49, 1993.

[729] Painton, K. and J. Campbell, Genetic algorithm in optimization of system reliability, *IEEE Transaction on Reliability,* vol. 44, no. 2, pp. 172–178, 1995.

[730] Ramachandran, V., V. Sivakumar, and K. Sathiyanarayanan, Genetics based redundancy optimization, *Microelectronics and Reliability,* vol. 37, no. 4, pp. 661–663, 1997.

[731] Sasaki, M. and M. Gen, Bicriteria knapsack problem with GUB structure by hybrid genetic algorithm, *Journal of Japan Society of Fuzzy Theory and Systems,* 1999 (forthcoming).

[732] Smith, A. E. and D. W. Coit, Reliability optimization of series-parallel systems using a genetic algorithm, *IEEE Transaction on Reliability,* vol. 45, no. 2, pp. 254–260, 1996.

[733] Taguchi, T., M. Gen, and K. Ida, Genetic algorithm for solving multiobjective nonlinear integer programming, *IEICE Trans. A,* vol. J79-A, no. 6, pp. 1221–1223, 1996, in Japanese.

[734] Taguchi, T., T. Yokota, and M. Gen, Reliability optimal design problem with interval coefficients using hybrid genetic algorithms, *Computer and Industrial Engineering,* vol. 35, pp. 373–376, 1998.

[735] Takeno, T., and G. Yamazaki, Improvement of local area transportation system scheduling using genetic algorithms, *Engg. Design and Auto.,* vol. 3, no. 2, pp. 191–200, 1997.

[736] Tsujimura, Y. and M. Gen, Evolutionary project scheduing under resource constraint, H. R. Parsaei, M. Gen, Y. Tsujimura, H. R. Leep and J. P. Wong eds., *Proc. of 2nd Inter. Conf. on Engg. Design and Auto.,* no. 308, 1998.

[737] Zhao, L., Y. Tsujimura, and M. Gen, Genetic algorithm for fuzzy clustering with application to manufacturing cell formation problem, in M. Jamshidi *et al,* eds.: *Intelligent Automation and Control; Recent Trends in Development and Applications,* vol. 4, pp. 799–804, 1997.

INDEX

2-connectivity, 210

Active schedule, 252
Adaptation of genetic algorithm, 13
  adaptive adaptation, 17
  deterministic adaptation, 17
  fuzzy control, 18
  fuzzy control controller, 18
  parameters adaptation, 14
  self-adaptive adaptation, 17
  structure adaptation, 15
Adaptive evaluation function, 86
Adaptive weight approach, 127
  adaptive penalty function, 130
  adaptive weights, 129
  hyperparallelogram, 127
Affine combination, 29
Airline crew scheduling problems, 56
Array-based methods, 394
  bond energy algorithm, 395
Augmented minimax problems, 180

Backtracking algorithm, 209
Beam search, 249
Bin-Packing Problem, 61
Binary relation, 101

Cauchy's method, 172
Centralized network, 363
  central site, 363
  centralized network design, 364
Combinatorial optimization, 53, 235
Complete alternative routing, 430
Compromise approach, 104
  compromise solution, 104
  lexicographic ordering, 105
  regret function, 104
Compromise-based fitness assignment, 109
Computer integrated manufacturing, 390
Computer network, 341
  bicriteria LAN design problems, 212
  bicriteria LAN topology design, 213
  bicriteria network topology design, 212
  expansion, 366
  network expansion, 368
  packet-switching networks, 352
Convex hull, 29
Crossover, 1, 171, 187
  affine, 29
  algorithm-based, 242
  arithmetic, 29, 171, 225, 228, 233, 415
  average, 29
  blend, 30
  cell-swap, 415
  cell-two-point, 416
  convex, 29
  cycle, 240
  direction-based, 33
  extended intermediate, 30
  intermediate, 29
  job-based order, 241
  linear, 29
  linear order, 240
  one-cut-point, 260, 308, 322, 334, 436
  order, 88, 239
  order-based, 239, 400
  partial schedule exchange, 241
  partial-mapped, 238
  partially matched crossover (PMX), 189
  path crossover, 355
  position-based, 239, 271, 349
  Prüfer number, 322
  random node field swap, 369
  simple, 415

Crossover (*Continued*)
 simple node field swap, 369
 sphere, 33
 subsequence exchange, 240
 substring exchange, 240
 two-cut-point, 408
 uniform, 208, 218
 unimodal normal distribution, 31

DataPacket, 357
Degree-based permutation, 86
DelayAnswer, 357
DelayEntry, 357
DelayRequest, 357
Direct search, methods, 226
 Hooke-Jeeves method, 226
 Powell's conjugate direction, 226
 simplex search, 226
Directed acyclic graph, 266
Distance method, 131
Dynamic programming, 194, 375

Efficient, 98
Efficient solutions, 101
Evolution strategies, 1
Evolutionary computation, 1
Evolutionary programming, 1
Exploitation, 16
Exploration, 16

Facility location, 378
 M/G/s queue, 378
 M/G/s queuing facility location, 378
 stochastic queue median, 378
Fitness sharing, 111
 genotypic, 111
 phenotypic, 111
 sharing along Pareto frontier, 111
 sharing function, 111
 sharing in multi-modal function, 111
 sharing on the objective space, 112
 sharing on the solution space, 112
Fixed alternative routings, 426
Flexible manufacturing systems, 390
Functional layout, 391, 421
Fundamental schedule, 258
Fusion operator, 58
Fuzzy linear programming, 143
 linear membership function, 146
Fuzzy mathematical programming, 142, 143
Fuzzy multiobjective integer programming, 178
Fuzzy nonlinear mixed integer goal programming, 165
 goal programming, 165
Fuzzy nonlinear programming, 156
 nonlinear programming, 156
Fuzzy numbers, 331

Genetic algorithm, 1
Genetic drift, 111
Genetic operators
 boundary, 32
 conventional, 29
 path genetic operators, 355
Genetic programming, 1
Geometric programming, 194
Goal programming, 109
Graph and network methods, 396
Graph partitioning, 396
Group technology, 253, 391
Grouping efficacy, 428

Height function, 290
Heuristic algorithms, 63, 82
 best fit heuristic, 64
 first fit heuristic, 64
 next fit heuristic, 64
Heuristic method, 194
Hybrid genetic algorithm, 226

Ideal point, 100
Inefficient, 99
Integer programming, 194
Intraprocessor precedence relation, 292
Inventory control, 297, 299
Job-shop layout, 391
Just-in-time, 390

Knapsack problem, 71
 0-1 Knapsack, 72
 bounded knapsack, 72
 multi-constrained knapsack, 72
 multiple-choice knapsack, 72
  generalized upper bounds (GUB), 73
  multiple-choice constraints, 73
 unbounded knapsack, 72

Lagrangean multiplier method, 194
LAN, 211
Linear combination, 29
Linear programming, 194
Local search, 7

Manufacturing cell design, 391
Mathematical programming methods, 395
 p-median model, 395
 p-median algorithm, 421
Memetic algorithms, 284

Minimax problem, 184
Minimum spanning tree, 81
Minimum spanning tree problem, 81
  bicriteria minimum spanning tree, 90
  capacitated minimum spanning tree, 363
  degree-constrained, 86
  quadratic, 81
Multi-stage process planning, 373
  dynamic recurrence expression, 375
Multiconstraint knapsack problem, 77
  multidimensional knapsack, 77
  multiple knapsack, 77
Multiple-choice knapsack problem, 72
Multiprocessor scheduling problem, 289
Mutation, 2, 88, 192
  boundary, 414
  directional, 33
  displacement, 242, 308, 322, 334
  Gaussian, 34
  heuristic mutation, 376
  insertion, 88, 242
  inversion, 242, 308, 322, 334
  local search-based, 271
  multi-non-uniform, 415
  multi-uniform, 414
  neighborhood search, 376
  neighborhood search-based, 243
  non-uniform, 33, 414
  path mutation, 356
  reciprocal exchange, 242
  shift, 242
  swap, 88, 218, 271, 349
  uniform, 172, 208, 218, 414

Negative ideal solution, 100
Network reliability design, 195
  all-terminal network reliability, 195
  all-terminal reliability, 196
  enumerative-based approach, 195
  exact calculation, 204
  GA-based approach, 195
  heuristic-based approach, 195
  Monte-Carlo simulation, 204
  network reliability, 335, 338
  reliability estimation, 204
  source-sink network reliability, 195
  tree-based network reliability, 211
Non-regular measure, 279
Nondominated solutions, 98
Nonlinear goal programming, 139
Nonlinear mixed integer programming, 221

Optimization
  multiobjective, 39
Optimizations
  combinatorial, 38
  constrained, 34
  global, 27

Pareto optimal solutions, 39, 98, 313, 333
Pareto ranking method, 116
  ranking selection, 116
Pareto-based approach, 108
  Pareto ranking, 108
  Pareto tournament, 108
Path growth, 345, 348
  dead node, 347
  dynamic adjacent matrix, 346
  eligible edge, 345
  eligible edge set, 346
  pendant node, 349
  reachability matrix, 348
Penalty function, 37
  constant penalty, 37
  dynamic penalty, 38
  problem-dependent, 38
  problem-independent, 38
  static penalty, 37
  variable penalty, 37
Plant location problems, 325
  capacitated plant location problem, 325
  uncapacitated plant location problem, 327
Positive ideal solution, 100
Power law function, 112
Preference structures, 101
Production plan, 298
Production scheduling, 297, 299
Prüfer number, 297
Pure literal string, 239

Random search, 7
Random weight approach, 125
  weighted-sum objective, 125
Reasonable schedule, 255
Regular measure, 279
Reliability design, 194, 195
  bicriteria, 194
  component reliability, 194
  with fuzzy goals, 195
Representation, 56, 74, 185
  bin-based, 65
  binary, 74
  binary encoding, 3
  binary string, 3, 78, 369
  building block, 8
  causality, 7
  cell number-based, 396, 407
  column-based representation, 56
  completeness, 6
  double string, 185

Representation (*Continued*)
epistasis, 8
group-based, 67
illegality, 4
infeasibility, 4
job sequence matrix encoding, 248
job-based representation, 249
Lamarckian property, 6
legality, 6
literal permutation encoding, 3
literal string encodings, 218
machine-based encoding, 249
matrix, 297
multi-dimensional encoding, 3
non-redundancy, 5
object-based, 66
one-dimensional encoding, 3
order, 74
order-based, 396
priority rule-based encoding, 247
priority-based encoding, 266, 343
Prüfer number, 297
real number encoding, 3
ringed double strings, 186
row-based, 58
spanning tree, 297
state permutation, 376
tree, 304
tree-based permutation, 365
variable-length, 74
Revised Hooke-Jeeves method, 229
Routing, 341
adaptive network routing, 352
broader gateway protocols, 353
exterior gateway protocol (EGP), 353
genetic-based adaptive routing, 353
interior gateway protocol, 352
open shortest path first protocol, 352
route entry, 354
routing information protocol, 352
routing table, 354
shortest path first protocol, 352

Scheduling problems
grouped jobs scheduling problem, 253
job partition, 279
job sequence, 279
job-shop scheduling, 235
literal permutation encodings, 238
multiprocessor scheduling, 288
operation sequence, 237
parallel machine scheduling, 278
precedence constraints, 237, 289
precedence relations, 290
resource-constrained project scheduling, 263
Selection, 171, 189
$(\mu + \lambda)$ 9, 323
deterministic, 437
elitist, 225, 274
elitist expected value, 187
elitist method, 171
elitist ranking, 187
elitist roulette wheel, 187
expected value, 187
proportional, 10
ranking, 187
roulette wheel, 9, 171, 187, 274, 287, 293, 323
sharing, 9
steady-state reproduction, 9
tournament, 9
Set-Covering Problem, 53
Shifting bottleneck heuristic, 249
Shortest path problem, 341
bicriteria shortest path problem, 342
Similarity coefficient methods, 394
Solution approaches, 102
adaptive weight, 109, 127
compromise, 104, 136
fixed weight, 109
generating, 102
goal programming, 138
nonpreemptive, 139
preemptive, 139
lexicographic ordering, 105, 118
Pareto approach, 105
preference-based, 102
random weight, 109, 125
utility function, 103
vector evaluation, 108
weighted-sums, 102
Spanning tree-based GA, 304, 315, 316

Topological sort, 266, 267, 270
Transportation problem, 297
balanced transportation problem, 300
bicriteria transportation problem, 330
capacitated transportation problem, 303
fixed charge transportation problem, 320
fuzzy coefficients, 330
generalized transportation problem, 303
linear transportation problem, 300
multi-objective transportation problem, 311
Tree-based network reliability
calculation, 216
reliability measure, 216

Triangular fuzzy number, 332

Vector evaluated genetic algorithm, 115

Weighted L_p—norm, 137
eighted gradient direction, 150, 162
Weighted-sum approach, 102
Work-in-progress, 391